STATISTICS: The Art and Science of Learning From Data

A basic understanding of statistics is essential for students to become informed consumers. Yet, many students fail to realize this fact until their college days are well behind them.

Alan Agresti and **Christine Franklin** have written a text that illustrates the power and value of statistics—while presenting concepts and techniques in a manner that is accessible, accurate, and honest. *Statistics: The Art and Science of Learning from Data* beckons students into the world of statistical decision making.

As respected teachers and innovative researchers, Alan and Christine are excited by statistics as a discipline and by statistical education. In creating this book, they worked with students, instructors, and other experts across all fields of statistics. These specialists shared their experiences and offered invaluable advice.

Statistics: The Art and Science of Learning from Data seeks to inspire students to ask and answer interesting questions, and presents numerous ways to help students grasp the use of statistics in the real world.

Are these objectives you'd like your course to meet? If so, we invite you to consider Agresti/Franklin's innovative new text. This Preliminary Edition invites you to learn more. It includes Chapters 1-8—the core of the course—in black-and-white format. The full-color textbook will be available in January 2006.

Innovative Example Format

"The examples are one of the strong points of this text..."
—Martin Lindquist, Columbia University

"The examples used in this text are excellent. ... A student ... is likely to believe that statistics is actually useful, and know how to use it, and also how not to use it."
—Alla Sikorskii, Michigan State University

"I think the writing style in these chapters is better than my current text..."
—Robert Price, East Tennessee State University

"Keep students' needs in mind and reduce confusion; This is an important distinguishing characteristic..."
—Rob Paige, Texas Tech

"Agresti/Franklin's use of current and interesting topics is very impressive..."
—Dawn White, CSU Bakersfield

Students learn from clear examples.

Agresti and Franklin's logical **5-step example format** encourages students to model their thinking to examine issues intelligently in statistics. By offering engaging, up-to-date, and diverse data, the authors provide examples compelling to both students and instructors. (See the following pages for a full example from the text.)

Applications include:

- Is smoking actually beneficial to your health?

- Random drug testing of air traffic controllers

- Are astrologers' predictions better than guessing?

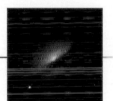

- How can we check for racial profiling?

- What proportion of the population won't pay higher prices to protect the environment?

- Are cell phones dangerous to your health?

Engaging Examples

THE 5 STEP EXAMPLE FORMAT

EXAMPLE 6

DID THE BUTTERFLY BALLOT COST AL GORE THE 2000 PRESIDENTIAL ELECTION?

Picture the Scenario

In the 2000 Presidential election in the U.S., the Democratic candidate was Al Gore and the Republican candidate was George W. Bush. In Palm Beach County, Florida, initial election returns reported 3407 votes for the Reform party candidate, Pat Buchanan. Some political analysts thought that this total seemed surprisingly large. They felt that most of these votes may have actually been intended for Gore (whose name was next to Buchanan's on the ballot) but wrongly cast for Buchanan because of the design of the "butterfly ballot" used in that county, which some voters found confusing. On the ballot (see the figure in the margin), Bush's name was the first entry in the left column, followed by Gore's. Buchanan's name appeared in the right column, across from Bush. Consequently, the second "punch hole", which ran down the center of the ballot, was for Buchanan, rather than Gore

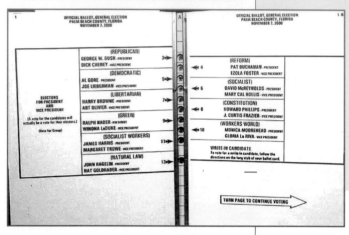

Question to Explore

For each of the 67 counties in Florida, the "Buchanan and the butterfly ballot" data file on the text CD includes the Buchanan vote and the vote for the Reform party candidate in 1996 (Ross Perot).

How can we explore graphically whether the Buchanan vote in Palm Beach County in 2000 was in fact surprisingly high, given the Reform party voting totals in 1996?

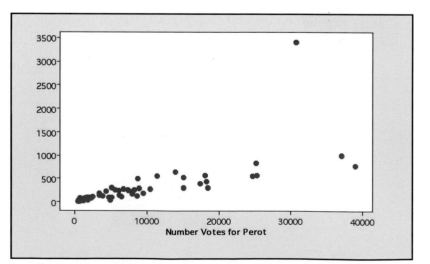

FIGURE 3.6: **MINITAB Scatterplot of Florida County-Wide Vote for Reform Party Candidates Pat Buchanan in 2000 and Ross Perot in 1996.** •
QUESTION: **Why is the top point, but not each of the two right-most points, considered an outlier relative to the overall trend of the data points?**

Think it Through

Figure 3.6 is a scatterplot of the county-wide vote for the Reform party candidates in 2000 (Buchanan) and in 1996 (Perot). Each point represents a county. This figure shows a strong positive association statewide: Counties with a relatively high Perot vote in 1996 tended to have a high Buchanan vote in 2000, and counties with a relatively low Perot vote in 1996 tended to have a low Buchanan vote in 2000. The Buchanan vote in 2000 was roughly only 3% of the Perot vote in 1996.

In Figure 3.6, one point falls well above the others. This severe outlier is the observation for Palm Beach County, the county that had the butterfly ballot. It is far removed from the overall trend for the other 66 data points, which follow approximately a straight line.

Insight

Alternatively you could plot the Buchanan vote against the Gore vote or against the Bush vote (Exercise 3.26). These and other analyses conducted by statisticians predicted that fewer than 900 votes were truly intended for Buchanan in Palm Beach County, compared to the 3407 votes he actually received. Bush won the state by 537 votes (and, with it, the Electoral College and the election), so this may have been a pivotal factor in determining the outcome of this election. Other factors that may have played a role were 110,000 disqualified "overvote" ballots in which people mistakenly voted for more than one Presidential candidate (with Gore marked on 84,197 ballots and Bush on 37,731) often because of confusion from names being listed on more than one page of the ballot, and 61,000 "undervotes" caused by factors such as "hanging chads" from manual punch-card machines in some counties. •

To practice this concept, try Exercise 3.26

STATISTICS: The Art and Science of Learning From Data

Table of Contents

Chapters 1–8 contained in the Preliminary Edition.

(NOTE: Each chapter concludes with Chapter Summary and Problems.)

STATISTICS: The Art and Science of Learning From Data

The Agresti/Franklin Review Team

Larry Ammann, University of Texas at Dallas

Pamela Arroway, North Carolina State University

Jeff Banfield, Montana State University

Sanjib Basu, Northern Illinois University

David Bauer, Virginia Commonwealth University

Jule Belock, Salem State University

Russel Carlson, University of Arizona

Ted Chang, University of Virginia

Ching-Yuan Chiang, James Madison University

Carrie Chmielarski, student - University of Georgia

John Cryer, University of Iowa

Phyllis Curtis, Grand Valley State University

Jacquelin Dietz, North Carolina State University

Bob Dobrow, Carlton College

William Duckworth, Iowa State University

Hans Engler, Georgetown University

Michael Evans, University of Toronto

Jindasa Gamage, Illinois State University

Ashis Gangopadhyay, Boston University

Steve Garren, James Madison University

Robert Goldman, Simmons College

Rob Gould, University of California, Los Angeles

Burke Grandjean, University of Wyoming

Shelby Haberman, Northwestern University

Stan Haines, University of South Carolina

Rebecca Head, Bakersfield College

Susan Herring, Sonoma State University

John Holcomb, Cleveland State University

Dawn Holmes, University of California, Santa Barbara

Debbie Hydorn, Mary Washington College

Martin Jones, College of Charleston

Colleen Kelly, San Diego State University

James Lang, Valencia Community College

Martin Lindquist, Columbia University

Jianguo Liu, University of North Texas

Robin Lock, St. Lawrence University

David Loewen, University of Manitoba

Richard Mahar, Loyola University, Chicago

David Mathiason, Rochester Institute of Technology

Darcy Mays, Virginia Commonwealth University

Eileen McDonald, Sonoma State University

Megan Meece, University of Florida

Xiao-Li Meng, Harvard University

Henry Mesa, Portland Community College-Rockcreek

Michael Monticino, University of North Texas

Tom Moore, Grinnell College

Linda Myers, Harrisburg Community College

Tereza Neocleous, University of Illinois at Urbana-Champaign

Rob Paige, Texas Tech University

Stephan Pelikan, University of Cincinnati

Stephanie Pickle, Virginia Polytechnic Institute and State University

Chandler Pike, University of Georgia

Cathy Poliak, Northern Illinois University

Robert Price, East Tennessee State University

R.V. Ramamoorthi, Michigan State University

Bob Raymond, University of St. Thomas

Diane Resnick, High School

Larry Ries, University of Missouri – Columbia

Maria Ripol, University of Florida

Neil Rogness, Grand Valley State University

Deb Rumsey, Ohio State University

Harold Sackrowitz, Rutgers, The State University of New Jersey

Steve Sawin, Fairfield University

Thomas Short, Villanova University

Alla Sikorskii, Michigan State University

Steve Wang, Harvard University

Dan Weiner, Boston University

Celia Welna, University of Hartford

Roger Woodard, North Carolina State University

Daming Xu, University of Oregon

Linda Young, University of Nebraska

Tian Zheng, Columbia University

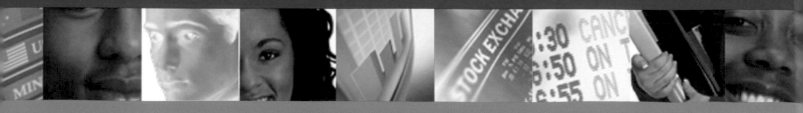

Class Test Sites

College of Charleston
University of Georgia
Georgia Southern University
University of Manitoba
Valdosta State University
Central Connecticut State University

STATISTICS: The Art and Science of Learning From Data

Alan Agresti, *University of Florida* · **Christine A. Franklin,** *University of Georgia*

PRELIMINARY EDITION

1 Lake Street
Upper Saddle River, NJ 07458

Executive Acquisitions Editor: Petra Recter
Executive Project Manager: Ann Heath
Project Manager: Michael Bell
Editor in Chief: Sally Yagan
Production Editor: Lynn Savino Wendel
Senior Managing Editor: Linda Mihatov Behrens
Executive Managing Editor: Kathleen Schiaparelli
Assistant Manufacturing Manager/Buyer: Alan Fischer
Art Director: Stacey Abraham
Writer: Elise Schneider
Director of Marketing: Patrice Jones
Marketing Assistant: Rebecca Alimena

PEARSON
Prentice
Hall

© 2006 by Pearson Education, Inc.
Pearson Prentice Hall
Pearson Education, Inc.
Upper Saddle River, New Jersey 07458

Pearson Prentice Hall™ is a trademark of Pearson
Education, Inc.

Printed in the United States of America

10 9 8 7 6 5 4 3 2 1

ISBN 0-13-185764-9

Pearson Education Ltd., London
Pearson Education Australia Pty. Limited, Sydney
Pearson Education Singapore Pte. Ltd.
Pearson Education North Asia, Ltd, Hong Kong
Pearson Education Canada, Ltd., Toronto
Pearson Educación de Mexico, S.A.,de C.V.
Pearson Education, Japan, Tokyo
Pearson Education Malaysia, Pte. Ltd.

Statistics: The Art and Science of Learning from Data
Alan Agresti and Christine Franklin

"Statistics ... the most important science in the whole world: for upon it depends the practical application of every other science and of every art; the one science essential to all political and social administration, all education, all organization based upon experience, for it only gives the results of our experience."

Florence Nightingale (1820 – 1910)

One of Florence Nightingale's achievements as a nurse was to promote improved health care by using the art and science of statistics to help people learn from data. She was one of the individuals who helped to develop a new way of looking at the world that is now vital in this modern information age. Today, everyone from students and teachers to workers and employers needs to be able to interpret data.

WHY DID WE WRITE THIS BOOK?

Having taught introductory statistics for more than 25 years, we have witnessed the welcome evolution from the traditional formula-driven watered-down mathematical statistics course to a concept-driven approach that places more emphasis on why statistics is important in the real world and less emphasis on probability. Our book's content and approach reflects our philosophy for teaching, our classroom experience, advances in technology, and student feedback on how they learn most easily. One of our main purposes in writing this book was to help make the conceptual approach more interesting and more readily accessible to the average college student. At the end of the course, we want more of these students to look back at their statistics course and feel they learned many worthwhile ideas that will serve them well the rest of their lives.

WHAT APPROACH DOES THIS BOOK TAKE?

Here are the main ways that *Statistics: The Art and Science of Learning from Data* attempts to provide an improved context for teaching introductory statistics:

Learn to ask and answer interesting questions: In presenting concepts and methods, we encourage students to think about the data and the appropriate analyses by continually posing questions. This is done in various ways, including (1) presenting a structured approach to examples that separates the question and the analysis from the scenario presented (2) providing lots of homework problems that encourage students to think and write, (3) requesting that students perform simulations and hands-on activities to help them learn by doing, and (4) posing information in the form of questions in the section titles and in the figure captions that are answered within the text.

Keep students' needs in mind and reduce confusion: In class tests, our students have told us that this book is more "readable" than many other introductory statistics texts. We have included a wide variety of interesting real-data examples and exercises that intrigue students. We have also attempted to simplify wherever possible, without sacrificing accuracy. A serious source of confusion for students is the multiplicity of inference methods resulting from the many combinations of confidence intervals and tests, means and proportions, large sample and small sample, variance known and unknown, two-sided and one-sided inference, independent and dependent samples, and so on. We've placed main emphasis on the most important cases for practical application of inference: large sample, variance unknown, two-sided inference, and independent samples. The many other cases are also covered (except for known variances), but more briefly, with the exercises focusing mainly on the way inference is commonly conducted in practice. We hope this will help to reduce student confusion.

Connect statistics to the real world through greater emphasis on proportions: Every day in the media, we see and hear about percentages and rates to summarize results of opinion polls or outcomes of medical studies or economic reports. Much less often do we hear about outcomes that are in any sense normally distributed or even nearly continuous, yet that's the main emphasis of most statistics texts. We've increased the attention paid to the analysis of proportions. For example, we first explain sampling distributions and inference methods using proportions for a binary outcome, and we use contingency tables to illustrate the concept of association and to show the potential influence of a lurking variable. We don't neglect quantitative data, but we believe that students should become comfortable with data in the form they'll most often see them in practice. This has pedagogical rewards. It's easier for a student to look at a sample of binary data or a distribution of sample proportions with a small sample size and understand what that distribution refers to than to do the same for a bell-shaped curve for a continuum of unobserved values.

HOW HAVE WE ORGANIZED THE CONTENTS?

We've shaped the organization and content using our beliefs about what works in the classroom and which topics merit inclusion in an introductory course. The main focus is on concepts and interpretation rather than blind plugging of numbers into formulas. We also want students to come to appreciate that in practice, assumptions are not perfectly satisfied, models are not exactly correct, distributions are not exactly normal, and all sorts of factors need to be considered in conducting a statistical analysis. As the title of our book reflects, data analysts soon realize that statistics is an art as well as a science.

Here's a brief chapter-by-chapter survey, summarizing the content of each but also pointing out ways in which our presentation may differ somewhat from other texts.

CHAPTER 1: Statistics: The Art and Science of Learning from Data

Our first chapter gives an overview to students about how data and statistical analyses surround us in our everyday lives. The chapter also gives students a sense of how data are used in an investigative process. An introductory overview of the field such as this chapter provides is missing from many statistics texts.

CHAPTER 2: Exploring Data with Graphs and Numerical Summaries

This first of two chapters on descriptive statistics introduces students to the univariate case--graphical and numerical methods for exploring and summarizing data on a single categorical or quantitative variable. The emphasis is on interpreting graphs and numerical summaries produced by computer or calculator output, rather than on constructing graphs and calculating numerical summaries by hand.

CHAPTER 3: Association: Contingency, Correlation, and Regression

This second of two chapters on descriptive statistics focuses on bivariate analyses. It introduces one of the most important topics of statistics--analyzing association between two variables. The chapter begins by exploring categorical variables through contingency tables. Students often see such summaries in their everyday lives, yet most books delay coverage of this material until near the end. Discussing contingency tables facilitates transition into Chapter 4 on producing data for topics such as survey sampling and comparisons in randomized experiments. It also provides background for the discussion of conditional probability in Chapter 5 on probability. The chapter next introduces correlation and regression methods. For many students, the first time that they see correlation is a "light-bulb" moment, the instant they understand the power and usefulness of statistics. The final section emphasizes important cautions, such as association not proving causation and the potential influence of lurking variables, two fundamental concepts that students need to see early in the course.

CHAPTER 4: Gathering Data

This chapter introduces students to good ways and poor ways to sample and to experiment. Why an entire chapter on producing data? Valid analysis and interpretation of data depends on having an appropriate study design. When is it appropriate to make cause and effect conclusions? How could other factors influence results? What determines the extent to which conclusions can be generalized? How important are issues such as non-response? Regardless of how sophisticated a statistical analysis is conducted, results may be nearly worthless if the study design is poor. However, even though randomized experiments are ideal, they are often not feasible or ethical. In practice, the types of studies we most often encounter are observational. Thus, we devote detailed coverage to types of observational studies, with emphasis on sample surveys.

CHAPTER 5: Probability in Our Daily Lives

This chapter introduces students to the basic ideas of probability and to how it can help us understand the forms that randomness takes in our lives. The unifying theme is how probability relates to our daily lives, using applications such as diagnostic disease testing and whether an apparent coincidence truly is coincidental. It's easy for students to get lost when probability is a major component of the course or is presented in a mathematical manner. So, we attempt to offer "just enough" to help students understand probability distributions and sampling distributions in the following chapter and conditional probability in analyses of contingency tables. Students are shown that simulation can estimate probabilities not easily calculated by simple rules.

CHAPTER 6: Probability Distributions

This chapter focuses on two of the most important probability distributions--the binomial for discrete random variables and the normal for continuous random variables -- and introduces the fundamental concept of sampling distributions. This is a topic that students find difficult, yet understanding it is crucial to their understanding of statistical inference. We believe sampling distributions are simplest to explain first for categorical data--using proportions in a small-sample example that identifies easily all the possible samples of binary data. We believe that students who work through these sections and also the applet-related activities will have a stronger understanding of sampling distributions than students usually achieve in basic statistics courses. The end of the chapter links sampling distributions to the basic ideas of statistical inference, the focus of the following two chapters.

CHAPTER 7: Statistical Inference: Confidence Intervals

In our opinion, in an introductory statistics course, more emphasis should be placed on interval estimation than significance testing. We learn less from testing that a parameter takes some specific value than from finding a range of plausible values for it. Statisticians increasingly share the belief that significance tests have been overemphasized. For instance, many psychology journals now forbid reporting P-values without accompanying information about effect sizes, such as confidence intervals. Moreover, confidence intervals are simpler than tests for students to understand, especially with the use of an applet for simulation. Because of this belief, we devote this first inference chapter solely to interval estimation. An activity using simulation helps students visualize the role of the sampling distribution in forming confidence intervals. We begin the chapter with forming a confidence interval for a proportion. We feel this is an easier transition for the students since the normal distribution is used with both the large sample case and the small sample method that we present. We then move to the confidence interval for a mean, introducing the *t*-distribution. We do not cover the case in which the population standard deviation is known, as this is rarely true in practice and increases the profusion of methods that leads to student confusion.

CHAPTER 8: Statistical Inference: Significance Tests of Hypotheses

This chapter emphasizes the P-value approach, which is used in practice and is more informative than simply making a dichotomous reject vs. not reject decision at an artificial alpha-level with a rejection region. As in the previous two chapters, we begin by introducing the significance test for a proportion, then move to the significance test for a mean. Both one-sided and two-sided alternatives are presented; however, we encourage more use of the two-sided alternative since this is commonly seen in practice and the results of using a two-sided alternative parallel (two-sided) confidence intervals. Also, *t*-methods are robust to violations of the normality assumption in that case.   In this chapter we also stress the limitations of significance tests. In doing so, we emphasize the distinction between practical and statistical significance, and we show the possible difficulties resulting when only "significant" results are reported.

CHAPTER 9: Comparing Two Groups

The chapter explains that a comparison of two groups is actually a bivariate analysis, in which the explanatory variable defines the groups to be compared. We avoid the case of known population standard deviation and we de-emphasize the analysis that assumes equal variances, showing how software can find the df value with the less restrictive analysis. We present methods for independent and for dependent samples (matched pairs), with main emphasis on the former. Since this text does not short-change categorical data, we also present methods for analyzing proportions using matched pairs. This topic is relatively simple to present but is rarely treated in introductory texts. This chapter also revisits the importance of taking into account potential lurking variables when comparing two groups, illustrating with an example satisfying Simpson's paradox.

CHAPTER 10: Analyzing the Association between Categorical Variables

In our opinion, introductory statistics books place too much emphasis on the chi-squared test. Like any significance test, it gives limited information. It should not be the sole or even the primary method taught in the inference chapter about bivariate categorical data. We have included follow-up analyses such as measuring association using the difference and the ratio of proportions, inspecting residuals to detect which cells are responsible for a large chi-squared statistic, and accounting for any natural ordering in the categories. This connects this material with regression methods presented in the following chapter (e.g., correlation, inspecting residuals) for bivariate quantitative data.

CHAPTER 11: Analyzing Association between Quantitative Variables: Linear Regression Analysis

This chapter expands on the concepts of simple linear regression introduced in Chapter 3. After reviewing the introductory material from Chapter 3 on straight-line prediction equations and the correlation, this chapter explains what is meant by a model, connects this with the notion of residual unexplained variation around a regression line (and the related ANOVA table and r-squared measure and F statistic), discusses "regression toward the mean," and presents inference methods. The chapter also discusses ways of checking model assumptions, and presents the exponential regression model as an alternative for some relationships that are not linear.

CHAPTERS 12-14: Advanced Topics

Chapters 12-14 contain more advanced topics for the first course or for a follow-up course. Chapter 12 covers multiple regression, including showing how to include categorical explanatory or response variables. Chapter 13 covers one-way and two-way ANOVA and shows the connection with multiple regression. Chapter 14 gives a brief introduction to nonparametric statistics.

HOW DO WE PROMOTE STUDENT LEARNING?

In an effort to motivate students to think about the material, to ask appropriate questions, and to develop good problem-solving skills, we have developed a number of special features that distinguish this text.

Chapter Opening Example
Each chapter begins with a high interest example that establishes some key questions and themes that are woven throughout the chapter. Illustrated with engaging photographs and illustrations, this example is designed to "grab" the student and draw them into the chapter. The issues discussed in the chapter's opening example are referenced and revisited in subsequent examples within the chapter.

Example Format
Recognizing that the worked examples are the key vehicle for engaging and teaching students, we have developed a unique structure to help students learn to model the question-posing and investigative thought process required to examine issues intelligently using statistics. The four components in this structure are:

- **Picture the Scenario:** Background information is presented so students can visualize the situation. This step places the data to be investigated in context and often provides a link back to previous examples.
- **Questions to Explore:** Questions are posed that follow directly from information in the scenario. This helps the student to focus on what is to be learned from the example and what types of questions are useful to ask.
- **Think It Through:** This step is the heart of each example as the questions posed are investigated and answered using appropriate statistical methods. Each solution is clearly matched to the questions so students can easily find the response to each posed Question to Explore.
- **Insight:** The overarching goal of the Insight step is to clarify the central ideas investigated in the example and to place them in a broader context that often states the conclusions in less technical terms. Many of the Insights also provide connections between seemingly disparate topics in the text by referring back to concepts learned previously and/or foreshadowing techniques and ideas to come later
- **To practice this concept, try . . . :** Following each example, students are directed to an end of the section exercise that allows immediate practice of a problem that requires the student to use similar reasoning to that employed in the example.

This example, from Chapter 6, demonstrates our example format:

EXAMPLE 16: EXIT POLL OF CALIFORNIA VOTERS REVISITED

Picture the Scenario
Example 1, at the beginning of the chapter discussed an exit poll of 3160 voters for the 2003 special California election called to determine whether the governor, Gray Davis, should be recalled from office. If the recall were successful, Davis would be removed from office and replaced as governor by the candidate who received the most votes in the election. The winning candidate was Arnold Schwarzenegger.

Questions to Explore
Suppose that exactly 50% of the population of all voters voted in favor of the recall. Describe the mean and standard deviation of the sampling distribution of:
a. the *number* in the sample who voted in favor of it.
b. the *proportion* in the sample who voted for it.

Think It Through
a. Each voter had two options (voting yes or voting no on the recall), so we have binary data. If the recall actually had 50% support in the population of voters, then the *number* of people in the exit-poll sample of 3160 voters who voted to recall is a binomial random variable with $n = 3160$ and $p = 0.50$. That random variable has mean $\mu = np = 3160(0.50) = 1580$. We expect about 1580 to vote for the recall and about 1580 to vote against it. The standard deviation is $\sigma = \sqrt{np(1-p)} = \sqrt{3160(0.50)(0.50)} = 28.1$. There is variability around the mean, and we would probably not observe *exactly* 1580 voters supporting the recall. For example, we might observe 1550, or perhaps 1610, both values falling about a standard deviation away from the mean.

b. The *proportion* of people in the sample who voted for the recall is the binomial random variable divided by 3160. If the population proportion $p = 0.50$, then the sampling distribution of the sample proportion has

$$\text{Mean} = p = 0.50$$

$$\text{Standard deviation} = \sqrt{\frac{p(1-p)}{n}} = \sqrt{\frac{(0.50)(0.50)}{3160}} = \sqrt{0.000079} = 0.0089.$$

There is variability around the mean, and we would probably not observe *exactly* a sample proportion of 0.50 supporting the recall. For example, we might observe a sample proportion of 0.49, or 0.51, both values falling about a standard deviation from the mean.

Insight
The standard deviation of the sampling distribution of the sample *proportion* is very small (0.0089). This means that with $n = 3160$, the sample proportion will probably fall close to the population proportion of 0.50.

◆

To practice this concept, try Exercise 6. 57a-c

Exercises

The text contains a strong emphasis on real data both in the examples and exercises. Nearly all the chapters contain more than 100 exercises. These exercises are realistic and ask students to provide interpretations of the data or scenario rather than merely to find a numerical solution. We show how statistics addresses a wide array of applications, from opinion polls, to market research, to health and human behavior. Our goal is to have students come to believe the comment by the author H.G. Wells (1866 – 1946) that "Statistical thinking will one day be as necessary for efficient citizenship as the ability to read and write."

Exercises are placed at the end of each section for immediate reinforcement, and a more comprehensive set of exercises is found at the end of the chapter. Each exercise has a descriptive title. Exercises for which technology is recommended are indicated with the icon, ⌨. Larger data sets used in examples and exercises are referenced in text and made available on the text CD. The exercises are divided into the following categories:

- *Practicing the Basics:* The section exercises and the first group of end-of-chapter exercises reinforce basic application of the methods.
- *Concepts and Investigations:* These exercises require that the student explore real data sets and carry out investigations for mini-projects, and explore concepts and related theory or extensions for the chapter's methods. This section also contains some multiple-choice and true/false exercises to help students check their understanding of the basic concepts and prepare for tests. It also contains a few more difficult, optional exercises (highlighted with the ♦ ♦ icon) that present some additional concepts and methods.
- *Class Explorations:* A brief third section provides exercises that are designed for group work based on individual investigations.

In Practice

One of our goals is to help students understand some of the differences between proper "academic" statistics and what is actually done in practice. As the title of the book suggests, data analysis in practice is as much of an art as a science. Although statistical theory has foundations based on precise assumptions and conditions, in practice the real world is not so simple. This makes it difficult to give general rules, always to be obeyed, for analyzing data. To that end, we have imbedded numerous in text references and separate *In Practice* boxes that alert students to the way statisticians actually analyze data in practice. These comments are based on our extensive consulting experience, observing what well-trained statisticians do in practice.

Activities and Simulations for Hands-on Learning

Activities are an important pedagogical tool for promoting understanding by helping the student to learn statistical concepts through hands-on exploration. Although we believe the activities help to facilitate learning, we have made them optional (and placed most of them at the ends of the chapters) so that instructors may use them at their discretion. Each chapter includes at least one activity. The activity often involves simulation, commonly using an applet available on the text Web site. Additional activities are available on the text CD.

Definition and Summary Boxes

We highlight key definitions, guidelines, procedures, "in practice" remarks, and other summaries in boxes throughout the text.

Immediate Help for the Student

Material presented in the margins of the text offer at-a-glance information to aid in comprehending material in the main text. The margin features include:

- **IN WORDS:** This box explains the definitions and symbolic notation found in the body of the text in a non-technical, less formal way to help the student understand "what it really means." In many cases students will get their "feel" for the concept more from this box than from the definition itself (which, for technical accuracy, must be more formal).

- **RECALL:** As the student progresses through the book, concepts are presented that call back to information learned in previous chapters. The Recall box directs the reader back to a previous presentation in the text to review and reinforce concepts and methods already covered.

- **ACTIVITY/APPLET:** The reader is directed to an activity or applet at the Prentice Hall Web site. Applets have great value, because they vividly show students certain concepts visually. For example, creating a sampling distribution is accomplished more readily with applets than with a static text figure. As noted, the activities and applets are presented as optional explorations in the text.

On the Shoulders of. . . History as Discovery

Many chapters highlight a significant contribution to statistics and the person behind the discovery. "On the Shoulders of…" boxes give students a historical perspective, showing the context in which major statistical ideas were developed, and showing that many such ideas were developed only within the last hundred years--very recent compared to other mathematical sciences.

Graphical Approach to Learning Statistics

Students are increasingly visual learners. Recognizing this fact, we have taken extra care to make the text figures informative. For instance, we've annotated many of the figures with labels that clearly identify the noteworthy aspects of the illustration. Further, each figure caption includes a question (answered at the end of the chapter) designed to challenge the student to actively interpret and think about the information being communicated by the graphic.

Technology

The ready availability of technology helps instructors to provide instruction that is less calculation based and more concepts oriented. Output from *computer software and calculators* is displayed throughout the textbook. The discussion focuses on interpretation of the output, rather than on the keystrokes needed to create the output. We use the generic-looking, popular, and reliable MINITAB statistical software for most of our output examples but also include selected screen captures from the *TI-83+/84* graphing calculator, as appropriate.

AN INVITATION RATHER THAN A CONCLUSION

We hope that students using this textbook will gain a lasting appreciation for the vital role the art and science of statistics plays in analyzing data and helping us make decisions in our lives. As evidenced by the approach we've outlined above, our major goals for this textbook are that students should learn how:

- To produce data that can provide answers to properly posed questions.
- To appreciate how probability helps us understand randomness in our lives, and the crucial concept of a sampling distribution and how it relates to inference methods.
- To choose appropriate descriptive and inferential methods for examining and analyzing data and drawing conclusions.
- To communicate the conclusions of statistical analyses clearly and effectively.
- To understand the limitations of most research, whether because it was based on an observational study rather than a randomized experiment or survey, or because a certain lurking variable was not measured that could have explained the observed associations.

We are excited about sharing through this text the insights that we have learned from our experience as teachers and from our students. Many students still enter statistics classes on the first day with dread because of its reputation as a dry, sometimes difficult, course. It is our goal to inspire a classroom environment that is filled with creativity, openness, realistic applications, and learning that students find inviting and rewarding. Many of our students have elected to take our second courses in statistics and even major or minor in statistics. We hope that this textbook will help the instructor and the students to experience a rewarding introductory course in statistics.

Alan Agresti, Gainesville, Florida
Chris Franklin, Athens, Georgia

Chapter 1
Statistics:
The Art and Science of Learning from Data

It's never been more the case than today: To *truly* make sense of the world around you, an understanding of statistics is essential.

In the business world, managers use statistics to analyze results of marketing studies about new products, to help predict sales, and to measure employee performance. Medical research studies use statistics to evaluate whether new ways to treat disease are better than existing ways. In fact, many, if not most, professional occupations today rely on statistical methods. In a competitive job market, an understanding of statistics provides an important advantage.

But it's important to understand statistics even if you will never use it in your job. Understanding statistics can help you navigate your world and make better choices. Why? Because every day you are bombarded with statistical information from news reports, advertisements, political campaigns, and surveys. How do you know what to heed and what to ignore? An understanding of the statistical reasoning underlying these pronouncements will help. For instance, this book will enable you to evaluate news stories of medical research studies more effectively and to know when you should be skeptical of their claims. Does that new diet you've seen promoted really help you lose weight and keep the weight off? Does regular moderate drinking of alcohol truly lessen the chance of heart disease?

We realize that you are probably not reading this book in hopes of becoming a statistician. (That's too bad, because there's a severe shortage of statisticians--more jobs than trained people. And with the ever-increasing ways in which statistics is being applied, it's an exciting time to be a statistician.) You may even suffer from math phobia and fear what lies ahead. Please be assured that to learn the main concepts of statistics, logical thinking and perseverance are more important than high-powered math skills. Don't be frustrated if learning comes slowly and you need to read about a topic a few times before it starts to make sense. Just as you would not expect to sit through a foreign language class session and be able to speak that language fluently, the same is true with the language of statistics. It takes time and practice. But we promise that work will be rewarded. Once you have completed even part of this text, you will understand much better how to make sense of statistical information, and hence, the world around you.

1.1 HOW CAN YOU INVESTIGATE USING DATA?

Does a low-carbohydrate diet result in significant weight loss? Are people more likely to stop at a Starbucks if they've seen a recent TV advertisement for their coffee? Information gathering is at the heart of investigating answers to such questions. The information we gather with experiments and with surveys is collectively called **data**.

For instance, consider an experiment designed to evaluate the effectiveness of a low-carbohydrate diet. The data might consist of measurements for the people participating in the weight study at the beginning of the study, weight at the end of the study, number of calories of food eaten per day, carbohydrate intake per day, body-mass index (BMI) at the start of the study, and gender. A marketing survey about the effectiveness of a TV ad for Starbucks could collect data on the percentage of people who went to a Starbucks since the ad aired and analyze how it compares for those who saw the ad and those who did not see it.

What Is Statistics?

You already have a sense of what the word "statistics" means. You hear statistics quoted about sports events (number of points scored by each player on a basketball team), statistics about the economy (median income, unemployment rate), and statistics about opinions and beliefs and behaviors (percentage of students who indulge in binge drinking). In this sense, a statistic is merely a number calculated from data. But we'll also use "statistics" in a broader sense, to refer to a field of study that deals with methods for producing and analyzing data.

Statistics

Statistics is the art and science of designing studies and analyzing the data that those studies produce. Its ultimate goal is translating data into knowledge and understanding of the world around us. In short, **statistics is the art and science of learning from data**.

Statistical methods help us investigate questions in an objective manner. The following examples ask questions that we'll learn how to answer using statistical investigations.

SCENARIO 1: PREDICTING AN ELECTION USING AN EXIT POLL

In elections, television networks often declare the winner well before all the votes have been counted. They do this using "exit polling," interviewing voters after they leave the voting booth. Using an exit poll, a network can often predict the winner after learning how several thousand people voted, out of possibly millions of voters.

In California in October 2003, a special election was held to determine whether Governor Gray Davis should be recalled from office. In the same election, voters were asked to vote for the candidate who would replace Davis as governor if he were recalled. The four candidates included the Hollywood actor Arnold Schwarzenegger. If a majority voted yes for the recall, the candidate who received the most votes (which was Schwarzenegger) would be the new governor. The exit poll on which TV networks relied for their projections found after sampling 3160 voters that 54% of the sample voted to recall Gray Davis and 46% voted not to recall him.[1] Was this enough evidence to predict whether the recall effort succeeded, even

---

[1] www.cnn.com/ELECTION/2003/recall

though information was available about the votes of only 3160 of the 8 million voters? We'll learn how to answer that question in this book.

SCENARIO 2: MAKING CONCLUSIONS IN MEDICAL RESEARCH STUDIES

Statistical reasoning is at the foundation of the analyses conducted in most medical research studies. Let's consider three examples of how statistics can be relevant.

Does regular aspirin intake reduce deaths from heart attacks? The Harvard Medical School conducted a landmark study to investigate. The people participating in the study regularly took either an aspirin tablet or a placebo tablet (a pill with no active ingredient). Of those who took aspirin, 0.9% suffered heart attacks during the study. Of those who took the placebo, 1.7% had heart attacks, nearly twice as many. Can you conclude that it's beneficial for people to take aspirin regularly? Or, could the observed difference be explained by how it was decided which people would receive aspirin and which would receive placebo? For instance, might those who took aspirin have had better results merely because they were healthier, on the average, than those who took placebo? Or, did those taking aspirin have better diet or did they exercise more regularly, on the average?

For years there has been controversy about whether regular intake of large doses of vitamin C is beneficial. Some studies have suggested that it is. But some scientists have criticized those studies' designs, claiming that the subsequent statistical analysis was meaningless. How do we know when we can trust the results of the statistics in a medical study reported in the media?

Suppose that you wanted to investigate whether smokers are more likely than non-smokers to get lung cancer. You could pick half of the students from your class and tell them to smoke a pack of cigarettes each day for the next fifty years, and tell the other half of the class not to smoke. Fifty years from now you could see whether more smokers than nonsmokers got lung cancer. Obviously such a study is not ethical. And who wants to wait 50 years to get the answer? Well, 50 years ago a statistician in Britain figured out how to study the effect of smoking on lung cancer using data that were already available. How did he do this?

This book will show you how to answer questions such as these. You'll learn how to be skeptical of results from many medical studies.

SCENARIO 3: USING A SURVEY TO INVESTIGATE PEOPLES' BELIEFS

How similar are your opinions and lifestyle to those of others? It's easy to find out. Every other year, the National Opinion Research Center at the University of Chicago conducts the General Social Survey (GSS). This survey of about 2000 adult Americans provides data about the opinions and behaviors of the American public. You can use it to investigate how adult Americans answer a wide diversity of questions, such as, "Do you believe in life after death?," "Would you be willing to pay higher prices in order to protect the environment?," "Do you think a preschool child is likely to suffer if his or her mother works?," "How much TV do you watch per day?," and "How many sexual partners have you had in the past year?" Similar surveys occur in other countries, such as the Eurobarometer survey for countries in

> Add picture in margin of GSS website, and then below a picture of part of the table showing the results from the GSS about life after death.

the European Union. We'll use data from such surveys to illustrate the proper application of statistical methods.

Why Use Statistical Methods?

The scenarios just presented illustrate the three main aspects of statistics:

- **Design:** Planning how to obtain data to answer the questions of interest

- **Description:** Summarizing the data that are obtained

- **Inference:** Making decisions and predictions based on the data

Design refers to planning how to obtain the data. How could you conduct an experiment, so that the results are trustworthy, to investigate whether regular large doses of vitamin C are beneficial? In a marketing survey, how do you select the people to survey so you'll get data that provide good predictions about future sales?

Description means exploring and summarizing patterns in the data. Files of raw data are often huge. For example, over time the General Social Survey has collected data about hundreds of characteristics on many thousands of people. Such raw data are not easy to assess--we simply get bogged down in numbers. It is more informative to use graphs or numbers that summarize the data, such as an average amount of TV watching or the percentage of people who believe in life after death.

IN WORDS
In English, the verb **"infer"** means to arrive at a decision or prediction by reasoning from known evidence.

Inference means making decisions or predictions based on the data. Usually the decision or prediction refers to a larger group of people, not merely those in the study. For instance, in the exit poll described in Scenario 1, of 3160 voters sampled, 54% said they voted to recall Gray Davis of California from the governor's job, whereas 46% said they voted against the recall. Using these data, we'll see we can predict that the recall would be successful out of the 8 million people who voted (opening the door for Arnold Schwarzenegger to become governor). Stating the percentages for the sample of 3160 voters is *description*, while predicting the outcome for all 8 million voters is *inference*.

Statistical description and inference are the ways of analyzing the data. You can use them to investigate and answer questions that are important to society. For instance, "Has there been global warming over the past decade?" "Is having the death penalty available for punishment associated with a reduction in violent crime?" "Does student performance in school depend on the amount of money spent per student, the size of the classes, or the teachers' salaries?" "How risky is it to take the smallpox vaccine?"

A topic that we have not mentioned yet but that is fundamental for developing statistical inference methods is **probability**. We'll study probability because it will help us to answer a question such as, "If the recall vote were actually supported by less than half of all voters, what's the chance it would get 54% support in an exit poll of 3160 voters?" If the chance were extremely small, we'd feel comfortable making the inference that the recall was supported by the majority of all 8 million voters.

<table>
<tr><td>Add **GSS** screen capture here for TV hours</td></tr>
</table>

ACTIVITY 1: DOWNLOADING DATA FROM THE INTERNET

It is simple to get descriptive summaries of data from the General Social Survey (GSS). We'll demonstrate, using one question it asked in recent surveys, "On a typical day, about how many hours do you personally watch television?"

- Go to the Web site www.icpsr.umich.edu/GSS/

- Click on the *Analyze* tab at the top of the screen.

- In the *Select an action* menu, place the dot at the *Frequencies and crosstabulation* option. Click on *Start*.

- The GSS name for the number of hours of TV watching is TVHOURS. Type TVHOURS as the *Row variable* name. Click on *Run the table*.

Now you'll see a table that shows the number of people and the percentage who made each of the possible responses. For all the years combined in which this question was asked, the most common response was 2 hours of TV a day (about 27% made this response).

What percentage of the people surveyed reported watching 0 hours of TV a day? How many people reported watching TV 24 hours a day?

Another question asked in the GSS is, "Taken all together, would you say that you are very happy, pretty happy, or not too happy?" The GSS name for this item is HAPPY. What percentage of people reported being very happy?

You might use the GSS to investigate what sorts of people are more likely to be very happy. Those who are happily married? Those who are in good health? Those who have lots of friends? Those who like their job? Those who attend religious services frequently? We'll see how to find out in this book.

IN WORDS

We'll see that the term "variable" refers to the characteristic being measured, such as number of hours per day that you watch TV.

To practice Activity 1, try Exercises 1.3 and 1.4.

SECTION 1.1: PRACTICING THE BASICS

1.1. Aspirin and heart attacks: The Harvard Medical School study mentioned in Scenario 2 used about 22,000 male physicians. The method of deciding whether a given person would receive aspirin or placebo was essentially made by flipping a coin. About 11,000 were assigned to take aspirin, and about 11,000 to take the placebo. The researchers described results by the percentage of each group that had a heart attack during the study. This was 0.9% for those taking aspirin and 1.7% for those taking placebo. They then used statistical methods to predict that if *all* male physicians could have participated in this study, the percentage having a heart attack would have been lower for those taking aspirin. Specify the aspect of this study that refers to (a) design, (b) description, and (c) inference.

1.2 Poverty and race: The **Current Population Survey (CPS)** is a monthly survey of households conducted by the U.S. Census Bureau for the Bureau of Labor Statistics. It provides a comprehensive body of data on the labor force, employment, and unemployment. A CPS of about 60,000 households indicated that of those households, 8.0% of the whites, 23.4% of the blacks, and 22.7% of the Hispanics had annual income below the poverty level (*Statistical Abstract of the United States, 2000*). Using these data, a statistical method makes the prediction that the percentage of *all* black households in the United States that had income below the poverty level was at least 22% but no greater than 25%. Specify the aspect of this study that refers to (a) description and (b) inference.

⌨ **1.3 GSS and heaven:** Go to the General Social Survey Web site www.icpsr.umich.edu/GSS/. By entering HEAVEN as the "row variable" name, find the percentages of people who said "yes, definitely," "yes, probably," "no, probably not," and "no, definitely not" when asked whether they believed in heaven.

⌨ **1.4 GSS and heaven and hell:** Refer to the previous exercise. You can obtain data for a particular survey year such as 1998 by entering YEAR(1998) in the "selection filter" option box before you click on *Run the Table*.
a. Do this for HEAVEN in 1998, giving the percentages for the four possible outcomes. (This question was asked only in 1991 and 1998.)
b. Summarize opinions in 1998 about belief in hell (row variable HELL). Was the percentage of "yes, definitely" responses higher for belief in heaven or in hell?

⌨ **1.5 GSS for subject you pick:** At the GSS Web site, click on *Subject* under Codebook Indexes. Find a subject that interests you and look up a relevant GSS code name to enter as the "row variable." Summarize the results that you obtain.

1.2 WE LEARN ABOUT POPULATIONS USING SAMPLES

We've seen that statistics consists of methods for **designing** investigative studies, **describing** (summarizing) data obtained for those studies, and making **inferences** (decisions and predictions) based on those data.

We Observe Samples But We Are Interested in Populations

The entities that we measure in a study are called the **subjects**. Usually the subjects are people, such as the individuals interviewed in a General Social Survey. But, they need not be. For instance, the subjects could be schools, countries, or days. We might measure characteristics such as:

- For each school: the per-student expenditure, the average class size, the average score of students on an achievement test

- For each country: the percentage of residents living in poverty, the birth rate, the percentage unemployed, the percentage who are computer literate

- For each day in an Internet café that you manage: the amount spent on coffee, the amount spent on food, the amount spent on Internet access

The **population** is the set of all the subjects of interest. In practice, we usually have data for only *some* of the subjects who belong to that population. These subjects are called a **sample**.

<table>
<tr><td>

IN WORDS:

Population: All subjects of interest

Sample: Subjects for whom we have data

</td><td>

Definition: Population and Sample

The **population** is the total set of subjects in which we are interested. A **sample** is the subset of the population for whom we have (or plan to have) data.

</td></tr>
</table>

In the 2002 General Social Survey (GSS), the sample was the 2765 people who participated in this survey. The population was the set of all adult Americans at that time (over 200 million).

EXAMPLE 1: WHAT'S THE SAMPLE AND THE POPULATION FOR AN EXIT POLL?

<table>
<tr><td>

EXAMPLE FORMAT

Examples in this book use the four parts shown in this example: **Picture the Scenario** introduces the context. **Question to Explore** states the question addressed. **Think It Through** shows the reasoning used to answer that question. **Insight** gives follow-up comments related to the example. Following the example, **To practice this concept** directs you to a similar "Practicing the Basics" exercise at the end of the section.

</td><td>

Picture the Scenario
Scenario 1 in the previous section discussed an exit poll. The purpose was to predict the outcome of the 2003 special election in California to consider whether the governor Gray Davis should be recalled from office. The exit poll sampled 3160 of the 8 million people who voted.

Question to Explore
For this exit poll, what was the population and what was the sample?

Think It Through
The population was the total set of subjects of interest, namely, the 8 million people who voted in this election. The sample was the 3160 voters who were interviewed in the exit poll. These are the people for whom the poll got data about their vote.

Insight
The ultimate goal of any study is to learn about the *population*. For example, the sponsors of this exit poll wanted to make an inference (prediction) about *all* 8 million voters, not just the 3160 voters sampled by the poll.

♦

</td></tr>
</table>

To practice this concept, try Exercises 1.7 and 1.8.

Occasionally data are available from an entire population. For instance, every ten years the U.S. Census Bureau gathers data from the entire population of Americans (or nearly all of them). But the census is an exception. Usually, it is too costly and time consuming to obtain data from an entire population. It is more practical to get data for a sample. The General Social Survey and polling organizations such as the Gallup poll usually select samples of about 1000 to 2500 Americans to learn about opinions and beliefs of the population of *all* Americans. The same is true for surveys in other parts of the world, such as Eurobarometer in Europe.

Descriptive Statistics and Inferential Statistics

Using the distinction between samples and populations, we can now tell you more about the use of **description** and **inference** in statistical analyses.

Description

Descriptive statistics refers to methods for summarizing the data. The summaries usually consist of graphs and numbers such as averages and percentages.

A descriptive statistical analysis usually combines graphical and numerical summaries. For instance, Figure 1.1 is a **bar graph** that shows the percentages of various types of households in the U.S. as of 2002. It summarizes a survey of 50,000 American households by the U.S. Census Bureau. The main purpose of descriptive statistics is to reduce the data to simple summaries without distorting or losing much information. Graphs and numbers such as percentages and averages are easier to comprehend than the entire set of data. It's much easier to get a sense of the data by looking at Figure 1.1 than by reading through the questionnaires filled out for the 50,000 sampled households. From this graph, it's readily apparent that the "traditional" household, defined as being a married man and woman with children in which only the husband is in the labor force, is no longer very common in the U.S. In fact, "Other" households, which include female-headed households and households headed by young adults or older Americans who do not reside with spouses are by far the most common.

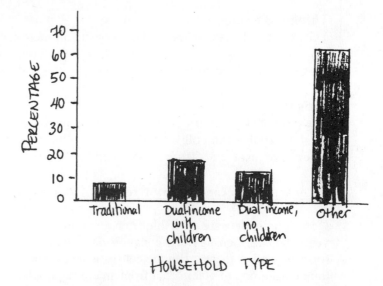

Figure 1.1: Types of U.S. Households, Based on Sample of 50,000 Households in the 2002 Current Population Survey.

Although data are usually available only for a sample, descriptive statistics are also useful when data are available for the entire population, such as in a census. By contrast, inferential statistics are used when data are available only for a sample but we want to make a decision or prediction about the entire population.

Inference

Inferential statistics refers to methods of making decisions or predictions about a population, based on data obtained from a sample of that population.

In most surveys, we have data for a sample, not for the entire population. We use descriptive statistics to summarize the sample data and inferential statistics to make predictions about the population.

EXAMPLE 2: OPINION ABOUT HANDGUN CONTROL

Picture the Scenario

Suppose we'd like to know what people think about controls over the sales of handguns. Let's consider how people feel who live in Florida, a state with a relatively high violent crime rate. The population of interest is the set of more than 10 million adult residents of Florida.

Since it is impossible to discuss the issue with all these people, we can study results from a recent poll of 834 residents of Florida conducted by the Institute for Public Opinion Research at Florida International University. In that poll, 54.0% of the sampled subjects said they favored controls over the sales of handguns. A newspaper article about the poll, using a method presented in this book, reports that the "margin of error" for how close this number falls to the population percentage is 3.4%. We'll see (in Chapter 7) that this means we can predict with high confidence (about 95% certainty) that the percentage of *all* adult Floridians favoring control over sales of handguns falls within 3.4% of the survey's value of 54.0%, that is, between 50.6% and 57.4%.

Question to Explore

In this analysis, what is the descriptive statistical analysis and what is the inferential statistical analysis?

Think It Through

The results for the sample of 834 Florida residents are summarized by the percentage, 54.0%, who favored handgun control. This is a descriptive statistical analysis. We're interested, however, not just in those 834 people but in the *population* of *all* adult Florida residents. The prediction that the percentage of *all* adult Floridians who favor handgun control falls between 50.6% and 57.4% is an inferential statistical analysis. In summary, we *describe* the *sample*, and we make *inferences* about the *population*.

Insight

The sample size of 834 was small compared to the population size of more than 10 million. However, because the values between 50.6% and 57.4% are all above 50%, the study concluded that a slim majority of Florida residents favored handgun control.

◆

To practice this concept, try Exercises 1.10 and 1.11.

An important aspect of statistical inference involves reporting the likely *precision* of a prediction. How close is the *sample* value of 54% likely to be to the true (unknown) percentage of the *population* favoring gun control? We'll see (in Chapters 4 and 6) why a well-designed sample of 834 people yields a sample percentage value that is very likely to fall within about 3 to 4% (the so-called *margin of error*) of the population value. In fact, we'll see that inferential statistical analyses can predict characteristics of entire populations quite well by selecting samples that are small relative to the population size. That's why most polls take samples of only about a thousand people, even if the population has millions of people. In this book, we'll see why this works.

Randomness and Variability

A sample tends to be a good reflection of a population when each subject in the population has the same chance of being included in that sample. That's the basis of **random sampling**. A simple example of random sampling is when a teacher puts each student's name on a slip of paper and places it in a hat and then draws names from the hat without looking. We'll see that random sampling allows us to make powerful inferences about populations.

Randomness is also crucial to performing experiments well. If, as in Scenario 2, we want to compare aspirin to placebo in terms of the percentage of people who later have a heart attack, it's best to randomly select who in the sample uses aspirin and who uses placebo. This tends to keep the groups balanced on all the factors that could affect the results. Without randomizing, for example, the people who decided to use placebo might have tended to be healthier than those who picked aspirin, which could produce misleading results.

People are different from each other, so, not surprisingly, the measurements we make on them vary from person to person. For the GSS question about TV watching in Activity 1, different people reported different amounts of TV watching. In the exit poll of Example 1, not all people voted the same way. If subjects did not vary, we'd need to sample only one of them (and it would be a rather uninteresting world). We learn more about this variability by sampling more people. If we want to predict the outcome of an election, we're better off sampling 100 voters than one voter, and our prediction will be even more reliable if we sampled 1000 voters.

Just as people vary, so do samples. Suppose you take an exit poll of 1000 voters to predict the outcome of an election. Suppose the Gallup organization also takes an exit poll of 1000 voters. Your sample will have different people than Gallup's. Consequently, the predictions will also differ. Perhaps your exit poll of 1000 voters has 480 voting for the Republican candidate, so you predict that 48% of all voters voted for that person. Perhaps their exit poll of 1000 voters has 440 voting for the

Republican candidate, so they predict that 44% of all voters voted for that person. Activity 2 at the end of the chapter shows, however, that with random sampling the amount of variability from sample to sample is quite predictable. Both of your predictions are very likely to fall within 5% of the actual population percentage who voted Republican.

The Basic Ideas of Statistics

Now that you know the purpose of descriptive statistics and inferential statistics, we can tell you about the key concepts of statistics that you'll learn in this book:

Chapter 2: Exploring Data with Graphs and Numerical Summaries--How can you present simple summaries of data? You replace lots of numbers with simple graphs and numerical summaries.

Chapter 3: Association: Contingency, Correlation, and Regression--How does annual income ten years after graduation correlate with college GPA? You can find out by studying the **association** between those characteristics.

Chapter 4: Gathering Data--How can you design an experiment or conduct a survey to get data to help you answer questions? You'll see why results may be misleading if you don't use **randomization**.

Chapter 5: Probability in Our DailyLives--How can you determine the chance of some outcome, such as your winning a lottery? Probability, the basic tool for evaluating chances, is also the key to how well inferential statistics work.

Chapter 6: Probability Distributions--You've probably heard of the **normal distribution** or "bell-shaped curve" that describes peoples' heights or IQs or test scores. Why does it occur so often? You'll see why, and you'll learn its key role in statistical inference.

Chapter 7: Statistical inference: Confidence Intervals--How can an exit poll of 2232 voters possibly predict well the results for millions of voters? You'll find out by applying the probability concepts of Chapters 5 and 6 to make **statistical inferences** that show how closely you can predict summaries such as population percentages.

Chapter 8: Statistical inference: Significance Tests About Hypotheses--How can a medical study make a decision about whether a new drug is better than a placebo? You'll see how you can control the chance that a statistical inference makes a correct decision about what works best.

Chapters 9-14: Applying Descriptive and Inferential Statistics to Many Kinds of Data--After Chapters 2-8 introduce you to the key concepts of statistics, the rest of the book shows you how to apply them in lots of situations. For instance, Chapter 9 shows how to compare two groups, such as using a sample of students from your university to make an inference about whether male and female students have different rates of binge drinking.

SECTION 1.2: PRACTICING THE BASICS

1.6 Description and inference:
a. Distinguish between *description* and *inference* as reasons for using statistics. Illustrate the distinction using an example.
b. You have data for a population, such as obtained in a census. Explain why descriptive statistics are helpful but inferential statistics are not needed.

1.7 EPA: The Environmental Protection Agency (EPA) uses a few new automobiles of each model every year to collect data on pollution emission and gasoline mileage performance. For the Honda Accord model, identify what's meant by the (a) subject, (b) sample, and (c) population.

1.8 Screening tests are fallible: A *New York Times* article (Feb. 17, 1999) about the PSA blood test for detecting prostate cancer stated: "The test fails to detect prostate cancer in 1 in 4 men who have the disease." Do you think this refers to a study in which the data set was obtained for a sample, or for a population? Explain.

1.9 Aspirin inference: The medical study mentioned in Scenario 2 and in Exercise 1.1 conducted an experiment with 22,000 male physicians, to investigate whether regular intake of aspirin reduces the chance of heart attack compared to taking placebo. The study concluded that for *all* male physicians, aspirin would be more effective than placebo. Identify (a) the sample, (b) the population, and (c) the inference.

1.10 Graduating seniors' salaries: The job placement center at your school surveys all graduating seniors at the school. Their report about the survey provides statistics such as the average starting salary and the percentage of students earning more than $30,000 a year. Are these descriptive, or inferential, statistical analyses? Explain.

1.11 At what age did women marry?: A historian wants to estimate the average age at marriage for women in New England in the early nineteenth century. Within her state archives she finds marriage records for the years 1800-1820. She takes a sample of those records, noting the age of the bride for each. The average age in the sample is 24.1 years. Using a statistical method, she finds a margin of error and estimates that the average age of brides for the population was between 23.5 and 24.7.
a. What part of this example is descriptive, giving a summary of the data?
b. What part of this example is inferential, making a prediction about a population?
c. To what population does the inference refer?

1.12 Age pyramids as descriptive statistics: The figure shown is a graph published by Statistics Sweden. It compares Swedish society in 1750 and in 1999 on the numbers of men and women of various ages, using "age pyramids." Explain how this indicates that:
a. In 1750, few Swedish people were old.
b. In 1999, Sweden had many more people than in 1750.
c. In 1999, of those who were very old, more were female than male.
d. In 1999, the largest five-year group included "baby boomers" born right after World War II.

Graphs of number of men and women of various ages, in 1750 and in 1999

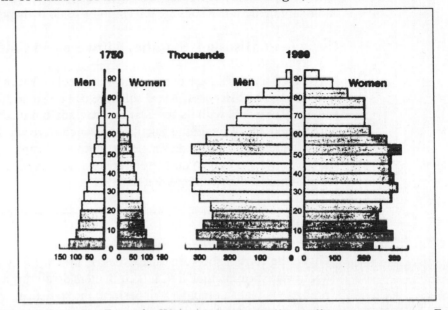

📖 **1.13 Gallup polls:** Go to the Web site for the Gallup poll, www.gallup.com. From reports listed on their homepage, give an example of (a) a descriptive statistical analysis and (b) an inferential statistical analysis.

1.14 National service: Consider the population of all students at your school. A certain proportion support mandatory national service (MNS) following high school. Your friend randomly samples 20 students from the school, and uses the sample proportion who support MNS to predict the population proportion at the school. You take your own, separate random sample of 20 students, and find the sample proportion that supports MNS.
a. For the two studies, are the populations the same?
b. For the two studies, are the sample proportions necessarily the same? Explain.

1.15 Samples vary less with more data: We'll see that the amount by which results vary from sample to sample depends on the sample size. In an election with two candidates, Smith and Jones, suppose Smith receives close to 50% of all votes. An exit poll is taken on the day of the election to predict the winner.
a. Which case would you find more surprising—taking an exit poll of 10 voters and finding that 0% or 100% of them voted for Smith, or taking an exit poll of 1000 voters and finding that 0% or 100% of them voted for Smith?
b. In exit polls, samples of size 1000 voters each will tend to give more similar percentages voting for a particular candidate than samples of 10 voters each. Using part (a), explain why you would expect this to be true.

1.3 WHAT ROLE DO COMPUTERS PLAY IN STATISTICS?

Today's researchers (and students) are lucky: Unlike those in the previous generation, they don't have to do complex statistical calculations by hand. Powerful user-friendly computing software and calculators are now readily available for statistical analyses. New statistical methods continue to be developed all the time, and many modern methods are too complex to do by hand.

Using (and Misusing) Statistics Software and Calculators

MINITAB and **SPSS** are two popular statistical software packages on college campuses. The **Microsoft Excel** software can do some statistical methods, sorting and analyzing data with its spreadsheet program, but its capabilities are limited. The **TI-83+** and **TI-84**[2] graphing calculators, which have similar output, are useful as portable tools for generating simple statistics and graphs. Throughout this text, we'll show examples of what you'll see if you use such products. We'll learn what to look for in output and how to interpret it.

IN PRACTICE

Given the current software capabilities, in practice why do we still have to learn about statistical methods? Can't computers do all this analysis for you? The problem is that a computer will perform the statistical analysis you request whether or not its use is valid in any given situation. Just knowing how to use software does not guarantee a proper analysis. You'll need a good background in statistics to understand which statistical method to use, which options to choose with that method, and how to interpret and make valid conclusions from the computer output. This text helps to give you this background.

Data Files

To make statistical analysis easier, large sets of data are organized in a **data file**. This file usually has the form of a spreadsheet. It is the way statistical software expects to receive the data.

Figure 1.2 is an example of part of a data file. It shows how a data file looks in **MINITAB**. The file shows data for eight students on the following characteristics:

> Gender (f = female, m = male)
> Racial-ethnic group (b = black, h = Hispanic, w = white)
> Age (in years)
> College GPA (scale 0 to 4)
> Average number of hours per week watching TV
> Whether a vegetarian (yes, no)
> Political party (dem = Democrat, rep = Republican, ind = independent)
> Marital status (1 = married, 0 = unmarried).

---

[2] We will use the shorthand, TI-83+/84, to indicate output from a graphing calculator

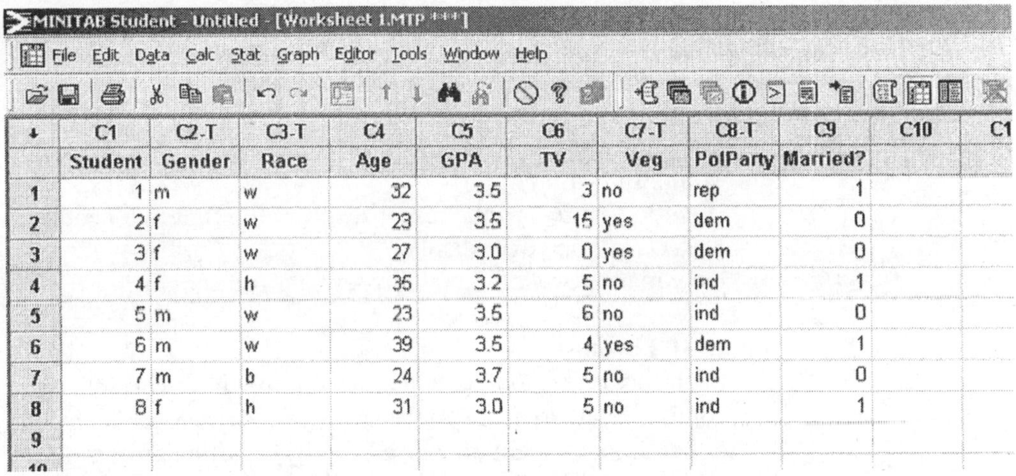

	C1	C2-T	C3-T	C4	C5	C6	C7-T	C8-T	C9	C10	C1
	Student	Gender	Race	Age	GPA	TV	Veg	PolParty	Married?		
1	1	m	w	32	3.5	3	no	rep	1		
2	2	f	w	23	3.5	15	yes	dem	0		
3	3	f	w	27	3.0	0	yes	dem	0		
4	4	f	h	35	3.2	5	no	ind	1		
5	5	m	w	23	3.5	6	no	ind	0		
6	6	m	w	39	3.5	4	yes	dem	1		
7	7	m	b	24	3.7	5	no	ind	0		
8	8	f	h	31	3.0	5	no	ind	1		
9											
10											

Figure 1.2: Part of a MINITAB Data File. Each row shows data for a particular student. Each column shows data for a particular characteristic.

Figure 1.2 shows the two basic rules for constructing a data file:

- Any one row contains measurements for a particular subject (for instance, person).
- Any one column contains measurements for a particular characteristic.

Some characteristics have numerical data, such as the values for hours of TV watching. Some characteristics have data that consist of categories or labels, such as the categories (yes, no) for whether a vegetarian or the labels (dem, rep, ind) for the categories of political affiliation.

Figure 1.2 resembles a larger data file from a questionnaire administered to a sample of students at the University of Florida. That data file, which includes other characteristics as well, is the one called "Florida student survey" on the CD that comes with this text. (You may want to find this file on the CD now, to practice accessing data files used in some homework exercises.)

To construct a similar data file for your class, try Activity 3 in the Chapter Explorations at the end of the chapter.

EXAMPLE 3: SHOULD YOUR COMPANY USE CATALOG SALES?

> To Come: Photo from "High Fidelity" of independent record shop

Picture the Scenario
You are the manager of a store that sells music CDs. Your sales have been shrinking because of competition from Internet companies (such as amazon.com) and the increasing practice of people downloading music from the Internet. You are considering trying to increase your sales by sending catalogs to potential customers. Because of the costs of sending the catalog, you would need to average at least $3 in orders per catalog mailed. To investigate whether you will be able to recover the costs and to find out which CDs are the most popular, you send a catalog to 500 potential customers. For the 10 CDs that you put in a special display on the cover of

the catalog, the first person who was sent the catalog ordered nothing, and the second person ordered one copy of CD 3 and two copies of CD 5, at a cost of $15 each. After a month you create a data file summarizing the orders of the 500 potential customers.

Questions to Explore

a. How do you record the responses from the first two people on the data file for the 10 CDs on the cover display?
b. How many rows of data will your data file contain?

Think It Through

a. Each row refers to a particular person to whom you sent the catalog. Each column refers to a CD on the cover of the catalog. Each entry of data is the dollar amount spent on a given CD by a particular person. So, the first two rows of the data file, which give the results for the first two people, are:

Person	cd1	cd2	cd3	cd4	cd5	cd6	cd7	cd8	cd9	cd10
1	0	0	0	0	0	0	0	0	0	0
2	0	0	15	0	30	0	0	0	0	0

b. The entire data file would consist of 500 lines of data, one line for each potential customer.

Insight

From the entire data file, you might determine that the average sales per person equaled $4. What could you then conclude about the average sales if you now send the catalog to tens of thousands of potential customers? Is it likely to be at least $3, so you'll break even? As we'll see (Chapters 4 and 8), you would need other information from the data file and about the method of sampling to answer this question. If you decide to start using mail-order marketing, you'll want to see how statistics can help you use information that you gather about customers' demographics and past purchases to develop and maintain a promising list of potential customers.

♦

To practice this concept, try Exercise 1.17 .

Databases

Most studies design experiments or surveys to collect data to answer the questions of interest. Often, though, it is adequate to take advantage of existing archived collections of data files, called **databases**. Many databases are available on the World Wide Web. By browsing various Internet addresses, you can obtain information on all sorts of things.

The General Social Survey, discussed in Scenario 3 and Activity 1, is one such database. The CD that comes with this text contains a database that has several data files that you'll use in some of the exercises. Here are some other databases that can be fun and informative to browse:

- Are you interested in pro baseball or basketball or football? Click on the statistics links at www.baseball.com or www.nba.com or www.nfl.com.

- Are you interested in what people believe around the world? Check out www.globalbarometer.org or www.europa.eu.int/comm/public_opinion or www.latinobarometro.org.

- Are you interested in results of Gallup polls about peoples' beliefs? See www.gallup.com.

Also very useful are search engines such as Google. Type in "data" in the search window at www.google.com, and it lists databases such as those maintained by the U.S. Census Bureau and Statistics Canada. A search engine also is helpful for finding data or descriptive statistics on a particular topic. For instance, if you want to find summaries of Canadians' opinions about the legalization of marijuana for medical treatment, try typing

poll legalize marijuana Canada medicine

in the search window.

Not all databases or reported data summaries give reliable information. For instance, many news organizations and search Web sites (such as cnn.com, abc.abcnews.go.com, netscape.com) ask you to participate in a "topic of the day" poll on their home page. Their summaries of how people responded are not reliable indications of how the entire population feels. They are biased, because it is unlikely that the people who respond are representative of the population. We'll explain why in Chapter 4.

Applets

IN WORDS
Applet = *appl*ication + *et*
An applet is a short *application* program for performing a specific task. We'll use them throughout the text.

Just like riding a bike, it's easier to learn statistics if you are actively involved, learning by doing. One way is to practice, by applying software or calculators to data files or data summaries that we'll provide. Another way is to perform activities that illustrate the ideas of statistics using **applets**. Using an applet, you can take samples from artificial populations and analyze them to discover properties of statistical methods applied to those samples. This is a type of **simulation**--using a computer to mimic what would actually happen if you selected a sample and used statistics in real life. So, let's get started with your active involvement.

ACTIVITY 2: LET'S USE AN APPLET TO SIMULATE RANDOMNESS AND VARIABILITY

Let's get a feel for randomness and variability, the concepts introduced on page 10 of this chapter, that we'll see are key ones in the study of statistics. To do this, we will simulate taking an exit poll of voters. We will do this using the "sampling applet" at www.prenhall.com/*TBA*. This applet generates samples resembling those we'd get with random sampling. Select from the menu for the parent population, *binary*. Each observation from an individual in the population then has two possible outcomes, as in sampling a person who voted in an election with two candidates.

Let's see what would happen with exit polling when 50% of the entire population (a proportion of 0.50) voted for each candidate. We'll first use a small poll, only 10 voters. First, represent this by taking a coin and flipping it ten times. Let the number of heads represent the number who voted for the Democrat. What proportion in your sample of size 10 voted for the Democrat?

Now let's do this with the applet. We'll regard outcome 1 as voting for the Democratic candidate and outcome 0 as voting for the Republican candidate. Go to the menu and set the population proportion for outcome 1 at 0.50. To take a sample of size 10, you go to *the sample size* menu and select 10 and then click on *animated*. You will see that a certain number of outcomes occurred for each type. When we did this, we got outcome 1 four times and outcome 0 six times. This simulates sampling 10 voters in the exit poll, in which 4 said they voted for the Democratic candidate and 6 for the Republican candidate. It corresponds to a sample proportion of 0.40 voting for the Democrat.

To illustrate how samples vary, take your own sample of size 10 using this applet. When you do this, you will probably get a proportion different than 0.40 voting for the Democrat, because the process is random. What did you get? Now collect another sample of size 10 and find the sample proportion. What did you get? Repeat taking samples of size 10 at least 5 times. Note the sample proportion for each sample and how the sample proportions compare to the population proportion, 0.50. Are they always close?

Now repeat this for a larger exit poll, taking a sample of 1000 voters instead of 10 (set *sample size* = 1000 on the applet menu). What proportion voted for the Democrat? Repeat taking samples of size 1000 at least 5 times. Note the sample proportion for each sample and how these sample proportions compare to the population proportion, 0.50. Do the sample proportions tend to be close to 0.50? We predict that all of your sample proportions fell between 0.45 and 0.55. Are we correct?

We would expect that some of the sample proportions you generated using a sample size of 10 fell much farther from 0.50 than the sample proportions you generated using a sample size of 1000. This illustrates by simulation that sample proportions tend to be closer to the population proportion when the sample size is larger. In fact, we will discover as we move forward in the textbook that we do much better in making inference about the population with larger sample sizes.

To practice this activity, try Exercises 1.21, 1.22, and 1.35.

SECTION 1.3: PRACTICING THE BASICS

1.16 Data file for friends: Construct (by hand) a data file of the form of Figure 1.2, for two characteristics with a sample of four of your friends. One characteristic should take numerical values, and the other should take values that are categories.

1.17 Shopping sales data file: Construct a data file describing the purchasing behavior of five people, described below, who visit a shopping mall. Enter purchase amounts they spent on clothes, sporting goods, books, and music CDs as the data. Customer 1 spent $49 on clothes and $16 on music CDs, customer 4 spent $92 on books, and the other three customers did not buy anything.

1.18 Internet poll: The sample of people who register their opinion in a poll at an Internet site is not a random sample. Why not? (*Hint*: For a random sample, every person in the population has the same chance of being in the sample.)

⌨ **1.19 Create a data file with software:** Your instructor will show you how to create data files using the software for your course. Use it to create the data file you constructed by hand in Exercise 1.16 or 1.17. Print a copy of the data file.

⌨ **1.20 Use a data file with software:** You will need to learn how to open a data file from the text CD or download one from the web for use with the software for your course. Do this for the "Florida student survey" data file on the text CD, from the survey mentioned following Figure 1.2. Print a copy of the data file.

⌨ **1.21 Simulate with the "sampling applet":** Refer to Activity 2 on page 18.
a. Repeat the activity using a population proportion 0.60: Take at least five samples of size 10 each, and observe how the sample proportions of the 1 outcome vary around 0.60, and then do the same thing with at least five samples of size 1000 each.
b. In (a), what seems to be the effect of the sample size on the amount by which sample proportions tend to vary around the population proportion, 0.60?
c. What is the practical implication of the effect of the sample size summarized in (b) with respect to making inferences about the population proportion when you collect data and observe only the sample proportion?

⌨ **1.22 Is a sample unusual?:** Suppose 70% of all voters voted for the Republican candidate. If an exit poll of 50 people were chosen randomly, would it be surprising if less than half that were sampled said they voted Republican? To reason an answer, use the applet described in Activity 2 to conduct at least ten simulations of taking samples of size 50 from a population with proportion 0.70. Note the sample proportions. Do you observe any sample proportion less than 0.50? What does this suggest?

CHAPTER SUMMARY

* Statistics consists of methods for conducting research studies and for analyzing and interpreting the data produced by those studies. **Statistics is the art and science of learning from data**.

- The first part of the statistical process involves **design**--planning an investigative study, such as how to obtain relevant data. The design often involves taking a **sample** from a **population** that contains *all* the **subjects** (usually, people) of interest. After we've collected the data, there are two types of statistical analyses:

    Descriptive statistics summarize the sample data with numbers and graphs.

    Inferential statistics make decisions and predictions about the entire population, based on the information in the sample data.

- With **random sampling**, each subject in the population has the same chance of being in the sample. This is desirable, because then the sample tends to be a good reflection of the population. Randomization is also important to good experimental design.

- The measurements we make of a characteristic **vary** from individual to individual. Likewise, results of descriptive and inferential statistics **vary**, depending on the sample chosen. We'll see that the study of **variability** is a key part of statistics. **Simulation** investigations generate many samples randomly, often using an **applet**. They provide a way of learning about the impact of randomness and variability from sample to sample.

- The calculations for data analysis are simple using computer software. The data are organized in a **data file**. This has a separate row of data for each subject and a separate column for each characteristic. However, you'll need a good background in statistics to understand which statistical method to use and how to interpret and make valid conclusions from the computer output.

CHAPTER PROBLEMS: PRACTICING THE BASICS

1.23 Charles and Camilla: An Associated Press (AP) story in 2002 reported that 55% of Britons believe that Prince Charles should marry his longtime consort, Camilla Parker Bowles. In articles such as this, do you think that the percentage reported was obtained using data from a sample, or from the entire population? Explain.

1.24 UW Student survey: In a University of Wisconsin (UW) study about alcohol abuse among students, 100 of the 40,858 members of the student body in Madison were sampled and asked to complete a questionnaire. One question asked was, "On how many days in the past week did you consume at least one alcoholic drink?"
a. Identify the population and the sample.
b. For the 40,858 students at UW, one characteristic of interest was the percentage who would respond "zero" to this question. For the 100 students sampled, suppose 29% gave this response. Does this mean that 29% of the entire population of UW students would make this response? Explain.

1.25 ESP: For several years, the General Social Survey asked subjects, "How often have you felt as though you were in touch with someone when they were far away from you?" Of 3887 sampled subjects who had an opinion, 1407 said never and

20

2480 said at least once. The proportion who had at least one such experience was 2480/3887 = 0.638.
a. Describe the population of interest.
b. Explain how the sample data are summarized using descriptive statistics.
c. For what population characteristic might we want to make an inference?

1.26 Presidential popularity: Each month the media report the U. S. president's current popularity rating. A CNN story in June 2004 about a Gallup poll of 1002 Americans reported, "47% of people polled said they approve of how Bush is handling the presidency. The margin or error is plus or minus 3 percentage points." Explain how this provides an *inferential* statistical analysis.

1.27 Bush vs. Kerry in other countries: A BBC story (September 9, 2004) about a poll in 35 countries about whether people favored George W. Bush or John Kerry in the 2004 U.S. Presidential election stated that Kerry was clearly preferred in 30 countries and Bush was clearly preferred in 3 countries. For instance, the percentage breakdown for (Kerry, Bush) was reported as (74, 7) in Norway (with the other 19% undecided or not reporting), (74, 10) in Germany, (64, 5) in France, (58, 14) in Italy, (45, 7) in Spain, (47, 16) in the UK, (61, 16) in Canada, but (26, 31) in Poland and (32, 57) in the Philippines. The company that performed the polling (Globescan) reported "The margin of error ranged from 2.3 to 5%."
a. Do these results summarize sample data, or population data?
b. Identify a descriptive aspect of this analysis.
c. Identify an inferential aspect of this analysis.

1.28 Reducing stress: Your school wants to make an inference about the percentage of students at the school who prefer having a several-day period between the end of classes and the start of final exams to help reduce the level of stress as students prepare for exams. A survey is taken of 100 students.
a. Identify the sample and the population.
b. For the study, explain the purpose of using (i) descriptive statistics and (ii) inferential statistics.

1.29 Marketing study: For the marketing study about CD sales in Example 3, identify the (a) sample and population, (b) descriptive and inferential aspects.

Select the correct response in the following two multiple-choice questions:

1.30 Believe in reincarnation?: In a survey of 750 Americans conducted by the Gallup organization, 24% indicated a belief in reincarnation. A method presented later in this book allows us to predict that for *all* adult Americans, the percentage believing in reincarnation falls between 21% and 27%. This prediction is an example of
a. descriptive statistics
b. inferential statistics
c. a data file
d. designing a study

1.31 Use of inferential statistics? : Inferential statistics are used
a. to describe whether a sample has more females or males.
b. to reduce a data file to easily understood summaries.

c. to make predictions about populations using sample data.

d. when we can't use statistical software to analyze data.

e. to predict the sample data we will get when we know the population.

1.32 True or false?: In a particular study, you could use descriptive statistics, or you could use inferential statistics, but you would rarely need to use both.

CHAPTER PROBLEMS: CONCEPTS AND INVESTIGATIONS

1.33 Statistics in the news: Pick up a recent issue of a national newspaper, such as *The New York Times* or *USA Today*. Identify an article that used statistical methods. Did it use descriptive statistics, or inferential statistics, or both? Explain.

1.34 What is statistics?: On a final exam that one of us recently gave, students were asked, "How would you define 'statistics' to someone who has never taken a statistics course?" One student wrote, "You want to know the answer to some question. There's no answer in the back of a book. You collect some data. Statistics is the body of procedures that helps you analyze the data to figure out the answer and how sure you can be about it." Pick a question that interests you, and explain how you might be able to use statistics to investigate the answer.

▣ 1.35 Surprising ESP data? : In Exercise 1.25, of 3887 sampled subjects, 63.8% said that at least once they felt as though they were in touch with someone when they were far away. Of 3887 sampled subjects who had an opinion, 1407 said never and 2480 said at least once. Suppose that only 20% of the entire population would report at least one such experience. If the sample of 3887 people were a random sample, would this sample proportion result of 0.638 be surprising? Investigate, using the applet described in Activity 2. Do this by simulating samples from the population with a proportion of 0.20. Use the sample size of 4000 as an approximation to the actual sample size of 3887.

▣ 1.36 Create a data file: Using the statistical software that your instructor has assigned for the course, find out how to enter a data file. Create a data file using the data in Figure 1.2 and print a copy.

CHAPTER PROBLEMS: CLASS EXPLORATIONS

▣ Activity 3: Your instructor will help the class create a data file consisting of the values for class members of characteristics based on responses to a questionnaire like the one on the following page. Alternatively, your instructor may ask you to use a data file of this type already prepared with a class of students at the University of Florida, the "Florida student survey" data file on the text CD. Using a spreadsheet program or the statistical software the instructor has chosen for your course, create a data file containing this information. Print the data. What are some questions you might ask about these data? Homework exercises in each chapter will refer to these data.

GETTING TO KNOW THE CLASS

Please answer the following questions. Do not put your name on this sheet. Skip any question that you feel uncomfortable answering. These data are being collected to learn more about you and your classmates and to form a database for the class to analyze.

(1)   What is your height (recorded in inches)? _____

(2)   What is your gender (M = Male, F = Female)? _____

(3)   How much did you spend on your last haircut? _____

(4)   Do you have a paying job during the school year at which you work on average at least 10 hours a week (y = yes, n = no)? _____

(5)   Aside from class time, how many hours a week, on average, do you expect to spend studying and completing assignments for this course? _____

(6)   Do you smoke cigarettes? (y = es, n = no) _____

(7)   How many different people have you dated in the last 30 days? _____

(8)   What was (is) your high school GPA (based upon a 4.0 scale)? _____

(9)   What is your current college GPA? _____

(10)   What is the distance (in miles) between your current residence and this class? _____

(11)   How many minutes each day, on average, do you spend browsing the Internet? _____

(12)   How many minutes each day, on average, do you watch TV? _____

(13)   How many hours each week, on average, do you participate in sports or have other physical exercise? _____

(14)   How many times a week, on average, do you read a daily newspaper? _____

(15)   Do you consider yourself a vegetarian? (y = yes , n = no) _____

(16)   How would you rate yourself politically? (1 = very liberal, 2 = liberal, 3 = slightly liberal, 4 = moderate, 5 = slightly conservative, 6 = conservative, 7 = very conservative) _____

(17)   What is your political affiliation? (D = Democrat, R = Republican, I = Independent) _____

CHAPTER 2
Exploring Data with Graphs and Numerical Summaries

EXAMPLE 1: WHAT CAN STATISTICS TEACH US ABOUT THREATS TO OUR ENVIRONMENT?

Picture the Scenario
As we've seen in Chapter 1, statistics plays a vital role in helping to inform us about issues in our daily lives.

One issue of major concern to many people is that of air pollution and the impact of the consumption of energy from fossil fuels on our environment. Most technologically advanced nations, such as the United States, use vast quantities of a diminishing supply of nonrenewable energy sources such as fossil fuels. We also see that countries fast becoming technologically advanced, such as China and India, are greatly increasing their levels of energy use. The result is increased emissions of carbon dioxide and other pollutants into the atmosphere. Emissions of carbon dioxide may also contribute to global warming, which is a potential danger to humankind in the future. Scientists use descriptive statistics to explore energy use, to study pollution, to learn whether or not pollution has an effect on climate, to compare different countries in how they contribute to the worldwide problems of pollution and climate change, and to measure the impact of global warming on the environment.

Questions to Explore
- How can we investigate which countries emit the highest amounts of carbon dioxide into the atmosphere or use the most non-renewable energy?
- Is climate change occurring, and how serious is it?

Thinking Ahead
In this chapter, we introduce **descriptive statistics** that help us to investigate questions like these. In Example 3 we'll use descriptive statistics to examine data on the dependence of the U.S. and Canada on fossil fuels for energy. Examples 11 and 17 explore carbon dioxide emissions for the world's largest nations and for nations in the European Union. We'll also use descriptive statistics to analyze some data regarding a big question—climate change and global warming. Example 9 presents an analysis of temperature change over time.

Chapter 1 distinguished between **descriptive statistics** (summarizing data) and **inferential statistics** (making a prediction or decision about a population, using sample data). Graphical and numerical summaries are the key elements of descriptive statistics.

Before we begin our journey into the art and science of analyzing data using descriptive statistics, we need to learn a few new terms. These pertain to the types of data we'll analyze. We'll then study ways of using graphs to describe data, followed by ways of summarizing the data numerically.

2.1 WHAT ARE THE TYPES OF DATA?

Statistical methods provide ways to measure and understand **variability**. For the characteristics observed in a study, variation occurs among the available subjects. For instance, there's variability among your schoolmates in weight, major, GPA, favorite sport, religious affiliation, and how long you spend studying today.

Variables

The characteristics observed to address the questions posed in a study are called **variables**.

Definition: Variable

A **variable** is any characteristic that is recorded for subjects in a study.

The terminology *variable* highlights that data values *vary*. For instance, a study of global warming might analyze the high temperature each day over the past century at each of several weather stations around the world. The variable is the high temperature. Examples of other variables for each day are the low temperature, whether it rained that day, and the number of centimeters of precipitation.

Variables Can Be Quantitative (Numerical) or Categorical (in Categories)

The data values that we observe for a variable are often referred to as the **observations.** Each observation can be numerical, such as number of centimeters of precipitation in a day. Or each observation can be a **category**, such as "yes" or "no" for whether it rained.

Definition: Categorical and Quantitative Variables

A variable is called **categorical** if each observation belongs to one of a set of categories.

A variable is called **quantitative** if observations on it take numerical values that represent different magnitudes of the variable.

IN WORDS

Consider your classmates. From person to person, there is variability in age, GPA, major, and whether he or she is dating someone. So, these are **variables**. Age and grade point average (GPA) take numerical values, so are **quantitative**. Major and dating status are **categorical**, since their values are categories, such as (psychology, business, history) for major and (yes, no) for dating status.

For weather stations, for instance, the nation in which a station is located and the daily observation of whether it rained (yes or no) are both categorical variables. The daily high temperature and the amount of precipitation are quantitative variables. For human subjects, examples of categorical variables include gender (with categories male and female), religious affiliation (with categories such as Catholic, Jewish, Muslim, Protestant, Other, None), place of residence (house, condominium, apartment, dormitory, other), and belief in life after death (yes, no). Other examples of quantitative variables are age, number of siblings, annual income, and number of years of education completed.

In the definition of a quantitative variable, why do we say that numerical values must *represent different magnitudes*? It's because we don't regard a variable such as telephone area code as quantitative, even though it is numerical. Quantitative variables measure "how much" there is of something (that is, *quantity* or *magnitude*). With them, you can find arithmetic summaries such as averages. Such summaries do not make sense for area codes, which are merely convenient numerical labels. The area codes do not differ in quantity, so a bank would not try to find the "average area code" of its customers. Other variables of this type are zip code and bankcard PIN numbers.

Graphs and numerical summaries describe the main features of a variable:

- For **quantitative** variables, key features to describe are the **center** and the **spread** (variability) of the data. For instance, what's a typical annual amount of precipitation? Is there much variation from year to year?

- For **categorical** variables, a key feature to describe is the relative number of observations (e.g., percentages) in the various categories.

Quantitative Variables Are Discrete Or Continuous

For a quantitative variable, each value it can take is a number. Quantitative variables can be either **discrete** or **continuous.**

IN WORDS	Definition: Discrete and Continuous Variables
A **discrete** variable is usually a count ("the number of …"). A **continuous** variable has a continuum of infinitely many possible values (such as time, distance, or a physical measurement such as weight).	A quantitative variable is **discrete** if its possible values form a set of separate numbers, such as 0, 1, 2, 3, …. A quantitative variable is **continuous** if its possible values form an interval.

Examples of discrete variables are the number of pets in a household, the number of children in a family, and the number of foreign languages in which a person is fluent. Any variable phrased as "the number of …" is discrete. The possible values are separate numbers such as {0, 1, 2, 3, 4, …}. The outcome of the variable is a count. *Any variable with a finite number of possible values is discrete.*

Examples of continuous variables are height, weight, age, and the amount of time it takes to complete an assignment. The collection of all the possible values of a continuous variable does not consist of a set of separate numbers, but rather an infinite region of values. The amount of time needed to complete an assignment, for example, could take the value 2.496631… hours. *Continuous variables have an infinite continuum of possible values.*

Which type of variable is "the number of people you have dated in the past month"? Consider the following number line.

| 0 | 1 | 2 | 3 | 4 | 5 | 6 | 7 | 8 | 9 | 10 |

When you mark the possible values for the variable on this number line, only the whole numbers will be marked. Each whole number is separated from other values on the line (such as 2.516), because no space between the whole numbers will be marked--It is not possible to date 2.516 persons. This is a discrete variable, since the possible outcomes are separate whole numbers--counts.

Which type of variable is "the distance (in miles) between your current residence and your statistics classroom"? The possible outcomes form an interval of real numbers on the number line, starting at 0. This interval includes values such as 0.5 miles, 2.4 miles, 5.38975 miles. This is a continuous variable.

You can distinguish between discrete and continuous variables by the way you can use your pencil to mark the possible values. If you cannot pick up your pencil to designate the possible range of measurements, the variable is continuous.

Frequency Tables

The first step in numerically summarizing data about a variable is to look at the possible values and count how often each occurs. For a categorical variable, each observation falls in one of the categories. We can use proportions or percentages to summarize the numbers of observations in the various categories. The category with the highest frequency is called the **mode**.

EXAMPLE 2: SHARK ATTACKS AROUND THE WORLD

Picture the Scenario
A sensational news item of recent years has been shark attacks. In the United States, shark attacks most commonly occur in Florida. There were 289 reported shark attacks in Florida between 1990 and 2002. What regions of the world besides Florida have experienced sharks attacks, and how many occur in each region? The International Shark Attack File (ISAF) provides data on shark attacks.

Table 2.1 classifies 735 shark attacks reported from 1990 through 2002, listing countries where attacks occurred as well as the three states in the U.S. where shark attacks were most common. The number of reported shark attacks for a particular region is a **frequency**, or count. It is easier to interpret frequencies after converting them to proportion or percentage form. The proportion is found by dividing the frequency by the total count of 735. The percentage is found by multiplying the proportion by 100.

IN PRACTICE

Why do we care whether a variable is *quantitative* or *categorical*, or whether a quantitative variable is *discrete* or *continuous*? We'll see that the method used to analyze a data set will depend on the type of variable the data represent.

Table 2.1: Frequency of Shark Attacks in Various Regions for 1990-2002.

Region	Frequency	Proportion	Percentage
Florida	289	0.393	39.3
Hawaii	44	0.060	6.0
California	34	0.046	4.6
Australia	44	0.060	6.0
Brazil	55	0.075	7.5
South Africa	64	0.087	8.7
Reunion Island	12	0.016	1.6
New Zealand	17	0.023	2.3
Japan	10	0.014	1.4
Hong Kong	6	0.008	0.8
Other	160	0.218	21.8
Total	735	1.00	100

Source: http://shark-gallery.netfirms.com/attack/statsw.htm

Questions to Explore
a. What is the variable? Is it categorical or quantitative?
b. Show how to find the proportion and percentage reported in Table 2.1 for Florida.
c. Identify the mode for these data.

Think It Through
a. Each observation (reported shark attack) identifies a region for the attack. Region is the variable. It is categorical, with categories shown in the first column of Table 2.1.

b. Of the 735 reported shark attacks, 289 were in Florida. This is a proportion of 289/735 = 0.393, nearly 4 of every 10 attacks. The percentage is 100(0.393) = 39.3%.

c. For the regions listed in this table, the greatest number of attacks was reported in Florida, with 39.3% of the reported worldwide attacks. Florida is the mode, because it has the greatest frequency of reported shark attacks

Insight
Two other regions of the U.S. (California and Hawaii) had about 10% of the reported attacks, so the U.S. had a total of about 50% of all attacks from 1990 through 2002.

The total proportion is 1.0, and the total percentage is 100%. In practice, the separate values may sum to a slightly different number (such as 99.9%) because of rounding.

♦

To practice this concept, try Exercise 2.7.

Here's a summary of the descriptive statistics we computed in Example 2:

> **Definition: Proportion and Percentage (Relative Frequencies)**
>
> The **proportion** of the observations that fall in a certain category is the frequency (count) of observations in that category divided by the total number of observations. The **percentage** is the proportion multiplied by 100. Proportions and percentages are also called **relative frequencies**.

A table of the form of Table 2.1 that lists the possible values of a variable and their frequencies and/or relative frequencies is called a **frequency table**.

> **Definition: Frequency Table**
>
> A **frequency table** is a listing of possible values for a variable, together with the number of observations for each value.

For a quantitative variable, the form of the frequency table depends on whether the variable is discrete or continuous. For a discrete variable, the frequency table usually lists each possible value. For a continuous variable, the frequency table divides the possible values into a set of intervals and displays the number of observations in each interval. We'll see examples for continuous variables in the next section.

SECTION 2.1: PRACTICING THE BASICS

2.1 Categorical/quantitative difference:
a. Explain the difference between categorical and quantitative variables.
b. Give an example of each.

2.2 Identify the variable type: Identify each of the following variables as categorical or quantitative.
a. Number of pets in family
b. County of residence
c. Choice of auto to buy (domestic or import)
d. Distance (in kilometers) commute to work

2.3 Categorical or quantitative?: Identify each of the following variables as categorical or quantitative.
a. Choice of diet (vegetarian, nonvegetarian)
b. Time spent in previous month attending place of religious worship
c. Ownership of a personal computer (yes, no)
d. Number of people you have known who have been elected to a political office

2.4 Discrete/continuous difference:
a. Explain the difference between a discrete variable and a continuous variable.
b. Give an example of each type.

2.5 Identify the type of variable: Identify each of the following variables as continuous or discrete.

 a) The length of time to run a marathon
 b) The number of people in line at a box office to purchase theater tickets
 c) The weight of a dog
 d) The number of pages in a book

2.6 Continuous-discrete: Repeat the previous exercise for the following:
a. The total playing time of a CD
b. The number of courses for which a student has received credit
c. The amount of money in your pocket (Hint: You could regard a number such as $12.75 as 1275 in terms of "the number of cents.")

2.7 Number of children: In the 2002 General Social Survey, respondents answered the question, "How many children have you ever had?" The results were

No. children	0	1	2	3	4	5	6	7	8+
Count	799	469	657	481	185	73	40	22	34

a. Is the variable, number of children, categorical or quantitative?
b. Is the variable, number of children, discrete or continuous?
c. Add proportions and percentages to this frequency table.
d. Which response is the mode?

2.2 HOW CAN WE DESCRIBE DATA USING GRAPHICAL SUMMARIES?

Looking at a graph often gives you more of a feel for a data set than looking at the raw data or a frequency table. In this section, we'll learn about graphs for categorical variables and then graphs for quantitative variables. We'll find out what we should look for in a graph to help us understand the data better.

Graphs for Categorical Variables

The two primary graphical displays for summarizing a categorical variable are the **pie chart** and the **bar graph**.

- A **pie chart** is a circle having a "slice of the pie" for each category. The size of a slice corresponds to the percentage of observations in the category.

- A **bar graph** displays a vertical bar for each category. The height of the bar is the percentage of observations in the category.

EXAMPLE 3: HOW MUCH ELECTRICITY USED COMES FROM RENEWABLE ENERGY SOURCES?

Picture the Scenario
On August 14, 2003, a major power outage hit the northeastern U.S. and parts of Canada. This major blackout brought attention to our dependence on energy and to

the diminishing stockpile of natural resources. As the worldwide demand for energy increases, conservation and alternative sources of energy become more important. The so-called *renewable* energy sources produce electricity without depleting natural resources.

Table 2.2 shows sources of electricity and their use in the United States and Canada. Coal and natural gas are significant fossil fuel contributors to the emission of carbon dioxide, which is a form of air pollution and may lead to climate change. Hydropower is a renewable source. The sources marked 'other' refer to other renewable sources such as solar power and wind power, which have received attention as energy sources that are less environmentally damaging.

Table 2.2: Sources of Electricity in the U.S. and Canada, and the Percentage of Electricity Use Generated by Each.

Source	U.S. Percentage	Canada percentage
Coal	51	16
Hydropower	6	65
Natural gas	16	1
Nuclear	21	16
Petroleum	3	1
Other	3	1
Total	100	100

(Source: *U.S. Department of Energy, Natural Resources Canada*)

Questions to Explore
a. Portray the information in Table 2.2 for the U.S. with a pie chart and a bar graph.
b. What percentage of electricity comes from renewable energy sources in the U.S. and in Canada?

Think It Through
a. Source of electricity is a categorical variable. Figure 2.1 is a pie chart of the U.S. data in Table 2.2. Sources of electricity with larger percentages have larger slices of the pie. The percentages are included in the labels for each slice of the pie.

Percentage use for sources of electricity

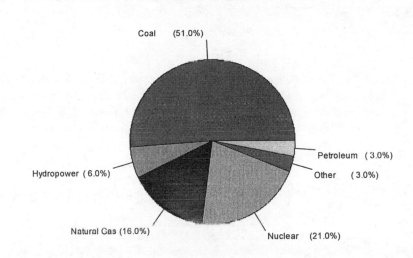

Figure 2.1: Pie Chart of Sources of Electricity in the United States. The label for each slice of the pie gives the category and the percentage of electricity generated by that source. The slice that represents the percentage generated by hydropower is 6% of the total area of the pic. **Question:** Why is it beneficial to label the pie wedges with the percent? (*Hint*: Is it always clear which of two slices is larger, and what percent a slice represents?)

Figure 2.2 shows the bar graph. The categories with larger percentages have higher bars. The scale for the percentages is shown on a vertical axis. The width is the same for each bar.

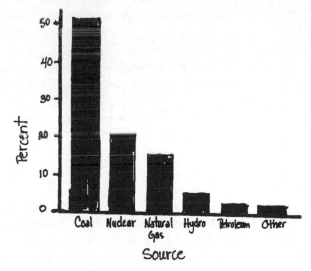

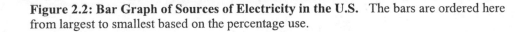

Figure 2.2: Bar Graph of Sources of Electricity in the U.S. The bars are ordered here from largest to smallest based on the percentage use.

b. The renewable sources provide only about 9% (6 + 3 = 9) of the electricity in the U.S. From either graph it is easy to see that the top source (the mode) is coal. By contrast, in Canada, Table 2.2 tells us that renewable sources provide about 66% of the electricity.

Insight
The pie chart and the bar graph both are simple to construct using software. The bar graph is more precise and more flexible. With a pie chart, when two slices are about the same size, it's often unclear which is larger. This is clearer from comparing heights of bars in a bar graph. When there are very many categories, it is not easy to view them all in a pie chart, but this is less a problem in the bar graph. We'll see that the bar graph can easily summarize how results compare for different groups (for instance, if we wanted to compare the U.S. and Canada for Table 2.2). If we just want to sketch a graph by hand, it's easier to sketch a bar graph.

◆

To practice this concept, try Exercise 2.8.

The bar graph in Figure 2.2 displayed the categories in decreasing order of the category percentages. This order made it easy to separate the categories with high percentages visually. In some applications, it is more natural to display them according to their alphabetical order or some other criterion. For instance, if the categories have a natural order, such as if we're summarizing the percentages of grades (A, B, C, D, F) for students in a course, we'd use that order in listing the categories on the graph.

Pareto charts

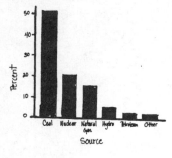

Figure 2.2 (shown again in the margin) is a special type of bar graph called a **Pareto chart**. Named after an Italian economist (Vilfredo Pareto, 1848-1923) who advocated its use, it is a bar graph with categories ordered by their frequency, from the tallest bar to the shortest bar. The Pareto chart is often used in business applications to identify categories that warrant the most attention. The chart helps to portray the **Pareto principle**, which states that a small subset of categories often contains most of the observations. For example, Figure 2.2 shows that three categories (coal, nuclear, and natural gas) were responsible for about 88% of U.S. electricity generation.

Graphs for Quantitative Variables

Now let's explore how to summarize *quantitative* variables graphically. We'll look at three types of displays-- the dot plot, stem-and-leaf plot, and histogram. We'll illustrate the graphs by analyzing data for a food product that many of us eat every day, as we continue our discussion of issues that influence our well-being.

Dot Plots

A **dot plot** shows a dot for each observation, placed just above the value on the number line for that observation. To construct a dot plot,

- Draw a horizontal line. Label it with the name of the variable, and mark regular values of the variable on it.
- For each observation, place a dot above its value on the number line.

EXAMPLE 4: EXPLORING THE HEALTH VALUE OF CEREALS: DOT PLOT

Picture the Scenario
Obesity, high blood pressure, high cholesterol, and heart disease are partially caused by a poor diet. Nutritional labels on packaged foods give us information about the amount of fat, cholesterol, sodium, vitamins, and carbohydrates contained in a serving of the food. Let's investigate the amount of sugar and salt in breakfast cereals, a popular food for kids and adults alike. Too much sugar contributes to weight gain. Too much sodium (the main ingredient of salt) contributes to high blood pressure. It is recommended that daily sodium consumption should not exceed 2400 milligrams (mg). There is no recommended limit for daily sugar consumption, but for a 2000-calorie-a-day diet, fewer than 50,000 mg is considered desirable.

[Photo: picture of some breakfast cereals and "Mikey_]

How much sugar and how much sodium are in breakfast cereals? Table 2.3 lists 20 popular cereals and the amounts of sodium and sugar contained in a single serving (usually ¾ cup), according to what's listed on the cereal boxes. The sodium and the sugar amounts are both quantitative variables. The variables are continuous, because they measure amounts that can take any positive real number value. They are measured in this table by rounding to the nearest number of milligrams (mg; 1000 mg = 1 gram).

TABLE 2.3: Sodium and Sugar Amounts in 20 Breakfast Cereals. The amounts refer to a single serving size. A third variable, Type, classifies the cereal as intended for adults (Type A) or children (Type C).

Cereal	Sodium (mg)	Sugar (mg)	Type
Frosted Mini Wheats	0	7,000	A
Raisin Bran	210	12,000	A
All Bran	260	5,000	A
Apple Jacks	125	14,000	C
Capt Crunch	220	12,000	C
Cheerios	290	1000	C
Cinnamon Toast	210	13,000	C
Crackling Oat Bran	140	10,000	A
Crispix	220	3,000	A
Frosted Flakes	200	11,000	C
Fruit Loops	125	13,000	C
Grape Nuts	170	3,000	A
Honey Nut Cheerios	250	10,000	C
Life	150	6,000	A
Oatmeal Raisin Crisp	170	10,000	A
Sugar Smacks	70	15,000	C
Special K	230	3000	A
Wheaties	200	3000	C
Corn Flakes	290	2000	A
Honeycomb	180	11,000	C

Source: *http://lib.stat.cmu.edu/DASL/Datafiles/Cereals.html*

Questions to Explore
a. Construct a dot plot for the sodium values of the 20 breakfast cereals. (We'll consider sugar amounts in the exercises.)
b. What does the dot plot tell us about the data?

Think It Through
a. Figure 2.3 shows a dot plot. Each cereal sodium value is represented with a dot above the number line. For instance, the labeled dot above 0 represents the sodium value of 0 mg for Frosted Mini Wheats.

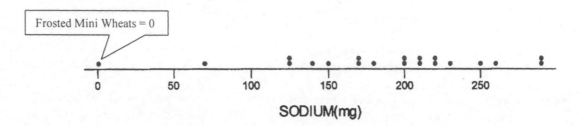

Figure 2.3: Dot Plot for Sodium Content of 20 Breakfast Cereals. The sodium value for each cereal is represented with a dot above the number line. **Question:** What does it mean when more than one dot appears above a value?

b. The dot plot gives us an overview of all the data. We see clearly that the sodium values fall between 0 and 290 mg, with most cereals falling between 125 and 250 mg.

Insight
The dot plot portrays the individual observations. The number of dots above a value on the number line represents the frequency of occurrence of that value. From a dot plot, we'd be able to reconstruct (at least approximately) all the data in the sample.

♦

To practice this concept, try Exercise 2.15.

Stem-and-Leaf Plots

Another type of graph, called a **stem-and-leaf plot**, is similar to the dot plot in that it portrays the individual observations.

- Each observation is represented by a **stem** and a **leaf**. Usually the stem consists of all the digits except for the final one, which is the leaf.

- Place the stems in a column, starting with the smallest. Place a vertical line to their right. On the right side of the vertical line, indicate each leaf (final digit) that has a particular stem. List the leaves in increasing order.

Cereal	Sodium
Frosted Mini Wheats	0
Raisin Bran	210
All Bran	260
Apple Jacks	125
Capt Crunch	220
Cheerios	290
Cinnamon Toast	210
Crackling Oat Bran	140
Crispix	220
Frosted Flakes	200
Fruit Loops	125
Grape Nuts	170
Honey Nut Cheerios	250
Life	150
Oatmeal Raisin Crisp	170
Sugar Smacks	70
Special K	230
Wheaties	200
Corn Flakes	290
Honeycomb	180

EXAMPLE 5: EXPLORING THE HEALTH VALUE OF CEREALS: STEM-AND-LEAF PLOT

Picture the Scenario
Let's re-examine the sodium values for the 20 breakfast cereals, shown again in the margin.

Questions to Explore
a. Construct a stem-and-leaf plot of the 20 sodium values.
b. How does the stem-and-leaf plot compare to the dot plot?

Think It Through
a. In the stem-and-leaf plot, we let the final digit of a sodium value form the leaf and the other digits form the stem. For instance, the sodium value for Sugar Smacks is 70. The stem is 7 and the leaf is 0. Each stem is placed to the left of the vertical bar. Each leaf is placed to the right of the bar. Figure 2.4 shows the stem-and-leaf plot. Notice that a leaf has only one digit, but a stem can have one or more digits.

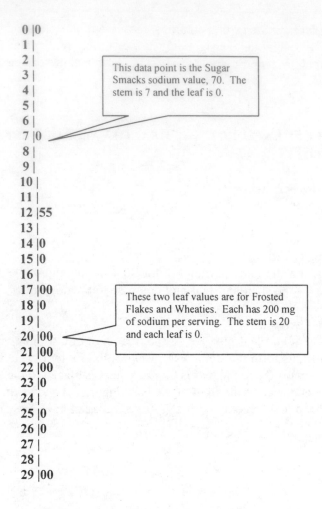

```
 0 |0
 1 |
 2 |
 3 |
 4 |
 5 |
 6 |
 7 |0
 8 |
 9 |
10 |
11 |
12 |55
13 |
14 |0
15 |0
16 |
17 |00
18 |0
19 |
20 |00
21 |00
22 |00
23 |0
24 |
25 |0
26 |0
27 |
28 |
29 |00
```

This data point is the Sugar Smacks sodium value, 70. The stem is 7 and the leaf is 0.

These two leaf values are for Frosted Flakes and Wheaties. Each has 200 mg of sodium per serving. The stem is 20 and each leaf is 0.

Figure 2.4: Stem-and-Leaf Plot for Cereal Sodium Values. The final digit of a sodium value forms the leaf, and the other digits form the stem. **Question:** Why do some stems not have a leaf?

The Sugar Smacks observation is labeled on the graph. Two observations have a stem of 12 and a leaf of 5. These are the sodium values of 125 for Fruit Loops and Applejacks.

b. The stem-and-leaf plot looks like the dot plot turned on its side, with the leaves taking the place of the dots. A stem has no leaf if there is no observation at that value. These are the values at which the dot plot has no dots. In summary, we get the same information from a stem-and-leaf plot as from a dot plot.

Insight
As with the dot plot, it's easy to use a stem-and-leaf plot to reconstruct the original data set, because they show all the individual observations. This becomes unwieldy for large data sets. We'll next look at a third way of graphing the data that can also handle large data sets.

♦

To practice this concept, try Exercise 2.14, parts (a) and (b).

Histograms

A more versatile way to graph the data, useful also for very large data sets, uses bars to summarize frequencies of outcomes.

> **Histogram**
>
> A **histogram** is a graph that uses bars to portray the frequencies or the relative frequencies of the possible outcomes for a quantitative variable.

EXAMPLE 6: HISTOGRAM OF TV WATCHING

Picture the Scenario
The 2002 General Social Survey asked, "On the average day, about how many hours do you personally watch television?" Figure 2.5 shows the histogram of the 905 responses.

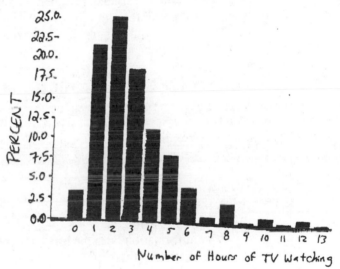

Figure 2.5: Histogram of GSS Responses about Number of Hours of TV Watching on the Average Day.

Questions to Explore
a. What was the most common outcome?
b. What percentage of people reported watching no more than 2 hours per day?

Think It Through
a. The most common outcome (the mode) is the value with the highest bar. This is 2 hours per day of TV watching.
b. To find the percentage for "no more than 2 hours per day," we need to look at the percentages for 0, 1 and 2 hours per day. They seem to be about 3, 22, and 26. Adding these percentages together tells us that about 51% of the respondents reported watching no more than 2 hours of TV per day.

Insight

In theory, TV watching is a continuous variable. However, the possible responses subjects were able to make here were 0, 1, 2, ..., so it was measured as a discrete variable. Figure 2.5 is an example of a histogram of a discrete variable.

◆

To practice this concept, try Exercise 2.23, parts (a) and (b)

Steps for Constructing a Histogram

- Divide the range of the data into intervals of equal width. For a discrete variable with few values, use the actual possible values.
- Count the number of observations (the frequency) in each interval, forming a frequency table.
- On the horizontal axis, label the values or the endpoints of the intervals. Draw a bar over each value or interval with height equal to its frequency (or percentage), values of which are marked on the vertical axis.

For a continuous variable, to construct a histogram you first need to divide the interval of possible values into smaller intervals. Then you can form a frequency table and graph the frequencies or percentages for those intervals.

EXAMPLE 7: EXPLORING THE HEALTH VALUE OF CEREALS: HISTOGRAM

Picture the Scenario

Let's reexamine the sodium values of the 20 breakfast cereals. Those values are shown again in the margin.

Cereal	Sodium
Frosted MW	0
Raisin Bran	210
All Bran	260
Apple Jacks	125
Capt Crunch	220
Cheerios	290
Cinnamon T	210
Crackling OB	140
Crispix	220
Frosted Flakes	200
Fruit Loops	125
Grape Nuts	170
Honey Nut C	250
Life	150
Oatmeal Raisin	170
Sugar Smacks	70
Special K	230
Wheaties	200
Corn Flakes	290
Honeycomb	180

Questions to Explore

a. Construct a frequency table.
b. Construct a histogram.
c. What information does the histogram not show that you can get from a dot plot or a stem-and-leaf plot?

Think it Through

a. To construct a frequency table, we divide the possible sodium values into separate intervals and count the number of cereals in each. The sodium values range from 0 to 290. The choice for the *number* of intervals is up to you, but they should each have the same width. We created Table 2.4 using eight intervals, each with a width of 40. For a continuous variable, 0 to 39 actually represents 0 to 39.999999..., 0 up to every number *below* 40. Sometimes you will see the intervals written as 0 to 40, 40 to 80, 80 to 120, and so on. However, for an observation that falls at an interval endpoint, then it's not clear in which interval it goes.

Statistical software can form intervals for you and find the counts. Frequency tables also usually show proportions and/or percentages for the intervals, as Table 2.4 illustrates.

Table 2.4: Frequency Table for Sodium in 20 Breakfast Cereals. The table summarizes the sodium values using eight intervals and lists the number of observations in each, as well as the proportions and percentages.

Interval	Frequency	Proportion	Percentage
0 to 39	1	0.05	5%
40 to 79	1	0.05	5%
80 to 119	0	0.00	0%
120 to 159	4	0.20	20%
160 to 199	3	0.15	15%
200 to 239	7	0.35	35%
240 to 279	2	0.10	10%
280 to 319	2	0.10	10%

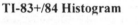

TI-83+/84 Histogram

b. Figure 2.6 shows the histogram for this frequency table. A bar is drawn over each interval of values, with height of the bar equal to the frequency. The histogram created using the TI83+/84 calculator is in the margin. The intervals axes are not labeled; however, the TRACE key may be used to view the endpoints of an interval and the frequency for an interval.

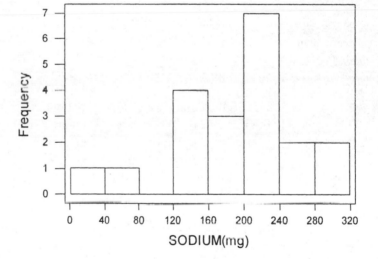

Figure 2.6: Histogram of Breakfast Cereal Sodium Values. The rectangular bar over an interval has height equal to the number of observations in the interval.

c. The histogram does not show the actual numerical values. For instance, we know that 1 observation falls below 40, but we do not know its actual value. In summary, with a histogram of a continuous variable, we lose the actual numerical values of individual observations, unlike a dot plot or a stem-and-leaf plot.

Insight

The histogram in Figure 2.6 labels the vertical axis with the frequencies. If it were labeled with proportions or percentages, the heights of the bars would be same. The histogram would be identical in appearance, and we would have the same graphical information about the sodium values. Either histogram would suggest that the cereal we choose to eat *does* matter if we wish to monitor our sodium intake.

♦

To practice this concept, try Exercise 2.19.

For a discrete variable, a histogram usually has a separate bar for each possible value rather than for intervals of values. An example is the TV watching data in Figure 2.5. When a discrete variable has a large number of possible values, such as a score on an exam, intervals can be formed with values grouped together.

Which Graph Should We Use?

We've now studied three graphs for quantitative variables--the dot plot, stem-and-leaf plot, and histogram. How do we decide which to use? Here are some guidelines:

<table>
<tr><td>

IN PRACTICE

Graphical displays are easily constructed using statistical software or graphing calculators. In practice, you won't have to draw them yourself. But it is important to understand *how* they are drawn and *how to interpret* them.

</td></tr>
</table>

- The dot plot and stem-and-leaf plot are more useful for small data sets, since they portray each individual observation. With large data sets, histograms work better.
- More flexibility is possible in defining the intervals with a histogram than in defining the stems with a stem-and-leaf plot.
- The histogram is usually the most compact of these displays.
- Data values are retained with the stem-and-leaf plot and dot plot but not with the histogram.

Unless the data set is small (say, about 50 or fewer observations), we usually prefer the histogram. When in doubt, create a histogram and the dot plot or stem-and-leaf plot, and then use whichever is clearer and more informative.

The Shape of a Distribution

For data on a quantitative variable, its graph or its frequency table is said to describe the "distribution of the data." Here are some things to look for in a graph showing a distribution:

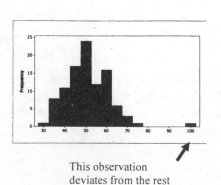

This observation deviates from the rest

- Look for the **overall pattern**. Do the data cluster together, or do one or more observations noticeably deviate from the rest, as in the histogram in the margin? We'll discuss such "outlier" observations later in the chapter.

- Do the data have a single mound? A distribution of such data is called **unimodal**. The highest point is at the **mode**. A distribution with *two* distinct mounds is called **bimodal.** A bimodal distribution results when a population is polarized on a controversial issue. For instance, each subject may be presented with ten scenarios in which a person found guilty may be subjected to the death penalty and asked in which of them a death sentence would be just. If we count the number of scenarios in which the subjects feel that the death penalty would be warranted, many responses would be close to 0 and many would be close to 10.

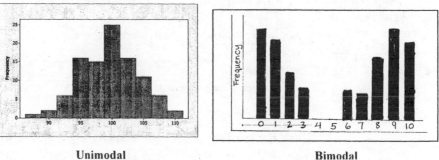

Unimodal   **Bimodal**

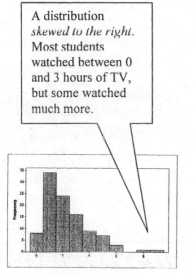

A distribution *skewed to the right*. Most students watched between 0 and 3 hours of TV, but some watched much more.

Skewed

- The **shape** of the distribution is often described as **symmetric** or **skewed.** A distribution is symmetric if the side of the distribution below a central value is a mirror image of the side above that central value. The distribution is skewed if one side of the distribution stretches out longer than the other side. For instance, the skewed distribution in the margin resulted from students in a class being asked how many hours they spent watching TV on the previous day.

In picturing features such as skew and symmetry, it is common to use smooth curves such as shown in Figure 2.7 to summarize the shape of a histogram. You can think of this as what can happen when you choose more and more intervals (making each interval narrower) and collect more data, so the histogram gets "smoother." The parts of the curve for the lowest values and for the highest values are called the **tails** of the distribution.

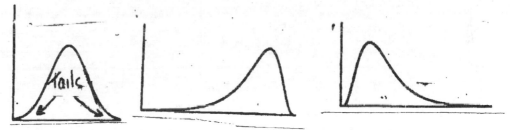

Symmetric   Skewed to the Left   Skewed to the Right

Figure 2.7: Curves for Data Distributions Illustrating Symmetry and Skew. Question: What does the longer tail indicate about the direction of skew?

IN WORDS

To "skew" means to pull in one direction.

Definition: Skewed Distribution

A distribution is **skewed to the left** if the left tail is longer than the right tail.

A distribution is **skewed to the right** if the right tail is longer than the left tail.

Identifying Skew

Let's consider some variables and think about what shape their distributions would have. How about IQ? Values cluster around 100, and tail off in a similar fashion in both directions. The appearance of the distribution on one side of 100 is roughly a mirror image of the other side, with tails of similar length. The distribution is approximately symmetric.

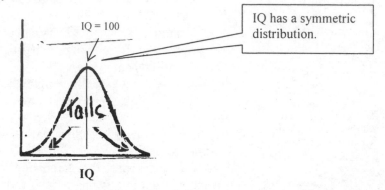

IQ = 100

IQ has a symmetric distribution.

IQ

How about life span for humans? Most people live to be at least about 60 years old, but some die at a very young age, so life span would probably be skewed to the left.

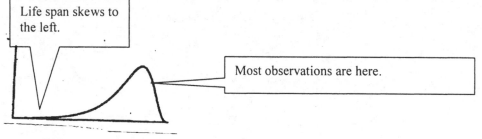

Life span skews to the left.

Most observations are here.

Life span

What shape would we expect for the distribution of annual incomes of adults? Probably there would be a long right tail, with some people having incomes much higher than the overwhelming majority of people. The distribution would be skewed to the right.

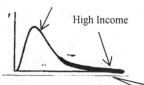

Most Income

High Income

Income

Income skews to the right. Relatively few are rich and have observations in this long right tail.

EXAMPLE 8: WHAT'S THE SHAPE OF THE DISTRIBUTION OF TV WATCHING?

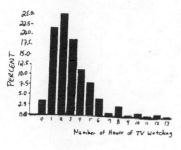

Number of Hours of TV Watching

Picture the Scenario
In Example 6, we constructed a histogram of the number of hours of TV watching reported in the GSS. It is shown again in the margin.

Question to Explore
How would you describe the shape of the distribution?

Think It Through
There appears to be a single mound of data clustering around the mode of 2. The distribution is unimodal. There also appears to be a long right tail, so the distribution is skewed to the right.

Insight
In a survey, an observation reported for a subject is not necessarily the "true" value. In this distribution, the percentage is quite a bit higher for 8 hours than for 7 or 9, and for 10 than for 9 or 11. Do subjects reporting high values tend to pick even numbers?

◆

To practice this concept, try Exercise 2.23, part(c).

Time Plots: Displaying Data over Time

We've been learning how to graph data collected for a variable at a particular time. For some variables, observations occur over time. Examples include the daily closing price of a stock, and the population of a country measured every decade in a census. A data set collected over time is called a **time series**.

We can display time-series data graphically using a **time plot**. This charts each observation, on the vertical scale, against the time it was measured, on the horizontal scale. A common pattern to look for is a **trend** over time, indicating either a long tendency of the data to rise or a long tendency to fall. To see a trend more clearly, it is beneficial to connect the data points in their time sequence.

To illustrate, Figure 2.8 is a time plot between 1995 and 2001 of the number of people worldwide who use the Internet (in millions). In 1995, relatively few people were using the Internet, but by 2001 about 500 million people were doing so. There is a clear increasing trend over time.

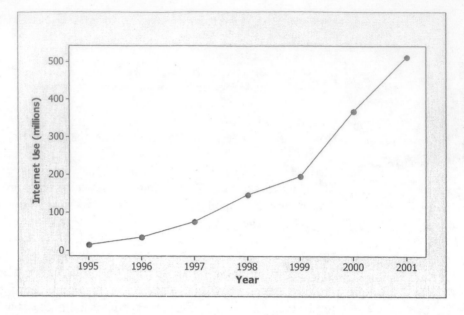

Figure 2.8: MINITAB Output for Number of People Using Internet between 1995 and 2001. (Source: www.netvalley.com/intvalstat.html)

Statistical software can easily construct a time plot for a time series. In practice, there's usually not such a clear trend as in Figure 2.8, as the next example illustrates.

EXAMPLE 9: IS THERE A TREND TOWARD WARMING IN NEW YORK CITY?

Picture the Scenario
Each day, weather stations around the world record the daily average temperature, defined as the average of the maximum and minimum temperature over a midnight-to-midnight time frame. In a given year, the annual average temperature is the average of the daily average temperatures for that year. Let's analyze data on annual average temperature (in degrees Fahrenheit) in Central Park, New York City from 1901--2000. This is a continuous, quantitative variable. The data are in the "Central Park Yearly Temps" data file on the text CD.

Question to Explore
What do we learn from a time plot of these annual average temperatures? Is there a trend toward warming in the U.S.?

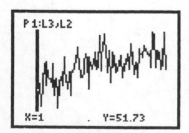

TI-83+/84 Time Plot

Think It Through
Figure 2.9 shows a time plot, constructed using Minitab software. The time plot constructed using the TI-83+/84 calculator is in the margin. The observations fluctuate considerably, but the figure does suggest there has been somewhat of an increasing trend in the annual average temperatures at this location.

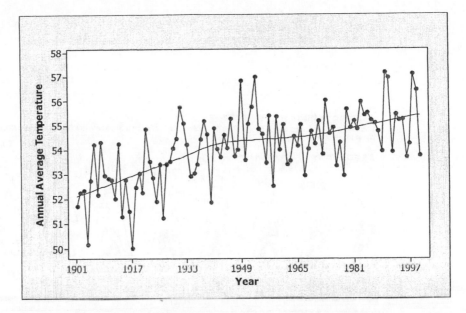

Figure 2.9: MINITAB Output for a Time Plot of Central Park, New York City Average Annual Temperatures. The annual average temperatures are plotted against the year from 1901-2000. A smoothing curve is superimposed. **Question:** Are the annual average temperatures tending to increase, decrease, or stay the same over time?

Insight

The short-term fluctuations in a time plot can mask an overall trend. It's possible to get a clearer picture by "smoothing" the data. This is beyond our scope here, but Exercise 2.130 shows one way to do this. Minitab presents the option of smoothing the data to portray a general trend (Click on "data view" and choose the "lowess" option under "smoother"). This is the smooth curve passing through the data points in Figure 2.9. This curve goes from a level of about 52 degrees in 1901 to about 55 degrees in 2000.

The data reported in Figure 2.9 refer to one location in the U.S. It would be important to explore other locations around the world to see if similar trends are evident there.

                                                                                    ◆

To practice this concept, try Exercises 2.27 and 2.29.

Beware of Poor Graphs

With modern computer-graphic capabilities, newspapers and other periodicals use graphs in an increasing variety of ways to portray quantitative information. This is not always done well, however, and you must look at graphs skeptically.

Figure 2.10 is an example. The graph is intended to show how enrollment has risen at the University of Georgia in recent years while the number of African American students has simultaneously decreased. The figure attempts to show a time plot for two racial groups, while using white and black human figures to portray the observations for the two groups. Do you see poor design aspects of this graph? Are the heights, or the areas, of the human figures proportional to the counts? Look at the relative sizes of the figures in 1996, compared to 2003 and 2004.

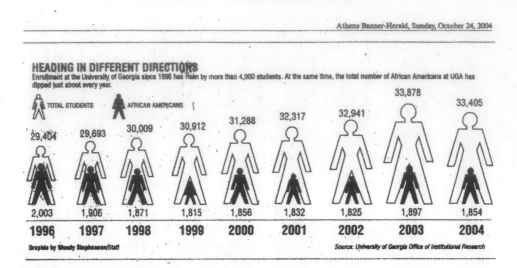

Figure 2.10: An Example of a Poor Graph. Question: What's misleading about the way the data are presented?

Here are some things to keep in mind when you construct a graph:

- Be cautious in using figures (such as people) in place of the usual bars or points. It can make a graph more attractive, but it is easy to get the relative percentages incorrect, as Figure 2.10 did.

- Label both axes and provide a heading to make clear what the graph is intended to portray. Figure 2.10 does not even provide a vertical axis.

- To help our eyes visually compare relative sizes accurately, the vertical axis should usually start at 0.

- It can be difficult to portray more than one group on a single graph when the variable values differ greatly. This is the case with Figure 2.10, in which one frequency is very small compared to the other.

Many books have been written in recent years showing innovative ways of clearly portraying quantitative information in graphs. Examples are *The Visual Display of Quantitative Information* and *Envisioning Information*, by Edward Tufte (Graphics Press).

On the Shoulders of ... FLORENCE NIGHTINGALE

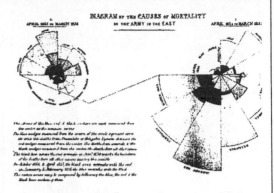

What relevance do graphs have to deaths from military combat?

During the Crimean War in 1854, the British nurse Florence Nightingale (1820-910) gathered data on the numbers of soldiers who died from various causes. She prepared graphical displays such as time plots and pie charts in an understandable format for policy makers. The graphs showed that more soldiers were dying from contagious diseases than from war-related wounds. The plots were revolutionary for her time. They helped her to promote her national cause of improving hospital conditions. After implementing sanitary methods, financed mostly with her own money, Nightingale showed with time plots that the relative frequency of soldiers' deaths from contagious disease decreased sharply and no longer exceeded that of deaths from wounds.

Throughout the rest of her life, Nightingale promoted the use of data for making informed decisions about public health policy. For example, she used statistical arguments to campaign for improved medical conditions in the U.S. during the Civil War in the 1860s (Franklin, 2002). Nightingale was once quoted as saying, "To understand God's thoughts we must study statistics, for these are the measure of His purpose" (Friendly 2000).

SECTION 2.2: PRACTICING THE BASICS

2.8　　Environmental protection: A 2001 Roper organization survey asked: "How far have environmental protection laws and regulations gone?" For the possible responses--not far enough, about right, and too far-- the percentages of responses were 44%, 35%, and 21%.
a. Sketch a bar chart to visually display the survey results.
b. Which is easier to sketch relatively accurately, a pie chart or a bar chart?
c. What is the advantage of using a graph to summarize the results instead of merely stating the percentages for each response?

2.9 What do alligators eat?: The bar chart (using Minitab) is from a study[1] investigating the factors that influence alligators' choice of food. For 219 alligators captured in four Florida lakes, researchers classified the primary food choice (in volume) found in the alligator's stomach in one of the categories--fish, invertebrate (snails, insects, crayfish), reptile (turtles, baby alligators), bird, or other (amphibian, mammal, plants).
a. Is primary food choice categorical or quantitative?
b. What is the mode for primary food choice?
c. About what percentage of alligators had fish as the primary food choice?
d. This type of bar chart has a special name. What is it?

---

[1] Data courtesy of Clint Moore.

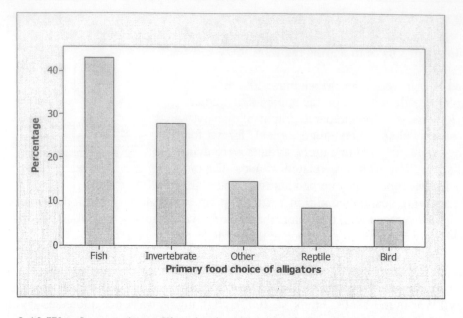

2.10 Weather stations: The pie chart (constructed using Minitab) shown portrays the regional distribution of weather stations in the U.S.
a. Do the slices of the pie portray (i) variables or (ii) categories of a variable?
b. Identify what the two numbers that are shown for each slice of the pie mean.
c. From looking at the graph without inspecting the numbers, would it be easier to identify the mode using this graph or using the corresponding bar graph? Why?

Regional Distribution of Weather Stations

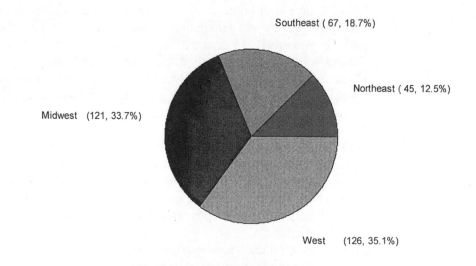

2.11 France's most popular holiday spot: Which countries are most visited by tourists from other countries? The table shows results according to www.aneki.com/visited.html.
   a.   Is country visited a categorical or quantitative variable?

b. In creating a bar graph of these data, would it be most sensible to list the countries alphabetically, or in the form of a Pareto chart? Explain.

c. Do a dot plot or stem-and-leaf plot make sense for these data? Explain.

Most Visited Countries, 2003

Country	Number of Visits (millions)
France	75.5
United States	50.9
Spain	48.2
Italy	41.2
China	31.2
United Kingdom	25.2
Russia	21.2
Mexico	20.6
Canada	20.4
Germany	19.0

2.12 U.S. married-couple households: According to the 2002 Current Population Survey, of U.S. married-couple households, 13% are traditional (with children and with only the husband in the labor force), 31% are dual-income with children, 25% are dual-income with no children, and 31% are other (such as older married couples whose children no longer reside in the household).

a. Is the variable of interest categorical or quantitative? Explain.

b. Which graph would you use to graphically summarize this information: a bar chart, dot plot, stem-and-leaf plot, or histogram? Why?

Region	Frequency
Florida	289
Hawaii	44
California	34
Australia	44
Brazil	55
South Africa	64
Reunion Island	12
New Zealand	17
Japan	10
Hong Kong	6
Other	160

2.13   Shark attacks worldwide: Table 2.1, part of which is shown again in the margin, summarized shark attacks for different regions of the world.

a. Why might a bar graph be preferable to a pie chart for graphing the information?

b. Using software or sketching, construct a bar graph, ordering the regions (i) alphabetically, (ii) as in a Pareto chart. Which do you prefer? Why?

2.14 Graphing exam scores: A teacher shows her class the scores on the midterm exam in the stem-and-leaf plot shown:

```
6 | 508
7 | 01136779
8 | 1223334677789
9 | 011234458
```

a. Identify the number of students in the class, and their minimum and maximum score.

b. Sketch how the data could be displayed in a dot plot.

c. Sketch how the data could be displayed in a histogram with four intervals.

2.15 Sugar dot plot: For the breakfast cereal data, a dot plot (constructed using Minitab) for the sugar values (in grams) is shown in the figure:

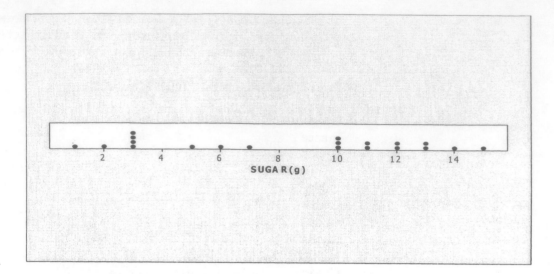

a. Identify the minimum and maximum sugar values.
b. Identify the sugar outcome that occurs most frequently. What is this called?

2.16 Leaf unit other than 1: When the observations are large numbers, their final digits are not shown in a stem-and-leaf plot. The plot specifies a **leaf unit** (sometimes also called a **stem unit**) by which to multiply each observation. For instance, for the breakfast cereal data with sugar measured in milligrams (as it is in Table 2.3, an excerpt of which is shown in the margin), software reports a figure such as shown here, but with the indication that "Leaf Unit = 1000." For instance, the observations of 14 and 15 in the final row of the plot actually represent observations of 14,000 and 15,000.

Cereal	Sugar
Frosted MW	7,000
Raisin Bran	12,000
All Bran	5,000

```
0 | 1
0 | 23333
0 | 5
0 | 67
0 |
1 | 00011
1 | 2233
1 | 45
```

a. In milligrams, what is the observation in the first row of the plot?
b. Identify the sugar outcome that occurs most frequently.
c. Suppose the plot shown above actually represented the ordered observations 10, 20, 30, 30, 30, 30, 50, ... , 150. Then, what would the leaf unit be for the plot?

2.17 Back-to-back stem-and-leaf plot: To compare two groups graphically, one can use the same stems for each and put leaves for one group on one side and for the other group on the other side. This is called a **back-to-back stem-and-leaf plot**. The figure shown compares sugar amounts for cereals listed in Table 2.3 according to whether they were intended either for adults or children.

Back-to-back stem-and-leaf plot of sugar for adult and children cereals

```
       Adult        Children
                | 100 | 0
          0     | 200 |
        0 0 0   | 300 | 0
                | 400 |
          0     | 500 |
          0     | 600 |
          0     | 700 |
                | 800 |
                | 900 |
         0 0    |1000| 0
                |1100| 0 0
           0    |1200| 0
                |1300| 0 0
                |1400| 0
                |1500| 0
```

Cereal	Type	Sodium
Frosted	A	0
Raisin	A	210
All Bra	A	260
Apple J	C	125
Capt C	C	220
Cheeri	C	290
Cinnast	C	210
Crackli	A	140
Crispix	A	220
Frosted	C	200
Fruit L	C	125
Grape	A	170
Honey	C	250
Life	A	150
Oatmea	A	170
Sugar S	C	70
Special	A	230
Wheatie	C	200
Corn Fla	A	290
Honeyco	C	180

a. Is the distribution of sugar values similar, or different, for the two cereal types? If they are different, describe the difference.

b. Construct a back-to-back stem-and-leaf plot for the sodium values, shown again in the margin. Is the distribution of sodium values similar, or different, for the two cereal types?

2.18 Truncating data and splitting stems: To make a stem-and-leaf plot more compact, we can **truncate** the data values: The last digit is removed. The figure shows this for the cereal sodium data shown in the margin. In constructing the plot, since the stems ranged only from 0 to 2, we listed each stem twice. They are then called **split stems**. The leaves on the first split stem range from 0 to 4 and the leaves on the second split stem range from 5 to 9. The splitting of the stems stops the display from getting too compact, since without splitting, the plot would have only three lines. The leaves are arranged in increasing order on each line.

Stem-and-leaf plot for truncated values of cereal sodium, using split stems

```
0 | 0
0 | 7
1 | 224
1 | 5778
2 | 0011223
2 | 5699
```

---

a. Explain why the range of data goes from 0 to 29 instead of from 0 to 290.

b. Identify on the plot the sodium value of 70.

c. Where on the plot is the observation of 230 for Special K?

d. What features of the data shown in a non-truncated plot are lost in this figure? (*Hint*: Do you have any detail of the large gap between the observations 0 and 70?

2.19 Histogram for sugar: For the breakfast cereal data, a histogram (constructed using Minitab) for the sugar values, in grams, is shown in the figure.

a. Identify the intervals of sugar values used for the plot.

b. Describe the shape of the distribution. What do you think might account for this unusual shape? (*Hint*: How else are the cereals classified in Table 2.3?)

c. What information can you get from the dot plot or stem-and-leaf plot of these data shown in Exercises 2.12 and 2.13 that you cannot get from this plot?

d. This histogram shows frequencies. If you were to construct a histogram using the *percentages* for each interval, how (if at all) would the shape of this histogram change?

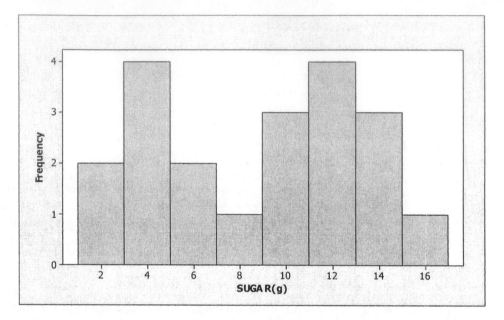

🖳 **2.20 Sugar plots:** Using software with the "Cereal" data set on the text CD, construct a (a) dot plot, (b) stem-and-leaf plot, (c) histogram. Explain how to interpret each plot.

2.21 Shape of the histogram: For each of the following variables, indicate whether you would expect its histogram to be symmetric, skewed to the right, or skewed to the left. Explain why.

a. Assessed value of houses in a large city

b. Number of times checking account overdrawn in the past year for the faculty in your school

c. IQ for the general population

d. The height of female college students

2.22 More shapes of histograms: Repeat the following exercise for:

a. The scores of students (out of 100 points) on a very easy exam in which most score perfectly or nearly so, but a few score very poorly

b. The weekly church contribution for all members of a congregation, in which the three wealthiest members contribute generously each week

c. Time needed to complete a difficult exam (maximum time is 1 hour)

d. Number of music CDs (compact discs) owned, for each student in your school.

2.23 How often do students read the newspaper?: Question 17 on the class survey (Exercise 1.37 in Chapter 1) asked, "Estimate the number of times a week, on average, that you read a daily newspaper."

a. Is this variable continuous, or discrete? Explain.

b. The histogram shown gives results of this variable when this survey was administered to a class of 36 University of Georgia students. Report the (i) minimum response, (ii) maximum response, (iii) number of students who did not read the newspaper at all, (iv) mode.

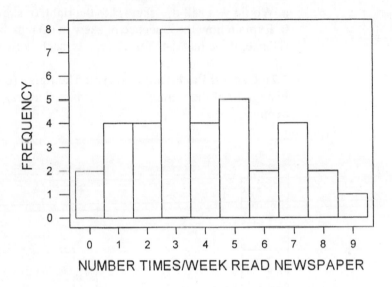

c. Describe the shape of the distribution.

2.24 SAT scores: The SAT is commonly used by colleges to help determine admission. Each year, the College Board reports the average SAT score for each state and the District of Columbia. The figure gives a histogram of the 2003 SAT state averages, from the "SAT state scores" data file on the text CD.

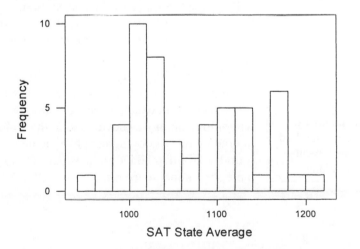

Histogram of the 2003 SAT State Averages

a. Is the distribution unimodal? Explain.

b. Identify the interval that has the highest frequency of values.

c. Identify the range of values for the (i) smallest interval, (ii) highest interval.

2.25 Shape of sodium values: Figure 2.6 showed a histogram of the 20 cereal sodium values, shown again in the margin.

a. Would you call this skewed to the right, or skewed to the left?

b. Explain how the evidence of skew in (a) may be due to only two observations. (Hence, if we had data for *all* cereals, such skew might not exist.)

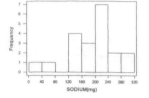

2.26 Central Park temperatures: The figure (constructed using MINITAB) shows a histogram of the Central Park, NY annual average temperatures for the Twentieth century.

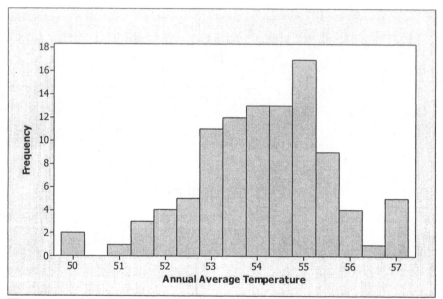

Histogram of Central Park Annual Average Temperatures

a. Describe the shape of the distribution.

b. What information can a time plot show that a histogram cannot provide?

c. What information does the histogram show that a time plot does not provide?

2.27 Is whooping cough close to being eradicated?: In the first half of the twentieth century, whooping cough was a frequently occurring bacterial infection that often resulted in death, especially among young children. A vaccination for whooping cough was developed in the 1940s. How effective has the vaccination been in eradicating whooping cough? One measure to consider is the **incidence rate** (number of infected individuals per 100,000 population) in the United States. The table shows incidence rates from 1925 to 1970.

Incidence Rates for Whooping Cough, 1925-1970[2]

Year	Rate per 100,000
1925	131.2
1930	135.6
1935	141.9
1940	139.6
1945	101.0
1950	80.1
1955	38.2
1960	8.3
1965	3.5
1970	2.1

a. Sketch a time plot. What type of trend do you observe?

b. Based on the trend from 1945 – 1970, was the whooping cough vaccination proving effective in reducing the incidence of whooping cough?

2.28 Whooping cough in recent years: Refer to the previous example. Has the incidence rate for whooping cough continued to decrease since 1970?

a.   The figure below gives a time plot of the data since 1980, reported in the MMWR.[3] Describe the trend.

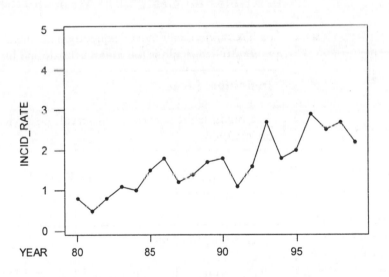

Time Plot of Whooping Cough Incidence Rates (per 100,000), 1980 – 1999

---

[2] Source: *Historical Statistics of the United States, Colonial Times to 1970, U.S. Department of Commerce, p.77*

[3] *www.cdr.gov/mmvr/*

b. What has the incidence rate been since about 1993? How does the incidence rate since 1993 compare with the incidence rate in 1970? Is the U.S. close to eradicating whooping cough?

c. Would a histogram of the incidence rates since 1935 address the question about the success of the vaccination for whooping cough? Why or why not?

⌨**2.29 Warming in Newnan, Georgia?:** Access the "Newnan Temps" file on the text CD, which reports the average annual temperatures during the twentieth century for Newnan, Georgia. Construct a time plot to investigate a possible trend over time. Is there evidence of climate change?

2.3   HOW CAN WE DESCRIBE THE CENTER AND LOCATION OF QUANTITATIVE DATA?

Section 2.2 introduced **graphical summaries** of data. For a quantitative variable, a graph shows the important feature of **shape**. It's always a good idea to look at the data first with a graph, to get a "feel" for the data. You can then consider **numerical summaries (statistics)**. For quantitative variables, we'll attempt to answer the questions, "What is a representative observation like?" and "Do the observations take similar values, or are they quite spread out?" Statistics that answer the first question describe the **center** of the distribution. Statistics that answer the second question describe the **spread**. We'll also study how the shape of the distribution influences the statistics and our choice of which are suitable for describing the data.

Describing the Center: The Mean and the Median

The best-known and most frequently used measure of the center of a distribution of a quantitative variable is the **mean**. It is found by averaging the observations.

Definition: Mean
The **mean** is the sum of the observations divided by the number of observations.

IN WORDS

The **mean** refers to "averaging," adding up the data points and dividing by how many there are. The **median** is the point that splits the data in two, half the data below it and half above it.

Another popular measure is the **median**. Half the observations are smaller than it, and half are larger.

Definition: Median
The **median** is the midpoint of the observations when they are ordered from the smallest to the largest (or from the largest to the smallest).

EXAMPLE 10: WHAT'S THE CENTER OF THE CEREAL SODIUM DATA?

Picture the Scenario
In Examples 4, 5, and 7 in Section 2.2, we investigated the sodium level in 20 breakfast cereals and saw various ways to graph the data. Let's return to the data on sodium levels and learn how to describe the center of the data. The observations (in mg) are

    0   70   125   125   140   150   170   170   180   200
   200   210   210   220   220   230   250   260   290   290

Questions to Explore
a. Find the mean
b. Find the median of the data.

Think It Through
a. We find the mean by adding all the observations and then dividing this sum by the number of observations, which is 20:

$$\text{Mean} = (0 + 70 + 125 + ...+ 290)/20 = 3710/20 = 185.5.$$

b. To find the median, we arrange the data from the smallest to the largest observation, as shown on the two data lines in the Picture the Scenario step. For the twenty observations, the smaller ten (on the first line) go from 0 to 200, and the larger ten (on the second line) go from 200 to 290. The midpoint is 200, which is the median. In summary, 200 is the midpoint of the ordered observations, and we get 185.5 by averaging all 20 of the observations.

Insight
The mean and median take different values. Why? The median measures the center by dividing the data into two equal parts, regardless of the actual numerical values above that point or below that point. The mean takes into account the actual numerical values of all the observations.

♦

To practice this concept, try Exercise 2.31.

In this example, what if the smaller ten observations go from 0 to 200, and the larger ten go from 210 to 290? Then, by convention, we take the median to be the average of the two middle observations, which is $(200 + 210)/2 = 205$.

How to Determine the Median

- Put the n observations in order of their size.
- When the number of observations n is odd, the median is the middle observation in the ordered sample.
- When the number of observations n is even, two observations from the ordered sample fall in the middle, and the median is their average.

A Closer Look at the Mean and the Median

Notation for the mean is used in a formula for the mean and in formulas for other statistics that describe spread around the mean.

Notation for a Variable and Its Mean

Variables are symbolized by letters near the end of the alphabet, most commonly x and y. The sample size is denoted by n. For a sample of n observations on a variable x, the mean is denoted by $\bar{x}$. Using the mathematical symbol $\sum$ for "sum," the mean has the formula

$$\bar{x} = \frac{\sum x}{n}.$$

This is short for "sum the values on the variable x and divide by the sample size."

For instance, the cereal data set has $n = 20$ observations. As we saw in Example 9,

$$\bar{x} = (\sum x)/n = (0 + 70 + 125 + ... + 290)/20 = 3710/20 = 185.5.$$

Here are some basic **properties of the mean**:

- The mean is the *balance point* of the data: If we were to place identical weights on a line representing where the observations occur, then the line would balance by placing a fulcrum at the mean.

The fulcrum shows the mean of the cereal data

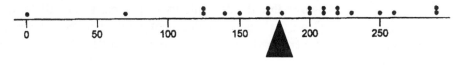

- For a skewed distribution, the mean is pulled in the direction of the longer tail, relative to the median. The next example illustrates this.

- The mean can be highly influenced by an **outlier**, which is an unusually small or unusually large observation.

Outlier

An **outlier** is an observation that falls well above or well below the overall bulk of the data.

EXAMPLE 11: CO_2 POLLUTION OF THE EIGHT LARGEST NATIONS

Picture the Scenario
The Pew Center on Global Climate Change[4] reports that possible global warming is largely a result of human activity that produces carbon dioxide (CO_2) emissions and other greenhouse gases. The CO_2 emissions from fossil fuel combustion are the result of the generation of electricity, heating, industrial processes, and gas consumption in automobiles. The *Human Development Report 2003*, published by the United Nations Development Programme[5], reported the per capita CO_2 emissions in 1999 by country. For the eight largest countries in population size (which make up more than half the world's population) these were, in metric tons per person:

> **DID YOU KNOW?**
>
> A metric ton is 1000 kilograms, which is about 2200 pounds.

China	2.3	Brazil	1.8
India	1.1	Russia	9.8
United States	19.7	Pakistan	0.7
Indonesia	1.2	Bangladesh	0.2

For these eight values, the mean is 4.6.

Question to Explore
a. What is the median of the data set?
b. Are any observations a potential outlier. Discuss its impact on how the mean compares to the median.
c. Using this data set, what is the effect an outlier can have on the mean?

Think It Through
a. The CO_2 values have $n = 8$ observations. The ordered values are

$$0.2, \ 0.7, \ 1.1, \ 1.2, \ 1.8, \ 2.3, \ 9.8, \ 19.7.$$

Since n is even, two observations are in the middle, the fourth and fifth ones in the ordered sample. These are 1.2 and 1.8. The median is their average, 1.5.

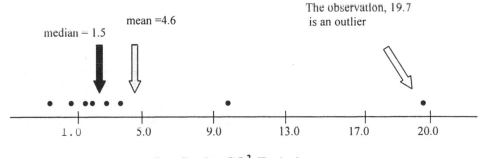

Per Capita CO^2 Emissions

---

[4] Source: http://www.pewclimate.org/global-warming-basics/facts_and_figures/

[5] hdr.undp.org

b. Let's consider a dot plot, as shown. The relatively high value of 19.7 falls well above the rest of the data. It is an outlier. This value as well as the value at 9.8 causes the mean, which is 4.6, to fall well above the median, which is 1.5.

c. The size of the outlier affects the calculation of the mean but not the median. If the observation of 19.7 for the U.S. had instead been the same as for China (2.3), the 8 observations would have had a mean of 2.4, instead of 4.6. The median would still have been 1.5. In summary, a single outlier can have a large impact on the value of the mean.

Insight

The mean may not be representative of where the bulk of the observations fall. This is fairly common with small samples when one observation is much larger or much smaller than the others. It's not surprising that the U.S. is an outlier, as the other nations are not nearly as economically advanced. Later in the chapter (Example 17) we'll compare carbon dioxide emissions for the U.S. to European nations.

♦

To practice this concept, try Exercise 2.32.

Comparing the Mean and Median

The shape of a distribution influences whether the mean is larger or smaller than the median. For instance, an extremely large value out in the right-hand tail pulls the mean to the right. It then usually falls above the median, as we just observed with the CO_2 data in Example 10. Generally, if the shape is

- perfectly symmetric, the mean equals the median.
- skewed to the right, the mean is larger than the median.
- skewed to the left, the mean is smaller than the median.

As Figure 2.11 illustrates, the mean is drawn in the direction of the longer tail.

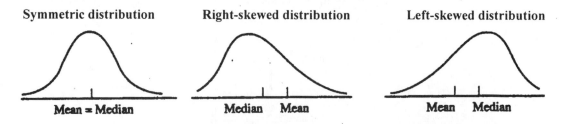

Symmetric distribution   **Right-skewed distribution**   **Left-skewed distribution**

Mean = Median   Median   Mean   Mean   Median

Figure 2.11: Relationship between the Mean and Median. Question: For skewed distributions, what causes the mean and median to differ?

When a distribution is close to symmetric, with the tails of similar length, the median and mean are similar. For skewed distributions, the mean lies toward the direction of skew (the longer tail) relative to the median, as Figure 2.11 shows. Extreme observations in a tail affect the balance point for the distribution, which is the mean.

The more highly skewed the distribution, the more the mean and median tend to differ.

Example 11 illustrated this property. The dot plot of CO_2 emissions shown there is skewed to the right. As expected, the mean of 4.6 falls in the direction of skew, above the median of 1.5. Another example is given by mean household income in the U.S. In 2001, the mean was \$58,208 and the median was \$42,228 (U.S. Bureau of the Census, *Current Population Reports*). This suggests that the distribution of household incomes in the U.S. is also skewed to the right.

Why is the median not affected by an outlier?

- How far an outlier falls from the middle of the distribution does not influence the median. The median is determined solely by having an equal number of observations above it and below it.

For the CO_2 data, for instance, if the value 19.7 for the U.S. were changed to 100, the median would still equal 1.5. The calculation of the mean uses *all* the numerical values, so unlike the median, it depends on how far observations fall from the middle. Because the mean is the balance point, an extreme value on the right side pulls the mean toward the right tail. Because the median is not affected, it is said to be **resistant** to the effect of extreme observations.

- A numerical summary of the observations is called **resistant** if extreme observations have little, if any, influence on its value.

The median is resistant. The mean is not.

These properties might lead you to believe that it's always better to use the median than the mean. That's not true. Often, there are advantages to having a measure use the numerical values of *all* the data. For instance, *for discrete data that take only a few values, quite different patterns of data can give the same result for the median*. It is then *too resistant*.

Let's see an example of this situation. This example also shows how to find the mean and median from a frequency table.

EXAMPLE 12: HOW MANY TIMES HAVE YOU BEEN MARRIED?

Picture the Scenario
A Census Bureau report[6] gave data on the number of times residents of the U.S. had been married, for subjects of various ages. Table 2.5 summarizes responses for subjects of age 20 to 24. The frequencies are actually *thousands* of people, for instance 7,074,000 men never married, but this does not affect calculations about the mean or median.

---

[6] Source: www.census.gov/prod/2002pubs/p70-80.pdf

Table 2.5: Number of Times Married, for Subjects of Age 20 to 24

	Frequency	
Number Times Married	Women	Men
0	5861	7074
1	2773	1541
2	105	43
Total	8749	8658

- -

Questions to Explore

a. Find the median and the mean for each gender.

b. Why is the median not particularly informative?

Think It Through

a. For the 8739 women, the ordered sample would be

$$0\,0\,0\,0\,0\,0\,.....\,0\ \ 1\,1\,1\,1\,1\,1\,...\,1\ \ 2\,2\,2\,2\,2\,2\,...\,2$$
$$\text{(5861 0s)} \qquad \text{(2773 1s)} \qquad \text{(105 2s)}$$

Since $n = 8739$, the observations would be listed from 1st to 8739th, and the middle observation would be the $(1 + 8739)/2 = 4370$th. However, only three distinct responses occur, and more than 4370 of them are 0. So, the median is 0. Likewise, the median is 0 for men.

To calculate the mean for women, it is unnecessary to add the 8739 separate observations to obtain the numerator of $\bar{x}$, because each value occurred many times. We can find the sum of the 8739 observations by multiplying each possible value by its frequency and then adding:

$$\sum x = 5861(0) + 2773(1) + 105(2) = 2983.$$

Since the sample size is 8739, the mean is $2983/8739 = 0.34$. Likewise, you can show that the mean for men is 0.19.

b. The median is 0 for women and for men. This makes it seem as if there's no difference between the genders on this variable. By contrast, the mean uses the numerical values of all the observations, not just the ordering. In summary, we learn that on the average, in this age group, women have been married more often than men.

Insight

The median ignores too much information when the data are highly discrete--that is, when a high proportion falls at only a few values. An extreme case occurs for **binary data**, which take only two values, 0 and 1. The median equals the more common outcome, but gives no information about the relative number of observations at the two levels. For instance, consider a sample of size 5 for the variable, number of times married. The observations $(1, 1, 1, 1, 1)$ and the observations $(0, 0, 1, 1, 1)$

both have a median of 1. The mean is 1 for (1, 1, 1, 1, 1) and 3/5 for (0, 0, 1, 1, 1). When observations take values of only 0 or 1, the mean equals the *proportion* of observations that equal 1. It is much more informative than the median.

♦

To practice this concept, try Exercise 2.50.

IN PRACTICE

- If a distribution is very highly skewed, the median is usually a more informative measure. It it usually preferred over the mean, because it better represents what is typical or representative.

- If the distribution is close to symmetric or only mildly skewed, the mean is usually the preferred measure.

- If the variable is discrete with few distinct values, the mean is usually preferred.

If you want to summarize the CO_2 emission data of Example 11, the median may be more relevant, because of the skew resulting from the extremely large value for the U.S. But for the marriage data in Example 12, the median discards too much information and the mean is more informative. In practice, both the mean and median are useful, and frequently *both* values are reported.

The Mode

We've seen that the **mode** is the value that occurs most frequently. It describes a typical observation in terms of the most common outcome. The mode is most often used to describe the category of a categorical variable that has the highest frequency. With quantitative variables, it is most useful with discrete variables taking a small number of possible values. For the marriage data of Table 2.5, for instance, the mode is 0 for each gender.

The mode need not be near the center of the distribution. It may be the largest or the smallest value, as in the marriage example. Thus, it is somewhat inaccurate to call the mode a measure of center, but often it is useful to report the most common outcome.

Measures of Location: The Quartiles and Other Percentiles

The median is a special case of a more general set of **measures of location** called percentiles.

Definition: Percentile

The ***p*th percentile** is the value such that *p* percent of the observations fall below or at that value.

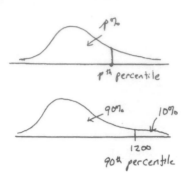

Suppose you're informed that your score of 1200 on the SAT college entrance exam falls at the 90th percentile. Set $p = 90$ in this definition. Then, 90% of those who took the exam scored between the minimum score and 1200. Only 10% of the scores were higher than yours.

Substituting $p = 50$ in this definition gives the 50th percentile. For it, 50% of the observations fall below or at it and 50% above it. But this is simply the median. The 50th percentile is usually referred to as the median.

Three useful percentiles are the **quartiles**. The **first quartile** has $p = 25$, the 25th percentile. The lowest 25% of the data fall below it. The **second quartile** has $p = 50$, the 50th percentile, which is the median. The **third quartile** has $p = 75$, the 75th percentile. The highest 25% of the data fall above it. The quartiles are denoted by Q1 for the first quartile, Q2 for the second quartile, and Q3 for the third quartile. The quartiles split the distribution into four parts, each containing one-fourth (25%) of the observations. See Figure 2.12.

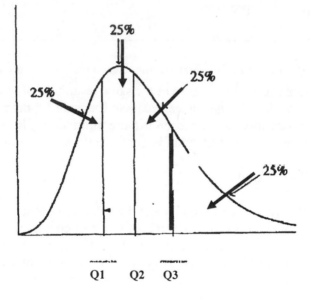

Figure 2.12: The Quartiles Split the Distribution into Four Parts. 25% is below the first quartile (Q1), 25% is between the first quartile and the second quartile (the median, Q2), 25% is between the second quartile and the third quartile (Q3), and 25% is above the third quartile. **Question:** Why is the second quartile also the median?

Notice that one quarter of the observations fall below Q1, two quarters (one half) fall below Q2 (the median), and three quarters fall below Q3. The middle 50% of the observations fall between the first quartile and the third quartile--25% from Q1 to Q2 and 25% from Q2 to Q3.

Finding Quartiles

- Arrange the data in order.
- Find the median (the midpoint). This is the **second quartile, Q2.**
- Consider the lower half of the observations. The median of these observations is the **first quartile, Q1.**
- Consider the upper half of the observations. Their median is the **third quartile, Q3.**

EXAMPLE 13: WHAT ARE THE QUARTILES FOR THE CEREAL SODIUM DATA?

Picture the Scenario

Knowing the quartiles not only tells you how the data split into four parts, but also gives information about the shape. Let's again consider the sodium values for the 20 breakfast cereals, from Table 2.3.

Questions to Explore

a. What are the quartiles for the 20 cereal sodium values?
b. Interpret the quartiles in the context of the cereal data. What does the distance between quartiles tell you about the distribution shape?

Think It Through

a. The sodium values, arranged in ascending order, are:

$$Q1 = 145$$

| 0 | 70 | 125 | 125 | **140** | **150** | 170 | 170 | 180 | **200** |

| **200** | 210 | 210 | 220 | **220** | **230** | 250 | 260 | 290 | 290 |

$$Q3 = 225$$

- The median of the 20 values is the average of the 10th and 11th observations, 200 and 200, which is **Q2 = 200 mg.**
- The first quartile Q1 is the median of the 10 smallest observations (in the top row), which is the average of 140 and 150, **Q1 = 145 mg.**
- The third quartile Q3 is the median of the 10 largest observations (in the bottom row), which is the average of 220 and 230, **Q3 = 225 mg.**

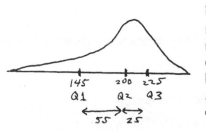

b. The sodium values range from 0 mg to 145 mg for the first quarter, 145 mg to 200 mg for the second quarter, 200 mg to 225 mg for the third quarter, and 225 mg to 290 mg for the fourth quarter. The distance of 55 from the first quartile to the median exceeds the distance of 25 from the median to the third quarter. This commonly happens when the distribution is skewed to the left, as shown in the margin figure. Although each quarter of a distribution may span different lengths, each quarter contains the same number (25%) of the observations.

Insight

The middle half of the data falls between Q1 = 145 and Q3 = 225. In the final section of the chapter we'll see a new type of graph that portrays the spread of the data using this information as well as the median and the extreme values.

To practice this concept, try Exercise 2.39.

IN PRACTICE

Percentiles other than the quartiles are reported usually only for large data sets. Software can do the computational work for you, and we won't go through the details. There are precise algorithms for the calculations involving interpolations, and different software often uses slightly different rules, even for the quartiles.

ACTIVITY 1: LET'S USE AN APPLET TO SIMULATE THE RELATIONSHIP BETWEEN THE MEAN AND MEDIAN

The Mean and Median Applet at the text Web site (www.prenhall.com/TBA) allows you to add and delete data points from a sample. Different colored arrows indicate the mean and median after each new point is added. At the applet, set the number line to have a range of values from 0 to 20.

- In Example 10 on CO_2 use, the per capita values (metric tons) for the 8 largest nations were (0.2, 0.7, 1.1, 1.2, 1.8, 2.3, 9.8, 19.7). Create this sample using the applet. Investigate what happens to the mean and median as you move the highest observation of 19.7 even higher, or move it lower. How does the outlier influence the relationship between the mean and median?

- Consider the effect of the sample size on the relationship between the mean and median, by adding to this distribution 20 values that are like the non-outliers. Does the outlier of 19.7 have as much effect on the mean when the sample size is larger?

To practice this concept, try Exercise 2.42, part (c), and Exercise 2.139.

SECTION 2.3: PRACTICING THE BASICS

2.30 Median versus mean: The mean and median describe the center.
a. Why is the median sometimes preferred? Give an example.
b. Why is the mean sometimes preferred? Give an example.

2.31 More on CO_2 emissions: The Pew Center on Global Climate Change reported the CO_2 emissions from fossil fuel combustion for the 7 countries in 1997 with the highest emissions. These values, reported as million metric tons of carbon equivalent, are 1490 (U.S.), 914 (China), 391 (Russia), 316 (Japan), 280 (India), 227 (Germany), and 142 (U.K.).

a. Find the mean and median.

b. The totals reported here do not take into account a nation's population size. Explain why it may be more sensible to analyze *per capita* values, as was done in Example 11.

2.32 Resistance to an outlier: Consider the following three sets of observations:

Set 1:   8, 9, 10, 11, 12
Set 2:   8, 9, 10, 11, 100
Set 3:   8, 9, 10, 11, 1000

a. Find the median for each data set.

b. Find the mean for each data set.

c. What do these data sets illustrate about the resistance of the median and mean to an outlier?

2.33 Income and race: According to the U.S. Bureau of the Census, *Current Population Reports*, in 2001 the median household income was $44,517 for whites and $29,470 for blacks, whereas the mean was $60,512 for whites and $39,248 for blacks. Does this suggest that the distribution of household income for each race is symmetric, or skewed to the right, or skewed to the left?   Explain.

2.34 Labor dispute: The workers and the management of a company are having a labor dispute. Explain why the workers might use the median income of all the employees to justify a raise but management might use the mean income to argue that a raise is not needed.

2.35 Cereal sodium center: The dot plot shows the 20 breakfast cereal sodium values from Example 4. What aspect of the distribution of values causes the mean to be less than the median?

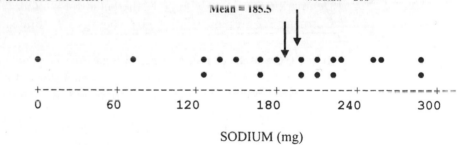

SODIUM (mg)

Dot Plot of Sodium Values for 20 Breakfast Cereals

2.36 Measures of center: The figure shows dot plots for three sample data sets.

a. For which, if any, data sets would you expect the mean and the median to be the same? Explain why.

b. For which, if any, data sets would you expect the mean and the median to differ? Which would be larger, the mean or the median? Why?

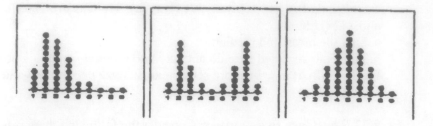

2.37 Vacation days: *National Geographic Traveler* magazine recently presented data on the annual number of vacation days averaged by residents of eight different countries. They reported 42 days for Italy, 37 for France, 35 for Germany, 34 for Brazil, 28 for Britain, 26 for Canada, 25 for Japan, and 13 for the United States.
a. Report the median.
b. By finding the median of the four values below the median, report the first quartile.
c. Find the third quartile.
d. Interpret the values found in parts (a)-(c) in the context of these data.

🖳 **2.38 European unemployment:** In recent years many European nations have suffered from relatively high unemployment. For the 15 nations that made up the European Union in 2003, the table shows the unemployment rates reported by Eurostat as of December 2003.
a. Find and interpret the median.
b. Find the first quartile (Q1) and the third quartile (Q3).
c. Find and interpret the mean.

European Union 2003 unemployment rates (www.europa.eu.int/comm/eurostat)

Belgium 8.3	Greece 9.3	Portugal 6.7	Ireland 4.6	Austria 4.5
Denmark 6.0	Spain 11.2	Netherlands 4.4	Italy 8.5	Sweden 6.0
Germany 9.2	France 9.5	Luxembourg 3.9	Finland 8.9	U.K. 4.8

2.39 Prices of stocks--measures of location: The data values below represent the prices per share of the 10 most actively traded stocks from the New York Stock Exchange (rounded to the nearest dollar) on Oct. 1, 2003.

2   4   11   12   13   15   31   31   37   47

a. Sketch a dot plot or construct a stem-and-leaf plot.
b. Find the median.
c. Find the first quartile (Q1) and the third quartile (Q3).
d. Based on the distances between Q1 and the median and between the median and Q3, what shape does this suggest the distribution has?

2.40 Prices of stocks-finding the mean: Using the data in the previous exercise:

a. Find the mean. Indicate the location of the mean on the plot. What feature of the distribution stands out with the location of the mean? (*Hint*: Are there gaps in the distribution?)

b. Is the mean similar to the median? Explain why it is or why it is not.

c. Write a short summary about the distribution of the stock prices using the statistical information from the previous exercise and this exercise.

2.41 Public transportation--center: The environmentally-conscious owner of a company in downtown Atlanta is concerned about the large use of gasoline by her employees due to urban sprawl, traffic congestion, and the use of energy inefficient vehicles such as SUVs. She would like to promote the use of public transportation. She decides to investigate how many miles her employees travel on public transportation during a typical day. The values for her ten employees (recorded to the closest mile) are:

0   0   4   0   0   0   10   0   6   0

a. Find and interpret the mean, median, and mode.

b. She has just hired an additional employee. He lives in a different city and travels 90 miles a day on public transport. Re-compute the mean and median. Describe the effect of this outlying observation.

💻**2.42 Public transportation--location:** Refer to the previous exercise.

a. Using software or by hand, find the quartiles of the original 10 observations.

b. Use the quartiles to describe the shape of the distribution.

c. Add the outlier of 90 to the data set, and use the Mean and Median applet to investigate what effect it has on the mean and median.

d. Now add 10 more data values that are near the mean of 2 for the original 10 observations. Does the outlier of 90 still have such a strong effect on the mean?

2.43 Female strength: The "high school female athletes" data file on the text CD has data for 57 high school female athletes on the maximum number of pounds they were able to bench press, which is a measure of strength. For these data, $\bar{x}$ = 79.9, Q1 = 70, median = 80, Q3 = 90.

a. Interpret each of these values.

b. Would you guess that the distribution is skewed, or roughly symmetric? Why?

2.44 Female body weight: The "college athletes" data file on the text CD has data for 64 college female athletes. The data on weight (in pounds) has $\bar{x}$ = 133, Q1 = 119, median = 131.5, Q3 = 144.

a. Interpret each of these values.

b. Would you guess that the distribution is skewed, or roughly symmetric? Why?

2.45 Student hospital costs: If you had data for all students in your school on the amount of money spent in the previous year on overnight stays in a hospital, probably the median and mode would be 0 but the mean would be positive.

a. Explain why.

b. Give an example of another variable that would have this property.

2.46 Net worth by degree: *Statistical Abstract of the United States 2000* reported that in 1998 for those with a college education, the median net worth was $146,400 and the mean net worth was $528,200. For those with a high school diploma only, the values were $53,800 and $157,800.
a.   Explain how the mean and median could be so different for each group.
b.   Which measure do you think gives a more realistic measure of a typical net worth, the mean or the median. Why?

2.47 Canadian income: According to Statistics Canada, in 2000 the median household income in Canada was $46,752 and the mean was $71,600. What would you predict about the shape of the distribution? Why?

2.48 Baseball salaries: According to sportingnews.com, in 2003 professional baseball players earned a mean salary of $2.5 million and a median salary of $800,000. What do you think causes these two values to be so different?

2.49 Number of marriages: Example 10 found the mean and median number of times married for subjects of age 20-24. For women, if 1400 observations shifted from 0 to 2, the distribution would be

Number Times Married	Frequency
0	4461
1	2773
2	1505

a. For this distribution, show that the median is still 0, as in Example 10, and the mean is 0.66 (it was 0.34 in Example 10).
b. Explain how the mean uses the numerical values of all the observations, not just the ordering.

2.50 Knowing homicide victims: The table summarizes responses of 4390 subjects in recent General Social Surveys to the question, "Within the past month, how many people have you known personally that were victims of homicide?"

Number of People You Have Known Who Were Victims of Homicide

Number of Victims	Frequency
0	3944
1	279
2	97
3	40
4 or more	30
Total	4390

----

a. To find the mean, it is necessary to give a score to the "4 or more" category. Find it, using the score 4.5. (In practice, you might try a few different scores, such as 4, 4.5, 5, 6, to make sure the result is not highly sensitive to that choice.)
b. Find the median. Note that the "4 or more" category is not problematic for it.
c. If 1744 observations shift from 0 to 4, show how the mean and median change.

d. Why is the median the same for (b) and (c), even though the data are quite different?

2.51 Accidents: One variable in a study measures how many serious motor vehicle accidents a subject has had in the past year. Explain why the mean would likely be more useful than the median for summarizing the responses of the 60 subjects.

2.4 HOW CAN WE DESCRIBE THE SPREAD OF QUANTITATIVE DATA?

A measure of the center is not enough to describe a distribution well. It tells us nothing about the spread of the data. With the cereal sodium data, if we report the median of 200 mg to describe the center, would the value of 250 mg for Honey Nut Cheerios be considered quite high, or are most of the data even farther from the median? To answer, we need numerical summaries for the spread of the distribution.

Measuring Spread: The Range

To see why a measure of the center is not enough, let's consider Figure 2.13. This figure graphically compares hypothetical income distributions of music teachers in public schools in Denmark and in the U.S. Both distributions are symmetric about $40,000 and have the same means. However, the annual incomes in Denmark (converted to U.S. dollars) go from $35,000 to $45,000, whereas those in the U.S. go from $20,000 to $60,000. Incomes are more uniform in Denmark and more spread out in the U.S. A simple way to describe this is with the **range**.

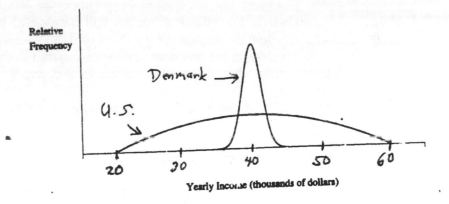

Figure 2.13: Income Distributions for Music Teachers in Denmark and in the U.S. The distributions have the same mean, but the one for the U.S. is more spread out. **Question:** How would the range for Denmark change if one teacher earned $100,000?

Definition: Range

The **range** is the difference between the largest and the smallest observations.

In Denmark the range is $45,000−$35,000 = $10,000. In the U.S. the range is $60,000−$20,000 = $40,000. The range is a larger value when the data are more spread out.

The range is simple to compute and easy to understand, but it ignores too much information. For instance, suppose half the music teachers in Denmark make $35,000 and half make $45,000. Then the range is the same as if only one teacher makes $35,000, one makes $45,000, and all the rest make $40,000, for which the data are much less spread out. The range would change dramatically in either case if one teacher made $100,000. It would change from $45,000−$35,000 = $10,000 to $100,000−$35,000 = $65,000. The range is affected dramatically by even a single outlier, regardless of how large the sample size is. It is not a resistant statistic. It shares the worst property of the mean, not being resistant, and the worst property of the median, ignoring the numerical values of nearly all the data.

Measuring Spread: The Interquartile Range

A much more resistant measure of spread summarizes the range for the *middle half* of the data. This middle half falls between the first and the third quartiles, Q1 and Q3. The distance from Q1 to Q3 is called the **interquartile range,**

IN WORDS

If the interquartile range of U.S. music teacher salaries equals $16,000, this means that for the middle half of the salaries, $16,000 is the distance between the largest and smallest salaries.

Definition: Interquartile Range (IQR)

The **interquartile range**, denoted by IQR, is the distance between the third and first quartiles,

$$IQR = Q3−Q1.$$

As with the range, the more spread out the data, the larger the IQR tends to be. But unlike the ordinary range, the IQR is resistant: It is not affected by any observations below the first quartile or above the third quartile. In contrast, the range depends solely on the minimum and the maximum values, the most extreme values, so the range changes as either extreme value changes.

For instance, for the breakfast cereal sodium data,

- Minimum value = 0
- First quartile Q1 = 145
- Median = 200
- Third quartile Q3 = 225
- Maximum value = 290

The range is 290−0 = 290 and the interquartile range is Q3−Q1 = 225−145 = 80. Suppose the highest sodium value were 1000 instead of 290. Then the range would change dramatically from 290 to 1000, due to this single observation. By contrast, the IQR would not change.

Measuring Spread: The Standard Deviation

The range and the interquartile range are simple ways of summarizing spread. Both use a limited amount of information, however. The range uses only the largest and the smallest observation, and the IQR uses only the first and third quartiles. The most popular summary of spread uses *all* the data. It describes a typical value of how far the observations fall from the mean. It does this by summarizing the **deviations** from the mean.

- The **deviation** of an observation x from the mean $\bar{x}$ is $(x - \bar{x})$, the difference between the two values.

For the cereal sodium values, the mean is $\bar{x} = 185.5$. The observation of 250 for Honey Nut Cheerios has a deviation of $250 - 185.5 = 64.5$. The observation of 70 for Sugar Smacks has a deviation of $70 - 185.5 = -115.5$. Figure 2.14 shows these deviations. Each observation has a deviation from the mean.

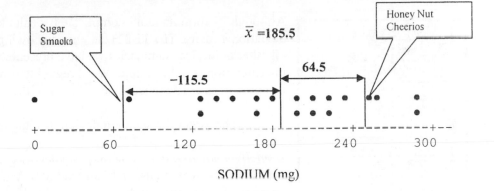

Figure 2.14: Dot Plot for Cereal Sodium Data, Showing Deviations for Two Observations. Questions: When is a deviation positive, and when is it negative?

- A deviation $x - \bar{x}$ is *positive* when the observation falls *above* the mean. A deviation is *negative* when the observation falls *below* the mean.
- The interpretation of the mean as the balance point implies that the positive deviations counterbalance the negative deviations. Because of this, the sum of the deviations always equals zero.
- Because the deviations always sum to zero, summary measures use either the squared deviations or their absolute values.
- The average of the squared deviations is called the **variance.** Because the variance uses the *square* of the units of measurement for the original data, its square root is easier to interpret. This is called the **standard deviation.**

IN WORDS

The symbol

$$\sum (x - \overline{x})^2$$

is called a **sum of squares**. It represents finding the deviation for each observation, squaring each deviation, and then adding them up. You can think of the standard deviation *s* as representing a *typical distance* of an observation from the mean.

Definition: The Standard Deviation *s*

The **standard deviation** *s* of *n* observations is

$$s = \sqrt{\frac{\sum (x - \overline{x})^2}{n-1}} = \sqrt{\frac{\text{sum of squared deviations}}{\text{sample size - 1}}}.$$

This is the square root of the **variance** s^2, which is an average of the squares of the deviations from their mean,

$$s^2 = \frac{\sum (x - \overline{x})^2}{n-1}.$$

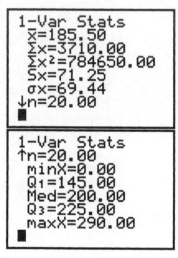

TI-83+/84 Output of Cereal Data

Although its formula looks complicated, a calculator can compute the standard deviation *s* easily. The TI-83+/84 output showing the sample standard deviation and all other numerical summaries we have presented is given in the margin. Its interpretation is quite simple: We'll regard it as a *typical distance* of an observation from the mean. The most basic property of the standard deviation is this:

- The larger the standard deviation *s*, the greater the spread of the data.

A small technical point. You may wonder why the denominator of the variance and the standard deviation use *n* – 1 instead of *n*. We said that the variance was an *average* of the *n* squared deviations, so should we not divide by *n*? In fact, that should be done if the data set has observations for an entire *population*. In practice, we rarely have data for the entire population and must instead rely on a *sample* of *n* observations to estimate the variability for the entire population. In that case, we tend to get better estimates by using *n* – 1 instead of *n* in the denominator. This book is not intended to present the mathematical theory of statistics; so, we'll have to ask you to accept this, even though it seems strange at first glance.

EXAMPLE 14: COMPARING WOMEN'S AND MEN'S IDEAL NUMBER OF CHILDREN

Picture the Scenario

Students in a small discussion class led by a teaching assistant for one of the authors were asked on a questionnaire at the beginning of the course, "How many children do you think is ideal for a family?" The observations, classified by the gender of the student, were:

        Men: 0, 0, 0, 2, 4, 4, 4
        Women: 0, 2, 2, 2, 2, 2, 4

Question to Explore

Both men and women have a mean of 2 and a range of 4. Are the distributions equally spread out? If not, which is more spread out?

Think It Through
Let's check dot plots for the data.

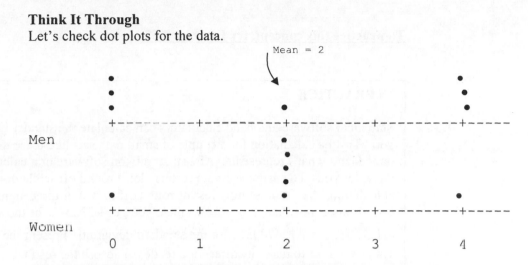

The average deviation from the mean for the male observations appears to be about 2. The observations for females mostly fall right at the mean, so their average deviation appears to be smaller.

Let's calculate the standard deviation for men. For their observations of 0, 0, 0, 2, 4, 4, 4, we have

Value	Deviation	Squared deviation
0	(0 - 2) = -2	4
0	(0 - 2) = -2	4
0	(0 - 2) = -2	4
2	(2 - 2) = 0	0
4	(4 - 2) = 2	4
0	(4 - 2) = 2	4
0	(4 - 2) = 2	4

The sum of squared deviations equals

$$\sum (x-\bar{x})^2 = 4+4+4+0+4+4+4 = 24.$$

The standard deviation of these $n = 7$ observations equals

$$s = \sqrt{\frac{\sum (x-\bar{x})^2}{n-1}} = \sqrt{\frac{24}{6}} - \sqrt{4} = 2.0.$$

This means that for men, a typical distance of an observation from the mean is 2.0. By contrast, you can calculate that the standard deviation for women is $s = 1.2$. Since $2.0 > 1.2$, the observations for males tended to be farther from the mean than those for females. In summary, the men's observations were more spread out.

Insight
The standard deviation is more informative than the range. For these data, it detects that the women were more consistent than the men in their viewpoints about the ideal number of children. The range does not detect this, as it equals 4 for each gender. ◆

To practice this concept, try Exercise 2.57.

IN PRACTICE

Statistical software and hand calculators can calculate the standard deviation s for you. Try the calculation for a couple of small data sets like these to help you understand what s represents. After that, rely on software or a calculator to do the labor for you. To ensure accurate results, don't round off while doing the calculations, but in presenting results round off to two or three significant digits. For instance, in calculating s for women, you get $1.3333\ldots$ for the variance, and $\sqrt{1.3333\ldots} = 1.1547005\ldots$ for the standard deviation. Present the result as $s = 1.2$ or $s = 1.15$ to make it easier for a reader to absorb the result.

EXAMPLE 15: WHAT'S A PLAUSIBLE STANDARD DEVIATION FOR EXAM SCORES?

Picture the Scenario
The first exam in your Statistics course is graded on a scale of 0 to 100. Suppose that the mean score in your class is 80.

Question to Explore
Which value is most plausible for the standard deviation s: 0, 10, or 50?

Think It Through
The standard deviation s is a *typical distance* of an observation from the mean. A value of $s = 0$ seems unlikely. For that to happen, every deviation would have to be 0. This means that every student must then score 80, the mean. A value of $s = 50$ is implausibly large, since 50 or -50 would not be a typical distance of a score from the mean of 80. We would instead expect to see a value of s such as 10. With $s = 10$, for instance, a typical distance is 10, as occurs with the scores of 70 and 90.

Insight
In summary, we've learned that s is a typical distance from the mean, larger values of s represent greater spread, and $s = 0$ means that all observations take the same value.

♦

To practice this concept, try Exercises 2.55 and 2.56.

Properties of the Standard Deviation, *s*

- The greater the spread of the data, the larger is the value of *s*.

- *s* = 0 only when all observations take the same value. For instance, if the reported ideal number of children for a sample of seven people is 2, 2, 2, 2, 2, 2, 2, then the mean equals 2, each of the seven deviations equals 0, and *s* = 0. This is the minimum possible spread for a sample.

- Since *s* measures spread around the mean, its use is appropriate only when the mean is appropriate for measuring the center. It can be influenced by an outlier, so it is a nonresistant measure.

Why is the standard deviation nonresistant? It uses the mean, which we know can be influenced strongly by outliers. Also, outliers have large deviations, and so they tend to have *extremely* large squared deviations. These can blow up the value of *s* and make it sensitive to outliers. So, it's often better to use the IQR to describe spread when a distribution is very highly skewed and has severe outliers.

To practice this concept, try Exercises 2.68 and 2.138.

Interpreting the Magnitude of *s*: The Empirical Rule

Bell Shaped Distribution

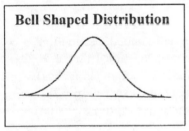

Suppose that a distribution is unimodal and approximately symmetric with a **bell shape**, as in the margin figure. The value of *s* then has a more precise interpretation. Using the mean and standard deviation, we can form intervals that contain certain percentages (approximately) of the data.

Empirical Rule

If a distribution is bell shaped, then approximately
- 68% of the observations fall within 1 standard deviation of the mean, that is, between $\bar{x} - s$ and $\bar{x} + s$ (denoted $\bar{x} \pm s$)
- 95% of the observations fall within 2 standard deviations of the mean ($\bar{x} \pm 2s$).
- All or nearly all observations fall within 3 standard deviations of the mean ($\bar{x} \pm 3s$).

IN WORDS

$\bar{x} - s$ denotes the value one standard deviation below the mean, $\bar{x} + s$ denotes the value one standard deviation above the mean, and $\bar{x} \pm s$ denotes all values falling *within* one standard deviation of the mean.

Figure 2.15 is a graphical portrayal of the Empirical Rule.

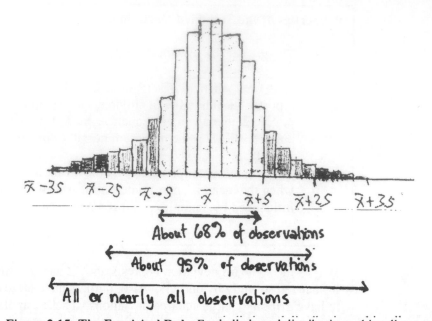

Figure 2.15: The Empirical Rule. For bell-shaped distributions, this tells us approximately how much of the data fall within one, two, and three standard deviations of the mean.
Question: About what percentage would fall *more than* two standard deviations from the mean?

EXAMPLE 16: DESCRIBING HEIGHTS FROM A STUDENT SURVEY

Picture the Scenario
Many human physical characteristics have approximately a bell-shaped distribution. Let's explore height, to see if it has a bell-shaped distribution. Question 1 on the student survey in Exercise 37 of Chapter 1 asked for the student's height (in inches). Figure 2.16 shows a histogram of the heights from responses to this survey by 378 students at the University of Georgia. Table 2.6 presents some descriptive statistics, using MINITAB.

Question To Explore
Can we use the Empirical Rule to describe the spread of these data? If so, how?

Sample of College Student Heights

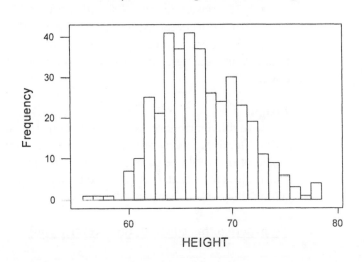

Figure 2.16: Histogram of Student Height Data. This summarizes heights of 378 college students. **Question:** How would you describe the shape, center, and spread of the distribution?

Table 2.6: Minitab Output for Descriptive Statistics of Student Height Data

Variable	N	Mean	Median
HEIGHT	378	67.032	67.000

Variable	StDev	Minimum	Maximum	Q1	Q3
HEIGHT	3.921	56.000	78.000	64.000	70.000

Think It Through

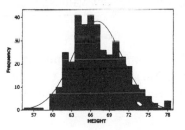

A curve superimposed to approximate Figure 2.16 has the shape shown in the margin. It is somewhat skewed to the right but is not too far from a symmetric bell shape. From Table 2.6, the mean and median are close, about 67 inches, Q1 and Q3 are both about 3 inches from the mean, and the minimum and maximum are both about 11 inches from the mean. Such equalities are reflective of an approximately symmetric distribution.

Since the mean is 67.0 and the standard deviation (labeled StDev in Table 2.6) is 3.9 inches, by the Empirical Rule, approximately

- 68% of the observations fall between
  $\bar{x} - s = 67.0 - 3.9 = 63.1$ and $\bar{x} + s = 67.0 + 3.9 = 70.9$,
  that is, within the interval (63.1, 70.9), about 63 to 71 inches.

- 95% of the observations fall within
  $\bar{x} \pm 2s$, which is $67.0 \pm 2(3.9)$, or (59.2, 74.8), about 59 to 75 inches.

- All or nearly all observations fall within $\bar{x} \pm 3s$, or (55.3, 78.7).

Of the 378 observations, by actually counting we find that
- 262 observations, 69%, fall within (63.1, 70.9).
- 363 observations, 96%, fall within (59.2, 74.8).
- 378 observations, 100%, fall within (55.3, 78.7).

In summary, the percentages predicted by the Empirical Rule are close to the actual ones.

Insight

Because the distribution was close to bell shaped, we could predict simple summaries quite effectively using only two numbers--the mean and the standard deviation. The distribution would be even closer to bell-shaped if we looked at the data only for the 117 males (for which it is centered around 71 inches) or only for the 261 females (for which it is centered around 65 inches).

◆

To practice this concept, try Exercise 2.60.

With a bell-shaped distribution for a large data set, the observations usually extend about 3 standard deviations below the mean and about 3 standard deviations above the mean. When the distribution is highly skewed, the most extreme observation in one direction may not be nearly that far from the mean. For instance, on a recent exam given to introductory statistics students, the scores ranged between 30 and 100, with median = 88, $\bar{x} = 84$ and $s = 16$. The maximum score of 100 was only one standard deviation above the mean (that is, $100 = \bar{x} + s = 84 + 16$). By contrast the minimum score of 30 was more than three standard deviations below the mean. This happened because the distribution of scores was highly skewed to the left. (See the margin figure).

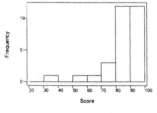

IN PRACTICE

The Empirical Rule may approximate the actual percentages poorly if the data are highly skewed or highly discrete (the variable taking relatively few values).

In Chapter 6, we'll see why the Empirical Rule "works." For a special family of smooth, bell-shaped curves (called the **normal distribution**), we'll see that we can find the percentage of the distribution in *any* particular region by knowing only the mean and the standard deviation.

Sample Statistics and Population Parameters

Of the descriptive measures introduced in this chapter, the mean $\bar{x}$ and the standard deviation s are the most commonly used in practice. We will refer to them frequently in the rest of the text. The formulas that define $\bar{x}$ and s refer to *sample* measurements. They are **sample statistics**.

RECALL

A **population** is the total group of individuals about whom you want to make conclusions. A **sample** is a subset of the population for whom you actually have data.

We will regularly distinguish between sample statistics and the corresponding values for the population. The population mean is the average of all observations in the population. The population standard deviation describes the spread of the population observations about the population mean. The term **parameter** is used for a summary measure of the population.

Definition: Parameter and Statistic

A **parameter** is a numerical summary of the population.

A **statistic** is a numerical summary of a sample taken from the population.

To distinguish further between sample statistics and population parameters, we'll use different notation for the parameter values. Often, Greek letters are used to denote the parameters. For instance, in later chapters we'll use μ (the Greek letter, mu) to denote the population mean and σ (the Greek letter lowercase sigma) to denote the population standard deviation.

Parameter values are usually unknown. The reason for taking a sample is to get sample statistics to estimate their values, because it's the population that we're interested in. Much of this text deals with ways of making inferences about unknown parameters using sample statistics.

SECTION 2.4: PRACTICING THE BASICS

2.52 Ways to measure spread: The standard deviation, the range, and the interquartile range (IQR) summarize the spread of the data.
a. Why is the standard deviation s usually preferred over the range?
b. Why is the IQR sometimes preferred to s?
c. What is an advantage of s over the IQR?

2.53 Variability of cigarette taxes: Here's a recent summary for the distribution of cigarette taxes (in cents) among the 50 states in the U.S.

Minimum = 2.5, Q1 = 36, Median = 60, Q3 = 100, Maximum = 205

a. About what proportion of the states have cigarette taxes (i) greater than 36 cents, (ii) greater than $1?
b. Between what two values are the middle 50% of the observations found?
c. Find the interquartile range. Interpret it.

2.54 Shape of cigarettes taxes: Refer to the previous exercise:
a. Based on the summary, do you think that this distribution was bell shaped? If so, why? If not, why not, and what shape would you expect?
b. Software reported $\bar{x} = 72.85$ and $s = 48$. Based on these values, do you think that this distribution is bell shaped? If so, why? If not, why not, and what shape would you expect?

2.55 Shape of home prices?: According to the National Association of Home Builders, the median selling price of new homes in the U.S. in 2002 was $187,100.
a. Would you expect the mean to be larger, smaller, or equal to $187,100? Explain.
b. Which of the following is the most plausible value for the standard deviation: (i) −15,000, (ii) 1000, (iii) 60,000, (iv) 1,000,000? Why?

2.56 Exam standard deviation: For an exam given to a class, the students' scores ranged from 35 to 98, with a mean of 74. Which of the following is the most realistic value for the standard deviation: -10, 0, 3, 12, 63? Clearly explain what's unrealistic about each of the other values.

2.57 Sick leave: A company decides to investigate the amount of sick leave taken by its employees. A sample of eight employees yields the following numbers of days of sick leave taken in the past year:

0   0   4   0   0   0   6   0

a. Find and interpret the range.
b. Find and interpret the quartiles and the interquartile range.
c. Find and interpret the standard deviation *s*.
d. Suppose the 6 was incorrectly recorded and is supposed to be 60. Redo (a)-(c) with the correct data and describe the effect of this outlier. Which measure of spread, the range, IQR, or *s*, is less affected by the outlier? Why?

2.58 Heights: For a sample of female and male college students, why would you expect the standard deviation for the overall distribution of student heights to be larger than the standard deviations for the separate male and female distributions?

2.59 Histograms and standard deviation: The figure shows histograms for three different samples, each with sample size *n* = 100.
a. Which sample has the (i) largest, (ii) smallest standard deviation?
b. To which sample(s) is the Empirical Rule relevant? Why?

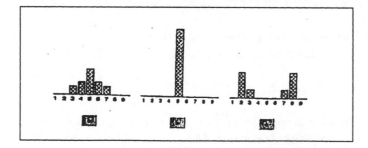

Histograms and Relative Sizes of Standard Deviations

⌨**2.60 Female strength:** The "high school female athletes" data file on the text CD has data for 57 female high school athletes on the maximum number of pounds they were able to bench press. The data are roughly bell shaped, with $\bar{x}$ = 79.9 and *s* = 13.3. Use the Empirical Rule to describe the distribution.

🖳**2.61 Female body weight:** The "college athletes" data file on the text CD has data for 64 female college athletes. The data on weight (in pounds) are roughly bell shaped with $\bar{x} = 133$ and $s = 17$.
a. Give an interval within which about 95% of the weights fall.
b. Identify the weight of an athlete who is three standard deviations above the mean In this sample? Would this be a rather unusual observation? Why?

No. Times	Count
0	7074
1	1541
2	43
Total	8658

2.62 Empirical Rule and skewed, highly discrete distribution: Example 10 gave data on the number of times married. For the observations for men, shown again in the margin, $\bar{x} = 0.19$ and $s = 0.40$.
a. Find the actual percentages of observations within one, two, and three standard deviations of the mean. How do these compare to the percentages predicted by the Empirical Rule?
b. How do you explain the results in (a)?

2.63 Computer use: During a recent semester at the University of Florida, students having accounts on a mainframe computer used a mean of 1921 and a standard deviation of 11,495 kilobytes of drive usage.
a. Would you expect this distribution to be symmetric, skewed to the right, or skewed to the left? Explain.
b. The minimum = 4, Q1 = 256, median = 530, Q3 = 1105, and maximum = 320,000. What does this suggest about the shape of the distribution? Why?
c. What could cause the standard deviation to be so large compared to the mean?
d. Does the Empirical Rule apply to this distribution? Why or why not?

2.64 How much TV?: The 2002 General Social Survey asked, "On the average day, about how many hours do you personally watch television?" Of 905 responses, the mode was 2, the median was 2, the mean was 2.98, and the standard deviation was 2.36. Based on these statistics, what would you surmise about the shape of the distribution? Why?

2.65 How many friends?: A recent General Social Survey asked respondents how many close friends they had. For a sample of 1467 people, the mean was 7.4 and the standard deviation was 11.0. The distribution had a median of 5 and a mode of 4. Based on these statistics, what would you surmise about the shape of the distribution? Why?

2.66 Judging skew using $\bar{x}$ and s: If the largest observation is less than one standard deviation above the mean, then the distribution tends to be skewed to the left. If the smallest observation is less than one standard deviation below the mean, then the distribution tends to be skewed to the right. A professor examined the results of the first exam given in her statistics class. The scores were

35   59   70   73   75   81   84   86

The mean and standard deviation are 70.4 and 16.8. Using these, determine if the distribution is either left or right skewed. Construct a dot plot to confirm your conclusion.

⌨**2.67 EU data file:** The "European Union unemployment" data file on the text CD contains unemployment rates in December 2003 for the 25 countries that were in the European Union in 2004. Using software:
a.   Construct a graph to describe these values.
b.   Find the range and IQR. Interpret.
c.   Find the standard deviation. Interpret.

⌨ **2.68 Create data with a standard deviation:** Use the standard deviation applet at www.prenhall.com/TBA to investigate how the standard deviation changes as the data change.
a. Create ten observations that have a mean of 5 and a standard deviation of about 2.
b. Create ten observations that have a mean of 5 and a standard deviation of about 4.
c. Placing ten values between 0 and 10, what is the largest standard deviation you can get? What are the values that have that standard deviation value?

2.5 HOW CAN WE GRAPHICALLY PORTRAY SPREAD AND DETECT OUTLIERS?

In any statistical analysis, you should examine the data for unusual observations, such as outliers. Can you think of a reason for why they differ so much from the other observations?

Detecting Potential Outliers

We've talked about outliers in an informal sense, but is there a formula for identifying them? One way to flag an observation as potentially being an outlier uses the interquartile range.

The $1.5 \times IQR$ Criterion for Identifying Potential Outliers

An observation is a potential outlier if it falls more than $1.5 \times IQR$ below the first quartile or more than $1.5 \times IQR$ above the third quartile.

From Example 13, the breakfast cereal sodium data has Q1 = 145 and Q3 = 225. So, IQR = Q3 – Q1 = 225 – 145 = 80. For those data:

- $1.5 \times IQR = 1.5 \times 80 = 120$.

- $Q1 - 1.5 \times IQR = 145-120 = 25$ (lower boundary, potential outliers below),
         and
  $Q3 + 1.5 \times IQR = 225+120 = 345$ (upper boundary, potential outliers above).

By the $1.5 \times IQR$ criterion, observations below 25 or above 345 are potential outliers. The only observation below 25 or above 345 is the sodium value of 0 mg for Frosted Mini Wheats. This is the only potential outlier.

Why do we identify an observation as a *potential* outlier rather than flagging it as a *definite* outlier? When a distribution has a long tail, some observations may be more than $1.5 \times IQR$ below the first quartile or above the third quartile even if they are not outliers, in the sense that they are not separated far from the bulk of the data. For instance, in a long tail, there need not be any long gaps between extreme observations and the rest of the data.

The Box Plot: Graphing a Five-Number Summary of Center and Spread

The quartiles and the highest and lowest values are five numbers often used as a set to describe center and spread.

The Five-Number Summary

The **five-number summary** of a dataset is the minimum value, first quartile Q1, median, third quartile Q3, and the maximum value.

The five-number summary is the basis of a graphical display of center and spread called the **box plot**. The **box** of a box plot contains the central 50% of the distribution, from the first quartile to the third quartile. A line inside the box marks the median. The lines extending from the box are called **whiskers**. These extend to encompass the rest of the data, except for potential outliers, which are shown separately.

Constructing a Box Plot

- A **box** goes from the lower quartile Q1 to the upper quartile Q3.
- A line is drawn inside the box at the median.
- A line goes from the lower end of the box to the smallest observation that is not a potential outlier. A separate line goes from the upper end of the box to the largest observation that is not a potential outlier. These lines are called **whiskers**. The potential outliers (more than 1.5 IQR below the first quartile or above the third quartile) are shown separately.

EXAMPLE 17: BOX PLOT FOR THE CEREAL SODIUM DATA

Picture the Scenario

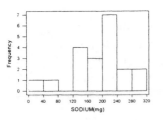

Example 7 constructed a histogram for the breakfast cereal sodium values. That figure is shown again in the margin. The two left-most bars resulted from sodium values of 0 and 70, which were considerably smaller than the others.

Questions to Explore

a. Figure 2.17 shows a box plot for those data. Labels are also given for the five-number summary. Explain how this box plot was constructed and how to interpret it.

b. Are the 0 and 70 values outliers?

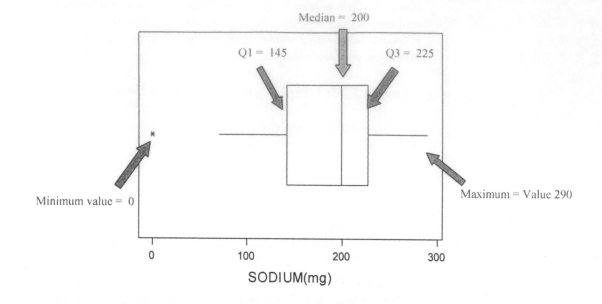

Figure 2.17: Box Plot for 20 Breakfast Cereal Sodium Values. The graph has a central box for the middle 50% of the distribution. The line in the box marks the median. Whiskers extend from the box to the smallest and largest observations, except to highlight potential outliers. The added labels show the five-number summary. The values refer to the horizontal number line. **Question:** Why is the left whisker drawn only down to 70 rather than to 0?

Think It Through

a. The five-number summary of sodium values for the 20 breakfast cereals was:

- Minimum = 0
- Q1 = 145
- Median = 200
- Q3 = 225
- Maximum = 290

The middle 50% of sodium values range from Q1 = 145 mg to Q3 = 225 mg, which are the two outer lines of the box. The median is 200 mg, shown by the center line in the box.

As we saw above, based on the lower and upper boundaries of 25 and 345 for identifying potential outliers using the $1.5 \times$ IQR criterion, the sodium value of 0 mg for Frosted Mini Wheats is the only potential outlier. On the box plot, an asterisk identifies it as a potential outlier. The whisker from Q1 is drawn down to 70, which is the smallest value that is not below the lower boundary of 25.

b. We've identified the observation of 0 as a potential outlier. In fact, there is a large gap between 0 and 70, the nearest observation. This indicates that 0 is an outlier.

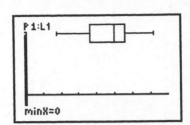

TI-83+/84 Box Plot

Insight

Most software identifies observations that are more than 1.5 IQR from the quartiles by a separate symbol, such as *. (Some software uses a separate symbol for observations more than 3 IQR from the quartiles.) The TI-83+/84 output for the boxplot is presented in the margin. The left whisker is drawn to 70 with the outlier at 0 on the *y*-axis. Why show potential outliers separately? One reason is to identify them for further study. Was the observation perhaps incorrectly recorded? Was that subject fundamentally different from the others in some way? Often it makes sense to repeat a statistical analysis without an outlier, to make sure the results are not overly sensitive to a single observation.

Another reason for identifying outliers is that they do not provide much information about the shape of the distribution, especially for large data sets. Some software also can provide a version of the box plot that extends the whisker to the minimum and maximum, even if outliers exist. However, an extreme value can give the impression of severe skew when actually the remaining observations are not at all skewed.

◆

To practice this concept, try Exercise 2.71.

The Box Plot Compared with the Histogram

A box plot does not portray the shape of a distribution as clearly as does a histogram. It does indicate skew from the relative lengths of the whiskers and the two parts of the box. In Figure 2.17 for the cereal sodium values, the left part of the box is wider than the right part, reflecting the greater distance between the first quartile and the median than between the third quartile and the median (55, compared to 25). The side with the larger part of the box and the longer whisker usually has skew in that direction.

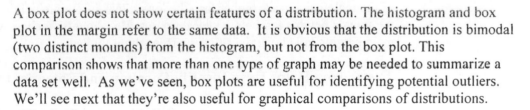

A box plot does not show certain features of a distribution. The histogram and box plot in the margin refer to the same data. It is obvious that the distribution is bimodal (two distinct mounds) from the histogram, but not from the box plot. This comparison shows that more than one type of graph may be needed to summarize a data set well. As we've seen, box plots are useful for identifying potential outliers. We'll see next that they're also useful for graphical comparisons of distributions.

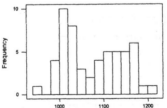

Side-By-Side Box Plots Help to Compare Groups

In Example 16 we looked at student heights. To compare heights for females and males, we could look at side-by-side box plots, as shown in Figure 2.18.

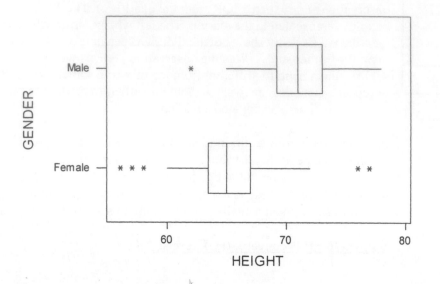

Figure 2.18: Box Plots of Male and Female Heights. The box plots are graphed using the same scale for heights. **Question:** What are approximate values of the quartiles for the two groups?

The box plots suggest that both distributions are approximately symmetric. The median (the center line in a box) is approximately 71 inches for the males and 65 inches for the females. Although the centers differ, the spread is similar, as indicated by the width of the boxes (which is the IQR) being similar. Both samples have heights that are unusually short or tall, flagged as potential outliers. The upper 75% of the male heights are higher than the lower 75% of female heights. That is, 75% of the female heights fall below their third quartile, about 67 inches, whereas 75% of the male heights fall above their first quartile, about 69 inches.

The *z*-Score Helps Us to Identify Location and Potential Outliers

The Empirical Rule tells us that for a bell-shaped distribution it is unusual for an observation to fall more than three standard deviations from the mean. An alternative criterion for identifying potential outliers then uses the standard deviation rather than the IQR.

- An observation in a bell-shaped distribution is regarded as a potential outlier if it falls more than about three standard deviations from the mean.

How do we know the number of standard deviations that an observation falls from the mean? When $\bar{x} = 84$ and $s = 16$, it is clear that a value of 100 is one standard deviation above the mean, since $(100 - 84) = 16$. Alternatively,

$$(100 - 84)/16 = 1.$$

Taking the difference between an observation and the mean and dividing by the standard deviation specifies the number of standard deviations that the observation falls from the mean. This number is called the **z-score**.

Definition: z-score

The **z-score** for an observation is the number of standard deviations that it falls from the mean. For sample data, the z-score is calculated as

$$z = \frac{\text{observation - mean}}{\text{standard deviation}}.$$

EXAMPLE 18: DETECTING COUNTRIES THAT ARE POLLUTION OUTLIERS

Picture the Scenario

In Example 11, we examined CO_2 emissions for eight of the world's most populous nations. We will now consider air pollution data for the 25 nations in the European Union. The "energy-eu" data file[7] on the text CD contains data for the 25 nations in the European Union (EU) on per capita carbon-dioxide (CO_2) emissions, in metric tons. The mean was 8.3 and the standard deviation was 3.3. Figure 2.19 shows a box plot of the data. The maximum value of 19.4 was recorded for Luxembourg, which is highlighted as a potential outlier.

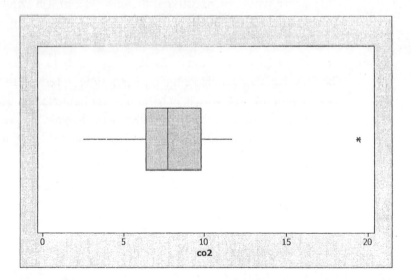

Figure 2.19: Minitab Box Plot of Per Capita Carbon Dioxide Emissions for European Union Nations. Question: Can you use this plot to approximate the five-number summary?

---

[7] Source: *Human Development Report 2004*, see hdr.undp.org/statistics/data

Questions to Explore

a. How many standard deviations from the mean was the CO_2 value for Luxembourg?

b. The CO_2 value for the U.S. was 19.8. According to the three standard deviation criterion, is the U.S. an outlier on carbon dioxide emissions relative to the EU?

Think It Through

a. Since $\bar{x} = 8.3$ and $s = 3.3$ inches, the z-score for the observation of 19.4 is

$$z = \frac{\text{observation-mean}}{\text{standard deviation}} = \frac{19.4 - 8.3}{3.3} = \frac{11.1}{3.3} = 3.4.$$

The per capita carbon dioxide emission for Luxembourg is 3.4 standard deviations above the mean. By the three standard deviation criterion, this is a potential outlier. Since it is well removed from the rest of the data, we'd regard it as an actual outlier. However, Luxembourg has only 350,000 people, so in terms of the amount of pollution it is not a major polluter in the EU.

b. The CO_2 value for the U.S. is even higher, so its z-score is higher. We find that $z = (19.9 - 8.3)/3.3 = 3.5$. The U.S. is also a CO_2 outlier, relative to values in the EU.

Insight

The z-scores of 3.4 and 3.5 are positive. This indicates that the observations are *above* the mean, because an observation above the mean results in a positive z-score. The z-score would be negative if the observation were *below* the mean. For instance, France has a CO_2 value of 6.2, which is below the mean of 8.3 and has a z-score of −0.6.

The U.S. has faced a challenge for some time to reduce its CO_2 emissions. Ten years ago, the President's Council on Sustainable Development[8] reported to then President Clinton that the U.S., with 5% of the world's population, used about 25% of global energy and was responsible for about 25% of the global emissions of CO_2.

♦

To practice this concept, try Exercises 2.69 and 2.72.

---

[8] In *Sustainable America*, released Feb. 1996. See clinton2.nara.gov/PCSD/Publications

---

On the Shoulders of: John Tukey (1915- 2000)

"The best thing about being a statistician is that you get to play in everyone's backyard." John Tukey

In the 1960s, John Tukey of Princeton University was concerned that statisticians were putting too much emphasis on complex data analyses and ignoring simpler ways to examine and learn from the data. Tukey developed new descriptive methods, under the title of **exploratory data analysis (EDA)**. These methods make few assumptions about the structure of the data and emphasize data display and ways of searching for patterns and deviations from those patterns. Two graphical tools that Tukey invented were the stem-and-leaf plot and the box plot.

Initially, few statisticians promoted EDA. However, in recent years, some EDA methods have become common tools for data analysis. Part of this acceptance has been inspired by the availability of technology (computer software and statistical calculators) that can implement some of Tukey's methods, although Tukey emphasized that his methods could be done with paper and pencil, especially for smaller data sets. Over his career, Tukey made many other important contributions to statistics and to science, including coining terms such as *software* and *bit*.

Tukey's work illustrates that statistics is an evolving discipline. Almost all of the statistical methods used today were developed in the past century, and new methods continue to be created, largely because of increasing computer power. John Tukey died recently, but he will be remembered for his unparalleled contributions to statistics and science and for his unusual creativity.

Photo to come

Figure 2.20: John Tukey, Who Developed Methods of Exploratory Data Analysis.

SECTION 2.5: PRACTICING THE BASICS

2.69 Energy statistics: The United Nations publication, *Energy Statistics Yearbook*, lists per capita consumption of energy. The 2001 data for the 25 nations that now make up the European Union are in the "energy-EU" data file on the text CD. The energy values (in kilograms) have a mean of 4998 and a standard deviation of 1824 and are roughly bell shaped, except for the value of 9814 for Luxembourg.

a. Using the MINITAB box plot shown, give approximate values for the five-number summary and indicate whether any countries were judged to be potential outliers according to that plot.

b. Italy had a value of 4222. How many standard deviations from the mean was it?

c. The U.S. is not in this data file, but its value was 11,067. Relative to the distribution for the European Union, how many standard deviations from the mean was it?

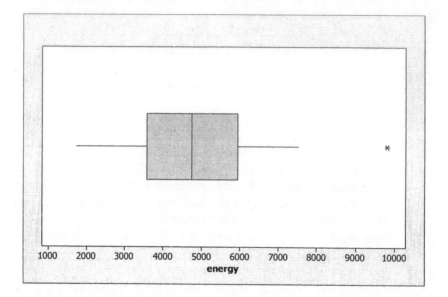

2.70 Computer use: During a recent semester at the University of Florida, students having accounts on a mainframe computer had drive use (in kilobytes) described by the five-number summary, minimum = 4, Q1 = 256, median = 530, Q3 = 1105, and maximum = 320,000.

a.   Would you expect this distribution to be symmetric, skewed to the right, or skewed to the left? Explain.

b.   Use the 1.5×IQR criterion to determine if any potential outliers are present.

2.71 SAT scores revisited: The statewide (and D.C.) average total SAT scores (math plus verbal) for 2003 are summarized in the box plot and histogram shown.

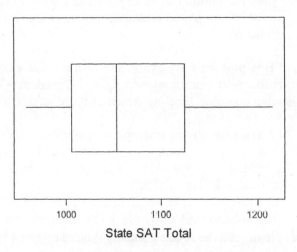

State SAT Total

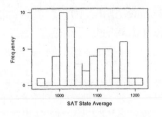

a. Exercise 2.24 considered the histogram for these data. It is shown again in the margin. Describe the shape of the histogram.

b. Explain how the box plot also gives you information about the distribution shape.

c. Compare the box plot to the histogram. What feature of the distribution would be missed by viewing only the box plot?

d. Using the box plot, give the approximate value for each component of the five-number summary.

2.72 Central Park temperature distribution revisited: Exercise 2.26 showed a histogram for the distribution of Central Park annual average temperatures for the 20th century. The box plot for these data is shown here.

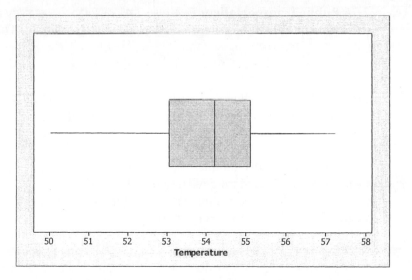

Temperature

a. According to the box plot, identify any potential outlier and give its approximate value.

b. Using the box plot, give the approximate values for each component of the five-number summary.
c. The distribution has mean = 54.1 and standard deviation = 1.45. Is the minimum of 50.04 a potential outlier according to the three standard deviation criterion? Explain.

2.73 Box plot for easy exam: The scores on an easy exam have mean = 88, standard deviation = 10, minimum = 65, Q1 = 77, median = 85, Q3 = 91, maximum = 100. Sketch a box plot, labeling which of these values are used in the plot.

2.74 European Union unemployment rates: Unemployment rates in December 2003 of European countries in the European Union (which were shown in Exercise 2.38) ranged from 3.9 to 11.2, with Q1 = 4.6, median = 6.7, Q3 = 9.2, a mean of 7.1, and standard deviation of 2.3.
a. In a box plot, what would be the values at the outer edges of the box, and what would be the values to which the whiskers extend?
b. Spain had the highest unemployment rate of 11.2. Is it an outlier according to the three standard deviation criterion? Explain.
c. What unemployment value for a country would have a z-score equal to 0?

2.75 Prices of stocks: The data values below represent the prices per share of the 10 most actively traded stocks from the New York Stock Exchange (rounded to the nearest dollar) on Oct. 1, 2003.

2   4   11   12   13   15   31   31   37   47

a. Construct a dot plot or stem-and-leaf plot.
b. Sketch or use software to construct a box plot. Interpret it.
c. What feature of the distribution displayed in the plot in (a) is not obvious in the box plot? (*Hint*: Are there any gaps in the data?)

2.76 Public transportation: Exercise 2.41 described a survey about how many miles per day employees of a company use public transportation. The sample values were:

0   0   4   0   0   0   10   0   6   0

Identify the five number summary, and sketch a box plot. Explain why Q1 and the median share the same line in the box. Why does the box plot not have a left whisker?

2.77 Air pollution: Example 18 discussed EU carbon dioxide emissions, which had a mean of 8.3 and standard deviation of 3.3.
a.   Canada's observation was 14.4. Find its z-score relative to the distribution of values for the EU nations, and interpret..
b. Sweden's observation was 5.3. Find its *z*-score, and interpret.

2.78 Female heights: For the 261 female heights shown in the box plot in Figure 2.18, the mean was 65.3 inches and the standard deviation was 3.0 inches. The shortest person in this sample had a height of 56 inches.
a. Find the *z*-score for the height of 56 inches.
b. What does the negative sign for the *z*-score represent?

c. Is this observation a potential outlier according to the three standard deviation distance criteria?   Explain.

2.79 Hamburger sales: The manager of a fast-food restaurant records each day the amount of money received from sales of food that day. Using software, he finds a bell-shaped histogram with a mean of $1165 and a standard deviation of $220. Would a day in which the sales equal $2000 be considered an unusually good day? Answer by providing statistical justification.

2.80 Who was Roger Maris: Roger Maris, who spent most of his professional baseball career with the New York Yankees, held the record for the most homeruns in one season (61) from 1961 until 1998, when the record was broken by Mark McGwire. Maris played in the major leagues from 1957 to1968. The number of homeruns he hit in each year that he played is summarized below.

Variable	N	Mean	Median	StDev
RMHR	12	22.92	19.50	15.98

Variable	Minimum	Maximum	Q1	Q3
RMHR	5.00	61.00	10.00	31.75

a.   Use the three standard deviation criterion to determine if any potential outliers are present.
b.   The criterion in (a) requires the distribution to be approximately bell-shaped. Is there any evidence here to contradict this? Explain.
c.   A sports writer commented that Roger Maris hit *only* 13 homeruns in 1966. Was this unusual for Maris? Comment, using statistical justification.

2.81 Florida students again: Refer to the "Florida student survey" data set on the text CD, and the data on weekly hours of TV watching.
a. Use software to construct a box plot. Interpret the information on the plot, and use it to describe the shape of the distribution.
b. Using a criterion for outliers, investigate whether there are any potential outliers.

CHAPTER SUMMARY

This chapter introduced **descriptive statistics**--graphical and numerical ways of **describing** data. The characteristics of interest that we measure are called **variables**, because values on them vary from subject to subject.

- A **categorical variable** has observations that fall into one of a set of categories.

- A **quantitative variable** takes numerical values that represent different magnitudes of the variable.

- A quantitative variable is **discrete** if it has separate possible values, such as the integers 0, 1, 2, ... for a variable expressed as "the number of ....". It is **continuous** if its possible values form an interval.

When we explore data, key features to describe are the **shape**, using graphical displays, and the **center** and **spread**, using numerical summaries.

Overview of Graphical Methods

- For categorical variables, data are displayed using **pie charts** and **bar graphs.** Bar graphs provide more flexibility and make it easier to compare categories having similar percentages.

- For quantitative variables, a **histogram** is a graph of a frequency table. It displays bars that specify frequencies or relative frequencies (proportions or percentages) for possible values or intervals of possible values. The **stem-and-leaf plot** (a vertical line dividing the final digit, the leaf, from the stem) and **dot plot** (dots above the number line) show the individual observations. They are useful for small data sets. These three graphs all show shape, such as whether the distribution is approximately bell-shaped, skewed to the right (longer tail pointing to the right), or skewed to the left.

- The **box plot** has a central box drawn between the first quartile and third quartile, with a line drawn in the box at the median. It has whiskers that extend to the minimum and maximum values, except for potential **outliers**. An **outlier** is an extreme value falling far below or above the bulk of the data.

- A **time plot** graphically displays observations for a variable measured over time. This plot can visually show **trends** over time.

Overview of Measures of the Center

Measures of the center attempt to describe a "typical" or "representative" observation.

- The **mean** is the sum of the observations divided by the number of observations. It is the balance point of the data.

- The **median** divides the ordered data set into two parts of equal numbers of observations, half below and half above that point. The median is the 50th percentile (second quartile). It is a more representative summary than the mean when the data are highly skewed.

- The lower quarter of the observations fall below the **first quartile (Q1)**, and the upper quarter fall above the **third quartile (Q3)**. These are the 25th percentile and 75th percentile. The quartiles and median split the data into four equal parts.

Overview of Measures of Spread

Measures of spread describe the variability of the data.

- The **range** is the difference between the largest and smallest observations.

- The **interquartile range (IQR)** is the difference between the third and first quartiles, which span the middle half of the data. It is a more **resistant** measure than the range, being unaffected by extreme observations.

- The **deviation** of an observation x from the mean is $x - \overline{x}$. The **variance** is an average of the squared deviations. Its square root, the **standard deviation**, is more useful, describing a typical distance from the mean.

- The **Empirical Rule** states that for a bell-shaped distribution:
  a. About 68% of the data fall within one standard deviation of the mean.
  b. About 95% of the data fall within two standard deviations, $\overline{x} \pm 2s$.
  c. Nearly all, if not all, the data fall within three standard deviations, $\overline{x} \pm 3s$.

- An observation is a potential outlier if it falls (a) more than $1.5 \times \text{IQR}$ below Q1 or above Q3, or (b) more than 3 standard deviations from the mean. The **z-score** is the number of standard deviations that an observation falls from the mean.

SUMMARY OF NOTATION

Mean $\overline{x} = \dfrac{\sum x}{n}$, where x denotes the variable, n is the sample size, and Σ indicates to sum the observations

Standard deviation $s = \sqrt{\dfrac{\sum (x - \overline{x})^2}{n-1}}$

$$z\text{-score} = \frac{\text{observed value - mean}}{\text{standard deviation}}$$

ANSWERS TO THE CHAPTER FIGURE QUESTIONS

Figure 2.1: *It is not always clear exactly what portion or percentage of a circle the slice represents, especially when a pie has many slices. Also, some slices may be representing close to the same percentage of the circle; thus, it is not clear which is larger or smaller.*

Figure 2.3: *The number of dots above a value on the number line represents the frequency of occurrence of that value.*

Figure 2.4: *There are no observations in the data set that have as their first two digits the particular stem value without a leaf.*

Figure 2.7: *The direction of the longer tail indicates the direction of the skew.*

Figure 2.9: *The annual average temperatures are tending to increase over time.*

Figure 2.10:   *The area of the figures does not accurately represent the frequencies being reported.*

Figure 2.11:   *The mean is pulled in the direction of the longer tail since extreme observations affect the balance point of the distribution. The median is not affected by the size of the observation.*

Figure 2.12:   *The second quartile has two quarters (25% and 25%) of the data below it. These two quarters are 50% of the data. Therefore, Q2 is the median since 50% of the data falls below the median.*

Figure 2.13:   *The range would change from $10,000 to $65,000.*

Figure 2.14:   *A deviation is positive when it falls above the mean and negative when it falls below the mean.*

Figure 2.15:   *Since about 95% of the data fall within 2 standard deviations of the mean, about 100%−95% = 5% fall more than 2 standard deviations from the mean.*

Figure 2.16:   *The shape is approximately symmetric, with a slight right skew. The center as measured by the mean and median is about 67 inches. A typical deviation form the mean is about 4 inches.*

Figure 2.17:   *The left whisker is drawn only to 70 since the sodium value of 0 is identified as a potential outlier.*

Figure 2.18:   *For males, the quartiles are about Q1 = 69 inches, Q2 = 71 inches, and Q3 = 73 inches. For females, the estimated quartiles are Q1 = 63 inches, Q2 = 65 inches, and Q3 = 67 inches.*

Figure 2.19:   *Yes. The left whisker extends to the minimum value, the left side of the central box is at Q1, the vertical line inside the central box is at the median, the right side of the central box is at Q3, and the outlier identified with a star is at the maximum value.*

CHAPTER PROBLEMS: PRACTICING THE BASICS

2.82 Categorical or quantitative?: Identify each of the following variables as categorical or quantitative.
a.   Number of children in family.
b.   Amount of time in football game before first points scored
c.   College major (English, history, chemistry, …)
d.   Type of music (rock, jazz, classical, folk, other)

2.83 Continuous or discrete?: Which of the following variables are continuous, when the measurements are as fine as possible?
a.   Age of mother
b.   Number of children in a family
c.   Cooking time for preparing dinner
d.   Latitude and longitude of a city
e.   Population size of a city

2.84 Immigration into U.S.: The table shows the number (in thousands) of the foreign-born population of the United States in 2002, by place of birth.

Forcign-born population in the U.S.

Place of Birth	Number (in Thousands)
Europe	4548
Caribbean	3102
Central America	11819
South America	2022
Asia	8281
Other	2680
Total	32452

Source: U.S. Statistical Abstract, 2003

a. Is "Place of birth" quantitative or categorical? Show how to summarize results by adding a column of percentages to the table.
b. By sketching or using software, plot the results in a bar graph labcling the horizontal axis with the categories in alphabetical order.
c. What is the mode?
d. How would the place of birth categories be ordered for a Pareto chart? What's its advantage over the ordinary bar graph?

2.85 Cool in China: A recent survey[9] asked 1200 university students in China to pick the personality trait that most defines a person as "cool." The possible responses allowed, and the percentage making each, were individualistic and innovative (47%), stylish (13.5%), dynamic and capable (9.5%), easygoing and relaxed (7.5%), other (22.5%).
a. Identify the variable being measured.
b. Classify the variable as categorical or quantitative.
c. Which of the following methods could you use to describe these data?: (i) bar chart, (ii) dot plot, (iii) box plot, (iv) median, (v) mean, (vi) mode, (vi) IQR, (vii) standard deviation.

2.86 Chad voting problems: The 2000 U.S. Presidential election had various problems in Florida. One was "overvotes" -- people mistakenly voting for more than one Presidential candidate. There were 110,000 overvote ballots, with Gore marked on 84,197 and Bush on 37,731. These ballots were disqualified. Was overvoting related to the design of the ballot? The figure shows dot plots (Minitab output) of the overvote percentages for 65 Florida counties organized according to the method of registering the vote – optical scanning, Votomatic (voters manually punch out chads)

---

[9] Source: Public relations firm Hill & Knowlton, as reported by *Time* magazine.

and Datavote (voter presses a lever that punches out the chad mechanically) and the number of columns on the ballot (1 or 2).

a. The overvote was highest (11.6%) in Gadsden County. Identify the number of columns and the method of registering the vote for that county.
b. Of the six ballot type and method combinations, which two seemed to perform best in terms of having relatively low percentages of overvotes?
c. How might these data be summarized further by a bar graph with six bars?

Overvote Percentages by Number of Columns On Ballot and Method Of Voting

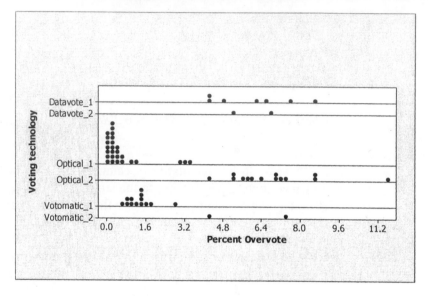

Source: A. Agresti and B. Presnell, *Statistical Science*, vol. 17, pp. 1-5, 2002.

2.87 Number of children: For the question "How many children have you ever had?" in the 2002 General Social Survey, the results were

No. children	0	1	2	3	4	5	6	7	8+
Count	799	469	657	481	185	73	40	22	34

a. Which is the most appropriate graph (dot plot, stem-and-leaf, or histogram) to use in displaying the data? Why?
b. Based on sketching or using software to construct the graph, characterize this distribution as skewed to the left, skewed to the right, or symmetric. Explain.

No. Times	Frequency
0	2
1	4
2	4
3	8
4	4
5	5
6	2
7	4
8	2
9	1

2.88 Newspaper reading: Exercise 2.23 gave results for the number of times a week a person reads a daily newspaper for a sample of 36 students at the University of Georgia. The frequency table is shown in the margin.

a. Construct a dot plot of the data.
b. Construct a stem-and-leaf plot of the data. Identify the stems and the leaves.
c. The mean is 3.94. Find the median.
d. Is this distribution skewed to the left, or skewed to the right, or symmetric? Why?

2.89 Match the histogram: Match the histogram with the appropriate description.
      (1) Skewed to the left

(2) Symmetric and bimodal
(3) Symmetric and unimodal
(4) Skewed to the right

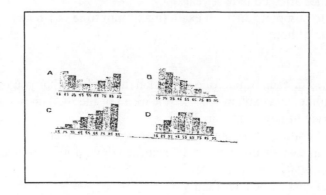

2.90 Enchiladas and sodium: How do sodium amounts vary in foods other than cereal? In the March 2000 *Consumer Reports*, comparisons were made among various brands of supermarket enchiladas in cost and sodium content.

Brand and Type	Cost (per serving)	Sodium content (mg)
Amy's Black Bean	$3.03	780
Patio Cheese	$1.07	1570
Banquet Cheese	$1.28	1500
El Charrito Beef	$1.53	1370
Patio Beef	$1.05	1700
Banquet Beef	$1.27	1330
Healthy Choice Chicken	$2.34	440
Lean Cuisine Chicken	$2.47	520
Weight Watchers Chicken	$2.09	660

a.   Construct a dot plot of the sodium amounts in the various enchilada brands.
b.   What is the advantage(s) of using the dot plot over constructing a histogram?
c.   Construct a stem-and-leaf plot of the sodium values. What does it show?
d.   Summarize your findings from these graphs.

2.91 Enchiladas and cost: Refer to the previous exercise. Repeat parts (a), (c) and (d) for the cost per serving. Summarize your findings.

2.92 Enchilada sodium revisited: In the previous two exercises, we examined the data graphically on cost and sodium content for nine enchilada brands. Let's take another look at those data. Use software for the following:
a.   Find the five-number summary for the sodium amounts, and interpret.
b.   Construct a box plot. What feature stands out about the central box? What is a possible explanation?
c.   Using the $1.5 \times IQR$ criterion, identify any potential outliers.
d.   Summarize the sodium values by their mean and standard deviation. What do you learn from the five-number summary that you do not learn from these?

2.93 What shape do you expect?: For the following variables, indicate whether you would expect its histogram to be bell shaped, skewed to the right, or skewed to the left. Explain why.
a. Number of times arrested in past year
b. Time needed to complete difficult exam (maximum time is 1 hour)
c. Assessed value of home
d. Age at death

2.94 Sketch plots: For each of the following, sketch roughly what you expect a histogram to look like, and explain whether the mean or the median would be greater.
a. The selling price of new homes in 2005
b. The number of children ever born per woman age 40 or over
c. The score on an easy exam (mean = 88, standard deviation = 10, maximum = 100, minimum = 50)
d. The number of cars owned per family
e. Number of months in which subject drove a car last year

2.95 Median vs. mean income: Based on the 2000 census, the U.S. Census Bureau reported that the median U.S. household income in 1999 was $41,994. Would you expect the mean household income to have been larger, or smaller? Explain.

2.96 Rich get richer?: A U.S. Federal Reserve study in 2000 indicated that for those families with annual incomes above $100,000, their median net worth was about $500,000 both in 1995 and in 1998, but their mean net worth rose from $1.4 million in 1995 to $1.7 million in 1998.
a. Is the distribution of net worth for these families likely to be symmetric, skewed to the right, or skewed to the left? Explain.
b. A newspaper story about this said that the mean uses "a calculation that captures the huge gains made by the wealthiest Americans." Why would the median not necessarily capture those gains?
c. The same report indicated that families earning less than $10,000 annually saw their median net worth fall by 25% over these three years to $3600. What was their median net worth in 1995?

2.97 Median and mean net worth by race: According to the U.S. Census Bureau, the median net worth in 2002 was $31,408 for black families and $47,199 for white families. For a given racial group, would you expect the mean net worth to be larger, or smaller, than the median? Explain.

2.98 Baseball salaries: During the strike of professional baseball players in 1994, two quite different numbers were reported for the central tendency of players' annual salaries. One was $1.2 million and the other was $500,000. One of these was the median and one was the mean. Which value do you think was the mean? Why?

2.99 Exams scores in a statistics class: A professor examined the results of the first exam given in her statistics class. The scores were

35   59   70   73   75   81   84   86

Find the interquartile range and the standard deviation. Interpret.

2.100 What does *s* equal?:
a. For an exam given to a class, the students' scores ranged from 35 to 98, with a mean of 74. Which of the following is the most realistic value for the standard deviation: −10, 1, 12, 60? Why?
b. The sample mean for a data set equals 80. Which of the following is an impossible value for the standard deviation? 200, 0, −20? Why?

2.101 Female heights: According to a recent report from the U.S. National Center for Health Statistics, females with age between 25 and 34 years have a bell shaped distribution for height, with mean of 65 inches and standard deviation of 3.5 inches.
a. Give an interval within which about 95% of the heights fall.
b. What is the height for a female who is three standard deviations below the mean? Would this be a rather unusual height? Why?

2.102 Is energy consumption bell shaped?: Residential electrical consumption in a recent month in Gainesville, Florida, had a mean of 780 and a standard deviation of 506 kilowatt-hours (Kwh). The minimum usage was 3 Kwh and the maximum was 9390 Kwh. (Data supplied by N. Todd Kamhoot, Gainesville Regional Utilities.) What shape do you expect this distribution to have? Why?

2.103 Shape of water use: Residential water consumption in a recent month in Gainesville, Florida, had a mean of 7.1 and a standard deviation of 6.2 (thousand gallons). What shape do you expect this distribution to have? Why? (Data supplied by N. Todd Kamhoot, Gainesville Regional Utilities.)

2.104 Dot plot of heights: The figures show dot plots of heights for female and male students from the "Heights" data file on the text CD, using the same horizontal scale.
a. What shape would you say that these have?
b. For males, the mean is 70.9 and the standard deviation is 2.9. Interpret using the Empirical Rule.
c. For females, the mean is 65.3 and the standard deviation is 3.0. Compare the center and spread of the height distributions for females and males.
d. Identify the lowest observation for males. How many standard deviations below the mean is it?
e. Is the observation highlighted in (d) a potential outlier according to its *z*-score? Is it an outlier based on the appearance of the dot plot?

Dot Plots of Male and Female Heights

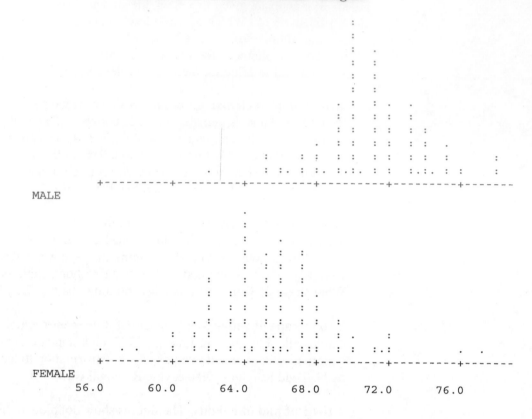

2.105 Cereal sugar values: Revisit the sugar data for breakfast cereals that are in the "cereal" data file on the text CD and shown in the margin in grams.

a. Construct a dot plot, stem-and-leaf plot, or histogram. Interpret.

b. The figure (Minitab output) shows a box plot of the sugar values. Interpret by giving approximate values for the five-number summary.

Cereal	Sugar (g)
Frosted MW	7
Raisin Bran	12
All Bran	5
Apple Jacks	14
Capt Crunch	12
Cheerios	1
Cinnamon T	13
Crackling OB	10
Crispix	3
Frosted Flakes	11
Fruit Loops	13
Grape Nuts	3
Honey Nut C	10
Life	6
Oatmeal Raisin	10
Smacks	15
Special K	3
Wheaties	3
Corn Flakes	2
Honeycomb	11

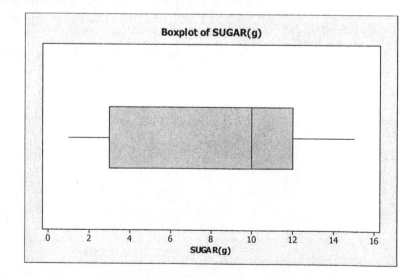

c. What does the box of the box plot suggest about possible skew?

d.   The mean is 8.20 and the standard deviation is 4.56. Find the z-score associated with the minimum sugar value of 1. Is such a sugar value a potential outlier according to its number of standard deviations from the mean?

▣2.106 Cigarette tax graphics: How do cigarette taxes per pack vary from one state to the next? The data set of 2003 cigarette taxes for all 50 states is in the "Cigarette tax" data file on the text CD.
a.   Is the variable, cigarette tax, quantitative or categorical?
b.   Use software to construct a histogram.
c.   Write a short description of the distribution, noting shape and possible outliers.

▣2.107 Cigarette tax center and spread: Refer to the previous exercise. Using software:
a.   Find the mean and median for the cigarette taxes. Which is larger, and why would you have expected that from the histogram in the previous exercise?
b.   Find the standard deviation, and interpret it.
c.   Does the Empirical Rule apply to this distribution? Why or why not? Compare the rule's predicted percentages to the actual percentages.

▣2.108 Temperatures in Central Park: Access the "Central Park Yearly Temps" on the text CD. Using software:
a.   Construct a histogram.
b.   Construct a box plot. Identify any potential outliers.
c.   Find and interpret the five-number summary.
d.   Find and interpret the mean and standard deviation.
e.   Summarize the distribution, noting shape, center, and spread.

2.109 Teachers' salaries: According to *Statistical Abstract of the United States, 2003*, average salary (in dollars) of secondary school classroom teachers in 2002 in the U.S. varied among states with a five-number summary of: minimum = 31,200, Q1 = 37,400, Median = 40,000, Q3 = 48,400, Maximum = 57,200.
a.   Find and interpret the range and interquartile range.
b.   In a box plot of these data, what would be the values at the (i) ends of the boxes, (ii) line in the middle of the box, (iii) lower end of the left whisker, (iv) upper end of the right whisker?
c.   Based on the five-number summary, predict the direction of skew for this distribution. Explain.
d.   If the distribution, although skewed, is approximately bell-shaped, which of the following would be the most realistic value for the standard deviation: (i) 100, (ii) 1000, (iii) 6000, or (iv) 25,000? Explain your reasoning.

2.110 Health insurance: In 2002, the five-number summary for the statewide percentage of people without health insurance had a minimum of 7.5% (Iowa), Q1 = 9.6, Median = 12.4, Q3 = 15.9, and maximum of 23.5% (Texas) (*Statistical Abstract of the United States, 2003*).
a.   Explain how a box plot would use these values.
b.   Do you think that the distribution is symmetric, skewed to the right, or skewed to the left? Why?
c.   Which of the following is the most plausible value for the standard deviation of this distribution: −16, 0, 4, 15, or 25? Why?

2.111 What box plot do you expect?: For each of the following variables, sketch a box plot that would be plausible for each variable.
a.　Exam score (minimum = 0, maximum = 100, mean = 87, standard deviation = 10)
b.　IQ (mean = 100 and standard deviation = 16)
c.　Weekly religious contribution (median = $10 and mean = $17)

2.112 High school graduation rates: The distribution of high school graduation rates in the United States in 2002 had a minimum value of 78.1 (Texas), first quartile of 81.6, median of 86.5, third quartile of 88.1, and maximum value of 92.2 (Alaska) (*Statistical Abstract of the United States, 2003*).
a.　Report and interpret the 50th percentile.
b.　Report the range and the interquartile range.
c.　Would a box plot show any potential outliers? Explain.

2.113 Blood pressure: A World Health Organization study (the MONICA project) of health in various countries reported that in Canada, systolic blood pressure readings have a mean of 121 and a standard deviation of 16. A reading above 140 is considered to be high blood pressure.
a.　What is the z-score for a blood pressure reading of 140? How is this z-score interpreted?
b.　The systolic blood pressure values have a bell-shaped distribution. Use the Empirical rule to report intervals of values within which (i) about 68% of the systolic blood pressure values fall, (ii) about 95% of the values fall, (iii) all or nearly all of the values fall.

2.114 Cereal outlier: The cereal sodium values have a mean of 185.5 and a standard deviation of 71.2.
a.　Find the z-score for the cereal that has a sodium value of 0.
b.　According to the three standard deviation criterion, is 0 a potential outlier?

CHAPTER PROBLEMS:　　CONCEPTS AND INVESTIGATIONS

⌨**2.115 Baseball's great homerun hitters:** The file "Baseball's HR Hitters" on the text CD contains data on the number of homeruns hit each season by some of baseball's great homerun hitters. Analyze these data using techniques introduced in this chapter to help judge statistically which player might be considered the best. Specify the criterion you use to compare the players.

⌨**2.116 Amount of TV viewing:** How much, on the average, do students watch TV each day? Access the "Georgia student survey" data file on the text CD or use your class data to explore this question using appropriate graphical and numerical methods. Prepare a short report describing your analyses and conclusions.

⌨**2.117 How much spent on haircuts?:** Is there a difference in how much males and females spend on haircuts? Access the "UGA student survey" data file on the text CD or use your class data to explore this question using appropriate graphical and numerical methods. Write a brief report describing your analyses and conclusions.

2.118 Comparing cereal sodium data: Revisit the cereal sodium data, contained in the "cereal" data file on the text CD. Conduct graphical and numerical descriptive analyses to compare the sodium amounts for the two types of cereal (adult and children). Write a brief description comparing the two distributions, addressing shape, center, spread, and possible outliers.

2.119 Controlling asthma: A study of 13 children suffering from asthma (*Clinical and Experimental Allergy,* vol. 20, pp. 429-432, 1990) compared single inhaled doses of formoterol (F) and salbutamol (S). Each child was evaluated using both medications. The outcome measured was the child's peak expiratory flow (PEF) eight hours following treatment. Is there a difference in the PEF level for the two medications? The data on PEF follow:

Child	F	S	Child	F	S	Child	F	S	Child	F	S
1	310	270	5	410	380	9	330	365	13	220	90
2	385	370	6	370	300	10	250	210			
3	400	310	7	410	390	11	380	350			
4	310	260	8	320	290	12	340	260			

a.   Construct plots to compare formoterol and salbutamol. Write a short summary comparing the two distributions of the peak expiratory flow.
b.   Consider the distribution of differences between the PEF levels of the two medications. Find the 13 differences and construct a plot of the differences. If on the average there is no difference between the PEF level for the two brands, where would you expect the differences to be centered? Write a short summary, noting shape, center, spread, and possible outliers.

2.120 Griffin, Georgia daily temperatures: Access the "Griffin GA Temps" data file on the text CD, which contains the daily temperatures for Griffin, Georgia from Jan. 1, 1931 to Dec. 31, 1997 in column 1 and for the year 1997 in column 2. Analyze the 1997 data, commenting on how daily temperatures compare to annual average temperatures with respect to their variability.

2.121 Mode versus median and mean: The mode is sometimes chosen as a measure of the typical value in a distribution.
a.   Give an example of a variable for which the mode applies, but not the mean or median.
b.   Suppose all the bars in a histogram have the same height. Explain why the mean and median apply, but not the mode.

2.122 You give examples: Give an example of a variable that you would expect to have a distribution that is
a.   Approximately symmetric
b.   Skewed to the right
c.   Skewed to the left
d.   Bimodal
e.   Skewed to the right, with a mode and median of 0 but a positive mean

2.123 Political conservatism and liberalism: Where do Americans tend to fall on the conservative-liberal political spectrum? The General Social Survey asks, "I'm going to show you a seven-point scale on which the political views that people might hold are arranged from extremely liberal, point 1, to extremely conservative, point 7. Where would you place yourself on this scale?" The table shows the seven-point scale and the distribution of 1331 responses for a recent survey (2002).

Score Category	Frequency
1. Extremely liberal	47
2. Liberal	143
3. Slightly liberal	159
4. Moderate, middle of road	522
5. Slightly conservative	209
6. Conservative	210
7. Extremely conservative	41

This is a categorical scale with ordered categories, called an **ordinal scale**. Ordinal scales are often treated in a quantitative manner by assigning scores to the categories and then using numerically summaries such as the mean and standard deviation.

a.   Using the scores shown in the table, the mean for these data equals 4.12. Using the reasoning from Example 12, set up the way this would be calculated.

b.   Identify the mode.

c.   In which category does the median fall?

♦♦ **2.124 Mean for grouped data:** Refer to the calculation of the mean in Example 12 and in the previous exercise. Explain why the mean for grouped data can be expressed as a sum, taking each possible outcome times the *proportion* of times it occurred.

2.125 Male heights: According to a recent report from the U.S. National Center for Health Statistics, for males with age 25-34 years, 2% of their heights are 64 inches or less, 8% are 66 inches or less, 27% are 68 inches or less, 39% are 69 inches or less, 54% are 70 inches or less, 68% are 71 inches or less, 80% are 72 inches or less, 93% are 74 inches or less, and 98% are 76 inches or less. These are called **cumulative percentages**.

a.   Which category has the median height? Explain why.

b.   Nearly all the heights fall between 60 and 80 inches, with fewer than 1% falling outside that range. If the heights are approximately bell-shaped, give a rough approximation for the standard deviation of the heights. Explain your reasoning.

For the multiple choice questions 2.126-2.128, select the best answer.

2.126 GRE scores: In a study of graduate students who took the Graduate Record Exam (GRE), the Educational Testing Service reported that for the quantitative exam, U.S. citizens had a mean of 529 and standard deviation of 127, whereas the non-U.S. citizens had a mean of 649 and standard deviation of 129.

a.   Both groups had about the same amount of variability in their scores, but non-U.S. citizens performed better, on the average, than U.S. citizens.

b.   If the distribution of scores was approximately bell-shaped, then almost no U.S. citizens scored below 400.

c. If the scores range between 200 and 800, then probably the scores for non-U.S. citizens were symmetric and bell-shaped.

d. A non-U.S. citizen who scored three standard deviations below the mean had a score of 200.

2.127 Facts about s: Which of these statements about the standard deviation s is false?

a. s can never be negative.

b. s can never be zero.

c. For bell-shaped distributions, about 95% of the data fall within $\bar{x} \pm 2s$.

d. s is a nonresistant (sensitive to outliers) measure of spread, as is the range.

2.128 Comparing grades: A study of grades at a local community college finds that the mean GPA for all students in the fall 2003 semester was 2.77. The Empirical Rule applies to this distribution of grades. A student with a GPA of 2.0 would like to know her relative standing in relation to the mean GPA and in relation to other students at the school. A numerical summary that would be useful for this is the

a. standard deviation.

b. median.

c. interquartile range.

d. number of students at the community college.

2.129 Soccer true-false: According to a story in the *Guardian* newspaper (football.guardian.co.uk) in the U.K., the mean wage for a Premiership player in 2001-2002 in the U.K. was 600,000 pounds. True or false: If the income distribution is skewed to the right, then the median salary was even larger than 600,000 pounds.

◆◆ 2.130 Smoothed temperatures. A method for smoothing out the jitters in a time plot finds **moving averages.** A particular moving average is the average of a certain number of observations next to each other in time. For instance, with moving averages of length 5, the first observation is the average of the first 5 data points, the second observation is the average of the data points in positions 2, 3, 4, 5, and 6, and so forth, until the last five data points are averaged. The figure is a time plot of these moving averages of the Central Park, New York annual average temperatures that had been plotted in Figure 2.9. There does seem to be somewhat of an increasing trend. Access the "Central Park Yearly Temps" data set on the text CD. This data set contains the average annual temperatures in Central Park, New York City from 1901 to 2000.

a. Using software, construct a time plot. Describe any patterns.

b. Construct a time plot using moving averages of size five. Describe any patterns observed in the time plot. Are the moving averages helpful in observing trends?

c. Compare the Central Park temperatures to the U.S annual average temperatures with respect to a possible warming trend. Write a short summary.

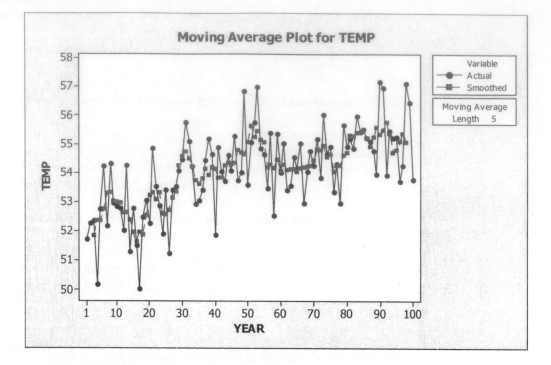

Minitab Output fFor the Time Plot of Moving Averages

♦♦ **2.131 Range and standard deviation approximation:** Use the Empirical Rule to explain why the standard deviation of a bell-shaped distribution for a large data set is often roughly related to the range by Range ≈ 6s. (For small data sets, one may not get any extremely large or small observations, and the range may be smaller, for instance about 4 standard deviations.)

♦♦ **2.132 Range the least resistant:** We've seen that measures such as the mean, the range, and the standard deviation can be highly influenced by outliers. Explain why the range is worst in this sense. (*Hint*: As the sample size increases, explain how a single extreme outlier has less effect on the mean and standard deviation, but can still have a large effect on the range.)

♦♦ **2.133 Using MAD to measure variability:** The standard deviation is the most popular measure of spread. It uses squared deviations, since the ordinary deviations sum to zero. An alternative measure is the **mean absolute deviation, MAD** = $\sum |x - \bar{x}| / n$.

a. Explain why greater spread tends to result in larger values of MAD.
b. Would you expect this to be more, or less, resistant than the standard deviation? Explain.

♦♦ **2.134 Rescale the data?:** The mean and standard deviation of a sample may change if data are rescaled (for instance, temperature changed from Fahrenheit to Celsius). For a sample with mean $\bar{x}$, adding a constant c to each observation changes the mean to $\bar{x} + c$, and the standard deviation s is unchanged. Multiplying each observation by $c > 0$ changes the mean to $c\bar{x}$ and the standard deviation to cs.

a. Scores on a difficult exam have a mean of 57 and a standard deviation of 20. The teacher boosts all the scores by 20 points before awarding grades. Report the mean and standard deviation of the boosted scores. Explain which rule above you used, and identify c.

b. Suppose that annual income for some group has a mean of $39,000 and a standard deviation of $15,000. Values are converted to British pounds for presentation to a British audience. If one British pound equals $2.00, report the mean and standard deviation in British currency. Explain which rule above you used, and identify c.

c. What is the effect on the quartiles of adding $c = 20$ to each observation?

d. What is the effect on the quartiles of multiplying each observation by $c = 2$?

e. Adding a constant and/or multiplying by a constant is called a **linear transformation** of the data. Do linear transformations change the *shape* of the distribution? Explain your reasoning.

CHAPTER PROBLEMS: CLASS EXPLORATIONS

2.135 The average student: Refer to the data file you created in Exercise 1.37. For variables chosen by your instructor, such as height, study time per week, number of people dated in last month, GPA, TV watching, hours a week in physical exercise, number of times reading a newspaper per week, describe the "average student" in your class. Prepare a one-page report summarizing your findings.

2.136 Analyze your data: Refer to the data file you created in Activity 3. For variables chosen by your instructor, conduct a descriptive statistical analysis, including graphics and measures of center and location. Prepare a two-page report summarizing your analyses, interpreting your findings.

2.137 Bad graphs: Search some publications and find an example of a graph that violates at least one of the principles for constructing good graphs discussed at the end of Section 2.2. The class will discuss the graphs found, summarizing what's wrong with them and how they could be improved.

2.138 Extreme observations: Using the "Standard Deviation Applet," create a set of 10 observations that is roughly symmetric.
a. Find the mean and the standard deviation.
b. Now add two extreme outliers, one on the left side and one on the right of the distribution. Explain how the outliers affect the (i) mean, (ii) standard deviation.
b) How does the effect of an outlier on the mean for this symmetric distribution compare to the effect of an outlier on the mean for a highly skewed distribution.

2.139 Create own data: Using the "Mean and Median Applet," create a data set to illustrate the effect of extreme observations on the mean and median. Write a short summary of your observations.

Bibliography

Franklin, Christine A. (2002). "The Other Life of Florence Nightingale," *Mathematics Teaching in the Middle School* 7(6): 337-339.

Friendly, Michael. (2000) "Florence Nightingale-Statistical Links."
www.math.yorku.ca/SCS/Gallery/flo.html

Tukey, J. W. (1977). *Exploratory Data Analysis*. Reading, MA: Addison-Wesley.

CHAPTER 3
Association: Contingency, Correlation, and Regression

EXAMPLE 1: IS SMOKING ACTUALLY BENEFICIAL TO YOUR HEALTH?

Picture the Scenario
There have been numerous studies over time investigating the potential harm of cigarette smoking. It is now generally believed that smoking is harmful to your health. However, one study found conflicting evidence. This study conducted a survey[1] of 1314 women in the United Kingdom during 1972-1974, asking each woman whether she was a smoker. Twenty years later, a follow-up survey observed whether each woman was deceased or still alive. The researchers studied the possible link between whether a woman smoked and whether she survived the 20-year study period. During that period, 24% of the smokers died and 31% of the nonsmokers died. The higher survival rate for the smokers is surprising.

Questions to Explore
- Is smoking actually beneficial to your health, since a smaller percentage of smokers died?
- What graphical and numerical methods can we use in exploring the data?
- If we observe a link between smoking status and survival status, is there something that could explain it?

Thinking Ahead
In Chapter 2 we distinguished between categorical and quantitative variables. In this study, we can identify two categorical variables. One is smoking status – whether a woman was a smoker (yes or no). The other is survival status – whether a woman survived the 20-year study period (yes or no). In practice, research investigations almost always need to analyze more than one variable. The link between two variables is often the primary focus.

This chapter presents descriptive statistics for examining data on two variables. We'll learn what these data suggest about the link between smoking and cancer. Even though fewer smokers died, we'll see that there might be some explanation for this. For example, perhaps the nonsmokers died in greater numbers because, on the average, they were older than the smokers. We will revisit this example in Section 3.4 (Example 16) and analyze the data while taking age into account as well.

---

[1] Described in article by D. R. Appleton et al., *American Statistician*, vol. 50, pp. 340-341 (1996).

Example 1 has categorical variables, but we'll learn also how to examine links between pairs of quantitative variables. For instance, we might want to answer questions such as "What's the relationship between the daily amount of gasoline use by automobiles and the amount of air pollution?" or "Do high schools that have higher per-student funding tend to have higher mean SAT scores for their students? ," which refer to quantitative variables.

Response Variables and Explanatory Variables

When we analyze data on two variables, our first step is to distinguish between the **response variable** and the **explanatory variable.**

IN WORDS	Response Variable and Explanatory Variable
The data analysis examines how the outcome on the **response** variable *depends on* or is *explained by* the value of the **explanatory** variable.	The **response variable** is the outcome variable on which comparisons are made. The **explanatory variable** defines the groups to be compared with respect to values on the response variable.

In Example 1, survival status (whether a woman is alive after 20 years) is the response variable. Smoking status (whether the woman was a smoker) is the explanatory variable. In a study of air pollution in several countries, the carbon dioxide (CO_2) level in a country's atmosphere might be a response variable, and the explanatory variable could be the country's amount of gasoline use for automobiles, or its gross domestic product, or its per capita energy use.

The main purpose of a data analysis with two variables is to investigate whether there is an **association** and to describe the nature of that association.

IN WORDS	Association between Two Variables
When there is an **association**, the likelihood of a particular value for one variable may depend on the value of the other variable. The chance of a college GPA above 3.5 is greater for those with high school GPA = 4.0 than for those with high school GPA = 3.0. So high school GPA and college GPA have an association.	An **association** exists between two variables if a particular value for one variable is more likely to occur with certain values of the other variable.

In Example 1, surviving the study period was more likely for smokers than for nonsmokers. So, there is an association between survival status and smoking status. For higher levels of per capita energy use, does the CO_2 level in the atmosphere tend to be higher? If so, then there is an association between energy use and CO_2 level.

This chapter presents methods for studying whether associations exist and for describing how strong they are. We explore associations between categorical variables in Section 3.1 and between quantitative variables in Sections 3.2 and 3.3.

3.1   HOW CAN WE EXPLORE THE ASSOCIATION BETWEEN TWO CATEGORICAL VARIABLES?

How would you respond to the question, "Taken all together, would you say that you are very happy, pretty happy, or not too happy?" We could summarize the percentage of people who have each of the three possible outcomes for this categorical variable using a table, a bar graph, or a pie chart. However, we'd probably want to know how the percentages depend on the values for other variables, such as a person's marital status (married, unmarried) or job satisfaction (high, low). For instance, is the percentage who report being very happy higher for people who are married than for people who are unmarried?

We'll now look at ways to form tables and graphs to summarize the association between two categorical variables. We'll illustrate for an example involving pesticide use in foods, and you can practice the concepts with some data on personal happiness in Exercises 3.2, 3.11, and 3.65.

EXAMPLE 2: ARE PESTICIDES PRESENT LESS OFTEN IN ORGANIC FOODS?

Picture the Scenario
Do you ever worry about whether pesticide residuals are on your foods? One appeal of eating organic foods is the belief that they are pesticide-free and thus healthier. However, relatively little fruit and vegetable acreage is cultivated as organic (only 2% in the U.S.), and consumers pay a premium for organic food.

How can we investigate how the percentage of foods carrying pesticide residues compares for organic foods and conventionally grown foods? The Consumers Union led a study based on sampling carried out by the United States Department of Agriculture (USDA) and the state of California.[2] The sampling was conducted as part of regulatory monitoring, in which organic and conventionally grown produce were analyzed for pesticide residues before being distributed for sale.

Table 3.1 is a summary of the data from this study. It displays the frequencies of foods that had and did not have pesticides, for organic produce and for conventionally grown produce. The table contains all possible category combinations of the two variables, "food type" and "pesticide status". To help us make educated decisions about which type of food to consume, we can examine the extent to which food type helps explain pesticide status.

Table 3.1: Frequencies for Food Type and Pesticide Status. The row totals and column totals are the frequencies for the categories of each variable. The counts inside the table give information about the association.

Pesticide Status

Food Type	Present	Not Present	Total
Organic	29	98	127
Conventional	19485	7086	26571
Total	19514	7184	26698

----

[2]Source: *Food Additives and Contaminants 200 ,Vol.19 No. ,427-446*

Questions to Explore

a. What is the response variable, and what is the explanatory variable?

b. How does the proportion of foods that had pesticides present compare for organic foods and for conventionally grown foods?

c. What proportion of all sampled produce items (including both the organic and conventionally grown) contained pesticide residuals?

Think It Through

a. Pesticide status, namely whether pesticide residuals are present, is the outcome of interest. The food type, organic or conventionally grown, is the variable that defines two groups to be compared on their pesticide status. So, pesticide status is the response variable and food type is the explanatory variable.

b. Table 3.1 shows that 29 out of 127 of the organic foods sampled contained pesticide residuals. The proportion with pesticides is 29/127 = 0.228. Likewise, 19,485 out of the 26,571 conventionally grown foods sampled contained pesticide residuals. The proportion is 19,485/26,571 = 0.733, much higher than for organic foods. Since 0.733/0.228 = 3.2, the relative occurrence of pesticide residues for conventionally grown produce is approximately 3 times that for organically grown produce.

c. The overall proportion of sampled produce that contained pesticide residuals is the total number that had pesticide residuals out of the total number of food items. This is

$$(29 + 19,485)/(127 + 26,571) = 19,514/26,698 = 0.731.$$

We can find this result using the column totals of Table 3.1. The value of 0.731 is close to the proportion containing pesticide residuals for conventionally grown foods alone, which we found to be 0.733. This is because conventionally grown foods make up a very high percentage of the sample. (In fact, using the row totals, we see that the proportion of the total sampled items that were conventionally grown was 26,571/26,698 = 0.995.) In summary, in this sample, pesticide residuals occurred in more than 73% of the sampled items, and they were much more common (about 3 times as common) for conventionally grown than organic foods.

Insight

Should we be alarmed that about 23% of organic food tested and 73% of conventionally grown food tested contained pesticide residues? The Consumers Union study stated that the level of pesticide residues found on both types of produce is usually far below the limits set by the Environment Protection Agency. Without pesticides, the amount of produce currently produced would not be possible. Scientists did find that the pesticide residues on organic produce were generally less toxic.

◆

To practice this concept, try Exercise 3.2.

Contingency Tables

Table 3.1 has two categorical variables: Food type (with categories organic and conventional), and pesticide status (with categories present and not present). With this table we can analyze the categorical variables separately, through the column totals for pesticide status and the row totals for food type. These totals are the category counts for the separate variables, for instance (19514, 7184) for the (present, not present) categories of pesticide status. We can also study the association between them, for instance by using the counts for the category combinations to find proportions, as in Example 2(b). Table 3.1 is an example of a **contingency table**.

<table>
<tr><td>

IN WORDS

A **contingency table** shows how many subjects are at each combination of categories of two categorical variables.

</td><td>

Definition: Contingency Table

A **contingency table** is a display for two categorical variables. Its rows list the categories of one variable and its columns list the categories of the other variable. Each entry in the table is the frequency of cases in the sample with certain outcomes on the two variables.

</td></tr>
</table>

Each row and column combination in a contingency table is called a **cell**. For instance, the first cell in the first row of Table 3.1 has the frequency 29, the number of observations in the organic category of food type and in the present category of pesticide status.

The process of taking a data file and finding the frequencies for the cells of a contingency table is referred to as **cross-tabulation** of the data. Table 3.1 is formed by cross-tabulation of food type and pesticide status for the 26,698 sampled cases. The cell frequencies are also called **joint frequencies**, because they refer to the two variables jointly (together).

Joint Proportions and Conditional Proportions

The joint frequencies in Table 3.1 can be expressed as **joint proportions**, by dividing each by the total number of foods sampled (26,698). Table 3.2 displays these joint proportions. For instance, the proportion of sampled foods that were both organic and that contained pesticide residuals was 0.001. The sum over all cells of the joint proportions is 1.0, except for possibly a slight difference due to rounding. Likewise, their row totals should add to 1.0, and their column totals should add to 1.0. The overall sample size n is listed in such tables, so the reader can determine the frequencies on which the joint proportions were based.

Table 3.2: Contingency Table of Joint Proportions for Food Type and Pesticide Status. The number in a cell is the proportion of the overall sample in that cell.

Pesticide Status

Food Type	Present	Not Present	Total	
Organic	0.001	0.004	0.005	
Conventional	0.730	0.265	0.995	
Total	0.731	0.269	1.000	$n = 26{,}698$

Consider the question, "Do organic and conventionally grown foods differ in the proportion of food items that contain pesticide residuals?" To answer, we find the proportions on pesticide status within each category of food type, and then compare them. From Example 2 with Table 3.1, the proportions that contain pesticide residuals are $29/127 = 0.23$ for organic foods and $19{,}485/26{,}571 = 0.73$ for conventionally grown foods. These proportions are called **conditional proportions**. The proportions are formed **conditional** upon (that is, given) food type. Given that the food type is organic, the proportion of food items with pesticides present equals 0.23. Table 3.3 shows the conditional proportions, given organic food in row 1 and given conventionally grown food in row 2.

Table 3.3: Contingency Table Showing Conditional Proportions on Pesticide Status, for Each Food Type. These treat pesticide status as the response variable. The sample size n in a row shows the total on which the conditional proportions in that row were based.

> **RECALL**
>
> Using Table 3.1, we obtain 0.23 from 29/127 and 0.77 from 98/127, the cell counts divided by the row total.

Pesticide Status

Food Type	Present	Not Present	Total	n
Organic	0.23	0.77	1.000	127
Conventional	0.73	0.27	1.000	26571

The conditional proportions in each row sum to 1.0. Usually, conditional proportions are more relevant than joint proportions. Whenever we distinguish between a response variable and an explanatory variable, it is natural to form conditional proportions for categories of the response variable.

We could instead have treated food type as the response variable. For instance, of those foods that had pesticide residue present, we could have found the proportion of food items that were organic and the proportion that were conventionally grown. However, it is more natural to analyze how pesticide status depends on food type than on how food type depends on pesticide status. That is, pesticide status is the more natural response variable.

EXAMPLE 3: HOW CAN WE COMPARE PESTICIDE RESIDUALS FOR THE FOOD TYPES GRAPHICALLY?

Picture the Scenario

For the data on food type and pesticide status, we've now seen three ways to display the data with a contingency table. Table 3.1 showed cell frequencies, Table 3.2 showed joint proportions, and Table 3.3 showed conditional proportions.

Questions to Explore

a. How can we use a graph to show the relationship between food type and pesticide status?
b. What does the graph tell us about this relationship?

Food Type	Pesticide Status	
	Present	Not Present
Organic	0.23	0.77
Conventional	0.73	0.27

Think It Through

a. We've seen that to compare the two food types on pesticide status, the response variable, we can find conditional proportions for the pesticide status categories. These conditional proportions are shown again in the margin. We can portray the association by graphing them.

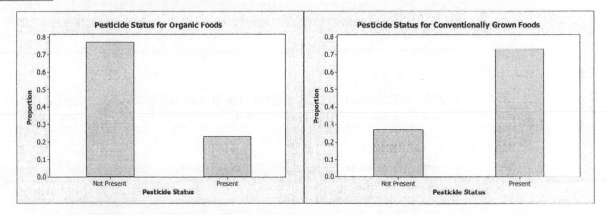

Figure 3.1: MINITAB Output of Conditional Proportions on Pesticide Status. In this graph, the conditional proportions are shown separately for each food type. **Question:** Can you think of a way to display the results that makes it easier to compare the food types on the pesticide status?

Figure 3.1 shows two bar graphs on pesticide status, one for organic foods and another for conventionally grown foods. More efficiently, we can construct a single bar graph that shows **side-by-side bars** to compare the conditional proportion of pesticide residues in conventionally grown and organic foods. Figure 3.2 shows this graph. The conditional proportion comparisons in this bar graph convey the same information as the two separate bar graphs in Figure 3.1. The advantage of the side-by-side bar graph is that it allows for easy comparison of the two food types on the pesticide status response variable.

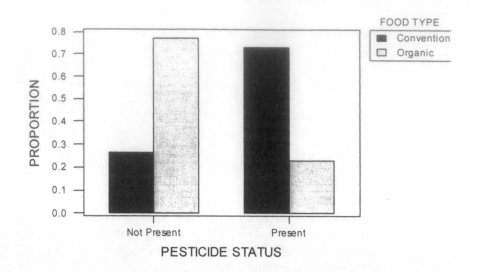

Figure 3.2: Conditional Proportions on Pesticide Status, Given the Food Type. For a particular pesticide status category, the side-by-side bars compare the two food types. **Question:** Comparing the bars, how would you describe the difference between organic and conventionally grown foods in the conditional proportion with pesticide residuals present?

b. Figure 3.2 clearly shows that that the proportion of foods having pesticides present is much higher for conventionally grown food than for organically grown food.

Insight
Chapter 2 used bar graphs to display proportions for a single variable. With two variables, bar graphs usually display conditional proportions, as in Figure 3.2. This is useful for making comparisons, such as the way Figure 3.2 compares organic and conventionally grown foods in terms of relative frequency of pesticide residuals.

♦

To practice this concept, try Exercise 3.4.

Is There an Association?

When you form a contingency table, first decide on which variable to treat as the response variable. In some cases, either variable could be the response variable, such as in cross-tabulating belief in heaven (yes, no) with belief in hell (yes, no). Then you can form conditional proportions in either or both directions. Studying the conditional proportions helps you judge whether there is an **association** between the variables.

Table 3.3 suggests that there is a reasonably strong association between food type and pesticide status, because the proportion of food items with pesticides present differs considerably (0.23 vs. 0.73) between the two food types. There would be no association if, instead, the proportion with pesticides present had been the same for each food type. For instance, suppose that for each food type, 60% had pesticides present and 40% did not have pesticides present, as shown in Table 3.4. Then, the food types have the same pesticide status distribution. If this had been the case, there would have been no association.

Table 3.4: Conditional Proportions on Pesticide Status for Each Food Type, Showing No Association. The conditional proportions for the response variable (pesticide status) categories are the same for each food type.

	Pesticide Status		
Food Type	Present	Not Present	Total
Organic	0.40	0.60	127
Conventional	0.40	0.60	26571

Figure 3.3 is a side-by-side display of the conditional proportions from Table 3.4. Compare Figure 3.3, no association, with Figure 3.2, which shows an association. In Figure 3.3, for any particular response category, the bars are the same height, indicating no association: Pesticide status does not depend on food type. By contrast, in Figure 3.2 the bars have quite different heights.

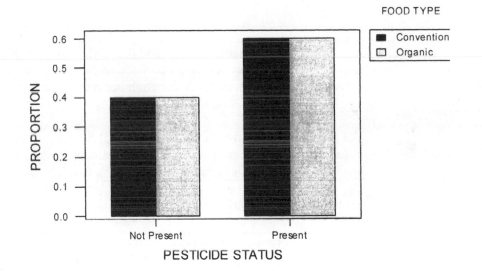

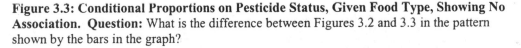

Figure 3.3: Conditional Proportions on Pesticide Status, Given Food Type, Showing No Association. Question: What is the difference between Figures 3.2 and 3.3 in the pattern shown by the bars in the graph?

SECTION 3.1: PRACTICING THE BASICS

3.1 Which is the response/explanatory variable?: For the following pairs of variables, which more naturally is the response variable and which is the explanatory variable?
a.   College grade point average (GPA) and high school GPA
b.   Number of children and mother's religion

c.   Happiness (not too happy, pretty happy, very happy) and whether married (yes, no)

d.   Monthly sales total for store and amount spent in that month on advertising

3.2 Does higher income make you happy?: Every General Social Survey includes the question "Taken all together, would you say that you are very happy, pretty happy, or not too happy?" The table uses the 2002 survey to cross-tabulate happiness with family income, measured as the response to the question, "Compared with American families in general, would you say that your family income is below average, average, or above average?"

Happiness and Family Income, from 2000 General Social Survey

	HAPPINESS			
INCOME	Not Too Happy	Pretty Happy	Very Happy	Total
Above average	17	90	51	158
Average	45	265	143	453
Below average	31	139	71	241
Total	93	494	265	852

a.   Identify the response variable and the explanatory variable.

b.   Construct the conditional proportions on happiness at each level of income. Interpret and summarize the association between these variables.

c.   What proportion of people reported being very happy?

3.3 Religious activities: In a recent General Social Survey, respondents answered the question, "In the past month, about how many hours have you spent praying, meditating, reading religious books, listening to religious broadcasts, etc.?" The responses on this variable were cross-tabulated with the respondent's gender. The table shows the results.

Religious Activity by Gender

	Number of Hours of Home Religious Activity					
Gender	0	1-9	10-19	20-39	40 or more	Total
Female	229	297	88	103	49	766
Male	276	243	59	40	16	634

a.   Identify the response variable and the explanatory variable.

b.   To summarize the difference between males and females with respect to number of hours of home religious activity, is it better to use the frequencies shown, or joint proportions, or conditional proportions? Why?

c.   Find the conditional proportions on the response variable, and interpret them.

⌨ **3.4 Religious activities revisited:** Refer to the previous exercise, and use software.

a. Create a side-by-side bar graph that compares males and females on the response variable. Summarize results.

b. What would be a disadvantage of using two pie charts to make this comparison?

3.5 Alcohol and college students: The Harvard School of Public Health, in its College Alcohol Study Survey, has surveyed college students in about 200 colleges in 1993, 1997, 1999, and 2001. The survey asks students questions about their drinking habits. Binge drinking is defined as 5 drinks in a row for males and 4 drinks in a row for females. The table shows results from the 2001 study, cross tabulating subjects' gender by whether they have participated in binge drinking.

Binge Drinking by Gender

Binge Drinking Status

Gender	Binge Drinker	Non-Binge Drinker	Total
Male	1908	2017	3925
Female	2854	4125	6979
Total	4762	6142	10904

a. Identify the response variable and the explanatory variable.

b. Find the joint proportion of subjects who were (i) male and a binge drinker, (ii) female and a binge drinker?

c. Can the joint proportions in (b) be used to answer the question, "Is there a difference between the proportions of male and female students who binge drink?" Why or why not?

d. Construct a contingency table that shows the conditional proportions of sampled students who do or do not binge drink, given gender. Interpret.

3.6 Revenues for public colleges: The table shows percentages for the sources of revenue for public degree-granting institutions.

a. Identify the response variable and the explanatory variable.

b. Why do the percentages add to 100% within columns rather than within rows?

c. What is the single greatest change from 1985-1986 to 1999-2000?

Percentages for Funding Sources during Two Time Periods

Source	1985-1986	1999-2000
Tuition and fees	14.5	18.5
Federal government	10.5	10.8
State governments	45.0	35.8
Local governments	3.6	3.8
Private gifts, grants, etc	3.2	4.8
Endowment income	0.6	0.7
Sales and services	20.0	21.6
Other sources	2.6	3.9

Source: National Center for Education Statistics (2002)

3.7 How to fight terrorism?: A survey of 1000 adult Americans (Rasmussen Reports, April 15, 2004) asked each whether the best way to fight terrorism is to let the terrorists know we will fight back aggressively or to work with other nations to find an international solution. The first option was picked by 53% of the men but by only 36% of the women in the sample.
a.   Identify the response variable and the explanatory variable, and their categories.
b.   Use a contingency table to display the results.
c.   Are the percentages reported here joint percentages, or conditional percentages? Explain.

3.8 Association between gender and political party?: In recent election years, there has been discussion about whether a "gender gap" exists in political beliefs and party identification. The table shows data collected from the 2002 General Social Survey on the categorical variables, gender and party identification (ID).

Party ID by Gender

Party Identification

Gender	Democrat	Independent	Republican	Total
Male	356	460	369	1185
Female	567	534	395	1496
Total	923	994	764	2681

a.   Identify the response and explanatory variables.
b.   What proportion of sampled individuals is (i) male and Republican, (ii) female and Republican? What types of proportions are these?
c.   What proportion is (i) male, (ii) Republican?
d.   The figure displays the proportion of individuals identifying with each political party, given gender. What are these proportions called? Is there a difference between males and females in the proportions that identify with a particular party? Summarize whatever gender gap you observe.

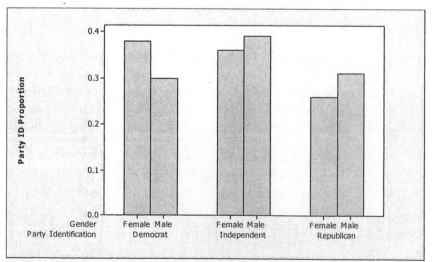

MINITAB Bar Graph of Party ID Proportions, by Gender

3.9 Unemployment by state: Look at the bar graph that shows percentage of the work force unemployed for several states as of August 2003.

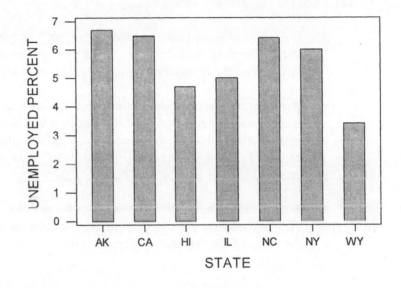

Bar Graph of Selected State Unemployment Percentages

a. Suppose the graph summarizes data in which each person sampled was categorized by their state of residence and by their employment status (employed or unemployed). Explain how this graphic could be regarded as summarizing 8 rows of a contingency table having 50 rows and 2 columns. Identify the two categorical variables for such a table, and identify the response variable.

b. Why don't the percentages shown in the graph add to 100? What type of percentages are they?

3.10 Types of proportions: Explain the difference between joint proportions and conditional proportions. Make up an example of a contingency table with frequencies, to illustrate.

📖 **3.11 Use the GSS:** Go to the GSS Web site, www.icpsr.umich.edu/GSS, click on "analyze," then "frequencies and crosstabulation," then "start," type "sex" for the row variable and "happy" for the column variable (without the quotation marks), put a check in the "row" box only for percentaging in the table options, and click on "Run the table."

a. Report the contingency table of counts.

b. Report the conditional proportions to compare the genders on reported happiness.

c. Are females and males similar, or quite different, in their reported happiness?

3.2   How Can We Explore the Association between Two Quantitative Variables?

In practice, research studies often focus on investigating the association between a response variable, which measures the outcome of interest, and an explanatory variable. The distribution of values on the response variable is analyzed at different values of the explanatory variable.

- The variables could both be *categorical*, such as food type and pesticide status. The data can then be displayed in a contingency table. We've seen we can explore the association by comparing conditional proportions.

- One variable could be *quantitative* and the other could be *categorical*, such as annual income and gender. We can then compare the categories (such as females and males) using summaries of center and spread for the quantitative variable (such as the mean and standard deviation of annual income) and graphics such as side-by-side box plots.

- Both variables could be *quantitative*, such as in the following example. We then analyze how the outcome on the response variable tends to change as the value of the explanatory variable changes. The rest of the chapter considers this case.

In exploring the relationship between two quantitative variables, we will use the principles introduced in Chapter 2 for a single variable. We first use graphics to look for an overall pattern. We follow up with numerical summaries. We check also for unusual observations that deviate from the overall pattern and that may affect the numerical summaries.

Example 4: Worldwide Use of the Internet

Picture the Scenario
In the past ten years the number of people who use the Internet has mushroomed. Although its use is now common in Westernized countries, in some countries few people use it. In 2001, for instance, about half of North Americans used the Internet, but only about 1% used it in Saudi Arabia, and only about 0.1% in Nigeria.

Table 3.5 shows recent data for 39 nations on Internet usage (Source: *Human Development Report 2003*, published by the United Nations Development Programme, http://hdr.undp.org). What other variables are likely to be associated with Internet use? One possibility is the degree of economic development of a nation. The per capita gross domestic product (GDP) is a summary description of a nation's wealth. For instance, Table 3.5 reports that GDP = $34,300 in the U.S., GDP = $27,100 in Canada, and GDP = $8400 in Mexico. Table 3.5 shows the data on Internet use and GDP as well as three other variables that could be associated with Internet use. The variables in the table are:

Internet: The percentage of adult residents who use the Internet

GDP: Gross domestic product, per capita, in thousands of U.S. dollars

CO_2: Carbon-dioxide emissions, per capita (a measure of air pollution, in metric tons, which is strongly associated with energy use)

Cellular: Percentage of adults who are cellular-phone subscribers

Fertility: Mean number of children per adult woman

The data are in the "Human development" data file on the text CD. This example uses only Internet use and GDP. We'll explore the other variables in exercises.

Table 3.5: Data for Several Nations from *Human Development Report, 2003*.

NATION	INTERNET	GDP	CO_2	CELLULAR	FERTILITY
Algeria	0.6	6.1	3.0	0.3	2.8
Argentina	10.1	11.3	3.8	19.3	2.4
Australia	37.1	25.4	18.2	57.4	1.7
Austria	38.7	26.7	7.6	81.7	1.3
Belgium	31.0	25.5	10.2	74.7	1.7
Brazil	4.7	7.4	1.8	16.7	2.2
Canada	46.7	27.1	14.4	36.2	1.5
Chile	20.1	9.2	4.2	34.2	2.4
China	2.6	4.0	2.3	11.0	1.8
Denmark	43.0	29.0	9.3	74.0	1.8
Egypt	0.9	3.5	2.0	4.3	3.3
Finland	43.0	24.4	11.3	80.4	1.7
France	26.4	24.0	6.1	60.5	1.9
Germany	37.4	25.4	9.7	68.2	1.4
Greece	13.2	17.4	8.2	75.1	1.3
India	0.7	2.8	1.1	0.6	3.0
Iran	1.6	6.0	4.8	3.2	2.3
Ireland	23.3	32.4	10.8	77.4	1.9
Israel	27.7	19.8	10.0	90.7	2.7
Japan	38.4	25.1	9.1	58.8	1.3
Malaysia	27.3	8.8	5.4	31.4	2.9
Mexico	3.6	8.4	3.9	21.7	2.5
Netherlands	49.0	27.2	8.5	76.7	1.7
New Zealand	46.1	19.2	8.1	59.9	2.0
Nigeria	0.1	0.8	0.3	0.3	5.4
Norway	46.4	29.6	8.7	81.5	1.8
Pakistan	0.3	1.9	0.7	0.6	5.1
Philippines	2.6	3.8	1.0	15.0	3.2
Russia	2.9	7.1	9.8	5.3	1.1
Saudi Arabia	1.3	13.3	11.7	11.3	4.5
South Africa	6.5	11.3	7.9	24.2	2.6
Spain	18.3	20.2	6.8	73.4	1.2
Sweden	51.6	24.2	5.3	79.0	1.6
Switzerland	30.7	28.1	5.7	72.8	1.4
Turkey	6.0	5.9	3.1	29.5	2.4
United Kingdom	33.0	24.2	9.2	77.0	1.6

United States	50.2	34.3	19.7	45.1	2.1
Vietnam	1.2	2.1	0.6	1.5	2.3
Yemen	0.1	0.8	1.1	0.8	7.0

Question to Explore

Use numerical summaries and graphics to describe the center and spread of the distributions of Internet use and GDP.

Think It Through

Using software, we find that for these 39 nations,

Internet (Percent use): Mean = 21.1, standard deviation = 18.5
GDP (1000$): Mean = 16.0, standard deviation = 10.6.

The standard deviation for Internet use is almost as large as the mean. This suggests skew to the right, since the smallest value (0.1) is only a bit more than a standard deviation below the mean. Figure 3.4 portrays the distributions graphically using box plots. The box plot for Internet use also suggests some skew to the right, since the whisker is much longer in the direction of the larger values.

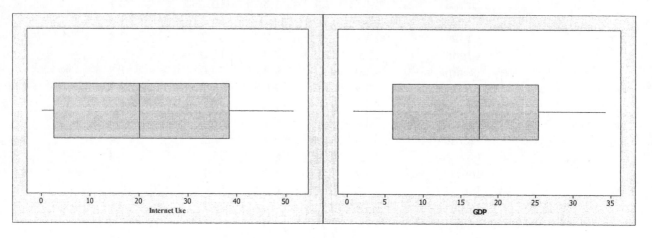

Figure 3.4: MINITAB Generated Box Plots for Internet Use and Per Capita Gross Domestic Product (GDP). Question: Approximate the five-number summary for Internet use, and indicate what the box plot suggests about the shape of its distribution.

Insight

The box plots portray each variable *separately*. How can we portray the *association* between Internet use and GDP, showing both variables at the same time? We'll study how to do that next.

♦

To practice this concept, try Exercise 3.14, part (a).

Looking for a Trend: The Scatterplot

With two quantitative variables it is common to denote the response variable by *y* and the explanatory variable by *x*. This is because graphical plots for examining the association use the *y*-axis (the vertical one) for values of the response variable and the *x*-axis (the horizontal one) for values of the explanatory variable. This graphical plot is called a **scatterplot.**

Definition: Scatterplot

A **scatterplot** is a graphical display for two quantitative variables. It uses the horizontal axis for the explanatory variable *x* and the vertical axis for the response variable *y*. The values of *x* and *y* for a subject are represented by a point relative to the two axes. The observations for the *n* subjects are *n* points on the scatterplot.

EXAMPLE 5: SCATTERPLOT FOR INTERNET USE AND GDP

Picture the Scenario
We return to the data from Example 4 for 39 nations on Internet use and the GDP (per capita gross domestic product).

Questions to Explore
a. Portray the relationship with a scatterplot.
b. What do we learn by inspecting it?

Think It Through
a. The first step is to identify the response variable and the explanatory variable. We'll study how Internet use depends on GDP. So, Internet use is the response variable. We use *y* to denote Internet use and *x* to denote GDP. We plot the values of Internet use on the vertical axis and the values of GDP on the horizontal axis. Any statistical software can construct a scatterplot. You use a data file such as Table 3.5 or the table excerpted in the margin with the values of Internet use in one column and the values of GDP in another column. You select from the data file the variable that plays the role of *x* and the variable that plays the role of *y*. Figure 3.5 shows the scatterplot that MINITAB produces for *x* = GDP and *y* = Internet use.

Nation	Internet	GDP
Algeria	0.6	6.1
Argentina	10.1	11.3
Australia	37.1	25.4
Austria	38.7	26.7
Belgium	31.0	25.5

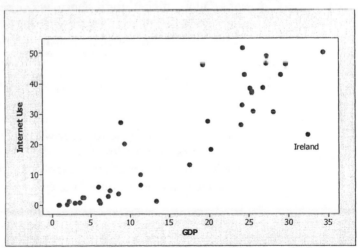

Figure 3.5: MINITAB Scatterplot for *y* = Internet Use and *x* = Gross Domestic Product (GDP), for 39 Nations. The point for Ireland is given by the ordered pair *x* = 32.4 and *y* = 23.3 and is labeled. **Question:** Is there any point that you'd identify as standing out in some way? Which country does it represent, and how is it unusual in the context of these variables?

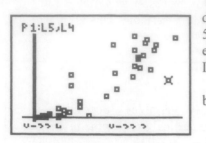

For example, consider the observation for the U.S. It has GDP $x = 34.3$ thousand dollars and Internet use $y = 50.2\%$. Its point has coordinate 34.3 for the x axis and 50.2 for the y axis. This is the top right point in Figure 3.5. The scatterplot is also easily generated by the TI-83+/84 calculator (see the margin, where the point for Ireland has been highlighted with the TRACE key).

b. Here are some things we learn by inspecting the scatterplot:
  • There is a clear trend. Nations with larger values of x (GDP) tend to have larger values of y (Internet use).
  • For countries with very low GDP (below about 8), there is little variability in Internet use. The Internet use is close to 0 for each such country.
  • The point for Ireland seems a bit unusual, in that its Internet use is relatively low (23.3%), considering its high GDP (32.4). Based on what we see for other countries with similarly high GDP, we might expect Internet use between about 40% and 50% rather than 23.3%. Likewise, the Internet use for Malaysia (27.3%) is perhaps a bit higher than we would expect for a country with such a relatively low GDP (8.8). Can you identify the point for Malaysia on the scatterplot?

Insight
There is a clear association. The countries with lower GDP values tend to have lower values of Internet use, and the countries with higher GDP values tend to have higher values of Internet use.

◆

To practice this concept, try Exercise 3.13, parts (a) and (b).

How to Examine a Scatterplot

We examine a scatterplot to study **association**. How do values on the response variable tend to change as values of the explanatory variable change? As the values of GDP get higher, for instance, we've seen that the values of Internet use tend to get higher. When there is a trend with the data points in a scatterplot, what's the direction? Is the association **positive,** or is it **negative**?

IN WORDS	**Positive Association and Negative Association**
Positive association: As x goes up, y tends to go up. **Negative association:** As x goes up, y tends to go down.	Two quantitative variables x and y are said to have a **positive association** when high values of x tend to occur with high values of y, and when low values of x tend to occur with low values of y. They are said to have a **negative association** when high values of one variable tend to pair with low values of the other variable, and low values of one pair with high values of the other.

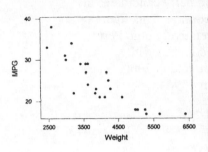

Figure 3.5 displays a positive association, because relatively high (low) values of GDP tend to occur with relatively high (low) values of Internet use. If we were to study the association between x = weight of car and y = gas mileage (MPG = miles per gallon), we would expect a negative association: Heavier cars would tend to get poorer gas mileage. The figure in the margin illustrates.

Sometimes there is no clear-cut choice of one variable as the response variable. For instance, with verbal SAT score and math SAT score, we could think of either as the response. We could label either as x or y. Regardless of our choice, we'll get the same information about the direction of the association. If there's a positive association when y = math SAT, we'd also get a positive association if we instead identified y = verbal SAT.

Here are some questions to explore when you examine a scatterplot:

- Does there seem to be a positive association, a negative association, or no clear evidence of any association?

- Can the trend in the data points be approximated reasonably well by a straight line? In that case, do the data points fall close to the line, or do they tend to scatter quite a bit?

- Are some observations unusual, falling well apart from the overall trend of the data points? What do the unusual points tell us?

EXAMPLE 6: DID THE BUTTERFLY BALLOT COST AL GORE THE 2000 PRESIDENTIAL ELECTION?

Picture the Scenario

In the 2000 Presidential election in the U.S., the Democratic candidate was Al Gore and the Republican candidate was George W. Bush. In Palm Beach County, Florida, initial election returns reported 3407 votes for the Reform party candidate, Pat Buchanan. Some political analysts thought that this total seemed surprisingly large. They felt that most of these votes may have actually been intended for Gore (whose name was next to Buchanan's on the ballot) but wrongly cast for Buchanan because of the design of the "butterfly ballot" used in that county, which some voters found confusing. On the butterfly ballot, Bush appeared first in the left column, followed by Buchanan in the right column, and Gore in the left column (see the figure in the margin).

Question to Explore

For each of the 67 counties in Florida, the "Buchanan and the butterfly ballot" data file on the text CD includes the Buchanan vote and the vote for the Reform party candidate in 1996 (Ross Perot). How can we explore graphically whether the Buchanan vote in Palm Beach County in 2000 was in fact surprisingly high, given the Reform party voting totals in 1996?

Think It Through

Figure 3.6 is a scatterplot of the county-wide vote for the Reform party candidates in 2000 (Buchanan) and in 1996 (Perot). Each point represents a county. This figure shows a strong positive association statewide: Counties with a relatively high Perot vote in 1996 tended to have a high Buchanan vote in 2000, and counties with a relatively low Perot vote in 1996 tended to have a low Buchanan vote in 2000. The Buchanan vote in 2000 was roughly only 3% of the Perot vote in 1996.

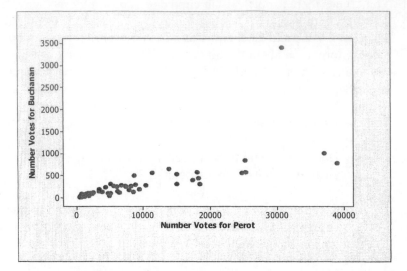

Figure 3.6: MINITAB Scatterplot of Florida County-Wide Vote for Reform Party Candidates Pat Buchanan in 2000 and Ross Perot in 1996. Question: Why is the top point, but not each of the two rightmost points, considered an outlier relative to the overall trend of the data points?

In Figure 3.6, one point falls well above the others. This severe outlier is the observation for Palm Beach County, the county that had the butterfly ballot. It is far removed from the overall trend for the other 66 data points, which follow approximately a straight line.

Insight

Alternatively you could plot the Buchanan vote against the Gore vote or against the Bush vote (Exercise 3.26). These and other analyses conducted by statisticians[3] predicted that fewer than 900 votes were truly intended for Buchanan in Palm Beach County, compared to the 3407 votes he actually received. Bush won the state by 537 votes (and, with it, the Electoral College and the election), so this may have been a pivotal factor in determining the outcome of this election. Other factors that may have played a role were 110,000 disqualified "overvote" ballots in which people mistakenly voted for more than one Presidential candidate (with Gore marked on 84,197 ballots and Bush on 37,731) often because of confusion from names being listed on more than one page of the ballot, and 61,000 "undervotes" caused by factors such as "hanging chads" from manual punch-card machines in some counties.

♦

To practice this concept, try Exercise 3.26.

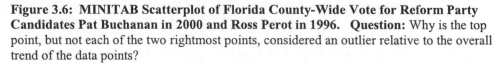

[3] For further discussion of these and related data, see Exercise 2.67 and the article by A. Agresti and B. Presnell, *Statistical Science*, vol. 17, pp. 1-5, 2002.

In practice, data points in a scatterplot sometimes fall close to a straight line trend, as we saw for all data except Palm Beach County in Figure 3.6. This represents a strong association, in the sense that we can predict the y-value quite well from knowing the x-value. The next step in analyzing a relationship is summarizing the strength of the association.

How Can We Summarize Strength of Association? The Correlation

When the data points follow roughly a straight line trend, the variables are said to have an approximately **linear** relationship. In some cases the data points fall close to a straight line, but more often there is quite a bit of variability of the points around the straight line trend. A summary measure called the **correlation** describes the strength of the linear association. It summarizes how close the data points fall to a straight-line trend.

Correlation

The **correlation** measures the direction of the association between two quantitative variables and the strength of its straight-line trend. Denoted by r, it takes values between -1 and $+1$.

- A positive value for r indicates a positive association and a negative value for r indicates a negative association.
- The closer r is to ± 1, the closer the data points fall to a straight line, and the stronger is the linear association. The closer r is to 0, the weaker is the linear association.

Let's begin our study of the correlation by looking at its values for some scatterplots. Figure 3.7 shows several scatterplots, and for each plot it shows the value of the correlation r

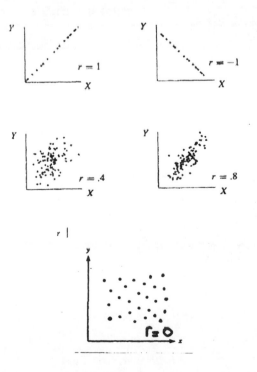

Figure 3.7: Some Scatterplots and their Correlations. The correlation gets closer to ± 1 when the data points fall closer to a straight line. **Question**: Why are the cases in which the data points are closer to a straight line considered to represent stronger association?

The correlation r takes its extreme values of +1 and -1 only when the data points follow a straight line pattern *perfectly*. These are the first two cases shown in Figure 3.7. Then, $r = +1$ occurs when the line slopes upward. The association is positive, since higher values of x tend to occur with higher values of y. The value $r = -1$ occurs when the line slopes downward, corresponding to a negative association.

In practice, don't expect the data points to fall perfectly on a straight line. However, the closer they come to that ideal, the closer the correlation is to 1 or -1. For instance, Figure 3.7 also shows a scatterplot in which the correlation $r = 0.8$. This is a stronger association than the scatterplot shown with correlation $r = 0.4$, for which the data points fall farther from a straight line.

Properties of the Correlation

RECALL
The **absolute value** of a number gives the distance the number falls from zero on the number line. The correlation values of -0.9 and 0.9 both have an absolute value of 0.9. They both represent a stronger association than correlation values of -0.6 and 0.6, for instance.

- The correlation measures the **direction** and **strength** of the **linear (straight-line) association** between two quantitative variables.
- A **positive correlation** indicates a **positive association**, and a **negative correlation** indicates a **negative association**.
- The correlation always **falls between -1 and $+1$**. The closer the value to 1, in absolute value (see the margin comments), the nearer the data points fall to a straight line.
- The value of the correlation **does not depend on the variables' units**. For example, suppose a variable measures the income of each subject in dollars. If we change the observations to units of euros or to units of thousands of dollars, we'll get the same correlation.
- Two variables have **the same correlation no matter which is treated as the response variable**.

The correlation can be calculated by all statistics software and by many calculators.

EXAMPLE 7: CORRELATION BETWEEN INTERNET USE AND GDP

Picture the Scenario
Example 5 displayed a scatterplot for Internet use and GDP (gross domestic product) for 39 nations. We observed a positive association.

Questions to Explore
Using software, what do we get for the correlation? How do we interpret the value?

Think It Through
Since the association is positive, we expect to find $r > 0$. If we input the columns of Internet use and GDP values into MINITAB and request the correlation from the Basic Statistics menu, we get

```
Correlations: gdp, internet

Pearson correlation of gdp and
internet = 0.888.
```

The correlation of $r = 0.888$ is not far below 1.0, the strongest possible correlation. This confirms the strong positive association that we observed in the scatterplot. In summary, a nation's extent of Internet use seems to be strongly associated with its GDP, with higher GDPs tending to correspond to higher Internet use.

Insight
The identifier *Pearson* for the correlation in the MINITAB output refers to the British statistician, Karl Pearson. In 1896 he provided the formula that is used to compute the correlation from sample data. This formula is shown next.

♦

To practice this concept, try Exercises 3.15 or 3.16.

What Formula Gives The Correlation?

Although software computes the correlation for us, it helps to understand it if you see the formula that defines it. For an observation x on the explanatory variable, let z_x denote the z-score that represents the number of standard deviations that x falls from the mean of x. That is,

RECALL
From Sec. 2.5, the *z-score* for an observation is the number of standard deviations that the observation falls from the mean. It equals the difference between the observation and the mean, divided by the standard deviation.

$$z_x = \frac{\text{observed value - mean}}{\text{standard deviation}} = \frac{(x - \bar{x})}{s_x},$$

where s_x denotes the standard deviation of the x-values. Similarly, let z_y denote the number of standard deviations that an observation y on the response variable falls from the mean of y. To obtain r, you calculate the product $z_x z_y$ for each observation, then find a typical value (a type of average) of those products.

Calculating the Correlation *r*

$$r = \frac{1}{n-1} \sum z_x z_y = \frac{1}{n-1} \sum \left(\frac{x - \bar{x}}{s_x} \right)\left(\frac{y - \bar{y}}{s_y} \right),$$

where *n* is the number of points, $\bar{x}$ and $\bar{y}$ are means, and s_x and s_y are standard deviations for *x* and *y*. The sum is taken over all *n* observations.

IN WORDS

A **quadrant** is any of the four regions into which a plane is divided by a horizontal line and a vertical line.

For *x* = GDP and *y* = Internet use, Example 7 found the correlation *r* = 0.888, using statistical software. To visualize how the formula works, let's revisit the scatterplot. This is reproduced in Figure 3.8, with a vertical line at the mean of *x* and a horizontal line at the mean of *y*. These lines divide the scatterplot into four quadrants. The summary statistics are:

$$\bar{x} = 16.0 \qquad \bar{y} = 21.1$$
$$s_x = 10.6 \qquad s_y = 18.5$$

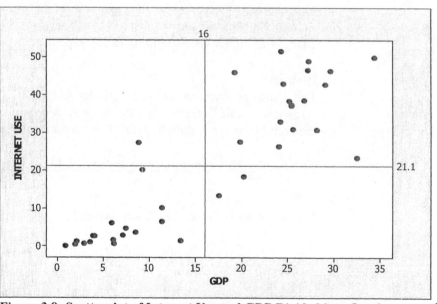

Figure 3.8: Scatterplot of Internet Use and GDP Divided into Quadrants at ($\bar{x}$, $\bar{y}$). 36 of the 39 data points lie in the upper-right quadrant (observations above the mean on each variable) or the lower-left quadrant (observations below the mean on each variable). **Question:** Do the points in these two quadrants make a positive, or a negative, contribution to the correlation value? (*Hint*: Is the product of z-scores for these points positive, or negative?)

The point for Ireland, (*x* = 32.4, *y* = 23.3) has as its z-scores (z_x = 1.55, z_y = 0.12). This point is labeled in Figure 3.8. Since *x* = 32.4 is to the right of the mean for *x* and *y* = 23.3 is above the mean of *y*, it falls in the upper-right quadrant. Both its z scores are positive. All except 3 of the 39 nations have points that fall in the upper-right and lower-left quadrants.

- The product of the z-scores for the points in the upper-right quadrant (such as Ireland) is positive. The product is also positive for points in the lower-left quadrant. Such points contribute to a positive correlation.
- The product of the z-scores for the points in the upper-left and lower-right quadrants is negative. Such points contribute to a negative correlation.

The overall correlation reflects the number of points in the various quadrants and how far they fall from the means.

IN PRACTICE

Hand **calculation of the correlation** r is tedious. We have not gone through an example, and you will not need to use the formula for r for calculation. You should rely upon software or a calculator. It's more important to understand how the correlation describes association, in terms of how it reflects the relative numbers of points in the four quadrants.

To get a feel for the correlation, use the "Regression by Eye" applet discussed in Exercise 3.127.

Graph the Data to Check Whether the Correlation Is Appropriate

The correlation is an efficient way to summarize lots of data points by a single number. But you must be careful to use it only when it is appropriate. Figure 3.9 illustrates why. It shows a scatterplot in which the data points follow a U-shaped curve. There is an association, because as x increases, y first tends to decrease, and then it tends to increase. For example, this might happen if x = age of person and y = annual medical expenses. Medical expenses tend to be high for newly born and young children, then they tend to be low until the person gets old, when they tend to become high again. However, $r = 0$ for the data in Figure 3.9.

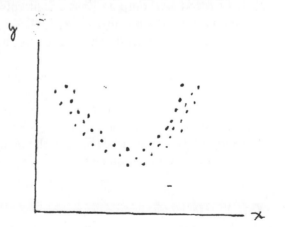

Figure 3.9: The Correlation Poorly Describes the Association when the Relationship Is Curved. For this U-shaped relationship, the correlation is 0 (or close to 0), even though the variables are strongly associated. **Question:** Can you use the formula for r, in terms of the way the points fall in the quadrants, to reason why the correlation would be close to 0?

The correlation is designed for straight line relationships. For Figure 3.9, $r = 0$ and it fails to detect the association. The correlation is not valid for describing association when the points cluster around a curve rather than around a straight line.

This figure highlights an important point to remember about *any* data analysis:

- Always plot the data.

If we merely used software to calculate the correlation for the data in Figure 3.9 but did not plot the data, we might mistakenly conclude that the variables have no association. They *do* have one, but it is not a straight line association.

IN PRACTICE

It is always important to **construct a scatterplot** to display the relationship. The correlation indicates the direction and strength only of an approximate *straight-line* relationship.

SECTION 3.2: PRACTICING THE BASICS

3.12 Used cars and direction of association: For the 100 cars on the lot of a used-car dealership, would you expect a positive association, negative association, or no association between each of the following pairs of variables? Explain why.
a.   The age of the car and the number of miles on the odometer
b.   The age of the car and the resale value
c.   The age of the car and the total amount that has been spent on repairs
d.   The weight of the car and the number of miles it travels on a gallon of gas

3.13 Cell phones and GDP: Table 3.5 showed data on the percent of the population in a country that uses cell phones, the variable called "cellular" in the table. Let $y =$ cell-phone use and $x =$ GDP.
a.   The MINITAB output shows a scatterplot. Describe this plot in terms of (i) the variability of cell-phone use values for nations that are close to 0 on GDP, (ii) identifying two nations that have less cell-phone use than you would expect, given their GDP.
b.   Give the approximate x and y coordinates for the nation that has highest (i) cell-phone use, (ii) GDP.
c.   Would the correlation be positive, or negative? Explain what it means for the correlation to have this sign.
d.   Suppose you considered the correlation only for those nations having GDP above 15. Would the correlation be stronger, or weaker, than for all 39 nations?

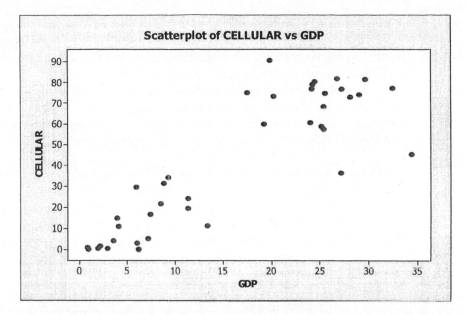

3.14 Economic development and air pollution: Table 3.5 also has data on the per capita carbon dioxide emissions in several countries, denoted by CO_2. This is a measure of the extent to which a country contributes to air pollution and to problems (such as global warming) that may be associated with air pollution. In this sample, its mean is 6.8, and its standard deviation is 4.7.

a.   The five-number summary for CO_2 is minimum = 0.3, Q1 = 3.0, median = 6.8, Q3 = 9.7, maximum = 19.7. Sketch a boxplot. Based on this and the mean and standard deviation, describe the shape of the distribution of CO_2 values.

b.   The MINITAB output shows a scatterplot relating CO_2 to a nation's per capita gross domestic product (GDP). Identify the approximate coordinates of the nation that is highest on each of these variables.

c.   The correlation between these variables for these data equals 0.786. Would you call this association strong, or weak? Explain.

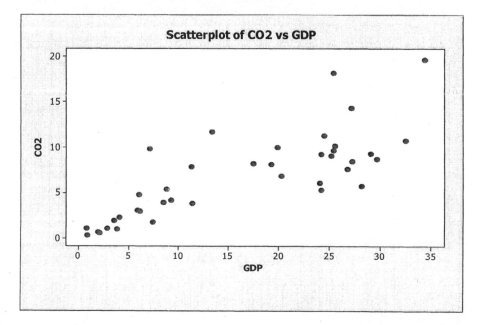

3.15 Politics and newspaper reading: For the "Florida student survey" dataset on the text CD, based on a survey of students in Florida, the correlation between y = political ideology (scored 1 = very liberal to 7 = very conservative) and x = number of times a week reading a newspaper is -0.066.
a.   Would you call this association strong, or weak? Explain.
b.   The correlation between political ideology and religiosity (how often attend religious services) is 0.580. For this sample, which explanatory variable, newspaper reading or religiosity, seems to have a stronger association with y? Explain.

3.16 Internet use correlations: For the "Human development" data file on the text CD, the correlation with Internet use is 0.888 for GDP, 0.818 for cellular-phone use, 0.669 for literacy, -0.551 for fertility, and 0.680 for CO_2 emissions.
a.   Which variable has the strongest linear association with Internet use.
b.   Which variable has the weakest linear association with Internet use.
c.   Interpret the correlation between Internet use and fertility (the mean number of children per adult woman).

⌨ **3.17 Air pollution and fertility:** Using the "Human development" data file on the text CD, analyze the data on y = CO_2 emissions with x = fertility.
a.   Construct a box plot for each variable. Summarize what you learn about the distributions.
b.   Construct a scatterplot. Based on it, do you expect the correlation to be positive, or negative? Explain. What can you learn from a scatterplot that you cannot learn from the box plots?
c.   Find the correlation. Interpret it.

3.18 Match the scatterplot with r: Match the scatterplot at the right with the correlation values.

1. $r = -0.9$

2. $r = -0.5$

3. $r = 0$

4. $r = 0.6$

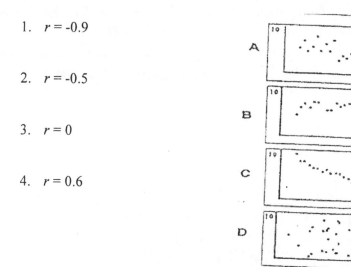

3.19 What makes $r = 1$?: Consider the data:

x		3	4	5	6	7
y		8	13	12	14	16

a. Sketch a scatterplot.
b. If one pair of (x,y) values is removed, the correlation for the remaining four pairs equals 1. Which pair is it?
c. If one y value is changed, the correlation of the five pairs equals 1. Identify the y value and how it must be changed for this to happen.

3.20 z-scores for $r = 1$: Refer to the previous exercise.
a. Find the z-scores on x and on y for the four remaining points in part (b). Comment on how z_x and z_y relate to each other for each point.
b. Compute r using the z-scores from part (a) for the four observations. Is this the value you expected to get for r? Why?

3.21 What makes $r = -1$? Repeat Exercise 3.19 with the following data, but such that removing or changing one point makes the correlation equal -1.

x	1	0	-1	-2	-3
y	0	2	4	9	8

3.22 $r = 0$: Sketch a scatterplot for which $r > 0$, but $r = 0$ after one of the points is deleted.

3.23 Correlation inappropriate: Describe a situation in which it is inappropriate to use the correlation to measure the association between two quantitative variables.

3.24 Which mountain bike to buy?:   Is there a relationship between the weight of a mountain bike and its price? A lighter weight bike is often preferred, but do lighter-weight bikes tend to be more expensive? The table below, from the "mountain bike" data file on the text CD, gives data on price, weight, and type of suspension (FU = full, FE = front end) for 12 brands.

Mountain Bikes

Brand and Model	Price($)	Weight (lb)	Type
Trek VRX 200	1000	32	FU
Cannondale SuperV400	1100	31	FU
GT XCR-4000	940	34	FU
Specialized FSR	1100	30	FU
Trek 6500	700	29	FE
Specialized Rockhop	600	28	FE
Haro Escape A7.1	440	29	FE
Giant Yukon SE	450	29	FE
Mongoose SX 6.5	550	30	FE
Diamondback Sorrento	340	33	FE
Motiv Rockridge	180	34	FE
Huffy Anorak 36789	140	37	FE

Source: Consumer Reports, June 1999

a.   You are shopping for a new bike. You are interested in whether and how weight affects the price. Which variable is the logical choice for the (i) explanatory variable, (ii) response variable?

b.   Sketch a scatterplot of price and weight. Does the relationship seem to be approximately linear? In what way does it deviate from linearity?

c.   The correlation equals −0.32. Interpret it in context. Does weight appear to affect the price strongly in a linear manner?

3.25 Enchiladas and sodium-revisited: Is there a relationship between the sodium content and the cost of enchiladas? Use software to analyze the data from *Consumer Reports* (March 2000) in the following table, contained in the "Enchiladas" file on the text CD.

Brand and Type	Cost (per serving)	Sodium content (mg)
Amy's Black Bean	$3.03	780
Patio Cheese	$1.07	1570
Banquet Cheese	$1.28	1500
El Charrito Beef	$1.53	1370
Patio Beef	$1.05	1700
Banquet Beef	$1.27	1330
Healthy Choice Chicken	$2.34	440
Lean Cuisine Chicken	$2.47	520
Weight Watchers Chicken	$2.09	660

a.   Identify the response variable and the explanatory variable.

b.   Construct a scatterplot. Does the association seem to be positive, or negative? Do you notice any unusual observations?

c.   What might explain the gap observed in the scatterplot? (*Hint:* How do the health-oriented products compare in sodium content to the others?)

d.   Obtain the correlation, r. Interpret this value in context.

3.26 Buchanan vote: Refer to Example 6 and the "Buchanan and the butterfly ballot" data file on the text CD. Let y = Buchanan vote and x = Gore vote.

a.   Construct a scatterplot. Identify any unusual points.

b.   For the county represented by the most outlying observation, about how many votes would you have expected Buchanan to get if the point followed the same pattern as the rest of the data?

c.   Repeat (a) and (b) using x = Bush vote and y = Buchanan vote.

3.3 HOW CAN WE PREDICT THE OUTCOME OF A VARIABLE?

We've seen how to explore the relationship between two quantitative variables graphically with a scatterplot. When the relationship follows approximately a straight-line pattern, the correlation describes it numerically. We can analyze the data further by finding an equation for the straight line that best describes that pattern. This equation predicts the value of the response variable using the value of the explanatory variable.

> ### Regression Line: An Equation for Predicting the Response Outcome
>
> The **regression line** predicts the value for the response variable y as a straight-line function of the value x of the explanatory variable. Let $\hat{y}$ denote the **predicted value** of y. The equation for the regression line has the form
>
> $$\hat{y} = a + bx.$$
>
> In this formula, a denotes the **y-intercept** and b denotes the **slope**.

> ### In Words
>
> The symbol $\hat{y}$, which denotes the predicted value of y, is pronounced *y-hat*.

EXAMPLE 8: HOW CAN ANTHROPOLOGISTS PREDICT HEIGHT USING HUMAN REMAINS?

Picture the Scenario
Anthropologists often try to reconstruct information using partial human remains at burial sites. For instance, after finding a femur (thighbone), they may want to predict how tall an individual was. An equation that allows them to do this is the regression line, $\hat{y} = 61.4 + 2.4x$, where $\hat{y}$ is the predicted height and x is the length of the femur, both in centimeters.

Questions to Explore
How can we graph the line that depicts how the predicted height depends on the femur length? A femur found at a particular site has length of 50 *cm*. What is the predicted height of the person who had that femur?

Think It Through
The formula $\hat{y} = 61.4 + 2.4x$ has y-intercept 61.4 and slope 2.4. It has the straight-line form $\hat{y} = a + bx$ with $a = 61.4$ and $b = 2.4$.

Each number x, when substituted into the formula $\hat{y} = 61.4 + 2.4x$, yields a value for $\hat{y}$. For simplicity in plotting the line, we start with $x = 0$, although in practice this would not be an observed femur length. The value $x = 0$ has $\hat{y} = 61.4 + 2.4(0) = 61.4$. Now, points on the y-axis have $x = 0$, so the line has height 61.4 at the point of its intersection with the y-axis. Because of this, the constant 61.4 in the equation is called the **y-intercept**. The line intersects the y-axis at the point with (x, y) coordinates (0, 61.4), which is 61.4 units up the y-axis.

The value $x = 50$ has $\hat{y} = 61.4 + 2.4(50) = 181.4$. When the femur length is 50 cm, the predicted height of the person is 181.4 cm. The coordinates are (50, 181.4). We can plot the line by connecting the points (0, 61.4) and (50, 181.4). Figure 3.10 plots the straight line for values of x between 0 and 50. In summary, the predicted height $\hat{y}$ increases from 61.4 to 181.4 as x increases from 0 to 50.

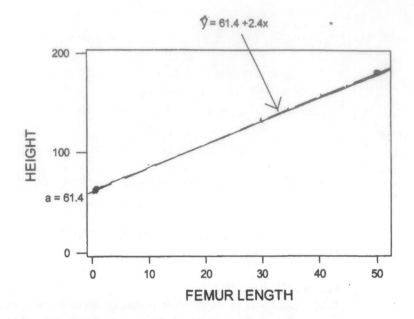

Figure 3.10: Graph of the Regression Line for x **= Femur Length and** y **= Height of Person. Questions:** At what point does the line cross the y-axis? How can you interpret the slope value of 2.4?

Insight

A regression equation is often called a **prediction equation**, since it provides predictions of the response variable y for any value of x. Sadly, this particular prediction equation had to be applied recently with bones found in mass graves in Kosovo, to help identify Albanians who had been executed by Serbians in 1998.[4]

To practice this concept, try Exercise 3.28, part (a), or 3.29, part (a).

◆

IN PRACTICE

The formula $\hat{y} = a + bx$ uses slightly different notation from the traditional formula for a line in mathematics textbooks, which is $y = mx + b$. In that equation, m = the slope (the coefficient of x) and b = y-intercept. Regardless of the notation, the interpretation of the y-intercept and slope are the same.

Interpreting the y-Intercept and Slope

The **y-intercept** is the predicted value of y when $x = 0$. This fact helps us plot the line, but it may not have any interpretative value if no observations had x values near 0. It does not make sense for femur length to be 0 cm, so the y-intercept for the equation $\hat{y} = 61.4 + 2.4x$ is not a relevant predicted height.

---

[4] "The Forensics of War," by Sebastian Junger in *Vanity Fair*, October 1999.

The **slope** b in the equation $\hat{y} = a + bx$ equals the amount that $\hat{y}$ changes when x increases by one unit. For two x-values that differ by 1.0, the $\hat{y}$ values differ by b. For the line $\hat{y} = 61.4 + 2.4x$, we've seen that $\hat{y} = 181.4$ at $x = 50$. If x increases by one unit to $x = 51$, we get $\hat{y} = 61.4 + 2.4(51) = 183.8$. The increase in $\hat{y}$ is from 181.4 to 183.8, which is 2.4, the value of the slope. For each 1-cm increase in femur length, height is predicted to increase by 2.4 cm. Figure 3.11 portrays the interpretation of the slope.

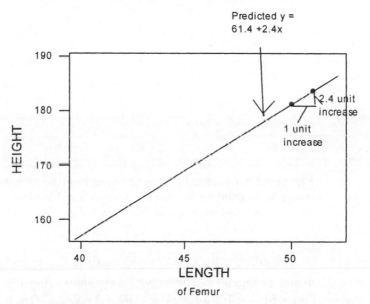

Figure 3.11: The Slope of a Straight Line. The slope is the change in the predicted value $\hat{y}$ of the response variable for a 1-unit increase in the explanatory variable x. For an increase in femur length from 50 cm to 51 cm, the predicted height increases by 2.4 cm. **Question:** What does it signify if the slope equals 0?

RECALL

Table 3.5 had entries like the following:

	GDP	Fertility
China	4.0	1.8
U.S.	34.3	2.1
Yemen	0.8	7.0

When the slope is negative, the predicted value $\hat{y}$ *decreases* as x increases. The straight line then goes downward, and the association is *negative*. For instance, for the data from Table 3.5 on y – fertility (mean number of children per woman) and x – GDP (gross domestic product), the regression equation is $\hat{y} = 3.5 - 0.07x$. This equation shows a negative association, with slope equal to -0.07. As the GDP of a country increases, the predicted fertility decreases.

When the slope = 0, the regression line is horizontal (parallel to the *x*-axis). The predicted value $\hat{y}$ of *y* stays constant at the *y*-intercept for any value of *x*. Then, the predicted value $\hat{y}$ does not change as *x* changes, and the variables do not exhibit an association. Figure 3.12 illustrates the three possibilities for the sign of the slope.

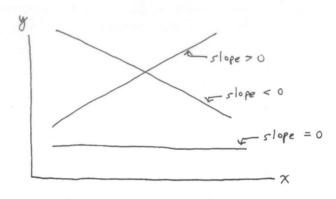

Figure 3.12: Regression Lines Showing Positive Association (Slope > 0), Negative Association (Slope < 0), and No Association (Slope = 0). Question: Would you expect a positive, or negative, slope when *y* = annual income and *x* = number of years of education?

The absolute value of the slope describes the *magnitude* of the change in $\hat{y}$ for a 1-unit change in *x*. The larger the absolute value, the steeper the regression line. A line with *b* = 4.2, such as $\hat{y} = 61.4 + 4.2x$, is steeper than one with *b* = 2.4. A line with *b* = −0.07 is steeper than one with *b* = −0.04.

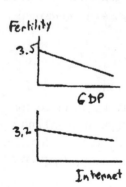

For instance, for the data from Table 3.5 on several nations, let *y* = fertility. When *x* = GDP, the regression equation is $\hat{y} = 3.5 - 0.07x$. When instead *x* = Internet use, the regression line is $\hat{y} = 3.2 - 0.04x$. This slope of −−0.04 is smaller in absolute value than the slope of −0.07 with GDP as the predictor. There is a slightly greater predicted decrease in fertility with an increase of one thousand dollars in per capita GDP than with an increase of 1% in Internet use. This does *not* mean that GDP has a stronger effect on fertility, because it does not make sense to equate a one thousand dollar increase in GDP with an increase of 1% in Internet use.

Depending on the units of measurement, a 1-unit increase in a predictor *x* could be a trivial amount, or it could be huge. However, the slope projects to any size of increase in *x*. For instance, in predicting *y* = fertility from *x* = Internet use by $\hat{y} = 3.2 - 0.04x$, a 1-unit increase in *x* is pretty small. If the percentage using the Internet is *50%* higher in one country than in another (e.g., 50% compared to 0%, as we see in comparing the U.S. to Yemen), we'll predict that its fertility changes by 50 times the amount for a 1-unit change. The predicted change in fertility is then $50(-0.04) = -2.0$, that is, two children fewer per woman.

Using Data to Find the Slope and *y*-Intercept of a Regression Line

How can we use the data to find the regression equation? We should first construct a scatterplot to make sure that the relationship follows approximately a straight line trend. If so, then statistical software or calculators can easily find the *y*-intercept and slope for the straight line that has the "best fit" to the data points.

EXAMPLE 9: HOW CAN WE PREDICT BASEBALL SCORING USING BATTING AVERAGE?

Picture the Scenario
In the sport of baseball, two statistics that can indicate a team's offensive ability are the team batting average (equal to the proportion of times the team's players get a hit, out of the times they get a hit or fail to get a hit) and team scoring--the mean number of runs scored per game by the team. Table 3.6 shows the values of these statistics at the end of the 2003 regular season for the teams in the American League. The data are in the "AL team statistics" data file on the textbook CD.

Table 3.6: Team Batting Average and Team Scoring (Mean Number of Runs Per Game) for American League Teams [5]

Team	Batting Average	Team Scoring
Boston	.289	5.93
Toronto	.279	5.52
Minnesota	.277	4.94
Kansas City	.274	5.16
Seattle	.271	4.91
New York	.271	5.41
Anaheim	.268	4.54
Baltimore	.268	4.59
Texas	.266	5.10
Tampa Bay	.265	4.41
Chicago	.263	4.88
Oakland	.254	4.74
Cleveland	.254	4.31
Detroit	.240	3.65

Scoring runs is a result of hitting, so team scoring is the response variable and team batting average is the explanatory variable. Figure 3.13 is the scatterplot for *y* = team scoring and *x* = team batting average, with the regression line superimposed. There is a straight-line trend summarized by a strong positive correlation, *r* = 0.875.

---

[5] *Source: www.usatoday.com/sports/baseball*

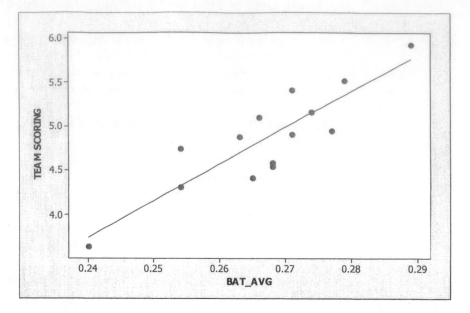

Figure 3.13: MINITAB Output for Scatterplot of Team Batting Average and Team Scoring (Mean Number of Runs per Game), with Regression Line Superimposed. **Question:** How can you find the prediction error that results when you use the regression line to predict the team scoring for a team?

Questions to Explore
    a. According to software, what is the regression equation?
    b. How do you interpret its slope?
    c. If a team has a batting average of 0.275 next year, what is their predicted mean number of runs per game?

Think It Through
    a. With the software package MINITAB, when we choose the regression option in the Regression part of the Statistics menu, part of the output tells us:

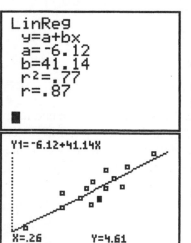

TI-83+/84 output

The regression equation is TEAM_SCORING = −6.13 + 41.2 BAT_AVG

The slope is the coefficient of the explanatory variable, which was denoted in the data file by BAT_AVG. It equals $b = 41.2$ The y-intercept is $a = -6.13$.   The regression equation is

$$\hat{y} = a + bx = -6.13 + 41.2x.$$

The TI-83+/84 calculator provides the regression equation along with the correlation r, as well as scatterplot with the regression line superimposed (see margin for output. The equation of the line is stored in Y_1, which automatically graphs on the scatterplot.)

b. Since the slope $b = 41.2$ is positive, the association is positive: The predicted team scoring increases as $x =$ team batting average increases. The slope refers to the change in $\hat{y}$ for a 1-unit change in x. However, batting average is a proportion. In Table 3.6, the batting averages fall between about 0.24 and 0.29, a range of 0.05. An increase of 0.05 in team batting average corresponds to a predicted increase of $(0.05)41.2 = 2.1$ in team scoring. That is, the mean number of runs scored is predicted to go up by 2.1 per game.

c. For the upcoming year, based on this regression equation we predict that an American League team with a team batting average of 0.275 will score an average of $\hat{y} = -6.13 + 41.2(0.275) = 5.2$ runs per game.

Insight

Figure 3.13 shows the regression line superimposed over the scatterplot. It applies only over the range of observed batting averages. For instance, it is not sensible to predict that a team with batting average of 0.0 will average -6.13 runs per game.

♦

To practice this concept, try Exercises 3.33 or 3.34. To get a feel for fitting a regression line, use the "Regression by Eye" applet discussed in Exercise 3.127.

Residuals Measure the Size of Prediction Errors

The regression equation $\hat{y} = -6.13 + 41.2x$ predicts team scoring for a given level of $x =$ team batting average. We can compare the predicted values to the actual team scoring to check the accuracy of those predictions.

For example, New York had $y = 5.41$ and $x = 0.271$. The predicted mean number of runs scored per game at $x = 0.271$ is $\hat{y} = -6.13 + 41.2x = -6.13 + 41.2(0.271) = 5.04$. The prediction error is the difference between the actual y value of 5.41 and the predicted value of 5.04, which is $y - \hat{y} = 5.41 - 5.04 = 0.37$. For New York, the regression equation under predicts y by 0.37 runs per game. For Tampa Bay, $x = 0.265$ and $\hat{y} = -6.13 + 41.2(0.265) = 4.79$. The actual value is $y = 4.41$, so the prediction is too high. The prediction error is $4.41 - 4.79 = -0.38$. The prediction errors are called **residuals**.

Definition: Residual

The **prediction error** for an observation, which is the difference $y - \hat{y}$ between the actual value and the predicted value of the response variable, is called the **residual** for that observation.

Each observation has a residual. Some are positive and some are negative. A positive residual occurs when the actual y is larger than the predicted value $\hat{y}$, so that $y - \hat{y} > 0$. A negative residual results when the actual y is smaller than the predicted value $\hat{y}$. The smaller the absolute value of a residual, the closer the predicted value is to the actual value, so the better is the prediction.

Graphically in the scatterplot, *the residual for an observation is the vertical distance between the point and the regression line*. Figure 3.14 illustrates this for the positive residual for New York. The residuals are vertical distances, because the regression equation predicts *y*, the variable on the vertical axis. Notice the parallel with what we did in analyzing contingency tables with categorical variables by studying values of the response variable, *conditional* on (given) values of the explanatory variable.

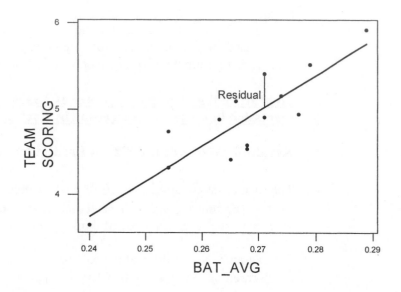

Figure 3.14: Scatterplot of Team Batting Average and Mean Number of Runs Scored, with the Residual for New York at the Point (0.271, 5.41). The residual is the prediction error, which is the vertical distance of the point from the regression line. **Question:** Why is a residual a *vertical* distance from the regression line?

EXAMPLE 10: HOW CAN WE DETECT AN UNUSUAL VOTE TOTAL?

Picture the Scenario

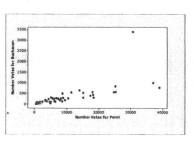

Example 6 investigated whether the vote total in Palm Beach County, Florida in the 2000 Presidential election was unusually high for Pat Buchanan, the Reform party candidate. We did this by plotting, for all 67 counties in Florida, Buchanan's vote against the Reform party candidate (Ross Perot) vote in the 1996 election.

Question to Explore

If we fit a regression line to *y* = Buchanan vote and *x* = Perot vote for the 67 counties, would the residuals help us detect any unusual vote totals for Buchanan?

Think It Through

A county with a large residual has its predicted vote far from the actual vote. This would indicate an unusual vote total. We can easily have software find the regression line and the residuals for the 67 counties. Then, we can quickly see whether any of the residuals are particularly large by constructing their histogram. Figure 3.15 shows it. The residuals cluster around 0, but one is very large, greater than 2000. Inspection of the results shows that this residual applies to Palm Beach county, for which the actual Buchanan vote was $y = 3407$ and the predicted vote was $\hat{y} = 1100$. Its residual is $y - \hat{y} = 3407 - 1100 = 2307$. In summary, in Palm Beach county Buchanan's vote was much higher than predicted.[6]

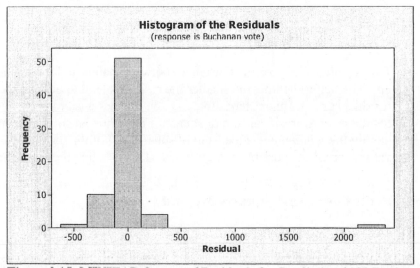

Figure 3.15: MINITAB Output of Residuals for Predicting 2000 Buchanan Presidential Vote in Florida Counties Using 1996 Perot Vote. Question: What does the right most bar represent?

Insight

As we'll discuss in the next section, an extreme outlier can pull the regression line toward it. Because of this, it's a good idea to fit the data *without* the Palm Beach county observation (that is, to the data for the other 66 counties alone) and see how well that line predicts results for Palm Beach county. We'd then get the regression equation $\hat{y} = 45.7 + 0.02414x$. Since the Perot vote in Palm Beach county was 30,739, this line would predict a Buchanan vote there of $\hat{y} = 45.7 + 0.02414(30,739) = 788$. This compares to the actual Buchanan vote of 3407 in Palm Beach county, for a residual of $3407 - 788 = 2619$. Again, the actual Buchanan vote seems very high.

♦

To practice this concept, try Exercise 3.36.

---

[6] A more complex analysis accounts for larger counts tending to vary more. However, our analysis is adequate for highlighting data points that fall far from the linear trend.

The Method of "Least Squares" Yields the Regression Line

We've seen that software finds the regression line for you. You could find your own line by placing a straightedge on the scatterplot to pass a line through the middle of the data points. To make the residuals (prediction errors) small, you would try to keep the vertical distances between the points and the line small.

Software chooses the optimal line to fit through the data points by making the residuals as small as possible. This involves compromise, since it can always find lines that perfectly predict one point but do poorly for many other points. The summary measure used to evaluate predictions is

$$\text{Residual sum of squares} = \sum (\text{residual})^2 = \sum (y - \hat{y})^2 .$$

This squares all the vertical distances between points and the line and then adds them up. The better the line, the smaller the residuals tend to be, and the smaller the residual sum of squares tends to be. Each line has a set of predicted values, a set of residuals, and a residual sum of squares. The line that software reports is the one having the *minimum* residual sum of squares out of the residual sum of squares values for all the possible straight lines. This is called the **least squares** method.

Definition: Least Squares Method

Among the possible lines that can go through data points in a scatterplot, the regression line results from the **least squares method**. This gives the line that has the smallest value for the residual sum of squares in using $\hat{y} = a + bx$ to predict y.

It's simple for software to use the least squares method to find the regression line for us. Besides making the errors as small as possible in this summary sense, this line:

- Has some positive residuals and some negative residuals, but their sum (and mean) equals 0.

- Passes through the point, $(\bar{x}, \bar{y})$.

The first property tells us that the too-low predictions are balanced by the too-high predictions. The second property tells us that the line passes through the center of the data.

There are formulas for the y-intercept and slope that the method of least squares yields, based on the sample data. They can be expressed in terms of summary statistics. Let $\bar{x}$ denote the mean of x, $\bar{y}$ the mean of y, s_x the standard deviation of the x values and s_y the standard deviation of the y values for the observations in the sample. The slope b is directly related to the correlation r.

Regression Formulas for *y*-Intercept and Slope

The **slope** equals $b = r\left(\dfrac{s_y}{s_x}\right)$.

The **y-intercept** equals $a = \bar{y} - b(\bar{x})$.

These formulas help us to interpret the regression line $\hat{y} = a + bx$ (for instance, see Exercises 3.122 and 3.123), but you should usually rely on software for the calculations.

For example, the baseball data in Example 9 have $\bar{x} = 0.267$ for batting average, $\bar{y} = 4.865$ for team scoring, $s_x = 0.0121$, and $s_y = 0.568$. The correlation is $r = 0.875$, so

$$b = r\left(\frac{s_y}{s_x}\right) = 0.875(0.568/0.0121) = 41.2,$$
$$\text{and } a = \bar{y} - b(\bar{x}) = 4.865 - 41.2(0.267) = -6.1.$$

The regression line to predict team scoring from batting average is $\hat{y} = -6.1 + 41.2x$.

The Slope, the Correlation, and the Units of the Variables

We've used the correlation to describe the strength of the association. Why can't we use the *slope* to do this, with bigger slopes representing stronger associations? The reason is that the numerical value of the slope depends on the units we use to measure the variables.

For example, the regression line between y = Internet use and x = GDP is $\hat{y} = -3.6 + 1.5x$. GDP was measured in *thousands* of U.S. dollars. For instance, $x = 19.2$ for New Zealand represents $19,200 per capita. Suppose we instead measure GDP in dollars, so $x = 19,200$ for New Zealand. A one-unit increase in GDP then refers to a single dollar per capita. This is only 1/1000 as much as a thousand-dollar increase, so the change in the predicted value of y would be 1/1000 as much, or $(1/1000)1.5 = 0.0015$. Thus, if x = GDP in dollars per capita, the slope of the regression equation is 0.0015 instead of 1.5. The strength of the association is the same in each case, since the variables and data base are identical. Only the units of measurement for one variable changed.

The slope b doesn't tell us whether the association is strong or weak, since we can make b as large or as small as we want by changing the units. By contrast, *the correlation does not change when the units of measurement change*. It is 0.888 between Internet use and GDP, whether we measure GDP in dollars per capita or in thousands of dollars per capita.

In summary, we've learned how the correlation describes the strength of the association. We've also seen how the regression line enables us to predict the response variable y using the explanatory variable x. Although correlation and regression methods serve different purposes, there are strong connections between them:

- They are both appropriate when the relationship between two quantitative variables can be approximated by a straight line.

- The correlation and the slope of the regression line have the same sign. If one is positive, so is the other one. If one is negative, so is the other one. If one is zero, the other is also zero.

There are some differences between correlation and regression methods. With regression we must identify response and explanatory variables. We get a different line if we use x to predict y than if we use y to predict x. By contrast, correlation does not make this distinction. We get the same correlation either way. Also, the values for the y-intercept and slope of the regression line depend on the units for the data, whereas the correlation does not. Finally, the correlation falls between -1 and $+1$, whereas the regression slope can equal any real number.

r-Squared (r^2)

When we predict a value of y, why do we use the regression line? We could instead predict y using a summary of the center of its distribution, such as the sample mean, $\bar{y}$. The reason for using the regression line is that if x and y have an association, then we can predict most y values more accurately by substituting x-values into the regression equation, $\hat{y} = a + bx$, than by using the sample mean $\bar{y}$ for prediction.

Another way to describe the strength of association refers to how well you can predict y using the regression equation. The square of the correlation summarizes how much less prediction error there is when you use the regression line to predict y, compared to using $\bar{y}$ to predict y. The prediction error is summarized by the sum of squared distances between the actual y-values and the predicted values. We use the notation r^2 for this measure because, in fact, it equals the square of the correlation r.

Specifically, r^2 is interpreted as the **proportional reduction in error**. For Internet use and GDP, $r = 0.888$, so $r^2 = (0.888)^2 = 0.79$. This means that the prediction error using the regression line to predict y is 79% smaller than the prediction error using $\bar{y}$ to predict y. Because the calculation of prediction error in r^2 refers to *squared* distances, it relates to comparing the *variance* of the y-values to the variability around the regression line. It is more difficult to interpret *variance* units than *standard deviation* units, so we wait until later in the book (Chapter 11) to introduce r^2 fully and explain it further. We mention it here merely because you'll see it listed on most software output for regression methods, and we wanted to give you a rough idea of what it represents.

Associations with Quantitative and Categorical Variables

In this chapter we've learned how to explore an association between two categorical variables and between two quantitative variables. With two quantitative variables x and y, identifying points in the scatterplot according to their values on a relevant categorical variable can help us to understand better the effects of x on y. This is done by using different symbols or colors on the scatterplot to portray the different categories of the categorical variable.

EXAMPLE 11: WHAT'S THE GENDER DIFFERENCE IN WINNING OLYMPIC HIGH JUMPS?

Picture the Scenario

The summer Olympic Games occurs every four years. One of the track and field events is the high jump. Men have competed in the high jump since 1896 and women since 1928. The "high jump" data file on the text CD contains the winning heights for each year, from *New York Times Almanac, 2004*, pp. 895-897.

Questions to Explore

a. How can we display the data on these two quantitative variables (winning height, year of the Olympics) and the categorical variable (gender) graphically?
b. How have the winning heights for the high jump changed over time? How different are the winning heights for men and women?

Think It Through

a. Figure 3.16 displays a scatterplot with $x =$ year and $y =$ winning height (in meters). The data points are displayed with a circle for men and a square for women. There were no Olympic games during World War II, so no observations appear for 1940 and 1944.

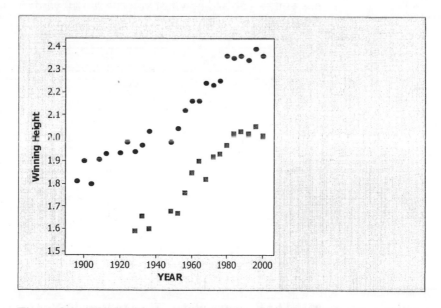

Figure 3.16: MINITAB Scatterplot for the Winning High Jumps (in Meters) in the Olympic Games. Different symbols (dots and squares) denote the two categories of the categorical variable (gender). The dots represent men and the squares represent women. **Question:** In a given year, what is the approximate difference between the winning heights for men and for women?

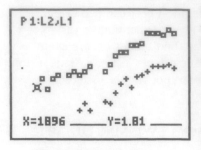

TI-83+/84 Scatterplot

b. The scatterplot shows that for each gender the winning heights have an increasing trend over time. Men have consistently jumped higher than women, between about 0.3 and 0.4 meters in any given year. The women's winning heights are similar to those for the men about 60 years earlier – for instance, about 2.0 meters around 1990-2000 for women and around 1930-1940 for men. A TI-83+/84 output of this scatterplot appears in the margin.

Insight

We could describe these trends by fitting a regression line to the points for men and a separate regression line to the points for women. However, note that in recent Olympic games, the winning distances have leveled off somewhat. We should be cautious in using such regression lines to predict future winning heights.

◆

To practice this concept, try Exercise 3.44.

SECTION 3.3: PRACTICING THE BASICS

3.27 Sketch plots of lines: Identify the values of the y-intercept a and the slope b, and sketch the following regression lines, for values of x between 0 and 10.

a. $\hat{y} = 7 + 0.5x$

b. $\hat{y} = 7 + x$

c. $\hat{y} = 7 - x$

d. $\hat{y} = 7$

3.28 Sit-ups and the 40-yard dash: Is there a relationship between how many sit-ups you can do and how fast you can run 40 yards? The figure shows the relationship between these variables for a study of female athletes to be discussed in Chapter 11.

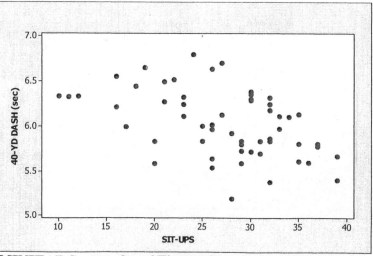

MINITAB Scatterplot of Time to Run 40-Yard Dash by Number of Sit-Ups

a.   The regression equation is $6.71 - 0.024x$. Find the predicted time in the 40-yard dash for a subject who can do (i) 10 sit-ups, (ii) 40 sit-ups. Based on these, sketch the regression line over this scatterplot.
b.   Interpret the y-intercept and slope of the equation in (a), in context.
c.   Based on the slope reported in (a), is the correlation positive, or negative? Explain.

⌨3.29 Home selling prices: The "house selling prices" file on the text CD contains a data set listing selling prices of homes in Gainesville, Florida in 2003 and some predictors for the selling price. For y = selling price (in thousands of dollars) and x = size of house (in thousands of square feet), $\hat{y} = 9.2 + 77.0x$.

a.   How much do you predict a house would sell for if it has (i) two thousand square feet, (ii) three thousand square feet.
b.   Using results in (a), explain how to interpret the slope.
c.   Is the correlation between these variables positive, or negative? Why?
d.   Suppose that a home that is three thousand square feet sold for $300,000. Find the residual, and interpret.

3.30 Internet use and GDP: For 39 nations, Example 7 found a correlation of 0.888 between y = Internet use (%) and x = per capita gross domestic product (GDP, in thousands of dollars). The regression equation is $\hat{y} = -3.61 + 1.55x$.

a.   Based on the correlation value, how did you know the slope would be positive?
b.   Ireland had a GDP of 32.4 thousand dollars and Internet use of 23.3%. Find its predicted Internet use based on the regression equation.
c.   Find the residual for Ireland. Interpret.

3.31 Cell phones and GDP: Table 3.5 showed data on the percent of the population in a country that uses cell phones, the variable called "cellular" in the table. Let y – cell-phone use and x = GDP. The regression equation is $\hat{y} = -0.13 + 2.62x$.

a.   Predict cell-phone use at the (i) minimum x-value of 0.8, (ii) maximum $x = 34.3$.
b.   Interpret the slope of the prediction equation. Is the association positive, or negative? Explain what this means.
c.   For the U.S., $x = 34.3$ and $y = 45.1$. Find the predicted cell-phone use and the residual. Interpret the large negative residual.

3.32 Air pollution and GDP: Exercise 3.14 analyzed a scatterplot relating per capita carbon dioxide emissions in several countries, denoted by CO_2, and per capita gross domestic product (GDP).

a.   Let $y = CO_2$ use and x = GDP. The regression equation is $\hat{y} = 1.26 + 0.346x$. Predict CO_2 at the (i) minimum x-value of 0.8, (ii) maximum $x = 34.3$.
b.   For the U.S., $x = 34.3$ and $y = 19.7$. Find the predicted CO_2 value and the residual. Interpret the large positive residual.
c.   Switzerland is nearly as economically advanced as the U.S., having $x = 28.1$. For it, $y = 5.7$. Find the predicted CO_2 value and the residual. Interpret.

3.33 Internet and fertility: For data from the "Human development" data file on the text CD, with y = Internet use with x = fertility (mean number of children per adult woman), MINITAB provides the result:

The regression equation is INTERNET = 40.6 - 8.17 FERTILITY

a.   What percent of the country is predicted to use the Internet if the mean number of children that a woman has is (i) 1, (ii) 3, (iii) 5?
b.   Explain how to interpret the slope of the prediction equation.
c.   Find and interpret the predicted value and residual for Canada, which has $x = 1.5$ and $y = 46.7$.

3.34 How much do seat belts help?: A study in 2000 by the National Highway Traffic Safety Administration estimated that failure to wear seat belts led to 9200 deaths in the previous year, and that that value would decrease by 270 for every 1 percentage point gain in seat belt usage. Let $\hat{y}$ = predicted number of deaths in a year and x = percentage of people who wear seat belts.
a.   Report the slope b for the equation $\hat{y} = a + bx$.
b.   If the y-intercept equals 28,910, then predict the number of deaths in a year if (i) no one wears seat belts, (ii) 73% of people wear seat belts (the value in 2000), (iii) 100% of people wear seat belts.

3.35 Regression between cereal sodium and sugar: The figure shows the result of using MINITAB to conduct a regression analysis of y = sodium and x = sugar for the breakfast cereal data considered in Chapter 2 (the "Cereal" data file on the text CD).
a.   Explain how to interpret the y-intercept and slope.
b.   Suppose you had just fitted a line to the scatterplot by eyeballing. In what sense would the line chosen by software be better than your line?
c.   Find the predicted value and residual for *Frosted Mini Wheats* ($x = 7000, y = 0$).
d.   Identify the observation and the residual from (c) on the scatterplot that shows the regression line. Does this residual appear to be large relative to the residuals for the other points on the scatterplot?

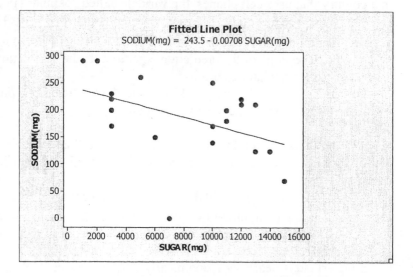

3.36 Looking at residuals: Refer to the previous exercise. The MINITAB figure shows a histogram of the residuals for this regression analysis. Explain what you learn from looking at this histogram, and explain how its leftmost bar relates to a point on the scatterplot.

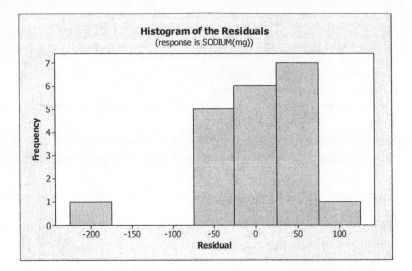

Histogram of the Residuals
(response is SODIUM(mg))

3.37 Regression and correlation between cereal sodium and sugar: Refer to the previous two exercises. Show the algebraic relationship between the correlation of −0.453 and the slope of the regression equation, using the fact that the standard deviations are 4561 mg for sugar and 71.2 mg for sodium.

3.38 TV and reading books: A high school student analyzes the relationship between x = number of books read for pleasure in the previous year and y = daily average number of hours spent watching television. For her three best friends, the observations are as shown in the table.
a.   Sketch a scatterplot.
b.   From inspection of the scatterplot, state the correlation. (*Note*: You should be able to figure it out without using software or formulas.)
c.   Find the mean and standard deviation for each variable.
d.   Using (b) and (c), find the regression line, using the formulas for the slope and the y-intercept. Interpret the y-intercept and the slope

Books	TV
0	5
5	3
10	1

3.39 Midterm-final correlation: For students who take Statistics 101 at Lake Wobegon College in Minnesota, both the midterm and final exams have mean = 75 and standard deviation = 10. The regression equation relating y = final exam score to x = midterm exam score is $\hat{y}$ = 30 + 0.60x.

a.   Find the predicted final exam score for a student who has (i) midterm score = 100, (ii) midterm score = 50. Note that in each case the predicted final exam score "regresses toward the mean" of 75. (This is a property of the regression equation that is the origin of its name, as Chapter 11 will explain.)
b.   Show that the correlation equals 0.60, and interpret it. (*Hint*: Use the relation between the slope and correlation.)

3.40 Predict final exam from midterm: In an introductory statistics course, x = midterm exam score and y = final exam score both have mean = 80 and standard deviation = 10. The correlation between the exam scores is 0.70.
a. Find the regression equation
b. Find the predicted final exam score for a student with midterm exam score = 80.

3.41 NL baseball: Example 9 related y = team scoring (per game) and x = team batting average for American League teams. For National League teams in 2004, $\hat{y}$ = −6.54 + 42.7x.
a. The team batting averages fell between 0.247 and 0.277. Explain how to interpret the slope.
b. In 2004, St. Louis had x = 0.277 and y = 5.28. Find their predicted mean number of runs per game, and the residual.

3.42 NL correlation and slope: Refer to the previous exercise.
a. The standard deviations were 0.01002 for team batting average and 0.501 for team scoring. The correlation between these variables was 0.854. Show how the correlation and slope of 42.7 relate, in terms of these standard deviations.
b. Software reports r^2 = 0.73. Explain how to interpret this measure.

3.43 Mountain bikes revisited: Is there a relationship between the weight and price of a mountain bike? This question was considered in Exercise 3.24. Analyze the "mountain bike" data file on the text CD.
a. Construct a scatterplot. Interpret.
b. Find the regression equation. Interpret the slope in context. Does the y-intercept have contextual meaning?
c. You decide to purchase a mountain bike that weighs 30 lb. What is the predicted price for the bike?

3.44 Mountain bike and suspension type: Refer to the previous exercise. The data file contains price, weight, and type of suspension system (FU = full, FE = front end) for 12 name brand bikes. You are interested in whether weight affects the price.
a. Does type of suspension make a difference in the relationship between price and weight? Below is a MINITAB scatterplot using different symbols for full and front end suspensions. Interpret.

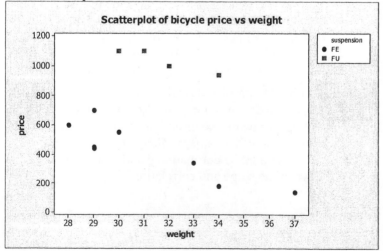

b.   Do you observe a linear relationship? Is the single regression line, which is $\hat{y} =$ 1896−40.45x, the best way to fit the data? How would you suggest fitting the data?
c.   Find separate regression equations for the two suspension types. Summarize your findings.
d.   The correlation for all 12 data points is $r = -0.32$. If the correlations for the full and front end suspension bikes are found separately, how do you believe the correlations will compare to $r = -0.32$? Find them, and interpret.
e.   You see a mountain bike advertised for $700 that weighs 28.5 lbs. The type of suspension is not given. Would you predict that this bike has a full or front end suspension? Statistically justify your answer.

⌨ **3.45 Olympic high jump**: The data discussed in Example 11 on winning high jumps in the Olympics are in the "high jump" data file on the text CD.
a.   Find the regression equation using the data for men alone. Interpret the slope.
b.   Repeat (a) using the data for women alone.
c.   Using the regression equations from (a) and (b), find the predicted winning high jumps for men and for women in the 2004 Olympic games.

3.4 WHAT ARE SOME CAUTIONS IN ANALYZING ASSOCIATIONS?

This chapter has introduced ways to explore **associations** between variables. We conclude the chapter by showing some things to be cautious about when analyzing associations.

Extrapolation Is Dangerous

Extrapolation refers to using a regression line to predict y-values for x-values outside the observed range of the data. This is riskier as we move farther from that range. If the trend changes, extrapolation will provide poor predictions.

Regression analysis is often used to fit a line to values of a quantitative variable observed over time. The line describes the time trend and allows us to make predictions for the future. But it is dangerous to extrapolate far into the future.

EXAMPLE 12: HOW CAN WE FORECAST FUTURE GLOBAL WARMING?

Picture the Scenario
In Chapter 2, we explored trends in mean annual temperatures over time using time plots. Let's consider the mean annual temperatures from 1901 – 2000 for Central Park, New York City[7] and use regression to explore a possible trend over time. These temperatures are in the "Central Park Mean Annual" data file on the text CD.

Questions to Explore
a. What does a regression line tell us about the trend over the twentieth century?
b. What does it extrapolate about the mean annual temperature in the year (i) 2005, (ii) 3000?

---

[7] *Source:http://www.ncdc.noaa.gov/oa/ncdc.html*

Think It Through

a.   Figure 3.17 shows a time plot of the Central Park mean annual temperatures. We see somewhat of an increasing trend over the century. For a regression analysis, the mean annual temperature is the response variable. Time (the year) is the explanatory variable. Suppose we set $x = (\text{year} - 1900)$, so $x = 0$ for 1900, $x = 1$ for 1901, $x = 2$ for 1902, and so forth. Then, x represents the number of years after 1900. Software tells us that the regression line is $\hat{y} = 52.5 + 0.031\,x$. Figure 3.17 shows the time plot with the regression trend line superimposed.

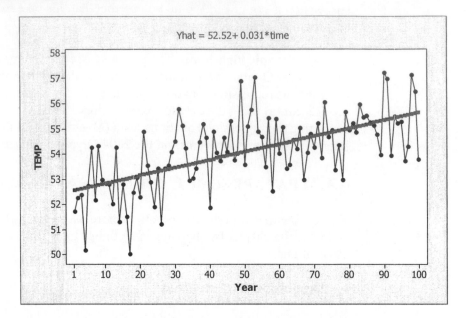

Figure 3.17: MINITAB Output for Time Plot of Central Park Mean Annual Temperature versus Time, Showing Fitted Regression Line. The years 1900 – 2000 are coded as times 0 through 100. **Question:** If the present trend continues, what would you predict for the mean annual temperature for 2005 (time $x = 105$)?

- The y-intercept, $a = 52.5$, indicates that for the year 1900 (coded as $x = 0$), the predicted mean annual temperature was 52.5 degrees Fahrenheit.
- The slope of the line, $b = 0.031$, indicates that for each 1 year increase, the predicted mean annual temperature increases by 0.031 degrees Fahrenheit.

The positive slope reflects an increasing trend: The mean annual temperature tended upward over the century. The slope of 0.031 seems close to 0, indicating that the warming was tiny. However, 0.031 is the predicted change per year. Over the century, the predicted change is $100(0.031) = 3.1$ degrees, which is quite significant.

b. The trend line can extrapolate the mean annual temperature for a future year. For example, for the year 2005 (coded as $x = 2005 - 1900 = 105$), the predicted mean annual temperature is

$$\hat{y} = 52.5 + 0.031(105) = 55.8 \text{ degrees Fahrenheit.}$$

Farther into the future we see more dramatic increases. Consider the next millennium, that is, the year 3000, which has $x = 3000 - 1900 = 1100$. The forecast is $\hat{y} = 52.5 + 0.031(1100) = 86.6$. This suggests that those living in New York then may never have the experience of seeing snow!

Insight

It is dangerous to extrapolate far outside the range of observed x values. There is no guarantee that the relationship will have the same trend outside that range. For instance, it seems reasonable to make the prediction above for 2005. That's not looking too far into the future from the last observation in 2000. However, it is foolhardy to predict for the year 3000. There is no basis for predicting that the same increasing, straight line trend will continue for the next 1000 years. As time moves forward, the mean annual temperatures may begin leveling off or even decreasing.

◆

To practice this concept, try Exercise 3.47.

Predictions about the future using time series data are called **forecasts**. When we use a regression line to forecast a trend for future years, we must make the assumption that the past trend will remain the same in the future. This is often risky to do.

Be Careful about Outliers that Are Potentially Influential

One reason to plot the data before you do a correlation or regression analysis is to check for an unusual observation. Such an observation can tell you something interesting, as in Example 6 about the Buchanan vote in the 2000 U.S. Presidential election. Furthermore, a data point that is an outlier on a scatterplot can have a substantial effect on the results, especially with small data sets.

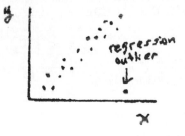

What's relevant here is not whether an observation is an outlier in terms of its x-value, relative to the other x-values, or in terms of its y-value, relative to the other y-values. Instead, we search for observations that are **regression outliers**, outliers in terms of being well removed from the trend that the rest of the data follow. The margin figure shows an observation that is a regression outlier, although it is not an outlier on x alone or on y alone.

When an observation has a large effect on the results of a regression analysis, it is said to be **influential**. Not all regression outliers are influential. For an observation to be influential on the correlation or slope, both of two conditions must hold:

- Its x value is relatively low or high compared to the rest of the data.
- The observation is a regression outlier, falling quite far from the trend that the rest of the data follow.

When both of these happen, the line tends to be pulled toward that data point and away from the trend of the rest of the points.

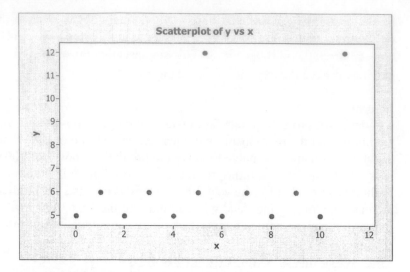

Figure 3.18: An Observation Is a Regression Outlier if it is Far Removed from the Trend that the Rest of the Data Follow. The top two points are regression outliers. Not all regression outliers are influential in affecting the correlation or slope. **Question:** Which of the regression outliers in this figure is influential?

Figure 3.18 shows two regression outliers. The correlation without the two regression outliers equals 0.00. The first regression outlier is near the middle of the range of *x*, so it does not have much potential for tilting the line up or down. It does not have much influence on the slope or the correlation. The correlation changes only to 0.03 when we add it to the data set. The second regression outlier is at the high end of the range of *x*-values. It is influential. The correlation changes to 0.47 when we add it to the data set.

EXAMPLE 13: IS HIGHER EDUCATION ASSOCIATED WITH HIGHER MURDER RATES?

Picture the Scenario

Table 3.7 shows data from *Statistical Abstract of the United States 2003* for the 50 states and the District of Columbia on several variables:

> **Violent crime rate:** The annual number of murders, forcible rapes, robberies, and aggravated assaults per 100,000 people in the populations
>
> **Murder rate:** The annual number of murders per 100,000 people in the population
>
> **Poverty:** Percentage of the residents with income below the poverty level
>
> **High school:** Percentage of the adult residents who have at least a high school education
>
> **College:** Percentage of the adult residents who have a college education
>
> **Single parent:** Percentage of families headed by a single parent

The data are in the "U.S. statewide crime" data file on the text CD. Let's look at the relationship between y = murder rate and x = college. We'll look at other variables in the exercises.

Table 3.7: Statewide Data on Several Variables

State	Violent crime	Murder	Poverty	High school	College	Single parent
Alabama	486	7.4	14.7	77.5	20.4	26
Alaska	567	4.3	8.4	90.4	28.1	23.2
Arizona	532	7	13.5	85.1	24.6	23.5
Arkansas	445	6.3	15.8	81.7	18.4	24.7
California	622	6.1	14	81.2	27.5	21.8
Colorado	334	3.1	8.5	89.7	34.6	20.8
Connecticut	325	2.9	7.7	88.2	31.6	22.9
Delaware	684	3.2	9.9	86.1	24	25.6
District of Columbia	1508	41.8	17.4	83.2	38.3	44.7
Florida	812	5.6	12	84	22.8	26.5
Georgia	505	8	12.5	82.6	23.1	25.5
Hawaii	244	2.9	10.6	87.4	26.3	19.1
Idaho	253	1.2	13.3	86.2	20	17.7
Illinois	657	7.2	10.5	85.5	27.1	21.9
Indiana	349	5.8	8.3	84.6	17.1	22.8
Iowa	266	1.6	7.9	89.7	25.5	19.8
Kansas	389	6.3	10.5	88.1	27.3	20.2
Kentucky	295	4.8	12.5	78.7	20.5	23.2
Louisiana	681	12.5	18.5	80.8	22.5	29.3
Maine	110	1.2	9.8	89.3	24.1	23.7
Maryland	787	8.1	7.3	85.7	32.3	24.5
Massachusetts	476	2	10.2	85.1	32.7	22.0
Michigan	555	6.7	10.2	86.2	23	24.5
Minnesota	281	3.1	7.9	90.8	31.2	19.6
Mississippi	361	9	15.5	80.3	18.7	30
Missouri	490	6.2	9.8	86.6	26.2	24.3
Montana	241	1.8	16	89.6	23.8	21.4
Nebraska	328	3.7	10.7	90.4	24.6	19.6
Nevada	524	6.5	10.1	82.8	19.3	24.2
New Hampshire	175	1.8	7.6	88.1	30.1	20
New Jersey	384	3.4	8.1	87.3	30.1	20.2
New Mexico	758	7.4	19.3	82.2	23.6	26.6
New York	554	5	14.7	82.5	28.7	26
North Carolina	498	7	13.2	79.2	23.2	24.3
North Dakota	81	0.6	12.8	85.5	22.6	19.1
Ohio	334	3.7	11.1	87	24.6	24.6
Oklahoma	496	5.3	14.1	86.1	22.5	23.5
Oregon	351	2	12.9	88.1	27.2	22.5
Pennsylvania	420	4.9	9.8	85.7	24.3	22.8
Rhode Island	298	4.3	10.2	81.3	26.4	27.4
South Carolina	805	5.8	12	83	19	27.1
South Dakota	167	0.9	9.4	91.8	25.7	20.7
Tennessee	707	7.2	13.4	79.9	22	27.9
Texas	545	5.9	14.9	79.2	23.9	21.5
Utah	256	1.9	8.1	90.7	26.4	13.6

Vermont	114	1.5	10.3	90	28.8	22.5
Virginia	282	5.7	8.1	86.6	31.9	22.2
Washington	370	3.3	9.5	91.8	28.6	22.1
West Virginia	317	2.5	15.8	77.1	15.3	22.3
Wisconsin	237	3.2	9	86.7	23.8	21.7
Wyoming	267	2.4	11.1	90	20.6	20.8

Questions to Explore

a. Construct the scatterplot between y = murder rate and x = college. Does any observation look like it could be influential in its effect on the regression line?
b. Use software to find the regression line. Check whether any observation is influential.

Think It Through

a. Figure 3.19 shows the scatterplot. The observation out by itself is D.C., having x = 38.3 and y = 41.8. This is the largest observation on both these variables. It satisfies both the conditions for a potentially influential observation: It takes a relatively extreme value on the explanatory variable (college), and it is a regression outlier, falling well away from the roughly horizontal linear trend of the other points. Including D.C. in the regression analysis will have the effect of pulling the slope of the regression line upward.

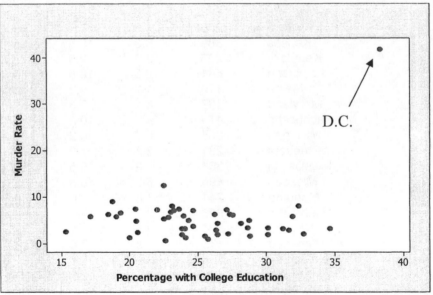

Figure 3.19: Scatterplot Relating Murder Rate to Percentage with College Education. Question: How would you expect the slope to change if D.C. is included in the regression analysis?

b. Using software, the regression line fitted to all 51 observations, including D.C., equals $\hat{y}$ = -3.1 + 0.33x. The slope is *positive*, as shown in the first plot in Figure 3.20. You can check that the predicted murder rates increase from 1.9 to 10.1 as the percentage with a college education increases from x = 15% to x = 40%, roughly the range of observed x-values. By contrast, when we fit the regression line only to the 50 states, ignoring the observation for D.C., $\hat{y}$ = 8.0 - 0.14x. The slope is −0.14, reflecting a *negative* trend, as shown in the second plot in Figure 3.20. Now, the predicted murder rate decreases from 5.9 to 2.4 as the percentage with a college education increases from 15% to 40%.

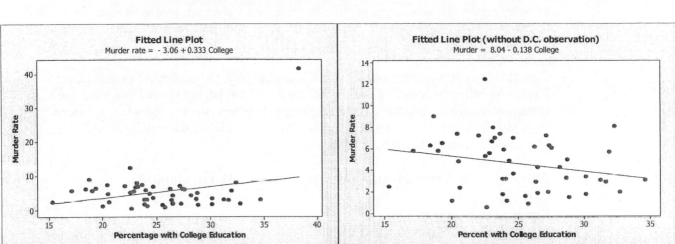

Figure 3.20: **MINITAB Scatterplots Relating Murder Rate to Percentage with College Education, with and without Observation for D.C.** **Question**: Which line better describes the trend for the 50 states?

Insight

The regression line including the D.C. observation makes it seem, misleadingly, as if the murder rate *increases* when the percentage with a college education increases. In fact, for the rest of the data the murder rate *decreases* somewhat. The observation for D.C. is highly influential in affecting the results of the regression analysis.

♦

To practice this concept, try Exercise 3.50.

The regression line including the D.C. observation distorts the relationship for the other 50 states. The regression line for the 50 states alone better represents the overall negative trend for the states. In reporting these results, we would use this line and note that the murder rate for D.C. is an outlier that falls well outside this trend.

This example shows yet once more that you should *construct a scatterplot* first before finding a correlation or a regression line. The correlation and the regression line are **non-resistant** measures: They are prone to distortion by outliers. *Investigate any regression outlier*. What's different about it? Was the observation possibly recorded incorrectly, or is it merely different from the rest of the data in some way? It is often a good idea to refit the regression line without it to see if it has a large effect, as we did in Example 13 with the D.C. observation.

To get a feel for influential points, use the "Correlation and Regression" applet discussed in Exercise 3.126.

Correlation Does Not Imply Causation

In a regression analysis, suppose we observe that as x goes up, so does y tend to go up (or go down). Can we conclude that there is a *causal* relationship, with changes in x causing changes in y?

Causality is central to science. We are all familiar with the concept of causality, at least in an informal sense. We know, for instance, that being exposed to a virus can cause the flu. But just observing an association between two variables is not enough to imply a causal relationship between the variables. There may well be some alternative explanation for the association.

EXAMPLE 14: DOES MORE EDUCATION CAUSE MORE CRIME?

Picture the Scenario

Figure 3.21 shows recent data (from the "Florida crime" data file on the text CD) from all 67 Florida counties on y = crime rate and x = education. Education was measured as the percentage of residents aged at least 25 in the county who had at least a high school degree. Crime rate was measured as the number of crimes in that county in the past year per 1000 residents.

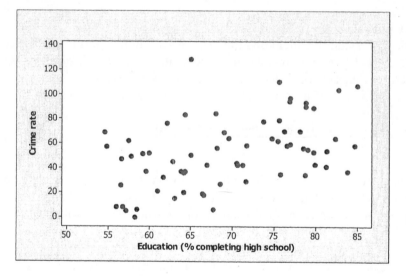

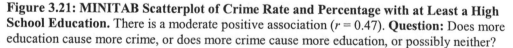

Figure 3.21: MINITAB Scatterplot of Crime Rate and Percentage with at Least a High School Education. There is a moderate positive association ($r = 0.47$). **Question:** Does more education cause more crime, or does more crime cause more education, or possibly neither?

As the figure shows, these variables have a positive association. The correlation is $r = 0.47$. Unlike the previous example, there is no obviously influential observation causing this positive correlation. Another variable measured for all counties is urbanization, measured as the percentage of the residents who live in metropolitan areas. It has a correlation of 0.68 with crime rate and 0.79 with education.

Question to Explore

From the positive correlation between crime rate and education, can we conclude that having a more highly educated populace causes the crime rate to go up?

170

Think It Through

The strong correlation of 0.79 between urbanization and education tells us that counties that are highly urban tend to have higher education levels. The moderately strong correlation of 0.68 between urbanization and crime rate tells us that the highly urban counties also tend to have higher crime. So, perhaps the reason for the positive correlation between education and crime rate is that education tends to be greater in more highly urbanized counties, but crime rates also tend to be higher in such counties. In summary, a correlation could occur without any causal connection.

Insight

For counties that have a common value on urbanization, the association between crime rate and education may look quite different. You may then see a *negative* correlation. Figure 3.22 is an idealized portrayal of how this could happen. It shows a negative trend between crime rate and education for counties having urbanization = 0 (i.e., none of the residents living in a metropolitan area), a separate negative trend for counties having urbanization = 50, and a separate negative trend for counties having urbanization = 100. If we ignore the urbanization values and look at the entire set of points, however, we see a positive trend-- higher crime rate tending to occur with higher education levels, as reflected by the positive correlation between these variables.

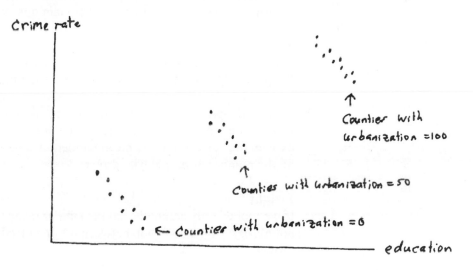

Figure 3.22: Hypothetical Scatter Diagram Relating Crime Rate and Education. The points are also labeled according to whether urbanization = 0, 50, or 100. **Question:** Sketch lines that represent (a) the overall positive relationship between crime rate and education, (b) the negative relationship between crime rate and education for the counties having urbanization = 0.

◆

To practice this concept, try Exercises 3.56 or 3.57.

Whenever an association occurs between two variables, other variables may have influenced that association. In Example 14, urbanization is a third variable influencing the association between crime rate and level of education. This illustrates an important point:

> **Correlation does not imply causation.**

In Example 14, crime rate and education were positively correlated, but that does not mean that having a high level of education in a county causes the crime rate to be high. Whenever we observe a correlation between variables x and y, there may be a third variable that is correlated with both x and y that may be responsible for their association. Here's another example to illustrate this point.

EXAMPLE 15: ICE CREAM POSITIVELY CORRELATED WITH DROWNING

| Photo - sunbather eating ice cream on the beach |

Picture the Scenario
The "Gold Coast" of Australia, south of Brisbane, is famous for its magnificent beaches. Because of strong rip tides, however, each year many people drown. Data collected each month of the year show a positive correlation between y = number of people who drowned in that month and x = number of gallons of ice cream sold in refreshment stands along the beach in that month.

Question to Explore
Identify another variable that could explain this association.

Think It Through
In the summer in Australia (especially January and February), the weather is hot. People tend to buy more ice cream in those months. They also tend to go to the beach and swim more in those months, and more people drown. In the winter, it is cooler. People tend to buy less ice cream, and fewer people go to the beach and fewer people drown. So, the mean temperature in the month is a variable that could explain the correlation. As mean temperature goes up, so do ice cream sales and so do the number of people who drown.

Insight
If we looked only at months having similar mean temperatures, probably we would not observe any association between ice cream sales and the number of people who drown.

♦

To practice this concept, try Exercise 3.56 or 3.57.

A third variable that is not measured in a study (or perhaps even known about to the researchers) but that influences the association between the response variable and the explanatory variable is referred to as a **lurking variable**.

> **Definition: Lurking Variable**
>
> A **lurking variable** is a variable, usually unobserved, that influences the association between the variables of primary interest.

In the example with a positive correlation between crime rate and education for counties in Florida, we would be remiss if we failed to recognize that that correlation could be explained by a lurking variable. This would happen if we observed only those two variables but not a variable such as urbanization, which is a lurking variable. Likewise, if we got excited by observing a positive correlation between ice cream sales and number of drownings, we'd be failing to recognize the mean temperature in the month as a lurking variable.

Simpson's Paradox

The statement that *correlation does not imply causation* can be expressed more generally as **association does not imply causation**. This is because this warning holds whether we are analyzing associations between quantitative variables or between categorical variables. Let's illustrate by revisiting Example 1, which presented a study indicating that smoking could apparently be beneficial to your health. Could a lurking variable be influencing the association?

EXAMPLE 16: IS SMOKING ACTUALLY BENEFICIAL TO YOUR HEALTH?

Picture the Scenario
Example 1 mentioned a survey[8] conducted of 1314 women in the United Kingdom during 1972-1974 that asked each woman whether she was a smoker. Twenty years later, a follow-up survey observed whether each woman was dead or still alive. Table 3.8 is a contingency table that shows the results. The response variable is survival status after 20 years. We find that 139/582, which is 24%, of the smokers died, and 230/732, or 31%, of the nonsmokers died. There was a greater survival rate for the smokers.

Table 3.8: Smoking Status and 20-Year Survival in Women

	Survival Status		
Smoker	Dead	Alive	Total
Yes	139	443	582
No	230	502	732
Total	369	945	1314

Could the age of the woman at the beginning of the study explain the association? Presumably it could if the older women were less likely than the younger women to be smokers. Table 3.9 shows a separate contingency table relating smoking status and survival status for the women separated into four age groups.

Table 3.9: Smoking Status and 20-Year Survival, for Four Age Groups

	Age Group							
	18-34		35-54		55-64		65+	
	Survival?		Survival?		Survival?		Survival?	
Smoker	Dead	Alive	Dead	Alive	Dead	Alive	Dead	Alive
Yes	5	174	41	198	51	64	42	7
No	6	213	19	180	40	81	165	28

---

[8] Described in article by D. R. Appleton et al., American Statistician, 50, 340-341 (1996).

Questions to Explore

a. Show that the counts in Table 3.9 are consistent with those in Table 3.8.

b. Use conditional percentages to describe the association in Table 3.9 between smoking status and survival status for each age group.

c. How can you explain the association in Table 3.8, whereby smoking seems to help you to live a longer life? How can this association be so different from the one shown in Table 3.9?

Think It Though

a. If you add the counts in the four separate parts of Table 3.9, you'll get the counts in Table 3.8. For instance, from Table 3.8, we see that 139 of the women who smoked died. From Table 3.9, we get this from $5 + 41 + 51 + 42 = 139$, taking the number of smokers who died from each age group. So, the counts in the two tables are consistent with each other.

b. Table 3.10 shows the conditional percentages of smokers and nonsmokers who died, for each age group. Figure 3.23 plots them. For each age group, a higher percentage of smokers than non-smokers died.

Table 3.10: Conditional Percentages of Deaths for Smokers and Nonsmokers, by Age. For instance, for smokers of age 18-34, from Table 3.10 the proportion who died was $5/(5+174) = 0.028$, or 2.8%.

	Age Group			
Smoker	18-34	35-54	55-64	65+
Yes	2.8%	17.2%	44.3%	85.7%
No	2.7%	9.5%	33.1%	85.5%

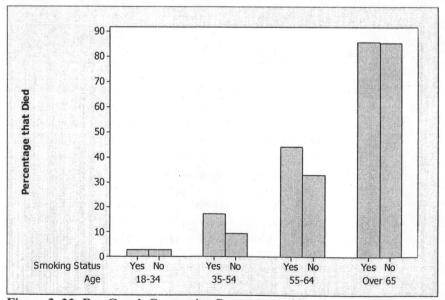

Figure 3. 23: Bar Graph Comparing Percentage of Deaths for Smokers and Nonsmokers, by Age. This side-by-side bar graph shows the conditional percentages from Table 3.10.

c. Could age explain the association? In Table 3.9, the proportion of women who smoked in 1972-1974 tended to be higher for the younger women. The percentage of smokers was 45% in the 18-34 age group (that is, $(5 + 174)/(5 + 174 + 6 + 213) = 0.45$), but only 20% in the 65+ age group. At the same time, younger women were less likely to die during the 20-year study period (perhaps due to natural causes of aging). For instance, the proportion who died was $(5 + 6)/(5 + 174 + 6 + 213) = 0.03$ in the 18-34 age group but 0.86 in the 65+ age group. In summary, the association in Table 3.8 could merely reflect younger women being more likely to be smokers, while the younger women were also less likely to die during this time frame.

Table 3.8 indicated that smokers had *higher* survival rates than nonsmokers. When we looked at the data separately in Table 3.9 for each age group, we saw the reverse: Smokers had *lower* survival rates than nonsmokers. The analysis using Table 3.8 did not account for age, which strongly influences the association. An association can look quite different after adjusting for the effect of a third variable by grouping the data according to its value.

Insight

This shows the dramatic influence that a lurking variable can have, unknown to researchers if they fail to include a certain variable in their study. Because of this reversal in the association after taking age into account, the researchers did *not* conclude that smoking was beneficial to your health -- quite the contrary.

♦

To practice this concept, try Exercise 3.60.

The fact that the direction of an association between two variables can change after we include a third variable and analyze the data at separate levels of that variable is known as **Simpson's paradox.** [9] We observed Simpson's paradox in Figure 3.22 in Example 14, in which a positive correlation between crime rate and education changed to a negative correlation when data were considered at separate levels of urbanization. This example serves as a warning that we've now seen repeatedly: *Be cautious about interpreting an association.* Always try to identify any lurking variable that may explain the association.

How Can Lurking Variables Affect Associations?

In interpreting associations, we must always be wary of possible lurking variables. They can affect associations in many ways. For instance, (1) a lurking variable may be a **common cause** of both the explanatory and response variable, or (2) a lurking variable and the explanatory variable may be two of **multiple causes** of the response variable.

---

[9] The paradox is named after a British statistician who in 1951 investigated conditions under which this flip-flopping of association can happen.

EXAMPLE 17: CAUSAL EFFECT OF HEIGHT ON VOCABULARY?

Picture the Scenario
Do tall students tend to have better vocabulary skills than short students? We might think so looking at a sample of students from grades 1, 6 and 12 of Lake Wobegon school district. The correlation was 0.81 between their height and their vocabulary test score: Taller students tended to have higher vocabulary test scores.

Question to Explore
Is there a causal relationship, whereby being taller gives you a better vocabulary, or is there a lurking variable that might explain this association?

Think It Through
This association might be explained by the sample having students of various ages. As age increases, both height and test score would tend to increase. So, age is a potential lurking variable. An analysis of the data separately for students in a particular grade (who have about the same age), would probably not find an association.

Insight
Height and vocabulary test score are probably correlated only because they both depend on a **common cause**, age. In Figure 3.24, each arrow depicts a causal dependency. As students get older, they simultaneously grow in height and learn a larger vocabulary. Because of this, height and vocabulary test performance seem to be positively correlated. In fact, they *are* correlated, but only because of a common dependence on age. They are not *causally* associated.

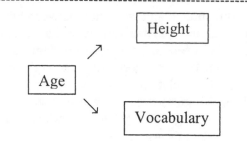

Figure 3.24: Graphical Depiction of a Common Cause. An arrow drawn from one variable to another denotes a causal influence. The positive correlation between height and vocabulary test score may be explained by the causal effect of age on both of them. **Question**: Would you still expect to see a positive correlation between height and the vocabulary test score if your sample included only 12-year-olds?

                                                                                    ◆

To practice this concept, try Exercises 3.58, part (a).

In practice, there's usually not a single variable that causally explains a response variable or the association between two variables. More commonly, there are multiple causes.

EXAMPLE 18: DOES BEING FROM A SINGLE-PARENT FAMILY CAUSE DELINQUENCY?

Picture the Scenario
Studies of crime have shown that teenagers from single-parent families are more likely to be juvenile delinquents.

Question to Explore
Does this imply that having only a single parent causes juvenile delinquency?

Think It Through
Realistically, probably lots of things contribute to juvenile delinquency. Many variables that you might think of as possible causes, such as the family income level and the quality of the neighborhood in which a child lives, are themselves likely to be associated with whether a family has a single parent. Perhaps single-parent families tend to be poor and live in bad neighborhoods, and perhaps being poor and living in a bad neighborhood makes a child more likely to be a juvenile delinquent. In summary, this association does not imply a causal relationship.

Insight
When there are **multiple causes** of a response variable, the association among them makes it difficult to study the effect of any single variable. Perhaps being from a single-parent family has a direct effect on delinquency but also an indirect effect through being more likely to be poor. Figure 3.25 depicts this possibility. The arrow between single-parent family and delinquency represents the direct effect, and the arrow between single-parent family and family income followed by the arrow from family income to delinquency represents the indirect effect.

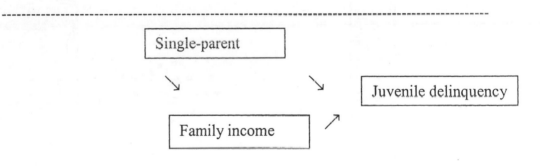

Figure 3.25: Graphical Depiction of Multiple Causes. Being from a single-parent family may have a direct effect on delinquency (represented by the arrow between the two) but also an indirect effect through the family being more likely to be poor (arrow from single-parent to family income), which also may affect delinquency (arrow from family income to delinquency). **Question:** Can you think of other variables that could affect delinquency but also be associated with whether a family has a single parent?

To practice this concept, try Exercise 3.58, part (b).

Associations Due to Time Trends

It's especially tricky to study cause and effect when two variables are measured over time. The variables may be associated merely because they both have a time trend. Suppose that both the divorce rate and the crime rate have an increasing trend over a 10 year period. They will then have a positive correlation: Higher crime rates occur in years that have higher divorce rates. Does this imply that an increasing divorce rate *causes* the crime rate to increase? Absolutely not. They would also be positively correlated with all other variables that have a positive time trend, such as annual average house price and the annual use of cell phones. There are likely to be other variables yet that are themselves changing over time and that have causal influences on the divorce rate and the crime rate.

SECTION 3.4: PRACTICING THE BASICS

3.46 Extrapolating murder: The MINITAB figure shows the data and regression line for the 50 states in Table 3.7 relating x = percentage of single-parent families to y = annual murder rate (number of murders per 100,000 people in the population).
a. The lowest x-value was for Utah and the highest was for Mississippi. Using the figure, approximate those x-values.
b. Find the predicted murder rate at $x = 0$. Does this prediction make sense? Why? What's wrong with making a prediction at $x = 0$ based on these data?

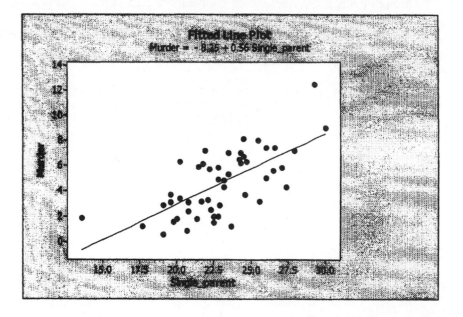

3.47 Women's Olympic high jumps: Example 11 discussed the winning height in the Olympic high jump. Using the "High Jump" data file on the text CD, MINITAB reports the regression line

```
Women_Meters = - 11.9 + 0.007 Year_Women
```

for predicting the women's winning height (in meters) using the year number.
a. Predict the winning Olympic high jump distance for women in the year (i) 2008, (ii) 3000.

b. Do you feel comfortable using this line to make either of the predictions in (a)? Explain.

3.48 Men's Olympic long jumps: The Olympic winning men's long jump distances (in meters) from 1986 to 2000 and the fitted regression line for predicting them using x = year are displayed in the MINITAB output below. (The data, in the "Long Jump" data file on the text CD, are from *2004 New York Times Almanac, pp. 895-897*.)
a.   Identify an observation that may influence the fit of the regression line. Why did you identify this observation?
b.   Which do you think is a more reasonable prediction for the winning Olympic long jump distance for men in the year 2008 – the sample mean of the y values in this plot, or $\hat{y} = -24.7 + 0.0168(2008)$? Why?
c.   Would you feel comfortable using the fitted line in the scatterplot to predict the winning Olympic long jump distance for men in the year 3000? Why or why not?

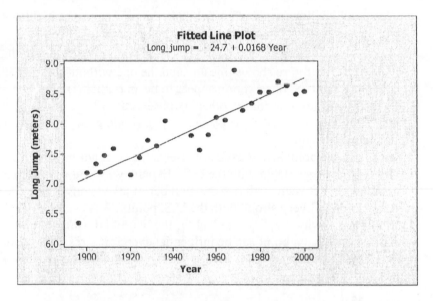

3.49 U.S. average annual temperatures: Use the "U.S. Temperatures" data file on the text CD.
a.   Fit a trend line, and interpret the slope.
b.   Predict the mean annual U.S. temperature for the year (i) 2010, (ii) 3000.
c.   In which prediction in (b) do you have more faith? Why?

3.50 Murder and education: Example 13 (page 166) found the regression line $\hat{y} = -3.1 + 0.33x$ for all 51 observations on y = murder rate and x = percent with a college education.
a.   Show that the predicted murder rates increase from 1.85 to 10.1 as percent with a college education increases from x = 15% to x = 40%, roughly the range of observed x-values.
b.   When the regression line is fitted only to the 50 states, $\hat{y} = 8.0 - 0.14x$. Show that the predicted murder rate decreases from 5.9 to 2.4 as percent with a college education increases from 15% to 40%.
c.   D.C. has the highest value for x (38.3) and is an extreme outlier on y (41.8). Is it a regression outlier? Why?

d.   What causes results to differ so according to whether D.C. is in the data set? Which line is more appropriate as a summary of the relationship? Why?

3.51 Murder and poverty: For Table 3.7 (page 167), the regression equation for the 50 states and D.C. relating y = murder rate and x = percent of people who live below the poverty level is $\hat{y} = -4.1 + 0.81x$. For D.C., $x = 17.4$ and $y = 41.8$.

a.   When the observation for D.C. is removed from the data set, $\hat{y} = 0.4 + 0.36x$. Does D.C. have much influence on this regression analysis? Explain.
b.   If you were to look at a scatterplot, based on the information above, do you think that the poverty value for D.C. would be relatively large, or relatively small? Explain.

3.52 TV watching and the birth rate: The figure shows recent data on y = the birth rate per 1000 people and x = the number of televisions per 100 people, for six African and Asian nations. The regression line, $\hat{y} = 29.8 - 0.024x$ applies to the data for these six countries. For illustration, another point is added at (81, 15.2) which is the observation for the United States. The regression line for all seven points is $\hat{y} = 31.2 - 0.195x$. The figure shows this line and the one without the U.S. observation.
a.   Does the U.S. observation appear to be an outlier (i) on x, (ii) on y, (iii) relative to the regression line for the other six observations?
b.   State the two conditions under which a single point can have such a dramatic effect on the slope.
c.   This one point also drastically affects the correlation, which is $r = -0.051$ without the U.S. but $r = -0.935$ with the U.S. Explain why you would conclude that the association between birth rate and number of televisions is (i) very weak without the U.S. point, (ii) very strong with the U.S. point.
d.   Explain why the U.S. residual for the line fitted using that point is very small. This shows that a point can be influential even if it's residual is not large.

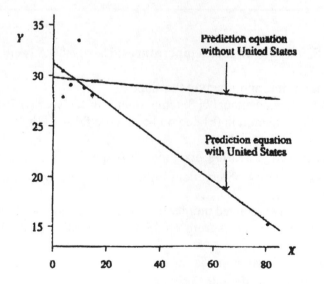

Regression Equations for Birth Rate and Number of TVs per 100 People

🖥 **3.53 Looking for outliers:** Using software, analyze the relationship between $x =$ college education and $y =$ percentage single-parent families, for the data in Table 3.7, which are in the "U.S. statewide crime" data file on the text CD.
a.   Construct a scatterplot. Based on your plot, identify two observations that seem quite different from the others, having y-value relatively very large in one case and somewhat small in the other case.
b.   Find the regression equation (i) for the entire data set, (ii) deleting only the first of the two outlying observations, (iii) deleting only the second of the two outlying observations.
c.   Is either observation influential on the slope? Summarize the influence.
d.   Including D.C., $\hat{y} = 21.2 + 0.089x$, whereas deleting that observation, $\hat{y} = 28.1 - 0.206x$. Find $\hat{y}$ for D.C. in the two cases. Does the predicted value for D.C. depend much on which regression equation is used?

🖥 **3.54 Regression between cereal sodium and sugar:** Consider the relationship between $x =$ sodium and $y =$ sugar with the breakfast cereal data in the "Cereal" data file on the text CD.
a.   Construct a scatterplot. Do any points satisfy the two criteria for a point to be potentially influential on the regression? Explain.
b.   Find the regression line and correlation using all the data points, and then using all except a potentially influential observation. Summarize the influence of that point.

🖥 **3.55 TV in Europe:** An article (by M. Dupagne and D. Waterman, *Journal of Broadcasting and Electronic Media,* vol. 42, pp. 208-220, 1998) studied variables relating to the percentage of TV programs in 17 Western European countries that consisted of fiction programs imported from the U.S. One explanatory variable considered was the percentage of stations in the country that are private. The data are in the "TV Europe" data file on the text CD.
a.   Construct a scatterplot. Describe the direction of the association.
b.   Find the correlation and the regression line. Draw the line on the scatterplot.
c.   Do you observe an outlier? If so, what makes the observation unusual? Would you expect it to be an influential observation?
d.   Delete the observation identified in part (c) from the data set. Find the regression line and the correlation. Compare your results to those in part (b). Was the observation influential? Explain.

3.56 Height and vocabulary: Refer to Example 17 on height and vocabulary. Sketch a hypothetical scatterplot (as we did in Figure 3.22 for the example on crime and education), labeling points by the child's grade (1, 6, and 12), such that overall there is a positive trend, but the slope would be about 0 when we consider children in a given grade. How does the age of the child play a role in the association between height and test score?

3.57 More firefighters cause worse fires?: Data are available for all fires in Chicago last year on $x =$ number of firefighters at the fire and $y =$ cost of damages due to the fire.
a.   Would you expect the correlation to be negative, zero, or positive? Why?
b.   If the correlation is positive, does this mean that having more firefighters at a fire causes the damage to be worse? Explain.

c.   Identify a third variable that could be considered a common cause of x and y. Construct a hypothetical scatterplot (like Figure 3.22 for crime and education), identifying points according to their value on the third variable, to help explain your argument.

3.58 Antidrug campaigns: An Associated Press story (6/13/2002) reported that "A survey of teens conducted for the Partnership for a Drug Free America found kids who see or hear anti-drug ads at least once a day are less likely to do drugs than youngsters who don't see or hear ads frequently? …. When asked about marijuana, kids who said they saw the ads regularly were nearly 15 percent less likely to smoke pot."

a.   Discuss at least one lurking variable that could affect these results.
b.   Explain how multiple causes could affect whether a teenager smokes pot.

3.59 Fluoride and AIDS: An Associated Press story (8/25/1998) about the lack of fluoride in most of the water supply in Utah quoted anti-fluoride activist Norma Sommer as claiming that fluoride may be responsible for AIDS, since the water supply in San Francisco is heavily fluoridated and since that city has an unusually high incidence of AIDS. Describe how you could use this story to explain to someone who has never studied statistics that association need not imply causation.

3.60 Death penalty and race: The table shows results of whether the death penalty was imposed in murder trials in Florida between 1976 and 1987. For instance, the death penalty was given in 53 out of 467 cases in which a white defendant had a white victim.

Death Penalty, by Defendant's Race and Victim's Race

Victim's Race	Defendant's Race	Death Penalty Yes	No	Total
White	White	53	414	467
	Black	11	37	48
Black	White	0	16	16
	Black	4	139	143

Source: M. Radelet and G. Pierce, *Florida Law Review*, Vol. 43, pp. 1-34, 1991

a.   First, consider only those cases in which the victim was white. Find the conditional proportions who got the death penalty when the defendant was white and when the defendant was black. Describe the association.
b.   Repeat (a) for cases in which the victim was black. Interpret.
c.   Now add these two tables together to get a summary contingency table that describes the association between the death penalty verdict and defendant's race, ignoring the information about the victim's race. Find the conditional proportions. Describe the association, and compare to (a) and (b).
d.   Explain how these data satisfy Simpson's paradox. How would you explain what is responsible for this, to someone who has not taken a statistics course?

3.61 NAEP scores: Eighth grade math scores on the National Assessment of Educational Progress had means of 277 in Nebraska and 271 in New Jersey (H.Wainer and L. Brown, *American Statistician*, Vol. 58, p. 119, 2004).
a. Identify the response variable and the explanatory variable.
b. For white students, the means were 281 in Nebraska and 283 in New Jersey. For black students, the means were 236 in Nebraska and 242 in New Jersey. For other nonwhite students, the means were 259 in Nebraska and 260 in New Jersey. Identify the third variable given here. Explain how it is possible for New Jersey to have the higher mean for each race, yet for Nebraska to have the higher mean when the data are combined. (This is another case of Simpson's paradox.)

CHAPTER SUMMARY

This chapter introduced descriptive statistics for studying the **association** between two variables. We explored how the value of a **response variable** (the outcome of interest) depends on the value of an **explanatory variable**.

- For two *categorical variables*, **contingency tables** summarize the counts of observations at the various combinations of categories of the two variables. Bar graphs can plot **conditional proportions** or percentages for the response variable, at different given values of the explanatory variable.

- For two *quantitative variables*, **scatterplots** display the relationship and show whether there is a **positive association** (upward trend) or a **negative association** (downward trend). The **correlation**, r, describes this direction and the strength of linear (straight line) association. It satisfies $-1 \leq r \leq 1$. The closer r is to -1 or 1, the stronger the linear association.

- When a relationship between two quantitative variables approximately follows a straight line, it can be described by a **regression line** $\hat{y} = a + bx$. The slope b describes the direction of the association (positive or negative, like the correlation) and gives the effect on $\hat{y}$ of a 1-unit increase in x.

- The correlation and the regression equation can be strongly affected by an **influential observation**. This takes a relatively small or large value on x and is a **regression outlier**, falling away from the straight-line trend of the rest of the data points. Be cautious of **extrapolating** a regression line to predict y-values at values of x that are far above or far below the observed x-values.

- **Association does not imply causation**. A **lurking variable** may affect the association. It is even possible that the association will reverse in direction after we adjust for a third variable. This phenomenon is called **Simpson's paradox**.

SUMMARY OF NOTATION

$$\text{Correlation } r = \frac{1}{n-1}\sum z_x z_y = \frac{1}{n-1}\sum\left(\frac{x-\bar{x}}{s_x}\right)\left(\frac{y-\bar{y}}{s_y}\right)$$

where n is the number of points, $\bar{x}$ and $\bar{y}$ are means, s_x and s_y are standard deviations for x and y. This shows that $r > 0$ when most data points fall in quadrants in which x and y are both above their means or both below their means.

Regression equation: $\hat{y} = a + bx$, for predicted response $\hat{y}$, y-intercept a, slope b.

Formulas for the slope and y-intercept are

$$b = r\left(\frac{s_y}{s_x}\right) \text{ and } a = \bar{y} - b(\bar{x}).$$

ANSWERS TO THE FIGURE QUESTIONS

Figure 3.1: Construct a single graph with side-by-side bars (as in Figure 3.2) to compare the conditional proportions of pesticide residues in the two types of food.

Figure 3.2: The proportion of foods having pesticides present is much higher for conventionally grown food than for organically grown food.

Figure 3.3: For a particular response, the bars in Figure 3.2 have different heights, in contrast to Figure 3.3 for which the bars have the same height.

Figure 3.4: The approximate five-number summary for the distribution of Internet use is minimum = 0, Q1 = 2, Median = 20, Q3 = 39, and maximum = 51. The box plot suggests some skew to the right.

Figure 3.5: The point labeled "Ireland" appears unusual. Its Internet use (about 23%) is lower than what's expected based on it's high GDP of about 32.

Figure 3.6: The top point is far above the overall straight-line trend of the other 66 points on the graph. The two rightmost points fall in the overall increasing trend.

Figure 3.7: If data points are closer to a straight line, there is a smaller amount of variability around the line; thus, we can more accurately predict one variable knowing the other variable, indicating a strong association.

Figure 3.8: These 36 points make a positive contribution to the correlation, since the product of their z-scores will be positive.

Figure 3.9: Yes, the points fall in a balanced way in the quadrants. The positive cross products in the upper-right quadrant are roughly counterbalanced by the negative cross products in the upper-left quadrant. The positive cross products in the lower-left quadrant are roughly counterbalanced by the negative cross products in the lower-right quadrant.

Figure 3.10: *The line crosses the y-axis at 61.4. For each femur length increase of 1 cm, the height is predicted to increase 2.4 cm.*

Figure 3.11: *The height is predicted to be the same for each femur length.*

Figure 3.12: *We would expect a positive slope.*

Figure 3.13: *The prediction error is represented by a vertical line from the prediction line to the observation. The distance between the point and line is the prediction error.*

Figure 3.14: *The residuals are vertical distances because the regression equation predicts y, the variable on the vertical axis.*

Figure 3.15: *The rightmost bar represents a county with a large positive residual. For it, the actual vote for Buchanan was much higher than what was predicted.*

Figure 3.16: *The difference in winning heights for men and women is typically about 0.3 to 0.4 meters.*

Figure 3.17: $\hat{y} = a + bx - 52.52 + 0.031(105) = 55.8.$

Figure 3.18: *The observation in the upper right corner will influence the tilt of the straight line fit.*

Figure 3.19: *The slope will become larger if D.C. is included in the regression analysis.*

Figure 3.20: *The fitted line to the data that does not include D.C.*

Figure 3.21: *Possibly neither. There may be another variable influencing the positive association between crime rate and amount of education.*

Figure 3.22: *Line (a) will pass through all the points, with a positive slope. In case (b), the lines passes only through the left set of points with urbanization = 0. It has the same negative slope that those points show.*

Figure 3.24: *No, the correlation would probably be close to 0.*

Figure 3.25: *The type of neighborhood in which the family lives.*

CHAPTER PROBLEMS: PRACTICING THE BASICS

3.62 Choose explanatory and response: For the following pairs of variables, identify the natural response variable and explanatory variable.
a.   Number of square feet in home and assessed value of home
b.   Political party preference (Democrat, Independent, Republican) and gender
c.   Annual income and number of years of education
d.   Number of pounds lost on diet and type of diet (low-fat, low-carbohydrate)

3.63 Graphing data: For each case in the previous exercise, describe the type of graph that could be used to portray the results.

3.64 Life after death for males and females: In the 1998 General Social Survey, respondents answered the question, "Do you believe in a life after death?" The table shows the responses cross-tabulated with gender.

Opinion About Life After Death by Gender

	Opinion about life after death			
Gender	Yes, Definitely	Yes, Probably	No, Probably Not	No, Definitely Not
Male	280	118	53	51
Female	380	129	57	59

a.   Construct a table of conditional proportions.
b.   Summarize results. Is there much difference between responses of males and females?

⌨ **3.65 God and happiness:** Go to the GSS Web site, www.icpsr.umich.edu/GSS, click on "analyze," then "frequencies and crosstabulation," then "start," type "god" for the row variable and "happy" for the column variable (without the quotation marks), and click on "Run the table."
a. Report the contingency table of counts.
b. Treating reported happiness as the response variable, find the conditional proportions. For which opinion about God are subjects most likely to be very happy?
c. To analyze the association, is it more informative to view the proportions in (b) or the frequencies in (a)? Why?

3.66 Degrees and income: The mean annual salaries earned in 1997-1999 by year-round workers with various educational degrees are given in the table:

Degree	Mean Salary
No diploma	$23,400
High school diploma	$30,400
Bachelor's degree	$52,200
Master's degree	$62,300
Doctoral degree	$89,400
Professional degree	$109,600

Source: U.S. Census Bureau and the U.S. Bureau of Labor Statistics, as reported in *The Gainesville Sun*, July 26, 2002.

a.   Identify the response variable? Is it quantitative, or categorical?
b.   Identify the explanatory variable? Is it quantitative, or categorical?
c.   Explain how a bar graph could summarize the data on the two variables.

3.67 Losing teachers: In the state of Georgia, finding that many teachers are leaving the field of education, the *Georgia Education Department* surveyed those leaving to find the reasons. The table shows summaries from 1999, 2000, and 2001.

	Percent Leaving		
Reasons	**1999**	**2000**	**2001**
Family reasons	7.2	7.3	6.8
Fired	2.5	2.1	1.9
Moved to new school system	15	14.2	15
Other	2.7	2.3	2.3
Resignation	56.6	59.4	62.1
Retirement	16	14.7	11.5

a. Identify the response variable and the explanatory variable.
b. What types of percentages are displayed? Why.
c. These results are based on about 10,000 replies each year. Assuming 10,000 responses in 2001, find the number who reported being fired.
d. Summarize any time trends you observe in these data.

3.68 Whooping cough: Consider the "whooping cough" data file on the text CD.
a. Identify the response variable and the explanatory variable.
b. Construct a bar graph to display the incidence rate of whooping cough contingent upon year. Interpret.

3.69 Women managers in the work force: The following side-by-side bar graph appeared in a 2003 issue of the *Monthly Labor Review* about women as managers in the work force. The graph summarized the percentage of managers in different occupations who were women, for the years 1972 and 2002.

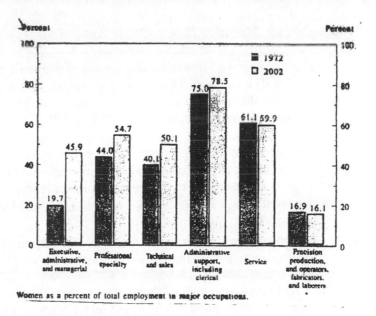

Source: *Monthly Labor Review(2003) Vol. 126 No.10, p.48*

a. Consider the first two bars in this graph. Identify the response variable and explanatory variable.
b. Express the information from the first two bars in the form of conditional proportions in a contingency table for two categorical variables.

3.70 Women as wage earners compared to men: The following side-by-side bar graph summarizes women's earnings as a percentage of men's earnings. The percentages are based on the median usual weekly earnings of full-time wage and salary workers. The results are broken down by age group.

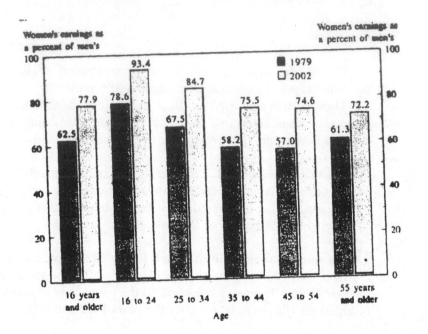

Source: *Monthly Labor Review (2003) Vol. 126 No 10, p.49*

a. The response variable is the assessment of women's earnings as a percent of men's. The graph also shows two explanatory variables. What are they?
b. What would you conclude about the wage earnings of women compared to men and how this has changed from 1979 to 2002?

3.71 Predict crime using poverty: A recent analysis of data for the 50 U.S. states on y = violent crime rate (measured as number of violent crimes per 100,000 people in the state) and x = poverty rate (percent of people in the state living at or below the poverty level) yielded the regression equation, $\hat{y} = 209.9 + 25.5x$.

a. Interpret the slope.
b. The state poverty rates ranged from 8.0 (for Hawaii) to 24.7 (for Mississippi). Over this range, find the range of predicted values for the violent crime rate.

3.72 Height and paycheck: The headline of an article in the *Gainesville Sun* (October 17, 2003) stated "Height can yield a taller paycheck." It described an analysis of four large studies in the U.S. and Britain by a University of Florida professor on subjects' height and salaries. The article reported that for each gender,

"… an inch is worth about $789 a year in salary. So, a person who is 6 feet tall will earn about $5,523 more a year than a person who is 5 foot 5."
a.   For the interpretation in quotes, identify the response variable and explanatory variable.
b.   State the slope of the regression equation, when height is measured in inches and income in dollars.
c.   Explain how the value $5,523 relates to the slope.

3.73 Predicting survey response rate   A study of mail survey response rate patterns of the elderly found a prediction equation relating x = age and y = percentage of subjects responding of to be $\hat{y} = 90.2 - 0.6x$, for ages between 60 and 90. (D. Kaldenberg et al., *Public Opinion Quarterly,* Vol.58, 1994, p.68).
a.   Interpret the slope in context.
b.   Does the y-intercept have a contextual interpretation? Why or why not.
c.   Find the predicted response for a 60-year-old and for a 70-year-old. Interpret.
d.   Find the difference in predicted response rates for any two ages that are ten years apart.

3.74 Predicting college GPA: An admissions officer claims that at his college the regression equation $\hat{y} = 0.5 + 7x$ approximates the relationship between y = college GPA and x = high school GPA, both measured on a four-point scale.
a.   Sketch this equation between x = 0 and 4, labeling the x and y axes. Is this equation realistic? Why or why not?
b.   Suppose that actually $\hat{y} = 0.5 + 0.7x$. Predict the GPA for two students having GPAs of 3.0 and 4.0. Interpret, and explain how the difference between these two predictions relates to the slope.

3.75 College GPA = high school GPA: Refer to the previous exercise. Suppose the regression equation is $\hat{y} = x$. Identify the y-intercept and slope. Interpret the line in context.

3.76 What's a college degree worth?: In 2002, the Associated Press reported a Census Bureau survey that the mean total earnings that a full-time worker in the U.S. can expect to earn between ages 25 and 64 is $1.2 million for those with only a high-school education and $2.1 million for those with a college degree but no advanced degree.
a.   Assuming four years for a college degree and a straight-line regression of y = total earnings on x = number years of education, what is the slope? Interpret it.
b.   If y instead measures earnings per year (rather than for 40 years), then what is the slope? Interpret.

3.77 Car weight and gas hogs: The table shows a short excerpt from the "car weight and mileage" data file on the text CD. That file lists several 2004 model cars with automatic transmission and their x = weight (in pounds) and y = mileage (miles per gallon of gas). The prediction equation is $\hat{y} = 45.6 - 0.0052x$.

Automobile Brand	Weight	Mileage
Honda Accord Sedan LX	3164	34
Toyota Corolla	2590	38
Dodge Grand Caravan	4218	25
Toyota Sienna XLE	4165	27
Dodge Dakota Club Cab	3838	22
Chevrolet Trail Blazer	4612	21
Jeep Grand Cherokee Laredo	3970	21
Hummer H2	6400	17

Source: auto.consumerguide.com, honda.com, toyota.com, landrover.com, ford.com

a.   Interpret the slope in terms of a 1000 pound increase in the vehicle weight.

b.   Find the predicted mileage and residual for a Hummer H2. Interpret.

3.78 Predicting Internet use from cell phone use: We now use data from the "human development" data file on cell-phone use and Internet use for 39 countries.

a.   The MINITAB output shows a scatterplot. Describe it in terms of (i) identifying the response variable and the explanatory variable, (ii) indicating whether it shows a positive or a negative association, (iii) describing the variability of Internet use values for nations that are close to 0 on cell phone use.

b.   Identify the approximate x and y coordinates for a nation that has less Internet use than you would expect, given its level of cell-phone use.

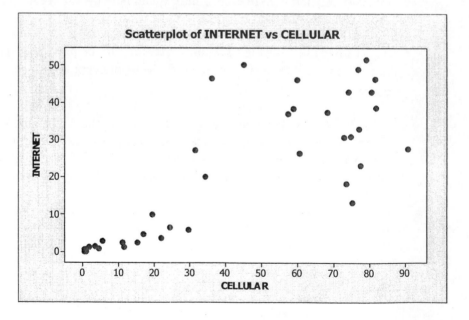

3.79 More on predicting Internet use: Refer to the previous exercise.

a.   The prediction equation is $\hat{y} = 1.27 + 0.475x$. Interpret the intercept and slope.

b.   Describe the relationship by noting how $\hat{y}$ changes as x increases from 0 to 90, which are roughly its minimum and maximum.

c. For the U.S., $x = 45.1$ and $y = 50.15$. Find its predicted Internet use and residual. Interpret the large positive residual.

3.80 Predicting cell-phone use from Internet use: Refer to the previous exercise.
a. When $y =$ cell phone use and $x =$ Internet use, $\hat{y} = 12.06 + 1.409x$. Describe the relationship by noting how $\hat{y}$ changes as x increases from 0 to 50, which are roughly its minimum and maximum.
b. For the U.S., $x = 50.15$ and $y = 45.1$. Find the predicted cell-phone use and the residual. Interpret the large negative residual.
c. The correlation between these variables equals 0.818. Based on the information in this exercise, what would the correlation be for the regression in the previous exercise, with Internet use as the response variable? Why?

3.81 Income depends on education?: For a study of counties in Florida, the table shows part of a printout for the regression analysis relating $y =$ median income (thousand of dollars) to $x =$ percent of residents with at least a high school education.
a. County A has 10% more of its residents than county B with at least a high school education. Find their difference in predicted median incomes.
b. Find the correlation. (*Hint* use the relation between the correlation and the slope of the regression line.) Interpret the (i) sign, (ii) strength of association.

Variable	Mean	Std Dev
INCOME	24.51	4.69
EDUCATION	69.49	8.86

The regression equation is
income = $-4.63 + 0.42$ education

3.82 Crime rate and urbanization: Refer to Example 14 on crime in Florida. For those data, the regression line between $y =$ crime rate (number of crimes per 1000 people) and $x =$ percentage living in an urban environment is $\hat{y} = 24.5 + 0.56x$.

a. Using the slope, find the difference in predicted crime rates between counties that are 100% urban and counties that are 0% urban. Interpret.
b. Interpret the correlation of 0.67 between these variables.
c. Show the connection between the correlation and the slope, using the standard deviations of 28.3 for crime rate and 34.0 for percentage urban.

💻 **3.83 Fertility and GDP:** Refer to the "Human development" data file on the text CD. Use $x =$ GDP and $y =$ fertility (mean number of children per adult woman).
a. Construct a scatterplot, and indicate whether regression seems appropriate.
b. Find the regression equation, and interpret the slope.
c. Using $x =$ percent using contraception, $\hat{y} = 6.7 - 0.065x$. Can you compare the slope of this regression equation with the slope of the equation with GDP as a predictor to determine which has the stronger association with y? Explain.
d. Contraception has a correlation of -0.887 with fertility. Which variable has a stronger association with fertility: GDP or contraception?

3.84 Women working and birth rate: Using data from several nations, a regression analysis of y = crude birth rate (number of births per 1000 population size) on x = women's economic activity (female labor force as a percentage of the male labor force) yielded the equation $\hat{y} = 36.3 - 0.30x$.

a.   Describe the effect by comparing the predicted birth rate for countries with $x = 0$ and countries with $x = 100$.

b.   The correlation is -0.55. Suppose that the correlation between the crude birth rate and the nation's gross national product (GNP) equals -0.35. Which variable, GNP or women's economic activity, seems to have the stronger association with birth rate? Explain.

💻 **3.85 SAT state averages and high school graduation rates:** Access the file "HS Graduation Rates" on the text CD. The file contains data for state high school graduation rates (2002) and SAT state averages (2003).

a.   Let HS graduation rates be the explanatory variable and SAT state averages be the response variable. Construct a scatterplot. Describe the relationship.

b.   Find the correlation. Interpret.

c.   Fit the regression equation to the data and show the line on the scatterplot.

d.   Summarize the relationship in a way that would be understandable to someone who has not taken a statistics course.

3.86 Education and income: The regression equation for a sample of 100 people relating x = years of education and y = annual income (in dollars) is $\hat{y} = -20,000 + 4000x$, and the correlation equals 0.50.

a.   The standard deviations were 2.0 for education and 16,000 for annual income. Show how to find the slope in the regression equation from the correlation.

b.   Show how to find the y-intercept.

3.87 Reverse education and income: Refer to the previous exercise. Suppose that now we let x = annual income and y = years of education. Will the correlation or the slope change in value? If so, show how.

3.88 Income in euros: Refer to the previous two exercises. Results in the regression equation $\hat{y} = -20,000 + 4000x$ for y = annual income were translated to units of euros, at a time when the exchange rate was \$1.25 per euro.

a.   Find the intercept of the regression equation. (Hint: What does 20,000 dollars equal in euros?)

b.   Find the slope of the regression equation.

c.   What is the correlation when annual income is measured in euros? Why?

💻 **3.89 Changing units for cereal data:** Refer to the "Cereal" data file on the text CD, with x = sugar and y = sodium (both in mg), for which $\hat{y} = 243.5 - 0.0071x$.

a.   Convert the amounts measured in milligrams to grams. Compare your results with the previous analysis. What changes, and what does not change? Explain why the results are what you would expect intuitively.

b.   Now suppose only y is changed to grams. What changes compared to measuring both in milligrams? Explain why you expect such a change.

c.   Now suppose only x is changed to grams. What changes compared to measuring both in milligrams? Explain why you expect such a change.

3.90 Murder and single-parent families: For Table 3.7 on the 50 states and D.C., the MINITAB figure shows the relationship between the murder rate and the percentage of single-parent families.

a.   For D.C., percentage of single-parent families = 44.7 and murder rate = 41.8. Identify D.C. on the scatterplot, and explain the effect you would expect it to have on a regression analysis.

b.   The regression line fitted to all 51 observations is $\hat{y} = -21.4 + 1.14x$. The regression line fitted only to the 50 states is $\hat{y} = -8.2 + 0.56x$.   Summarize the effect of including D.C. in the analysis.

c.   Identify the approximate coordinates of one other point that could have some influence on the analysis. What would you expect its effect on the slope to be?

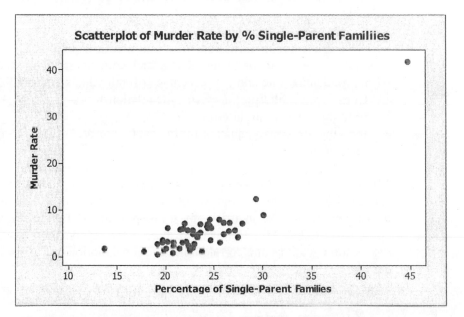

3.91 Violent crime and college education: For the "U.S. statewide crime" data file from the text CD, let y = violent crime rate and x = percent with a college education.

a.   Construct a scatterplot. Identify any points that you think may be influential in a regression analysis.

b.   Fit the regression line using all 51 observations. Interpret the slope.

c.   Fit the regression line after deleting the observation identified in (a). Interpret the slope, and compare results to (b).

3.92 Violent crime and high school education: Repeat the previous exercise using x = percent with at least a high school education. This shows that an outlier is not especially influential if its x-value is not relatively large or small.

3.93 Crime and urbanization: For the "U.S. statewide crime" data file on the text CD, using MINITAB to analyze y = violent crime rate and x = urbanization (percentage of the residents living in metropolitan areas) gives the results shown:

```
Variable   N   Mean   StDev  Minimum     Q1  Median     Q3  Maximum
violent   51  441.6  241.4     81.0  281.0   384.0  554.0   1508.0
urban     51  68.36  20.85     27.90  49.00   70.30  84.50   100.00
The regression equation is violent = 36.0 + 5.93 urban
```

a.   Using the five-number summary, sketch a box plot for *y*. What do this and the reported mean and standard deviation of *y* tell you about the shape of the distribution of violent crime rate?
b.   Report the regression equation, and find the correlation (*Hint*: Use what you know about the relation between the correlation and the slope).

3.94 Crime and urbanization influence: Refer to the previous exercise.
a.   Construct a scatterplot. Does this show any potentially influential observations? Would you predict that the slope would be larger, or smaller, if you delete this observation? Why?
b.   Fit the regression without the observation highlighted in (a). Describe the effect on the slope.

3.95 High school graduation rates and health insurance: Access the "HS Graduation Rates" file on the text CD, which contains statewide data on *x* = high school graduation rate and *y* = percentage of individuals without health insurance.
a.   Construct a scatterplot. Describe the relationship.
b.   Find the correlation. Interpret.
c.   Find the regression equation for the data. Interpret the slope, and summarize the relationship.

3.96 African droughts and dust: Is there a relationship between the amount of dust carried over large areas of the Atlantic and the Caribbean and the amount of rainfall in African regions? In an article (by J. Prospero and P. Lamb, *Science*, vol. 302, p. 1024, Nov. 7, 2003) the following scatterplots were given along with corresponding regression equations and correlations. The precipitation index is a measure of rainfall deficit.

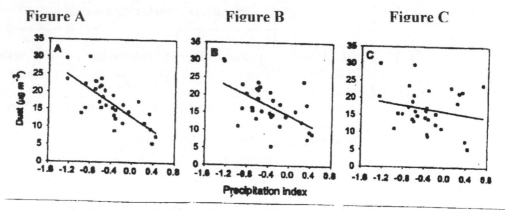

a.   Match the following regression equations and correlations with the appropriate graph.

(i) $\hat{y} = 14.05 - 7.18x$. $r = -0.75$
(ii) $\hat{y} = 16.00 - 2.36x$. $r = -0.44$
(iii) $\hat{y} = 12.80 - 9.77x$. $r = -0.87$

b.   Based on the scatterplots and information in part (a), what would you conclude about the relationship between dust amount and rainfall amounts?

3.97 Income and height: A survey of adults revealed a positive correlation between the height of the subject and their income in the previous year. Explain how gender could be a potential lurking variable that could explain this association.

3.98 More TV watching goes with fewer babies?: For United Nations data from several countries, there is a strong negative correlation between the birth rate and the per capita television ownership.
a.   Does this imply that having higher television ownership causes a country to have a lower birth rate?
b.   Identify a lurking variable that could provide an explanation for this association.

3.99 More sleep causes death?: An Associated Press story (2/15/2002) quoted a study at the University of California at San Diego that reported, based on a nationwide survey, that those who averaged at least 8 hours sleep a night were 12 percent more likely to die within six years than those who averaged 6.5 to 7.5 hours of sleep a night.
a.   Explain how the subject's age could be positively associated both with time spent sleeping and with an increased death rate, and hence could be a lurking variable responsible for the observed association between sleeping and the death rate.
b.   Explain how the subject's age could be a common cause of both variables.

3.100 Ask Marilyn: Marilyn vos Savant writes a column for *Parade* magazine to which readers send questions, often puzzlers or questions with a twist. In one column, a reader posed the following question *(Parade Magazine, 28 April 1996, p 6):* "A company decided to expand, so it opened a factory generating 455 jobs. For the 70 white-collar positions, 200 males and 200 females applied. Of the females who applied, 20% of the females and only 15% of the males were hired. Of the 400 males applying for the blue-collar positions, 75% were hired. Of the 100 females applying, 85% of were hired. A federal Equal Employment enforcement official noted that many more males were hired than females, and decided to investigate. Responding to charges of irregularities in hiring, the company president denied any discrimination, pointing out that in both the white collar and blue collar fields, the percentage of female applicants hired was greater than it was for males. But the government official produced his own statistics, which showed that a female applying for a job had a 58% chance of being denied employment while male applicants had only a 45% denial rate. As the current law is written, this constituted a violation....Can you explain how two opposing statistical outcomes are reached from the same raw data?"
a.   Construct two contingency tables giving counts relating gender to whether hired (yes or no), one table for white-collar jobs and one table for blue-collar jobs.
b.   Construct a single contingency table for gender and whether hired, combining all 900 applicants into one table. Verify that the percentages not hired are as quoted above by the government official.
c.   Comparing the tables in (a) and (b), explain why this is an example of Simpson's paradox.

CHAPTER PROBLEMS: CONCEPTS AND INVESTIGATIONS

⌨ **3.101 Cereal data:** Refer to the "Cereal" dataset on the text CD. Letting $x =$ sodium content and $y =$ sugar content, construct a scatterplot, including different symbols for the Type of Cereal. Analyze the relationship among these three variables, and summarize in a short report.

⌨ **3.102 Crime and education:** Refer to the "U.S. statewide crime" data file on the course CD. Analyze the relationship between violent crime rate and percentage having at least a high school education. Write a two-page report summarizing your analyses and interpreting your findings. Include the key parts of the software output as an appendix.

⌨ **3.103 NL baseball team ERA and number of wins:** Is a baseball team's earned run average (ERA = the average number of earned runs they give up per game) a good predictor of the number of wins that a team has for a season? The data for the National League teams in 2003 are available in the "NL team statistics" file on the text CD. Conduct a correlation and regression analysis, including graphical and numerical descriptive statistics. Summarize results in a short report.

⌨ **3.104 SAT scores:** Let's explore the association between state average SAT score and percentage of students in the state who take the exam. Analyze the data in the "SAT state scores" data file on the text CD, choosing an appropriate response variable, using graphics for the individual variables and for both variables together, and finding and interpreting the correlation and the regression equation.

⌨ **3.105 ACT scores:** Repeat the analyses of the previous exercise, using the state ACT averages from the "ACT state scores" file found on the text CD.

⌨ **3.106 Time studying and GPA:** Is there a relationship between the amount of time a student studies and a student's GPA? Access the "Georgia Student Survey" file on the text CD or use your class data to explore this question using appropriate graphical and numerical methods. Write a brief report summarizing results.

⌨ **3.107 Warming in Newnan, Georgia:** Access the "Newnan GA Temps" file on the text CD, which contains data on average annual temperatures for Newnan, Georgia during the 20th century. Fit a regression line to these temperatures and interpret the trend. Compare the trend to the trend found in Example 12 for Central Park, New York temperatures.

3.108 Regression for dummies: You have done a regression analysis for the catalog sales company you work for, using monthly data for the last year on $y =$ total sales in the month and $x =$ number of catalogs mailed in preceding month. You are asked to prepare a 200-word summary of what regression does under the heading "Regression for Dummies," to give to fellow employees who have never taken a statistics course. Do this, being careful not to use any technical jargon with which the other employees may not be familiar.

3.109 Fish fights Alzheimer's: An AP story (July 22, 2003) described a study conducted over four years by Dr. Martha Morris and others from Chicago's Rush-

Presbyterian-St. Luke's Medical Center involving 815 Chicago residents aged 65 and older (*Archives of Neurology*, July 21, 2003). Those who reported eating fish at least once a week had a 60% lower risk of Alzheimer's than those who never or rarely ate fish. However, the story also quoted Dr. Rachelle Doody of Baylor College of Medicine as warning, "Articles like this raise expectations and confuse people. Researchers can show an association, but they can't show cause and effect." She said it is not known whether those people who had a reduced risk had eaten fish most of their lives, and whether other dietary habits had an influence. Using this example, describe how you would explain to someone who has never taken statistics (a) what a lurking variable is, (b) how there can be multiple causes for any particular response variable, and (c) why they need to be skeptical when they read new research results such as in this story.

3.110. What's wrong with regression?: Explain what's wrong with the way regression is used in each of the following examples:

a. Winning times in the Boston marathon (at www.bostonmarathon.org) have followed a straight line decreasing trend from 160 minutes in 1927 (when the race was first run at the Olympic distance of about 26 miles) to 130 minutes in 2004. After fitting a regression line to the winning distances, you use the equation to predict that the winning time in the year 2300 will be about 13 minutes.

b. For data for several cities on x = % of residents with a college education and y = median price of home, you get a strong positive correlation. You conclude that having a college education causes you to be more likely to buy an expensive house.

c. A regression between x = number of years of education and y = annual income for 100 people shows a modest positive trend, except for one person who dropped out after 10th grade but is now a multimillionaire. It's wrong to ignore any of the data, so we should report all results including this point, such as the correlation r = -0.28.

Answer the following multiple-choice questions 3.111-3.118:

3.111 Correlate GPA and GRE: In a study of graduate students who took the Graduate Record Exam (GRE), the Educational Testing Service reported a correlation of 0.37 between undergraduate grade point average (GPA) and the graduate first-year GPA[10]. This means that:

a.   As undergraduate GPA increases by one unit, graduate first-year GPA increases by 0.37 units.

b.   Since the correlation is not 0, we can predict a person's graduate first-year GPA perfectly if we know their undergraduate GPA.

c.   The relationship between undergraduate GPA and graduate first-year GPA follows a curve rather than a straight line.

d.   As one of these variables increases, there is a weak tendency for the other variable to increase also.

3.112 Regression and correlation: If your analysis of two quantitative variables produced the regression equation, $\hat{y}$ = 0.40 − 0.008x, then

a.   the y-intercept is negative.

b.   the y-intercept has a greater effect on y than does the slope.

---

[10] Source: *GRE Guide to the Use of Scores*, 1998-1999, www.gre.org

c.   there is an inverse relationship between x and the predicted value of y; as one increases, the other decreases.

d.   the correlation equals 0.40 when $x = 0$.

3.113 Properties of r : Which of the following is <u>not</u> a property of r?

a. r is always between -1 and 1.

b. r depends on which of the two variables is designated at the response variable.

c. r measures the strength of the linear relationship between x and y.

d. r does not depend on the units of y or x.

e. r has the same sign as the slope of the regression equation.

3.114 Interpreting r : One can interpret $r = 0.30$ as

a.   a weak, positive association

b.   30% of the time $\hat{y} = y$.

c. $\hat{y}$ changes 0.30 units for every one-unit increase in x.

d.   a stronger association than two variables with $r = -0.70$.

3.115 Correct statement about r : Which one of the following statements is correct?

a.   The correlation is always the same as the slope of the regression line.

b.   The mean of the residuals from the least-squares regression line is 0 only when $r = 0$.

c.   The correlation is the percentage of points that lie in the quadrants where x and y are both above the mean or both above the mean.

d.   The correlation is inappropriate if a U-shaped relationship exists between x and y.

3.116 Describing association between categorical variables: You can summarize the data for two categorical variables x and y by

a.   drawing a scatterplot of the x and y values.

b.   constructing a contingency table for the x and y values.

c.   calculating the correlation between x and y.

d.   fitting a regression line between x and y.

e.   constructing a box plot for each variable.

3.117 Correlate income with education: Based on recent data from the *Statistical Abstract of the United States* compiled for the 50 states, for the explanatory variable x = percent of adult residents with at least 16 years of education, the correlation equals about 0.70 with y = per capita income and -0.50 with y = percent of residents living below the poverty level.

a.   As x increases, the predicted percent of people below the poverty level increases.

b.   The correlation between per capita income and percent living below the poverty level is $0.70 - 0.50 = 0.20$.

c.   D.C. is higher than all 50 states on x and on percent of residents living below the poverty level. If we included D.C. in the analysis, the correlation between these two variables would probably not be as strongly negative.

d.   As x increases 1-unit, the predicted per-capita income increases by 0.70 thousand dollars.

3.118 Slope and correlation: The slope of the regression equation and the correlation are similar in the sense that

a.   they do not depend on the units of measurement.

b. they both must fall between -1 and +1.
c. they both have the same sign.
d. neither can be affected by severe regression outliers.

3.119 True-false: The variables y = annual income (thousands of dollars), x_1 = number of years of education, and x_2 = number of years experience in job are measured for all the employees having city-funded jobs, in Knoxville, Tennessee. Suppose that the following regression equations and correlations apply:
i) $\hat{y} = 10 + 1.0\,x_1$, $r = 0.30$.

ii) $\hat{y} = 14 + 0.4\,x_2$, $r = 0.60$.

The correlation is -0.40 between x_1 and x_2. Which of the following statements are true?
a. The strongest association is between y and x_2.

b. The weakest association is between x_1 and x_2.

c. The regression equation using x_2 to predict x_1 has negative slope.

d. Each additional year on the job corresponds to a $400 increase in predicted income.
e. The predicted mean income for employees having 20 years of experience is $4000 higher than the predicted mean income for employees having 10 years of experience.

♦♦ 🖳 3.120 Asthma: The "asthma" data file on the text CD shows data for a study of 13 children suffering from asthma. This study compared single inhaled doses of formoterol (F) and salbutamol (S). Each child was evaluated using both medications. The outcome measured was the child's peak expiratory flow (PEF) eight hours following treatment. The purpose of the study was to see if there was evidence to support a difference in the PEF level for the two brands. A graphical way to compare the PEF levels is to construct a scatterplot. Let the horizontal axis represent the formoterol PEF levels and the vertical axis represent the salbutamol PEF levels. What would be an appropriate line to use in the scatterplot to judge if the two brands differ with respect the PEF levels? Do you believe there is a difference based on your finding from the scatterplot? Give a short description of your findings.

♦♦ 3.121 Correlation does not depend on units: Suppose you convert y = income from British pounds to dollars, and suppose a pound equals 2.00 dollars.
a) Explain why the y-values double, the mean of y doubles, the deviations $(y - \bar{y})$ double, and the standard deviation s_y doubles.
b) Using the formula for calculating the correlation, explain why the correlation would not change value.

♦♦ 3.122 When correlation = slope: Consider the formula $b = r(s_y / s_x)$ that expresses the slope in terms of the correlation. Suppose the data are equally spread out for each variable. That is, suppose the data satisfy $s_x = s_y$. Show that the correlation and the slope are the same. (In practice, the standard deviations are not usually identical. However, explain how this provides an interpretation for the correlation as representing what we would get for the slope of the regression line if the two variables were equally spread out.)

♦♦ **3.123 Center of the data:** Consider the formula $a = \bar{y} - b\bar{x}$ for the y-intercept.

a. Show that $\bar{y} = a + b\bar{x}$. Explain why this means that the predicted value of the response variable is $\hat{y} = \bar{y}$ when $x = \bar{x}$.

b. Show that an alternative way of expressing the regression model is as $(\hat{y} - \bar{y}) = b(x - \bar{x})$. Interpret this formula.

♦♦ **3.124 Final exam scores "regress toward mean" of midterm scores**: Let $y =$ final exam score and $x =$ midterm exam score. Suppose that the correlation is 0.70 and that the standard deviation is the same for each set of scores.

a. Using part (b) of the previous exercise and the relation between the slope and correlation, show that $(\hat{y} - \bar{y}) = 0.70(x - \bar{x})$.

b. Explain why this means that the predicted difference between your final exam grade and the mean for the class is 70% of the difference between your midterm exam score and the mean for the class. Your score is predicted to "regress toward the mean." The concept of *regression toward the mean*, which is responsible for the name of regression analysis, will be discussed in Section 11.2.

CHAPTER PROBLEMS: CLASS EXPLORATIONS

⌨ **3.125 Analyze your data**: Refer to the data file the class created in Activity 3. For variables chosen by your instructor, conduct a regression and correlation analysis. Prepare a one-page report summarizing your analyses, interpreting your findings.

⌨ **3.126 Activity: Effect of moving a point:** The 'Correlation and Regression' Applet at www.prenhall.com/TBA lets you add and delete points on a scatterplot. The correlation and the regression line are automatically calculated for the points you provide.

a. Your instructor will give the students five data points that have an approximate linear relation between x and y but a slope and correlation near 0. Plot these in a scatterplot

b. Add a sixth observation that is influential. What did you have to do to get the slope and correlation to change greatly? Show how to place the influential point so the correlation (i) increases, (ii) decreases, (iii) becomes negative.

c. Now consider the effect of sample size on the existence of influential points. Start by creating 20 points with similar pattern as in (a) that have a slope and correlation near 0. Is it now harder to add a single point that makes the correlation and regression change greatly? Why? Your class will discuss the results.

⌨**3.127 Activity: Guess the correlation and regression:** The "Regression by Eye" applet at www.prenhall.com/ allows you to guess the correlation r and the regression line for a scatterplot with randomly generated data points. Your instructor will give you instructions on how to use this applet to practice matching the correct value of r to a scatterplot and to estimate the correct regression line.

CHAPTER 4
Gathering Data

EXAMPLE 1:   ARE CELL PHONES DANGEROUS TO YOUR HEALTH?

Picture the Scenario

Cell phones have become the must-have communication gadget of the new millennium. There's no doubt about it: The use of cell phones has become a routine part of our lives. But how safe are they? Cell phones emit electromagnetic radiation, produced in the form of non-ionizing radio-frequency energy. A cellular phone's antenna is the main source of this energy. The closer the antenna is to the user's head, the greater the exposure to cell phone radiation.

With the increase in the popularity of cell phones has come a growing concern that heavy use of cell phones may increase a person's risk of getting cancer. Several studies have explored whether there's an association between cell phone use and the occurrence of cancer. For instance:

Study 1: A German study (Stang, 2001) compared 118 patients with a rare form of eye cancer called uveal melanoma to 475 healthy patients who did not have the eye cancer. The patients' cell phone use was measured using a questionnaire. The eye cancer patients used cell phones more often, on the average.

Study 2: A U.S. study (Muscat et al., 2000) compared 469 patients with brain cancer to 422 patients who did not have brain cancer. The patients' cell-phone use was measured using a questionnaire. The two groups' use of cell phones was similar.

Study 3: An Australian study (Repacholi, 1997) conducted an experiment with 200 transgenic mice, specially bred to be susceptible to cancers of the immune system. One hundred mice were exposed for two half-hour periods a day to the same kind of microwaves with roughly the same power as the kind transmitted from a cell phone. The other 100 mice were not exposed. After 18 months, the brain tumor rate for the mice exposed to cell phone radiation was twice as high as the brain tumor rate for the unexposed mice.

Studies 1 and 3 found an association between cell phone use and cancer. Study 2 did not.

Questions to Explore
- Why is it that results of different medical studies sometimes disagree?
- What is the best study design for gathering data to explore whether there is an association between cell-phone use and the occurrence of cancer? Can such a study design establish a direct causal link between cell phone use and cancer? Is this study design ethically feasible to conduct?

Thinking Ahead

A knowledge of different **study designs for gathering data** can help us understand how contradictory results can happen in scientific research studies and help us determine which studies deserve our trust. In this chapter, we'll learn that the study design can have a major impact on its results. Unless the study is well designed and implemented, the results may be meaningless, or worse yet, misleading. We'll see how to answer the questions posed above regarding cell phone studies in Example 2, in the two subsections following Example 3, and in Example 11.

Chapters 2 and 3 introduced graphical and numerical summaries for describing data. We described shape, center, and spread, looked for patterns and unusual observations, and explored associations between variables. For these and other statistical analyses to be useful, we must have "good data." But what's the best way to gather data to ensure they are "good"? The studies described in Example 1 used two different methods. The first two studies merely *observed* subjects, whereas the third study *conducted an experiment* with them. We'll now learn about such methods, we'll study their pros and cons, and we'll see both good ways and bad ways of using them. This will help us understand when we can trust the conclusions of a study and when we need to be skeptical.

4.1 SHOULD WE EXPERIMENT OR SHOULD WE MERELY OBSERVE?

We use statistics to learn about a **population**. The German and the U.S. studies mentioned in Example 1 examined subjects' cell phone use and whether or not the subjects had brain or eye cancer. Since it would be too costly and time consuming to seek out *all* individuals in the populations of Germany or the U.S., the researchers used **samples** to gather data. These studies, like many, have two variables of primary interest--a **response variable** and an **explanatory variable**. The response variable is whether or not a subject has cancer (eye or brain). The explanatory variable is the amount of cell phone use.

Types of Studies: Experimental and Observational

Many studies, such as the Australian study in Example 1, perform an **experiment**. Subjects are assigned to experimental conditions, such as receiving radiation or not receiving radiation, that we want to compare on the response outcome. These experimental conditions are called **treatments**. They correspond to different values of the explanatory variable, such as whether the subject receives radiation.

Definition: Experiment

A researcher conducts an **experiment** by assigning subjects to certain experimental conditions and then observing outcomes on the response variable. The experimental conditions, which correspond to assigned values of the explanatory variable, are called **treatments**.

For example, the Australian study used mice as the subjects. Each mouse was assigned to one of two treatments, receiving radiation or not receiving radiation. These are the categories of the explanatory variable--whether or not the subject is exposed to cell phone radiation. The purpose of the experiment was to examine the association between this variable and the response variable--whether or not the mouse developed a brain tumor.

An experiment assigns each subject to a treatment and then observes the response. By contrast, many studies merely *observe* the values on the response variable and the

IN WORDS

An **observational study** merely *observes* rather than *experiments* with the study subjects. An **experimental study** assigns to each subject a treatment (such as exposure or nonexposure to cell phone radiation) and then observes the outcome on the response variable.

explanatory variable for the sampled subjects without doing anything to them. Such studies are called **observational studies.**

Definition: Observational Study

In an **observational study**, the researcher observes values of the response variable and explanatory variables for the sampled subjects, without anything being done to the subjects (such as imposing a treatment).

EXAMPLE 2: DIFFERENT TYPES OF CELL PHONE STUDIES

Picture the Scenario
Example 1 described three studies about whether there's an association between cell phone use and cancer. We've seen that the Australian study was an experiment, using mice. The German and the U.S. studies both observed the amount of cell phone use for cancer patients and for non-cancer patients using a questionnaire.

Questions to Explore
a. Were the German and U.S. studies experiments or observational studies?
b. How were these studies fundamentally different from the Australian study?

Think It Through
a. In both the German and the U.S. studies, the explanatory variable was the amount of cell phone use. Information on the amount of cell phone use was gathered by giving a questionnaire to the sampled subjects. The subjects (people) decided how much they would use a cell phone and, thus, their amount of radiation exposure. The studies merely observed this exposure. No experiment was performed. The studies were therefore *observational* studies.

b. In the Australian study, the researcher rather than the subject (the mouse) decided how much radiation each subject would receive. The researchers did not merely observe the subjects but determined which treatment each would receive. The study was therefore an *experimental* study rather than an observational study.

Insight
One reason that results of different medical studies sometimes disagree is that they are not the same *type* of study. An experimental study with mice is not directly comparable to an observational study with humans, and as we'll see, there are different types of observational studies, some more trustworthy than others.

♦

To practice this concept, try Exercise 4.2, part (a), and Exercise 4.3.

Let's consider another study. From its description, let's see if we can figure out the response and explanatory variables and whether it was an experiment or an observational study.

EXAMPLE 3: DOES DRUG TESTING REDUCE STUDENTS' DRUG USE?

Picture the Scenario

"Student Drug Testing Not Effective in Reducing Drug Use" proclaimed the headline in a May 19, 2003 news release from the University of Michigan. It reported results from a study of 76,000 students in 497 high schools and 225 middle schools nationwide.[1] Each student in the study filled out a questionnaire. One question asked whether the student used drugs. The study found that drug use was similar in schools that tested for drugs and in schools that did not test for drugs. For instance, the table in the margin shows the conditional proportions on drug use for the sampled twelfth graders from the two types of schools.

Conditional proportions on drug use for Twelfth grade students

Drug Tests?	Drug Use Yes	No	n
Yes	0.37	0.63	5653
No	0.36	0.64	17,437

Questions to Explore

a. What were the response and explanatory variables?
b. Was this an observational study, or was it an experiment?

Think It Through

a. The purpose of the study was to compare the percentage of students who used drugs in schools that tested for drugs and in schools that did not test for drugs. Whether the student used drugs was the response variable. The explanatory variable was whether the student's school tested for drugs. Both variables were categorical, with categories "yes" and "no." For each grade, the data were summarized in a contingency table, as shown above in the margin for twelfth graders.

b. For each student, the study merely observed whether his or her school tested for drugs and whether he or she used drugs. So this was an observational study.

Insight

An experiment would have assigned schools to use or not to use drug testing, rather than leaving the decision to the schools. For instance, the study could have "randomly" selected half the schools to do drug testing and half not to do it, and then a year later measured the student drug use in each school. As we will soon discuss, an experimental study gives the researcher more control over outside influences. This control allows more accuracy in studying the effect of an explanatory variable on a response variable.

♦

To practice this concept, try Exercise 4.2, part (a), and Exercise 4.3.

Advantage of Experiments over Observational Studies

RECALL

From Section 3.4, a **lurking variable** is a variable, not observed in the study, that influences the association between the response and explanatory variables due to its own association with each of those variables.

In an observational study, lurking variables can affect the results. The German study in Example 1 found an association between cell phone use and eye cancer. However, there could be lifestyle, genetic, or health differences between the subjects with eye cancer and those without it, and between those who use cell phones a lot and those who do not. This could affect the association. For example, a possible lifestyle lurking variable is computer use. Perhaps those who use cell phones a lot also use computers a lot. Perhaps high exposure to computer screens increases the chance of

---

[1] Study by R. Yamaguchi et al., reported in *Journal of School Health*, vol. 73, pp. 159-164, 2003.

eye cancer. In that case, the higher prevalence of eye cancer for heavier users of cell phones could be due to their higher use of computers, not their higher use of a cell phone.

By contrast, an experiment reduces the potential for lurking variables to affect the results. Why? We'll see that with a type of "random" selection to determine which subjects receive each treatment, the groups of subjects receiving the different treatments are "balanced": That is, the groups have similar distributions on other variables, such as lifestyle, genetic, or health characteristics. For instance, if we randomly decide which mice receive radiation and which do not, the two treatment groups of mice have similar distributions on health. One group will not have mice that are much healthier, on the average, than the other group. When the groups are balanced on a potential lurking variable, there is no association between the lurking variable and the explanatory variable (for instance, between health and whether the mouse receives radiation). When this is true, the lurking variable will not affect how that explanatory variable is associated with the response variable. One treatment group will not tend to have more cancer cases simply because its mice have poorer health, on the average.

Establishing **cause and effect** is central to science. But it's not possible to establish cause and effect definitively with observational studies. There's always the possibility that some lurking variable could be responsible for the association. If people who make greater use of cell phones have a higher rate of eye cancer, how do we know it's not because of some variable that we failed to measure in our study, such as computer use? As we learned in Section 3.4, **association does not imply causation**.

Because it's easier to adjust for lurking variables in an experiment than in an observational study, *we can study the effect of an explanatory variable on a response variable more accurately with an experiment than with an observational study*. With an experiment, the researcher has control over which treatment (e.g., radiation or no radiation) each subject receives. If we find a higher rate of brain cancer for those exposed to radiation, we can be more sure that we've discovered a causal relationship than if we merely find an association in an observational study. Because of this, the best method for determining causality is to conduct an experiment.

What Type of Study Is Possible?

If experiments are preferable, why ever conduct an observational study? Why bother to measure whether cell phone usage was greater for those with cancer than for those not having it? Why not instead conduct an experiment, such as the following: Pick half the students from your school "at random" and tell them to use a cell phone each day for the next 50 years. Tell the other half of the student body not to ever use cell phones. Fifty years from now, analyze whether cancer was more common for those who used cell phones.

You can see obvious difficulties with such an experiment. First, there is an ethical issue. It's not ethical for a study to expose part of the sample to something (such as cell phone radiation) that may potentially be harmful. Second, it's difficult in practice to make sure that the subjects do as told. How can you monitor them to ensure that they adhere to their treatment assignment over the 50-year experimental period? Third, who wants to wait 50 years to get an answer?

For these reasons, medical experiments are often performed over a short time period, and often with animals rather than humans. In the Australian study, mice were exposed to radiation similar to that emitted from a cell phone. Because inferences about human populations are more trustworthy when we use samples of humans than when we use samples of animals, however, scientists often resort to observational studies. We'll study methods for designing a good observational study. This can yield useful information when an experiment is not practical.

Finally, another reason observational studies are common is that many questions of interest do not involve trying to assess causality. For instance, if we want to gauge the public's opinion on some issue, or if we want to conduct a marketing study to see how people rate a new product, it's completely adequate to use an observational study that samples (appropriately) the population of interest.

Using Data Already Available

Of course, you yourself will not conduct an experiment or an observational study every time you want to answer some question, such as whether cell phone use is dangerous. It's human nature to rely instead on data that are already available. The most readily available data come from your personal observations. Perhaps a friend recently diagnosed with brain cancer was a frequent user of cell phones. Is this strong support for concluding that frequent cell phone use increases the likelihood of getting brain cancer?

Informal observations of this type are called **anecdotal evidence**. Unfortunately, there is no way to tell if they are representative of what happens for an entire population. Sometimes you hear people give anecdotal evidence to attempt to disprove causal relationships. "My Uncle Geoffrey is 85 years old and his entire adult life he's smoked a pack of cigarettes a day, yet he's as healthy as a horse." An association does not need to be perfect, however, to be causal. Not all people who smoke a pack of cigarettes each day will get lung cancer, but a much higher proportion of them will do so than people who are nonsmokers. Perhaps Uncle Geoffrey is lucky to be in good health, but that should not encourage you to tempt the fates by smoking regularly.

Instead of using anecdotal evidence to draw conclusions, you should rely on data collected in reputable research studies. You yourself can find research results on topics of interest by entering keywords in search engines on the Internet. This directs you to published results, such as in medical journals for medical studies.[2] Results from well-designed experiments and observational studies are more trustworthy than anecdotal evidence.

The previous chapters have already mentioned many sources of available data. For instance, Activity 1 in Chapter 1 introduced the General Social Survey (GSS), a

---

[2] Surveys of research about cell phones suggest that there is no convincing evidence yet of adverse radiation effects (e.g., D. R. Cox, *J. Roy. Statist. Soc., Ser. A*, vol. 166, pp. 241-246, 2003). The primary danger appears to be people using cell phones while driving!

popular observational database for tracking opinions and behaviors of the American public. Likewise, federal governments collect an enormous amount of demographic and economic data. Some central sites for such data are www.fedstats.gov for the U.S., www.statcan.ca for Canada (Statistics Canada), www.inegi.gob.mx for Mexico (INEGI, the Instituto Nacional de Estadistica Geografia e Informatica), www.statistics.gov.uk for the United Kingdom, and www.abs.gov.au for Australia. The U.S. site has a link for the Census Bureau (www.census.gov). Their publications include the annual *Statistical Abstract of the United States*.

The Census and Other Sample Surveys

The General Social Survey (GSS) is an example of a **sample survey**. It gathers information by interviewing a sample of subjects from the U.S. adult population to provide a snapshot of that population.

Definition: Sample Survey

A **sample survey** selects a sample of people from a population and interviews them to collect data.

A sample survey is a type of observational study. The subjects provide data on the variables measured. There is no assignment of subjects to different treatments. For instance, Example 3 on student drug use in schools that tested or did not test for drugs was a sample survey.

Most countries conduct a regular **census**. This is a survey that attempts to count the number of people in the population and to measure certain characteristics about them. It is different from most surveys, which sample only a small part of the entire population.

In Article 1, Section 2, the U.S. Constitution states that a complete counting of the U.S. population is to be taken every 10 years. The first census was taken in 1790 and the most recent was in 2000. The first census counted 3.9 million people, while today the U.S. population is estimated to be nearly 300 million. Other than counting the population size, here are three key reasons for conducting the U.S. census:

- The Constitution mandates that seats in the House of Representatives be apportioned to states based on their portion of the population measured by the census. When the 1910 census was completed, Congress fixed the number of seats at 435. With each new census, states may gain or lose seats depending on how their population size compares with other states.
- Census data are used in the drawing of boundaries for electoral districts.
- Census data are used to determine the distribution of federal dollars to states and local communities.

Methods for taking the U.S. census have varied since horseback riders served as enumerators (the people who make the counts) in the first censuses. From the 1930s to the 1960s, women who were homemakers served as the enumerators. But as women increasingly took jobs outside the home, too few enumerators were available, so the Census Bureau began to mail forms to households in 1970. The mail-back rate for the forms was 90%. However, this decreased to 65% for the 1990 and 2000 censuses. The 35% who did not respond required follow-ups. The Census Bureau sends a reminder requesting that the form be returned. If this is unsuccessful, enumerators make personal visits to the household address.

Although it's the intention of a census to sample *everyone* in a population, in practice this is not possible. Some people do not want to be counted. Some do not have known addresses for the Census Bureau to send a census form: Some people are homeless; others are transient. Practical problems make any census less than perfect. Fortunately, the U.S. founding fathers realized that it was impractical to attempt a census every year. Yet data are needed regularly on economic variables, such as the unemployment rate. Thus, the U.S. Census Bureau and other government agencies continually take smaller samples of the population, rather than relying solely on the complete census. An example is the monthly Current Population Survey of about 50,000 households.

It is usually much more practical to take a sample rather than to try to measure everyone in a population. In fact, in the late 1990s statisticians at the Census Bureau proposed using sampling methods to estimate numbers of people missed by the ordinary census. Their proposal became a political battleground between Republicans (who did not want sampling used) and Democrats (who supported sampling). In 1999, the Supreme Court ruled 5 to 4 that sampling could not be used to adjust the counts for apportioning seats in Congress but that sampling could be used to adjust counts for federal funding.

For a sample survey to be informative, it is important that the sample reflect the population well. As we will discuss next, "random" selection--letting chance determine which subjects are in the sample--is the key to getting a good sample.

SECTION 4.1: PRACTICING THE BASICS

4.1 Cell phones: Consider the U.S. cell phone study described in Example 1.
a.   Identify the response variable and the explanatory variable.
b.   Was this an observational study or an experiment? Explain why.

4.2 Acoustic neuroma: A study identified nearly all cases in Sweden in 1999-2000 of acoustic neuroma, a benign tumor on the auditory nerve (S. Lonn et al., *Epidemiology*, vol. 15, p. 653, 2004). It matched these subjects with others of similar age, gender, and residential area who did not have acoustic neuroma and compared the two groups on their mobile phone use. When considering tumors on the same side of the head as the phone was normally used, those with tumors were more likely to have used mobile phones for at least 10 years than those without tumors.
a. Was this study an experiment or an observational study? Explain.
b. Explain why this does not prove that greater use of mobile phones causes subjects to be more likely to develop acoustic neuroma.

4.3 GSS: Activity 1 in Chapter 1 described the General Social Survey. Is the GSS an experiment or an observational study? Explain.

4.4 Experiments vs. observational studies: When either type of study is feasible, an experiment is usually preferred over an observational study. Explain why, using an example to illustrate.

4.5 School testing for drugs: Example 3 discussed a study comparing high schools that tested for drugs with high schools that did not test for drugs, finding similar levels of student drug use in each. State a potential lurking variable that could affect the results of such a study. Describe what the effect could be.

4.6 Hormone therapy and heart disease: Since 1976 the Nurses' Health Study (conducted by researchers at Harvard University) has followed more than 100,000 nurses. Every two years, the nurses fill out a questionnaire about their habits and their health. Results from this study indicated that post-menopausal women have a reduced risk of heart disease if they take a hormone replacement drug.
a.   Suppose the hormone-replacement drug actually has no effect. Identify a potential lurking variable that could explain the results of the observational study. (*Hint*: Suppose that the women who took the drug tended to be more conscientious about their personal health than those who did not take it.)
b.   Recently a randomized experiment called the Women's Health Initiative was conducted by the National Institutes of Health to see if hormone therapy is truly helpful. The study was planned to last for eight years but was stopped after five years when it was noticed that women who took hormones had 30% *more* heart attacks. This study suggested that rather than reducing the risk of heart attacks, hormone replacement drugs can actually increase the risk.[3] Explain how it is that two studies about the same subject could reach such different conclusions. (For attempts to reconcile the studies, see a story by Gina Kolata in *The New York Times*, April 21, 2003.)
c.   Explain why randomized experiments, when feasible, are preferable to observational studies.

4.7 Smoking affects lung cancer?: You would like to investigate whether smokers are more likely than non-smokers to get lung cancer. For the students in your class, you pick half "at random" and tell them to smoke a pack of cigarettes each day, and you tell the other half not to ever smoke. Fifty years from now, you will analyze whether more smokers than nonsmokers got lung cancer.
a.   Is this an experiment or an observational study? Why?
b.   Summarize at least three practical difficulties with this planned study.

4.8 Hairdressers at risk: In a study by Swedish researchers (*Occupational and Environmental Medicine* 2002;59: 517-522), 2410 women who had worked as hairdressers and given birth to children were compared to 3462 women from the general population who had given birth. The hairdressers had a slightly higher percentage of infants with a birth defect.
a.   Identify the response variable and the explanatory variable.
b.   Is this study an observational study or an experiment? Explain.
c.   Can we conclude that there's something connected with being a hairdresser that
d.   causes higher rates of birth defects? Explain.

---

[3] See article by H. N. Hodis et al., *New England Journal of Medicine*, August 7, 2003.

4.9 Helping low-income smokers quit: A study (*Journal of Family Practice* 2000; 50:138-144) conducted in three Michigan health centers evaluated the effectiveness of usual care for smokers wishing to quit to the usual care enhanced by computer assisted telephone counseling sessions by nurses and counselors. The study used a sample of 233 low-income adult smokers. Each subject was assigned randomly either to the usual care (physician advice and follow-up) or to the usual care plus counseling, and their smoking status (still smoking or quit smoking) was observed after 3 months. The percentage who had quit smoking was higher for the group receiving counseling.
a.   Identify the response variable and the explanatory variable.
b.   Was this study an observational study, or an experiment? Explain.
c.   How does the design of this study take into account potential lurking variables?

4.10 Experiment or observe?: Explain whether an experiment or an observational study would be more appropriate to investigate the following:
a.   Whether or not smoking has an effect on coronary heart disease.
b.   Whether a Honda Accord or a Toyota Camry gets better gas mileage.
c.   Whether or not higher SAT scores tend to be positively associated with higher college GPAs.
d.   Whether or not a special coupon attached to the outside of a catalog makes recipients more likely to order products from a mail-order company.

4.11 Physician likes "cold turkey": A physician who has practiced for more than 20 years believes that going "cold turkey'" is more effective than nicotine patches and other smoking cessation methods in helping smokers to quit. The physician gives as evidence his observation of patients in his practice.
a.   Explain the limitations of such anecdotal evidence.
b.   Mention a lurking variable that could affect the association the physician observed.

4.12 Seat belt anecdote: Andy once heard about a car crash victim who died because of being pinned in the wreckage by a seat belt he could not undo. Because of this, Andy refuses to wear a seat belt when he rides in a car. How would you explain to Andy the fallacy behind relying on this anecdotal evidence?

4.13 Census every 10 years?: Give at least two reasons why you think the U.S. takes a census only every 10 years. What are reasons for taking the census at all?

⌨ **4.14 Canadian census:** Using the Internet, find the date that the Canadian census was first conducted. Find the Canadian population size then and now according to the most recent Canadian census.

4.2 WHAT ARE GOOD WAYS AND POOR WAYS TO SAMPLE?

The sample survey is a very common type of observational study. The first step of a sample survey is to define the population targeted by the study. Sometimes, this is straightforward. For instance, the Gallup organization (www.gallup.com) conducts a monthly survey of about 1000 adult Americans to report the percentage of those sampled who respond "approve" when asked, "Do you approve or disapprove of the way [the current president of the United States] is handing his job as president?" The population consists of all adults living in the U.S.

Other times, you must specify guidelines to identify the population. For instance, suppose a sports magazine plans to survey the opinions of basketball fans about ways to improve professional basketball. Then it might target the population of people of age 18 and older who watch at least three pro basketball games each season.

Sampling Frame and Sampling Design

Once you've identified the population, the second step is to compile a list of subjects in it, so you can sample from it. This list is called the **sampling frame**.

Definition: Sampling frame

The **sampling frame** is the list of subjects in the population from which the sample is taken.

Ideally, the sampling frame lists the entire population of interest. In practice, as in a census, it's usually hard to identify every subject in the population.

Suppose you plan to sample students at your school, asking students to complete a questionnaire about various issues. The population is all students at the school. One possible sampling frame is the student directory. Another one is a list that the registrar keeps of registered students.

Once you have a population and a sampling frame, you need to specify a method for selecting subjects from the sampling frame. The method you use is called the **sampling design**.

Here's one possible sampling design for sampling students at your school: You sample all students in your statistics class. It's easy to conduct such a sample survey, because the class is a readily available sample. But do you think your class is necessarily reflective of the entire student body? Do you have a representative mixture of freshmen through seniors, males and females, athletes and nonathletes, working and nonworking students, political party affiliations, and so on? With this sampling design, it's doubtful.

When you pick a sample merely by convenience, the results may not be representative of the population. Some response outcomes may occur much more than in the population, and some may occur much less. Consider a survey question about the number of hours a week that you work at an outside job. If your class is primarily juniors and seniors, they may be more likely to work than freshmen or sophomores. The mean or median response in the sample may be much larger than in the overall student population. Information from the sample may be misleading.

Simple Random Sampling

You're much more likely to obtain a representative sample if you let *chance*, rather than *convenience*, determine the sample. The sampling design should give each student an equal chance to be in the sample. It should also enable the data analyst to figure out how likely it is that descriptive measures (such as sample means) fall close to corresponding values we'd like to make inferences about for the entire population. These are reasons for using **random sampling**.

You have probably many times been part of a selection process in which your name was put in a box or hat along with many other names, and someone blindly picked a name for a prize. If the names were thoroughly mixed before the selection, this emulates a *random* type of sampling, called a **simple random sample**.

Add photo of someone drawing a paper slip from a hat

RECALL

As in Chapters 2 and 3, *n* denotes the number of observations in the sample, called the **sample size**.

Definition: Simple Random Sample

A **simple random sample** of *n* subjects from a population is one in which each possible sample of that size has the same chance of being selected.

A *simple random sample* is often just called a **random sample**. The "simple" adjective distinguishes this type of sampling from more complex random sampling designs presented in Section 4.4.

EXAMPLE 4: SAMPLING CLUB OFFICERS FOR A NEW ORLEANS TRIP

Picture the Scenario

A campus club can select two members to attend the club's annual conference in New Orleans, with all expenses paid by the national organization. The club members decide that two of the five club officers should attend. The five officers are President (P), Vice-President (V), Secretary (S), Treasurer (T), and Activity Coordinator (A). They make up the population and the sampling frame. The club members want the selection process to be fair, so they decide to select a simple random sample of size *n* = 2. The five names are written on identical slips of paper, placed in a hat, mixed, and then a neutral person blindly selects two slips from the hat.

Questions to Explore
a. What are the possible samples of two officers?
b. What is the chance that a particular sample of size 2 will be drawn?
c. What is the chance that the Activity Coordinator will be chosen?

Think It Through
a. Using the letters that denote each officer, the possible samples of two officers are:

(P,V)	(P,S)	(P,T)	(P,A)	(V,S)
(V,T)	(V,A)	(S,T)	(S,A)	(T,A)

b. There are 10 possible samples. This process of blindly selecting two slips from a hat ensures that each sample has an equal chance of occurring. Since there are 10 possible samples, the chance of any one sample being selected by randomly drawing slips is 1 out of 10.

c. The Activity Coordinator (A) occurs in 4 of the samples. So the chance that the Activity Coordinator is selected for a free trip is 4 out of 10, or 2/5.

Insight
The chance of 2/5 for the Activity Coordinator is the same as for the President or any other officer. In practice, especially with larger populations, it is difficult to mix the slips so that each sample truly has an equal chance of selection. There are better ways of selecting simple random samples, as we'll learn next.

♦

To practice this concept, try Exercise 4.15.

How Can We Select a Simple Random Sample?

<table>
<tr><td>

Did You Know:

The *random number applet* at www.prenhall.com/TBA can also be used to generate random numbers.

</td><td>

What's a better way to take a simple random sample than blindly drawing slips of paper out of a hat? You first number the subjects in the sampling frame. You then generate a set of these numbers randomly. Finally, you sample the subjects whose numbers were generated.

</td></tr>
</table>

One way you can generate numbers randomly is using a **random number table**. This is a table containing a sequence of digits such that any particular digit is equally likely to be any of the numbers 0, 1, 2, …, 9 and does not depend on the other numbers generated. If a particular digit is a 6, for instance, the next digit is just as likely to be a 6 (or a 0) as any other number. The numbers fluctuate according to no set pattern. Table 4.1 shows part of a random number table.

Table 4.1: A Portion of a Table of Random Numbers

Line/Col.	(1)	(2)	(3)	(4)	(5)	(6)	(7)	(8)
1	10480	15011	01536	02011	81647	91646	69179	14194
2	22368	46573	25595	85393	30995	89198	27982	53402
3	24130	48360	22527	97265	76393	64809	15179	24830
4	42167	93093	06243	61680	07856	16376	39440	53537
5	37570	39975	81837	16656	06121	91782	60468	81305

Using Random Numbers to Select a Simple Random Sample

To select a simple random sample,
- Number the subjects in the sampling frame, using numbers of the same length (number of digits).
- Select numbers of that length from a table of random numbers or using software or a calculator with a random number generator.
- Include in the sample those subjects having numbers equal to the random numbers selected.

EXAMPLE 5: AUDITING THE ACCOUNTS OF A SCHOOL DISTRICT

Picture the Scenario

Local school districts must be prepared for annual visits from state auditors whose job is to verify that accounts within the school district are in financial compliance. The auditors must verify the actual dollar amount of the accounts and verify that the money is being spent appropriately. A school district may have many accounts, sometimes 100 or more. It is too time-consuming and expensive for auditors to review all accounts, so they typically review only some of them. So that a school district cannot anticipate which accounts will be reviewed, the auditors often take a simple random sample of the accounts.

Questions to Explore

a. How can the auditors use random numbers to select 10 accounts to audit in a school district that has 60 accounts?

b. Why is it important for the auditors not to use personal judgment in selecting the accounts to audit?

Think It Through

a. The sampling frame consists of the 60 accounts. We first number the accounts 01 through 60. In a random number table, we select two digits at a time until we have 10 different two-digit numbers between 01 and 60. (We number the accounts 01 through 60 rather than 1 through 60 because we select two-digit numbers from the table.) Any pair of random digits has the same chance of selection.

Choose a random starting place in the table. We will illustrate by starting with row 1, column 1 of Table 4.1. The random numbers in this row are

　　10480　15011　01536　02011　81647　91646　69179　14194

The first 10 two-digit numbers are:

　　10　48　01　50　11　01　53　60　20　11

The first account chosen is the one numbered 10. The second chosen is the one numbered 48, and so forth. We observe that 01 and 11 occur twice. We discard the two repeats, and we need two more numbers between 01 and 60. The next ten twodigit numbers are:

　　81　64　79　16　46　69　17　91　41　94

We skip the numbers 81, 64, 79, since no account in the sampling frame of 60 accounts has an assigned number that large. The last two accounts sampled are those numbered 16 and 46. In summary, the auditors should audit the accounts numbered 01, 10, 11, 16, 20, 46, 48, 50, 53, and 60.

b. By using simple random sampling, the auditors hold the school district responsible for all accounts. If the auditors personally chose the accounts to audit, they may tend to select certain accounts each year. A school district would soon learn which accounts they need to have in order and which accounts can be given less attention.

Likewise, the Internal Revenue Service (IRS) randomly selects tax returns for audit, so taxpayers cannot predict ahead of time whether they will be audited.

Insight
The auditors may be concerned that a simple random sample will not guarantee enough "large" accounts to be audited. Section 4.4 introduces a modification of simple random sampling that can ensure a more balanced representation of large and small accounts.

♦

To practice this concept, try Exercise 4.16.

IN PRACTICE

You can obtain **random numbers** on the Internet (see, for instance, random.org, or the *random number applet* at www.prenhall.com/TBA) or you can use a random integer function with a statistical calculator or software to generate random numbers. You can instruct Minitab to generate randomly any number of integers you would like between a minimum of 01 and maximum of 60, as shown in Figure 4.1, or other ranges depending on the numbers you've assigned to subjects in the sampling frame. These days, such options are used more often than random number tables.

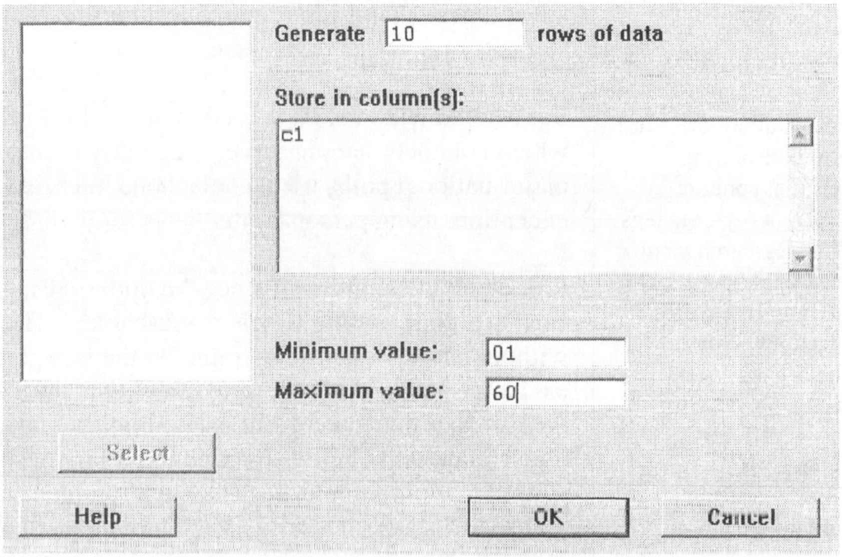
Generate
10
rows of data
Store in column(s):
c1
Minimum value:
01
Maximum value:
60
Select
Help
OK
Cancel

Figure 4.1: This Minitab screen is used to randomly generate 10 integers between 01 and 60, inclusively. The 10 random numbers will be scored in column 1 of the Minitab spreadsheet.

Methods of Collecting Data in Sample Surveys

In Example 5, the subjects sampled were accounts. More commonly, they are people. For any polling organization, it is difficult to get a good sampling frame, since a list of all adults in a population typically does not exist. So, to conduct a survey, they must pick a place where almost all people can be found. This is not a shopping mall or a rock concert or a college campus but instead a person's place of residence.

Once we identify the desired sample for a survey, how do we contact the people to collect the data? The three most common methods are:

- Personal interview
- Telephone interview
- Self-administered questionnaire

Picture of the census short form used in 2000

Personal interview
In a personal (face-to-face) interview, an interviewer asks prepared questions and records the subject's responses. An advantage is that subjects are more likely to agree to participate. A disadvantage is the cost. Also, some subjects may not answer

sensitive questions pertaining to opinions and to lifestyle, such as alcohol and drug use, that they might answer on a printed questionnaire.

Telephone interview

A telephone interview is like a personal interview but conducted over the phone. A main advantage is lower cost, since no travel is involved. A disadvantage is that the interview might have to be short. Subjects aren't as patient on the phone and may hang up before starting or completing the interview.

Self-administered questionnaire

Subjects are requested to fill out a questionnaire mailed to them by post or e-mail. An advantage is that it is cheaper than a personal interview. A disadvantage is that more subjects may fail to participate.

So, which method is most used in practice? Let's consider the Gallup Organization, which conducts sample surveys regularly to measure public opinion. They, like most major national polls, use the telephone interview. The General Social Survey is an exception, using personal interviews for their questionnaire, which is quite long.

For telephone interviews, since the mid-1980s, many sample surveys have used **random digit dialing** to select households. The survey can then obtain a sample without having a sampling frame. In the U.S., typically the area code and the 3-digit exchange are randomly selected from the list of all such codes and exchanges. Then, the last four digits are dialed randomly, and an adult is selected randomly within the household. Although this sampling design incorporates randomness, it is not a simple random sample, because each sample is not equally likely to be chosen. (Do you see why not? Does everyone have a telephone, or exactly one telephone?)

How Accurate Are Results from Surveys with Random Sampling?

The most common use of data from sample surveys is to estimate population percentages. For instance, a Gallup poll recently reported that 30% of Americans worried that they might not be able to pay health-care costs during the next 12 months. How good is such a sample estimate? When you read results of surveys, you'll often see a statement such as "The **margin of error** is plus or minus 3 percentage points." This means that it's very likely that the reported sample percentage is no more than 3% lower or 3% higher than the population percentage. So, if Gallup reports that 30% worry about health-care costs, it's very likely that in the entire population the percentage that worry about health-care costs is between about 27% and 33% (that is, within 3% of 30%). "Very likely" means that 95 times out of 100 such statements are correct.

Chapters 6 and 7 will show details about the margin of error and how to calculate it. For now, we'll use a rough approximation for the margin of error. The more precise formulas in Chapters 6 and 7 will give a somewhat smaller value when the estimated percentage is far from 50%. When using a simple random sample of n subjects, the margin of error is approximately

$$\frac{1}{\sqrt{n}} \times 100\%.$$

EXAMPLE 6: FINDING THE MARGIN OF ERROR FOR A GALLUP POLL

Picture the Scenario

In April 2003, there was worldwide concern about the spread of the viral respiratory illness called *severe acute respiratory syndrome* (SARS). It is spread by close person-to-person contact and has symptoms of high fever and difficulty in breathing. The majority of reported cases of SARS were in Asian countries, mainly China, but cases were reported in other countries, including the U.S. and Canada. The World Health Organization (WHO) implemented measures to control its spread, so that an epidemic would not occur. The outbreak was contained, with WHO reporting 8098 cases resulting in 774 deaths.

A Gallup poll conducted for CNN and *USA Today* in three successive weeks monitored Americans' concern about SARS. The poll questioned 1001 adults using telephone interviews. Figure 4.2 shows bar graphs that the Gallup organization used to report results.

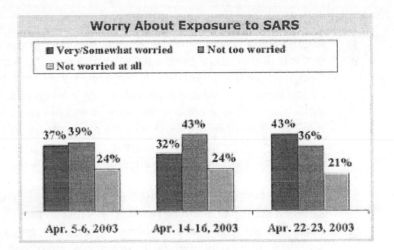

Figure 4.2: Results[4] of Gallup Polls about SARS

Questions to Explore

Find an approximate margin of error for these results. How is it interpreted?

Think It Through

The sample size was $n = 1001$. The Gallup poll was a random sample, but not a simple random sample. It used random digit dialing, which can give results nearly as accurate as with a simple random sample. So the margin of error was approximately

$$\frac{1}{\sqrt{n}} \times 100\% = \frac{1}{\sqrt{1001}} \times 100\% = 0.03 \times 100\% = 3\%.$$

In the final poll, Gallup reported that 43% were somewhat or very worried about exposure to SARS. This refers to the *sample*. The margin of error of 3% suggests

---

[4] Source: http://www.gallup.com/poll/releases/pr030425.asp

that in the *population* of adult Americans, roughly between about 40% and 46% were worried about SARS exposure.

Insight

You may be surprised or skeptical that a sample of only 1001 people out of a huge population (such as 200 million adult Americans) can provide such precise inference. That's the power of random sampling. We'll see *why* this works in Chapter 6.

♦

To practice this concept, try Exercises 4.21 and 4.22 and try using the : "Sampling" applet as described in Exercise 4.124.

Be Wary of Sources of Potential Bias in Sample Surveys.

A variety of problems can cause responses from a sample to tend to favor some parts of the population over others. Then, results from the sample are not representative of the population and are said to exhibit **bias**.

Bias may result from the sampling method. The main way this occurs is if the sample is not random. We'll see examples later in the section. Another way it can occur is due to **undercoverage**--having a sampling frame that lacks representation from parts of the population. A telephone survey will not reach homeless people or prison inmates or other people not having a telephone. If its sampling frame consists of the names in a telephone directory, it will not reach those having an unlisted number. Responses by those who are not in the sampling frame might tend to be quite different from those who are in the frame. Bias resulting from the sampling method, such as nonrandom sampling or undercoverage, is called **sampling bias**.

Sampling bias is merely one type of possible bias in a sample survey. When some sampled subjects cannot be reached or refuse to participate, there is **nonresponse bias.** The subjects who are willing to participate may be different from the overall sample in some way, perhaps having strong, emotional convictions about the issues being surveyed. Even those who do participate may not respond to some questions, resulting in nonresponse bias due to **missing data**. All major surveys suffer from some nonresponse bias. The General Social Survey has a nonresponse rate of about 20-30%. The nonresponse rate is much higher for many telephone surveys. By contrast, government conducted surveys often have lower nonresponse rates. The Current Population Survey, which measures factors related to employment, has a nonresponse rate of only about 7%. To reduce nonresponse bias, investigators try to make follow-up contact with the subjects who do not return questionnaires.

Results have dubious worth when there is substantial nonresponse. For instance, in her book *Women and Love* (Knopf, 1987), author Shere Hite presented results of a survey of adult women in the United States. One of her conclusions was that 70% of women who had been married at least five years have extramarital affairs. She based this conclusion on responses to questionnaires returned from a sample of 4500 women. This sounds impressively large. However, the questionnaire was mailed to about 100,000 women. We cannot know whether this sample of 4.5% of the women who responded is representative of the 100,000 who received the questionnaire, much less the entire population of adult American women.

Besides sampling bias and nonresponse bias, a third type of potential bias is in the actual responses made. This is called **response bias**. An interviewer might ask the questions in a leading way, such that subjects are more likely to respond a certain way. Or, subjects may lie because they think their response is socially unacceptable, or they may give the response that they think the interviewer prefers.

<table>
<tr><td>

Cartoon of a poorly worded question

</td></tr>
</table>

If you design an interview or questionnaire, you should strive to construct questions that are clear and understandable. *Avoid questions that are confusing, long, or leading*. The wording of a question can greatly affect the responses. A Roper Poll was designed to determine the percentage of Americans who express some doubt that the Holocaust occurred in World War II. In response to the question, "Does it seem possible or does it seem impossible to you that the Nazi extermination of the Jews never happened?" Twenty-two percent said it was possible the Holocaust never happened. The Roper organization later admitted that the question was worded in a confusing manner. When they asked, "Does it seem possible to you that the Nazi extermination of the Jews never happened, or do you feel certain that it happened?" only 1% said it was possible it never happened.[5]

Even the order in which questions are asked can dramatically influence results. One study[6] asked, during the Cold War, "Do you think the U.S. should let Russian newspaper reporters come here and send back whatever they want?" and "Do you think Russia should let American newspaper reporters come in and send back whatever they want?" For the first question (about Russian reporters), the percentage of yes responses was 36% when it was asked first and 73% when it was asked second.

Types of Bias in Sample Surveys

Sampling bias occurs from using nonrandom samples or having undercoverage.

Nonresponse bias occurs when some sampled subjects cannot be reached or refuse to participate or fail to answer some questions.

Response bias occurs when the subject gives an incorrect response (perhaps lying), or the question wording or the way the interviewer asks the questions is confusing or misleading.

Volunteer and Other Convenience Samples: Poor Ways to Sample

Many surveys are conducted using **convenience** samples--samples that are easy to obtain. Have you ever been stopped on the street or at a shopping mall to participate in a survey? It is easy for the interviewer to obtain data relatively cheaply. But the sample is unlikely to be representative of the population. Severe biases may result because of the time and location of the interview and the judgment of the interviewer about whom to interview. For example, working people might be underrepresented if

---

[5] *Newsweek*, July 25, 1994.
[6] Described in Crossen (1994).

the interviews are conducted on workdays between 9 a.m. and 5 p.m. Poor people may be underrepresented if the interviewer conducts interviews at an upscale shopping mall. Beware of convenience samples. They usually give a biased view of the population of interest.

Have you ever answered a survey you've seen posted on the Internet, such as at the home page for a news organization? A sample of this type, called a **volunteer sample**, is the most common type of convenience sample. As the name implies, subjects volunteer for the sample. People who volunteer to participate may have opinions that are not representative of the entire population. Volunteer samples introduce sampling bias into the results. For instance, a survey by the Pew Research Center (1/6/2003) estimated that 46% of Republicans said they like to register their opinions in online surveys, compared with only 28% of Democrats. (Why? One possible lurking variable is personal income: On the average, Republicans are weathier than Democrats, and wealthier people are more likely to have access to and use the Internet.) Thus, results of online surveys may be more weighted in the direction of Republicans' beliefs (for instance, more conservative) than the general population.

Convenience samples are undesirable. Sometimes, however, they are necessary, both in observational studies and in experiments. This is often true in medical studies. Suppose we want to investigate how well a new drug performs compared to a standard drug, for subjects who suffer from severe mental depression. We're not going to be able to find a sampling frame of all depression patients and take a simple random sample of them. We may, however, be able to sample depression patients at certain medical centers. Even then, randomization should be used wherever possible. For the study patients, we can randomly select who is given the new drug and who is given the standard one.

Some samples are poor not only because of their convenience but also because they use an inappropriate sampling frame. An example of this was a poll in 1936 to predict a Presidential election.

EXAMPLE 7: TAKING A POOR SAMPLE--THE *LITERARY DIGEST* POLL

Picture the Scenario

The magazine called *Literary Digest* conducted a poll to predict the result of the 1936 Presidential election between Franklin Roosevelt (Democrat and incumbent) and Alf Landon (Republican). At the time, their poll was famous, because they had correctly predicted three successive elections. In 1936 they mailed questionnaires to 10 million people and asked how they planned to vote. The sampling frame was constructed from telephone directories, country club memberships, and automobile registrations. Approximately 2.3 million people returned the questionnaire. The *Digest* predicted that Landon would win, getting 57% of the vote. Instead, Landon actually got only 36%, and Roosevelt won in a landslide.

Questions to Explore

a. What was the population?

b. How could the *Literary Digest* poll make such a large error, especially with such a huge sample size? What type of bias could have occurred in this poll?

Think It Through

a. The population was all registered voters in the U.S. in 1936.

b. This survey had two severe problems:

- *Sampling bias due to undercoverage of the sampling frame and a nonrandom sample*: In 1936, the U.S. was in the Great Depression. Those who had cars and country club memberships and thus received questionnaires tended to be relatively wealthy. The wealthy tended to be primarily Republican, the political party of Landon. Many potential voters were not on the lists used for the sampling frame. There was also no guarantee that a subject in the sampling frame was a registered voter.
- *Nonresponse bias*: Of the 10 million people who received questionnaires, 7.7 million did not respond. As might be expected, those individuals who were unhappy with the incumbent (Roosevelt) were more likely to respond.

Insight

For this same election, a pollster who was getting his new polling agency off the ground surveyed 50,000 people and predicted that Roosevelt would win. Who was this pollster? George Gallup. The Gallup Organization is still with us today. However, the *Literary Digest* went out of business soon after the 1936 election.

♦

To practice this concept, try Exercise 4.25.

A Large Sample Size Does Not Guarantee an Unbiased Sample

Many people think that as long as the sample size is large, it doesn't matter how the sample was selected. This is incorrect, as illustrated by the *Literary Digest* poll. A sample size of 2.3 million did not prevent poor sample results. Many Internet surveys have thousands of respondents, but a volunteer sample of thousands is not as good as a random sample, even if that random sample is much smaller. *We're almost always better off with a simple random sample of 100 people than with a volunteer sample of thousands of people.*

Summary: Key Parts of a Sample Survey

- Identify the **population** of all the subjects of interest.
- Construct a **sampling frame**, which attempts to list all the subjects in the population.
- Use a **random sampling design**, implemented using random numbers, to select *n* subjects from the sampling frame.
- Be cautious about **sampling bias** due to nonrandom samples (such as volunteer samples) and sample undercoverage, **response bias** from subjects not giving their true response or from poorly worded questions, and **nonresponse bias** from refusal of subjects to participate.

SECTION 4.2: PRACTICING THE BASICS

4.15 Choosing officers: In Example 4, "Sampling from a Population of Club Officers," two of the officers denoted by (P, V, S, T, A) were selected to attend a conference in New Orleans. Suppose instead that *three* officers can get fully paid trips. For a simple random sample of three of the five officers:
a. Show all the possible samples.
b. What is the chance that a particular sample of size three will be drawn?
c. What is the chance that the Activity Coordinator will be chosen?

4.16   Sample students: A class has 50 students. Use the second row of digits in the random number table (Table 4.1) to select a simple random sample of three students. If the students are numbered 01 to 50, what are the numbers of the three students selected?

4.17 Random number applet: Use the random number applet at www.prenhall.com/??? or use software to (a) answer the previous question, (b) select a simple random sample of three students out of a class of 500 students.

4.18 Auditing accounts - applet: Use the *Random Number* applet at www.prenhall.com/TBA to select 10 of the 60 school district accounts described in Example 5. Explain how you did this, and identify the accounts that will be in the sample.

4.19 Auditing accounts – software: Use either statistical software or a statistical calculator to select 10 of the 60 school district accounts described in Example 5. Explain how you did this, and identify the accounts that will be in the sample.

4.20 Sampling from a directory: A local telephone directory has 50,000 names, 100 a page for 500 pages. Show how you could choose a simple random sample of 10 names. Explaining all steps of how you found and used random numbers, select 10 numbers to identify subjects for the sample.

4.21 Comparing polls: The table shows the result of the 2000 Presidential election along with the vote predicted by several organizations in the days before the election.

The sample sizes were typically about 1000 to 2000 people. The percentages for each poll do not sum to 100 because of voters who indicated they were undecided.
a.   Treating the sample sizes as 1000 each, find the approximate margin of error.
b.   Do most of the predictions fall within the margin of error of the actual vote percentages? Considering the relative sizes of the sample and the population and the fact that some people were undecided, would you say that these polls had good accuracy?

	Predicted Vote		
Poll	Gore	Bush	Nader
Gallup/CNN/USA	46	48	4
Harris	47	47	5
ABC/Wash Post	45	48	3
CBS	45	44	4
NBC/WSJ	44	47	3
Pew Research	47	49	4
Actual vote	**48**	**48**	**3**

Source: *www.ncpp.org/1936-2000.htm*

4.22 Margin of error and n: The Gallup poll in Example 6 reported that 43% of Americans worried that they would be exposed to SARS . Find the approximate margin of error if (a) $n = 100$, (b) $n = 400$, (c) $n = 1600$. Explain how the margin of error changes as n increases.

4.23 Bias due to perceived race: A political scientist at the University of Chicago studied the effect of the race of the interviewer.[7] Following a phone interview, respondents were asked whether they thought the interviewer was black or white (all were actually black). Perceiving a white interviewer resulted in more conservative opinions. For example, 14% agreed that "American society is fair to everyone" when they thought the interviewer was black, but 31% agreed to this statement when posed by an interviewer that the respondent thought was white. Which type of bias does this illustrate: Sampling bias, nonresponse bias, or response bias? Explain.

4.24 Confederates: Some southern states in the U.S. have wrestled with the issue of a state flag that is sensitive to African Americans and not divisive. Suppose a survey asks, "Do you oppose the present state flag that contains the Confederate symbol, a symbol of past slavery in the South and a flag supported by extremist groups?"
a.   Explain why this is an example of a leading question.
b.   Explain why a better way to ask this question would be, "Do you favor or oppose the current state flag containing the Confederate symbol?"

4.25 Call in your opinion: Many TV news and entertainment programs ask viewers to offer their opinions on an issue of the moment by calling an 800 or 900 phone number or voting on the Internet. For instance, one night the ABC program *Nightline* asked viewers whether the United Nations should continue to be located in the United States. Of more than 186,000 callers, 67% wanted the United Nations out of the United States. At the same time, a scientific poll using a random sample of about 500 respondents estimated the true percentage wanting the United Nations out of the United States to be about 28% (D. Horvitz et al., *Chance*, vol. 8, pp. 16-25, 1995).

---

[7]   Study by Lynn Sanders, as reported by *Washington Post*, June 26, 1995.

a.   The random sample is much smaller. Do you think it is more trustworthy, or less trustworthy, than the results of the call-in poll for *Nightline*? Explain your reasoning.

b.   Is the call-in poll a random sample? If not, what type of sample is it?

4.26 Job market for MBA students: An article from the Feb 2, 2003 *Atlanta Journal Constitution* about the bleak job market for MBA students graduating in 2003 described an opinion survey conducted by a graduate student at a major state university. The student polled 1500 executive recruiters, asking their opinions on the industries most likely to hire. He received back questionnaires from 97 recruiters, of whom 54 indicated that health care was the industry most likely to see job growth.

a.   What is the population for this survey?

b.   What was the intended sample size? What was the sample size actually observed? What was the percentage of nonresponse?

c.   Describe two potential sources of bias with this survey.

4.27   Gun control: More than 75% of Americans answer *yes* when asked, "Do you favor cracking down against illegal gun sales?" but more than 75% say *no* when asked, "Would you favor a law giving police the power to decide who may own a firearm?"

a.   Which statistic would someone who opposes gun control prefer to quote?

b.   Explain what is wrong with the wording of each of these statements.

4.28 Stock market associated with poor mental health: An Internet survey of 545 Hong Kong residents suggested that close daily monitoring of volatile financial affairs may not be good for your mental health (*J. Social and Clinical Psychology* 2002;21:116-128). Subjects who felt that their financial future was out of control had the poorest overall mental health, whereas those who felt control over their financial future had the best mental health.

a.   What is the population of interest for this survey?

b.   Describe why this is an observational study.

c.   Briefly discuss the potential problems with the sampling method used and how these problems could affect the survey results.

4.29   "What rots beneath": This was the headline in a *New York Times* article (May 19, 2003). The Gowanus Canal in Brooklyn has become famous for its contamination from sewage and industrial waste. Scientists and Army Corps of Engineers technicians have used augurs, drills, and a split-spoon, which sucks up the muck of the canal bottom, to analyze what is 'living' in the canal.

a.   Describe the population under study.

b.   Explain why a census is not practical for this study. What advantages does sampling offer?

4.30 Are outgoing people good employees?: *Yahoo News* (June 28, 2002) reported that a new study suggests that the most outgoing employees tend to be the biggest troublemakers. The study looked at 105 employees of one private company whose work was mostly projectbased. Employees completed questionnaires that measured for extroversion and conscientiousness. It also conducted personal interviews with the supervisors about employees' counterproductive behavior. The study concluded that whereas extroverted, conscientious employees were least likely to be rated counterproductive, the less-than-conscientious extroverts stirred up the most trouble. Introverts, on the other hand, tended not to get into much trouble.

a.   What data collection methods were used by the researcher to obtain information from the employees and supervisors?

b.   What are potential biases that could occur with the methods described in (a)?

c.   This survey is an example of an observational study. What are the explanatory and response variables?

d.   Do you feel comfortable with generalizing the conclusion made based on this study to the population of all employees for all companies in the country in which this study was conducted? Why or why not?

4.31 Identify the bias: A newspaper in Los Angeles designs a survey to estimate the proportion of the city's adult residents who favor a proposal to legalize casinos in the city. It takes a list of the 1000 people who have subscribed to the paper the longest, and sends each of them a questionnaire that asks, "Do you think it is a good idea to legalize casinos in Los Angeles, which will broaden the tax base and contribute money to education?" After analyzing results from the 50 people who reply, they report that 90% of the local citizens are in favor of the proposal. Identify the bias that results from:

a.   Sampling bias due to undercoverage.

b.   Sampling bias due to the sampling design.

c.   Nonresponse bias.

d.   Response bias due to the way the question was asked.

4.32 Types of bias: Give an example of a survey that would suffer from:

a.   Sampling bias due to the sampling design.

b.   Sampling bias due to undercoverage.

c.   Response bias.

d.   Nonresponse bias

4.3: WHAT ARE GOOD WAYS AND POOR WAYS TO EXPERIMENT?

We've just learned about good ways and poor ways to gather a sample in an observational study. Likewise, there are good ways and poor ways to conduct an experiment. We now discuss key aspects of planning an experiment well.

First, let's recall from Section 4.1 what's meant by an experiment: We assign each subject to an experimental condition, called a **treatment**. We then observe the outcome on the response variable. The goal of the experiment is to investigate the association--how the treatment affects the response. An advantage of an experiment over an observational study is that it provides stronger evidence for causation.

In an experiment, the subjects are sometimes called **experimental units**. This is merely to emphasize that the objects measured need not be human beings. They could, for example, be schools, stores, mice, or computer chips.

The Australian study described in Example 1 conducted an experiment on 200 mice. The mice were the experimental units (the subjects). The explanatory variable was "whether received radiation." The treatments were its categories, *radiation* and *no radiation*. The researchers assigned 100 mice to each treatment. After 18 months, the mice were examined for brain tumors. Whether a mouse developed brain tumors was the response variable.

The Elements of a Good Experiment

Let's consider another example to help us learn what makes a good experiment. It is common knowledge that smoking is a difficult habit to break. Studies have reported that regardless of what smokers do to quit, most relapse within a year. Some scientists have suggested that smokers are less likely to relapse if they take an antidepressant regularly after they quit. How can you design an experiment to study whether antidepressants help smokers to quit?

For this type of study, as in most medical experiments, it is not feasible to randomly sample the population (all smokers who would like to quit). It is necessary to use a convenience sample. For instance, a medical center might advertise to attract volunteers from the smokers in a community who would like to try to quit.

Control comparison group

Suppose you have 400 volunteers who would like to quit smoking. You could ask them to quit, starting today. You could have each start taking an antidepressant, and then a year from now check how many have relapsed. Perhaps 42% of them would relapse. But this is not enough information.[8] You need to be able to compare this to the percentage who would relapse if they were *not* taking the antidepressant.

An experiment normally has a primary treatment of interest, such as receiving an antidepressant. But it should also have a second treatment for comparison to help you analyze the effectiveness of the primary treatment. So the volunteers should be split into two groups: One group receives the antidepressant, and the other group does not. You could give the second group a pill that looks like the antidepressant but that does not have any active ingredient--a placebo. This second group is called the **control group.** After subjects are assigned to the treatments and observed for a certain period of time, the relapse rates are compared.

Why bother to give the placebo to the control group, if the pill doesn't contain any active ingredient? This is partly so that the two treatments appear identical to the subjects. (As we'll discuss below, subjects should not know which treatment they are receiving.) This is also because people who take a placebo tend to respond better than those who receive nothing, perhaps for psychological reasons. This is called the **placebo effect.** For instance, of the subjects not receiving the antidepressant, perhaps 75% would relapse within a year, but if they received a placebo pill perhaps 55% of them would relapse. When you compare this to a relapse rate of 42% for subjects who received antidepressants, it makes a big difference how you define the second treatment.

---

[8] Chapter 11 explains another reason a control group is needed. The "regression effect" implies that, over time, poor subjects tend to improve and good subjects tend to get worse, in relative terms.

In some experiments, a control group may receive an existing treatment rather than a placebo. For instance, a smoking cessation study might analyze whether an antidepressant works better than a nicotine patch in helping smokers to quit. It may not be necessary to include a placebo group if the nicotine patch has already been shown in previous studies to be more effective than a placebo. Or the experiment could compare all three treatments: antidepressants, nicotine patch, and a placebo.

Randomizing in an experiment

In a smoking cessation experiment, how should the 400 study subjects be assigned to the treatment groups? Should you personally decide which treatment each subject receives? This could result in bias. If you are conducting the study to show that the antidepressant is effective, you might consciously or subconsciously place smokers who you believe will be more likely to succeed into the group that receives the antidepressant.

It is better to use **randomization** to assign the subjects: Pick a simple random sample of 200 of the 400 subjects to receive the antidepressant. The other 200 subjects will form the control group. Randomization helps to prevent bias from one treatment group tending to be different from the other in some way, such as having better health or being younger. In using randomization, we attempt to *balance the treatment groups* by making them similar with respect to their distribution on potential lurking variables. This enables us to attribute any difference in their relapse rates to the treatments they are using, not to lurking variables or to researcher bias.

You might think you can do better than randomization by using your own judgment to assign subjects to the treatment groups. For instance, when you identify a potential lurking variable, you could try to balance the groups according to its values. If age is that variable, every time you put someone of a particular age in one treatment group, you could put someone of the same age in the other treatment group. There are two problems with this. First, *many* variables are likely to be important, and it is difficult to balance groups on *all* of them at once. Second, you may not have thought of other relevant lurking variables. Even if you can balance the groups on the variables you identified, the groups could be unbalanced on these other variables, causing the overall results to be biased in favor of one treatment.

When you hear about new research findings, you can feel more confident about their worthiness if they come from a randomized experiment with a control group than if they come from an experiment without a control group or from an observational study. An analysis[9] of published medical studies about treatments for heart attacks indicated that the new therapy provided improved treatment 58% of the time in studies without randomization and control groups but only 9% of the time in studies having randomization and control groups. Although this does not prove anything (as this analysis itself was an observational study), it does suggest that studies conducted without randomization or other ways to reduce bias produce results that tend to be overly optimistic.

In summary, use randomization to assign subjects to the treatments:
- To eliminate bias that may result if you (the researchers) assign the subjects.

---

[9] See Crossen (1994), p. 168.

- To balance the groups on variables that you know affect the response.
- To balance the groups on lurking variables that may be unknown to you.

Blinding the study

It is important that the treatment groups be treated as equally as possible. Ideally, the subjects are **blind** to the treatment to which they are assigned. In the smoking cessation study, the subjects should not know whether they are taking the antidepressant or a placebo. Whoever has contact with the subjects during the experiment, including the data collectors who record the subjects' response outcomes, should also be blind to the treatment information. Otherwise they could intentionally or unintentionally provide extra moral support to one group. When neither the subject nor those having contact with the subject know the treatment assignment, the study is called **double-blind.** That's ideal.

EXAMPLE 8: DESIGN THE STUDY TO ASSESS ANTIDEPRESSANTS FOR QUITTING SMOKING

Picture the Scenario
We've mentioned that some scientists believe it helps smokers to quit smoking if they take antidepressants. To investigate this, a study[10] followed 429 men and women who were 18 or older and had smoked 15 cigarettes or more per day for the previous year. The subjects were highly motivated to quit and in good health. They were assigned to one of two groups: One group took 300 mg daily of bupropion, an antidepressant that has the brand name Zyban. The other group did not take an antidepressant. At the end of a year, the study observed whether each subject had successfully abstained from smoking or had relapsed.

Questions to Explore
a. Identify the response and explanatory variables, treatments, and experimental units.
b. How should the researchers assign the subjects to the two treatment groups?
c. Without knowing more about this study, what would you identify as a potential problem with the study design?

Think It Through
a. This experiment has:

Response variable: Whether the subject abstains from smoking for 1 year
Explanatory variable: Whether the subject received Zyban™ (yes or no)
Treatments: Zyban, no Zyban
Experimental units: The 429 volunteers who are the study subjects

b. The researchers should randomize to assign subjects to the two treatments. They could use random numbers (Table 4.1, Random Number Applet, or software) to select a simple random sample of 215 (half) of the subjects to form the group that uses Zyban. The procedure would be:
- Number the study subjects from 001 to 429.

---

[10] *Annals of Internal Medicine* 2001; 135:423-33

- Pick a three-digit random number between 001 and 429. If the number is 392, then the subject numbered 392 is put in the Zyban group.
- Continue to pick three-digit numbers until you've picked 215 distinct values between 001 and 429. This determines the 215 subjects who will receive Zyban.

c. The description of the experiment in "Picture the Scenario" did not say whether the subjects who did not receive Zyban were given a placebo or whether the study was blinded. If not, these would be potential sources of bias for the study.

Insight

In the actual reported study, the subjects were randomized to receive Zyban or a placebo for 45 weeks. The study *was* double-blinded. At the end of 1 year, 55.1% of the subjects receiving Zyban were not smoking, compared with 42.3% in the placebo group. After 18 months, 47.7% if the Zyban subjects were not smoking compared to 37.7% for the placebo subjects. However, after 2 years, the percentage of non smokers was similar for the two groups, 41.6% versus 40%.

♦

To practice this concept, try Exercise 4.28.

DID YOU KNOW?	**IN PRACTICE**
A randomized experiment comparing medical treatments is often referred to as a **clinical trial**.	In medicine, the randomized experiment (clinical trial) has become the "gold standard" for evaluating new medical treatments. The Cochrane Collaboration (www.cochrane.org) is an organization devoted to synthesizing evidence from medical studies all over the world. According to this organization, there have now been hundreds of thousands of randomized experiments comparing medical treatments (Senn, 2003, p. 68). In most countries, a company cannot get a new drug approved for sale unless it has been tested in a randomized experiment.

Sample Size and Statistical Significance

Now suppose you've conducted an experiment, and it's a year later. You find that 44% of the subjects taking Zyban have relapsed, whereas 55% of the control group relapsed. Can you conclude that Zyban was effective in helping smokers quit?

RECALL

For simple random sampling, the margin of error is approximately

$$\frac{1}{\sqrt{n}} \times 100\%,$$ which is 10% for $n = 100$, 7% for $n = 200$, and 3% for $n = 1000$.

Not quite yet. You must convince yourself that this difference between 44% and 55% cannot be explained by the variation that occurs naturally just by chance. Even if the effect of Zyban is no different from the effect of placebo, the sample relapse rates would not be exactly the same for the two groups. Just by ordinary chance, in the random assignment of subjects to treatment groups, on the average one group may be slightly more committed to quitting smoking than the other group.

In a randomized experiment, the variation that could be expected to occur just by chance alone is roughly like the margin of error with simple random sampling. So if a treatment has $n = 215$ observations, this is about $(1/\sqrt{215}) \times 100\% = 7\%$ for a percentage. If the population percentage of people who relapse is 50%, the sample percentage of 215 subjects who relapse is very likely to fall between 43% and 57% (that is, within 7% of 50%). The difference in a study between 44% relapsing and 55% relapsing could be explained by the ordinary variation expected for this size of sample.

By contrast, if each treatment had $n = 1000$ observations, the ordinary variation we'd expect due to chance is only about 3% for each percentage. Then, the difference between 44% and 55% could not be explained by ordinary variation: There is no plausible common percentage for the two treatments such that 44% and 55% are both within 3% of its value. Zyban would truly seem to be better than placebo. The difference expected due to ordinary variation is smaller with larger samples. You can be more confident that sample results reflect a true effect when the sample size is large than when it is small. Obviously, we cannot learn much by using only $n = 1$ subject for each treatment. The process of assigning *several* experimental units to each treatment is called **replication**.

When the difference between the results for the two treatments is so large that it would be rare to see such a difference by ordinary variation (for instance, 55% relapse vs. 44% relapse when $n = 1000$ for each group), we say that the results are **statistically significant**. We can then conclude that, in the population, the response truly depends on the treatment. How to determine this and how it depends on the sample size are topics we'll study in later chapters.[11] Suffice it to say for now, the larger the sample size, the better.

When a study has statistically significant results, it's still possible that the observed effect was merely due to chance. Even if the treatments are identical, one may seem better just because, by sheer luck, random assignment gave it many subjects who tend to respond better, perhaps being healthier on the average. For this reason, another type of replication is also important. You can feel more confident if other researchers perform similar experiments and get similar results.

Generalizing Results to Broader Populations

We've seen that random samples are much preferable to convenience samples, yet convenience samples are the only realistic way to conduct many experiments. When an experiment uses a convenience sample, be cautious about the extent to which results generalize to a larger population. Look carefully at the characteristics of the sample. Do they seem representative of the overall population? If not, the results have dubious worth.

Many medical studies use volunteers at *several* medical centers to try to obtain a broad cross-section of subjects. But some studies mistakenly try to generalize to a broader population than the one from which the sample was taken. A psychologist may conduct an experiment using a sample of students from an introductory psychology course. For the results to be of wider interest, however, the psychologist might claim that the conclusions generalize to *all* college students, to all young adults, or even to all adults. Such generalizations may well be wrong, since the

---

[11] It's a bit more complicated than the reasoning shown above. The margin of error for comparing two percentages differs from the one for estimating a single percentage.

sample may differ from those populations in fundamental ways, such as in average socioeconomic status or the distribution of race or gender.

Carefully assess the scope of conclusions in research articles, the mass media, and advertisements. Evaluate critically the basis for the conclusions by noting the experimental design or the sampling design upon which the conclusions are based.

Summary: Key Parts of a Good Experiment

- A good experiment has a **control comparison group**, **randomization** in the assignment of experimental units to treatments, **blinding**, and **replication**.
- The **experimental units** are the subjects--the people, animals, or other objects to which the treatments are applied.
- The **treatments** are the experimental conditions imposed on the experimental units. One of these may be a **control** (for instance, a placebo) that provides a basis for determining if a treatment is truly effective. The treatments correspond to values of an explanatory variable.
- **Randomize** in assigning the experimental units to the treatments. This tends to balance the comparison groups with respect to lurking variables.
- **Replicating** the treatments on many experimental units helps, so that observed effects are not due to ordinary variability but instead are due to the treatment. Repeat studies to increase confidence in the conclusions.

IN WORDS

A **factor** is a categorical explanatory variable (such as whether the subject takes an antidepressant) having experimental conditions (the **treatments**, such as Zyban or no Zyban) as categories of that variable.

Multifactor Experiments

Categorical explanatory variables in an experiment are often referred to as **factors**. The experiment described in Example 8 had a factor measuring whether a subject used the antidepressant called Zyban.

Many experiments have more than one factor. Suppose the researchers also wanted to use a second factor with two treatments, using a nicotine patch versus not using one. The experiment could then have four treatment groups that result from cross classifying the two factors. See Table 4.2. The four treatments are Zyban alone, nicotine patch alone, Zyban and nicotine patch, neither Zyban nor nicotine patch.

Table 4.2: An Experiment Can Use Two (or More) Factors at Once. The treatments are the combinations of categories of the factors.

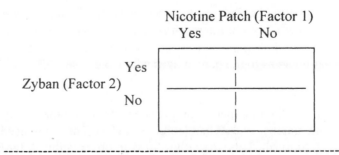

Why use two factors in an experiment? Why not do one experiment about Zyban and a separate experiment about the nicotine patch? The reason is that we can learn more from a two-factor experiment. For instance, the combination, using both a nicotine patch and Zyban, may be more effective than using either method alone.

EXAMPLE 9: DO ANTIDEPRESSANTS AND/OR NICOTINE PATCHES HELP SMOKERS QUIT?

Picture the Scenario

In the previous example, we analyzed a study about whether Zyban helps a smoker quit cigarettes. Let's now consider nicotine patches as another possible cessation aide.

Question to Explore

How can you design a single study to investigate the effects of nicotine patches and Zyban on whether a subject relapses into smoking?

Think It Through

You could use the two factors with four treatment groups shown in Table 4.2. Figure 4.3 portrays the design of a randomized experiment, with placebo alternatives to Zyban and to the nicotine patch.

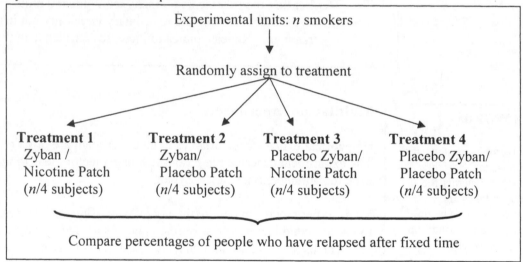

Figure 4.3: Diagram of a Randomized Experiment with Two Factors (Whether Use Zyban, and Whether Use a Nicotine Patch). Question: A 3-factor design could also incorporate whether a subject receives counseling to discourage smoking. How many treatments would such a design have?

The study should be double-blind: Subjects should not know whether or not the pill is active or whether or not the patch is active, and neither should the individuals who evaluate them during the experiment. After a fixed amount of time, you compare the four treatments as to the percentages who have relapsed.

Insight

In fact, a two-factor experiment *has* been conducted that compared the effectiveness of nicotine patches and Zyban (Jorenby, 1999). The study used $n = 893$ smokers with the design described above, with follow-up observations after 6 weeks. A slight departure is that the study used fewer subjects for the double-placebo treatment than the others. Although it is common to make treatment groups similar in size, it is not necessary to have them exactly equal. The most effective treatments were Zyban alone or in combination with a nicotine patch. We will explore the results in more detail in the exercises.

◆

To practice this concept, try Exercise 4.33.

The experiments described in Examples 8 and 9 used **completely randomized designs**: Once the subjects were selected for the experiment, they were randomly assigned to one of the treatments. Example 8 used a one-factor completely randomized design and Example 9 used a two-factor completely randomized design. If we include a third factor of whether or not the subject receives counseling, we would have a three-factor completely randomized design with $2 \times 2 \times 2 = 8$ treatments.

Block Designs

With a completely randomized design, the goal of the randomization is to balance the treatment groups with respect to other variables. Just by chance, however, the groups may have some lack of balance, especially when sample sizes are small. An alternative experimental design identifies a key explanatory variable that could influence the response outcome and then groups together experimental units that take the same or similar values on this variable. Each such group is called a **block**. A **block design** attempts to **control** the effects of the variable whose levels form the blocks (called the **blocking variable**), by comparing the treatments within blocks.

For instance, suppose we think that how long a person has been smoking may affect whether an antidepressant is successful for helping smokers to quit. We might consider three blocks of subjects: Those who have smoked for less than two years, those who have smoked for between two and ten years, and those who have smoked for more than ten years. We could then assign the subjects to the two treatments randomly within each block. We might find out that Zyban is effective for those who are recent smokers but not effective for long-term smokers.

Blocks and a Block Design

A **block** is a collection of experimental units that have the same (or similar) values on a key variable expected to influence the response outcome. An experimental design that identifies blocks before the start of the experiment and assigns subjects to treatments within those blocks is called a **block design**.

EXAMPLE 10: USING A BLOCK DESIGN IN THE SMOKING CESSATION STUDY

Picture the Scenario
Let's return to Example 8, "Design the Study to Assess Antidepressants for Quitting Smoking." In this study, one factor was considered, using the antidepressant Zyban versus using a placebo.

Question to Explore

Smokers may have a more difficult time quitting smoking if they live with another smoker. How can an experiment use a block design to explore this possibility in a study to compare Zyban with placebo?

Think It Through

The researchers could treat whether a subject lives with another smoker as a blocking variable. The subjects would then be split into two blocks: Those who live with another smoker, and those who do not live with smokers. Subjects within a block are then similar in terms of whether they live with a smoker. Within each block, the subjects can then be randomly assigned to take Zyban or a placebo.

Figure 4.4 shows a flow chart of this block design, when 250 of the 429 study subjects live with nonsmokers and 179 live with another smoker. If, for instance, Zyban is more effective than placebo only when the subject lives with a nonsmoker, we can learn that from a study using this design.

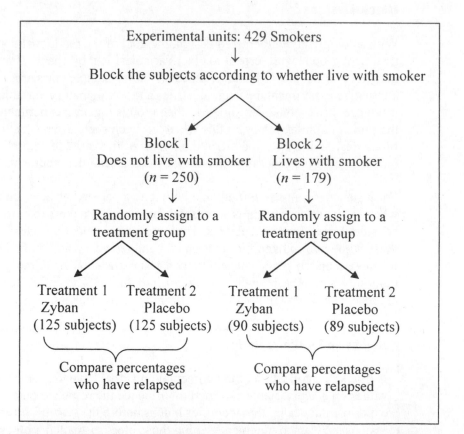

Figure 4.4: Diagram of an Experiment Using a Block Design.
Question: What is the advantage of this design over a completely randomized design (that is, without blocks), in which whether the subject lives with a smoker is ignored?

Insight

This design is not a *completely* randomized design: Although subjects are randomized to Zyban or placebo, they are not randomized to live or not live with a smoker. Subjects are not randomized to levels of blocking variables. They naturally occur within a particular block. This type of block design is called a **randomized block** design, because randomization of units to treatments occurs within blocks.

♦

To practice this concept, try Exercise 4.37.

Matched-Pairs Designs

In the randomized block design of Example 10, subjects within a block were randomly assigned to the two treatments. In another type of block design, each subject serves as a block. Both treatments are then observed for each subject. For instance, an experiment to evaluate a diet might take measurements on each subject before using the diet and then again after using the diet for some period of time. This type of block design is called a **matched-pairs design.**

Medical experiments often use a matched-pairs design to compare treatments for chronic conditions that do not have a permanent cure. For instance, suppose a study plans to compare an oral drug with a placebo for treating migraine headaches. The subjects could take one treatment the first time they get a migraine headache and the other treatment the second time they get a migraine headache. The response is whether the subject's pain is relieved. For instance, the first three subjects might contribute the results to the data file:

Subject	Drug	Placebo
1	Relief	No relief
2	Relief	Relief
3	No relief	No relief

For the entire sample, we would compare the percentages of subjects who have pain relief with the drug and with the placebo.

This type of matched-pairs design is called a **crossover design**. Subjects cross over from using one treatment to the other. It provides the benefit that it helps to remove certain sources of potential bias. Using the same subjects for each treatment keeps potential lurking variables from affecting the results, because those variables take the same values for each treatment. For instance, any difference observed between the drug and the placebo responses is not due to subjects taking the drug having better overall health, because each subject received both treatments.

In a block design, each block can have more than two treatments. For instance, the migraine headache crossover study could investigate relief for subjects who take a low dose of an oral drug, a high dose of that drug, and a placebo. Where possible, randomization is used within each block to assign experimental units to treatments. In the migraine headache study, to reduce possible bias due to the order in which the three treatments are taken, the order would be randomized. Otherwise, bias could occur due to subjects always receiving one treatment before another. An example of potential bias is a positive "carry-over" effect from the first drug taken, if that drug has lingering effects that can help improve results for the second drug taken.

SECTION 4.3: PRACTICING THE BASICS

4.33 Never leave home without duct tape: There have been anecdotal reports of the ability of duct tape to remove warts. In an experiment conducted at the Madigan Army Medical Center in the state of Washington (*Archives of Pediatric and Adolescent Medicine* 2002; 156: 971-974), 51 patients between the ages of 3 and 22 were randomly assigned to receive either duct-tape therapy (covering the wart with a piece of duct tape) or cryotherapy (freezing a wart by applying a quick, narrow blast of liquid nitrogen). After two months, the percentage successfully treated was 85% in the duct tape group and 60% in the cryotherapy group.

a.   Identify the response variable, the explanatory variable, the experimental units, and the treatments.

b.   Describe the steps of how you could randomize in assigning the 51 patients to the treatment groups.

4.34: Rats and cell phones: The Science Daily website (*www.sciencedaily.com*) described a study at Washington University that exposed rats to two common types of cell phone radiation for four hours a day, five days a week, for two years. One third of the rats were exposed to analog cell phone frequency, one third to digital cell phone frequency, and one third served as controls and received no radiation. At the end of two years, the brain, spinal cord and other organs were examined for cancerous tumors. No statistically significant difference was found among groups in the percentage of tumors.

a.   Use a diagram similar to Figure 4.4 to portray this study. Clearly label the experimental units, explanatory and response variables, and treatments.

b.   What is a difference between this study and the Australian cell phone study discussed in Example 1 that could explain the different conclusions? This illustrates the importance of not relying on the results of one study.

4.35 No statistical significance: A randomized experiment investigates whether an herbal treatment is better than a placebo in treating subjects suffering from depression. Unknown to the researchers, the herbal treatment has no effect: Subjects have the same score on a rating scale for depression (for which higher scores represent worse depression) no matter which treatment they take.

a. The study will use eight subjects, numbered 1 to 8. Using random numbers, pick the four subjects who will take the herbal treatment. Identify the four who will take the placebo.

b. After taking the assigned treatment for three months, subjects' results on the depression scale are as follows:

Subject	Response	Subject	Response	Subject	Response	Subject	Response
1	85	2	60	3	44	4	95
5	69	6	78	7	50	8	75

Based on the treatment assignment in (a), find the sample mean response for those who took the herbal supplement and for those who took the placebo.

c. Using the means in (b), explain how (i) sample means can be different even when there is "no effect" in a population of interest, and (ii) a difference between two sample means may not be "statistically significant" even though those sample means are not equal.

4.36 Blind study: Why do you think that, ideally, subjects should be blind to the treatment to which they are assigned? Why should it matter whether they know?

4.37 Two factors helpful?: A two-factor experiment designed to compare two diets and to analyze whether results depend on gender randomly assigns 20 men and 20 women to the two diets, 10 of each to each diet. After three months the sample mean weight losses are as shown in the table.

Sample Mean Weight Loss by Diet Type and Gender

Diet	Gender Female	Male	Overall
Low carb	12	0	6
Low fat	0	12	6
Overall	6	6	6

a. Identify the two factors and the response variable.
b. What would the study have concluded if it did only a one-factor analysis on diet?
c. What could the study learn from the two-factor study that it would have missed by doing a one-factor study on diet alone or a one-factor study on gender alone?

4.38 Assigning subjects to two-factor study: Refer to Example 9, "Do Antidepressants and/or Nicotine Patches Help Smokers Quit?" Explain steps to assign 800 subjects randomly to the four treatments, 200 to each.

4.39 Zyban and nicotine patch study results: The 893 subjects described in Example 9 were evaluated for abstinence from cigarette smoking at the end of 12 months. The table shows the percentage in each group that were still abstaining.

Group	Abstinence Percentage	Sample Size
Nicotine Patch Only	16.4	244
Zyban Only	30.3	244
Nicotine Patch with Zyban	35.5	245
Placebo Only	15.6	160

---

a. Find the approximate margin of error for the abstinence percentage in each group. Explain what a margin of error means.
b. Based on the results in (a), does it seem as if the treatments Zyban only and Placebo only are "significantly different"? Explain.
c. Based on the results in (a), does it seem as if the treatments Zyban only and Nicotine patch with Zyban are "significantly different"? Explain.
d. Based on the results in parts (a), (b), and (c), how would you summarize the results of this experiment?

4.40 Prefer Coke or Pepsi?: You would like to conduct an experiment with your class to see if students tend to prefer Coke or Pepsi.

a. Explain how you could do this, incorporating ideas of blinding and randomization, (i) with a completely randomized design, (ii) with a matched-pairs design.
b. Which design would you prefer? Why?

4.41 Caffeine jolt: A recent study (*Psychosomatic Medicine* 2002; 64: 593-603) claimed that people who consume caffeine regularly may experience higher stress and higher blood pressure. In the experiment, 47 regular coffee drinkers consumed 500 milligrams of caffeine in a pill form (equivalent to four 8 oz. cups) during one workday, and a placebo pill during another workday. The researchers monitored the subjects' blood pressure and heart rate, and the subjects recorded how stressed they felt.
a. Identify the response variable(s), explanatory variable, experimental units, and the treatments.
b. Is this an example of a completely randomized design, or a crossover design? Explain.
c. Describe how blocking was implemented in this experimental design. What can the researchers hope to control by doing this?

4.42 Comparing gas brands: The marketing department of a major oil company wants to investigate whether cars get better mileage using their gas (Brand A) than from an independent one (Brand B) that has cheaper prices. The department has 20 cars available for the study.
a. Identify the response variable, the explanatory variable, and the treatments.
b. Explain how to use a completely randomized design to conduct the study.
c. Explain how to use a matched-pairs design to conduct the study. What are the blocks for the study?
d. Give an advantage of using a matched-pairs design.

4.43 Allergy relief: An experiment is being designed to compare relief from hay fever symptoms given by a low dose of a drug, a high dose of the drug, and a placebo. Each subject who suffers from hay fever and volunteers for the study is observed on three separate days, with a different treatment used each day. There are two days between treatments, so that a treatment does not have a carry-over effect for the next treatment assigned.
a. What are the blocks in this block design? What is this type of block design called?
b. Suppose the study is conducted as a double-blind study. Explain what this means.
c. Explain how randomization within blocks could be incorporated into the study.

4.44 Colds and vitamin C: For some time there has been debate about whether regular large doses of vitamin C reduce the chance of getting a common cold.
a. Explain how you could design an experiment to test this. Describe all parts of the experiment, including (i) what the treatments are, (ii) how you assign subjects to the treatments, (iii) how you could make the study double-blind.
b. An observational study indicates that people who take vitamin C regularly get fewer colds, on the average. Explain why these results could be misleading.

4.4   WHAT ARE WAYS TO PERFORM OBSERVATIONAL STUDIES?

We've seen that experiments assign subjects to values of the explanatory variable (the treatments), but observational studies merely observe the value of the explanatory variable. An experiment can better investigate cause and effect. With an observational study, it's possible that an association is due to some lurking variable.

However, observational studies do have advantages. It's often not practical to conduct an experiment. For instance, how could you design one to study the effect of cell phone use on getting cancer, with human subjects? Observational studies can observe subjects in a realistic setting. Recall the three cell phone studies from Example 1. The two observational studies observed people in their ordinary lives, without attempting to manipulate whether or not they could use a cell phone. The one experimental study used mice. The results suggested that cell phone radiation increases the occurrence of brain tumors, but we don't know whether results from an artificial laboratory environment for mice can be extrapolated to humans.

It is possible to have a well-designed and informative observational study. Sample surveys that select subjects randomly are an example. Let's now learn more about how to conduct good sample surveys.

Sample Surveys: Other Random Sampling Designs Are Useful in Practice

Simple random sampling gives every possible sample the same chance of selection. In practice, more complex random sampling designs are often easier to implement. Sometimes they are even preferable to simple random sampling.

Cluster sampling

To use simple random sampling, we need a reliable sampling frame--the list of all, or nearly all, subjects in the population. Unfortunately, this is often not available. If you work in the marketing department of a company in Toronto and want to select a simple random sample of all adult Canadians, do you think you would be able to get a list of this population? Even if a sampling frame is available, the task of locating the subjects for the sample may be difficult.

It is often easier to identify **clusters** of subjects. A study of residents of a country can identify counties or census tracts. A study of students can identify schools. A study of the elderly living in institutions can identify nursing homes. We can then obtain a sample by randomly selecting the clusters and then observing each subject in the clusters chosen.

For instance, suppose you wanted to sample about 1% of the families in your city. You could use city blocks as clusters. Using a map to label and number city blocks, you could select a simple random sample of 1% of the blocks and then select every family on each of those blocks for your observations.

Cluster Random Sample

Divide the population into a large number of **clusters**, such as city blocks. Select a simple random sample of the clusters. Use the subjects in those clusters as the sample. This is a **cluster random sample**.

Government agencies often use cluster samples. The clusters are usually geographical locations, such as city blocks or census tracts. For personal interviews, when the subjects within a cluster are close geographically, cluster sampling is less expensive per observation than simple random sampling. By interviewing every family in a particular city block, you can obtain many observations quickly and with little travel.

In summary, cluster random sampling is a preferable sampling design if
- A reliable sampling frame is not available, or
- The cost of selecting a simple random sample is excessive.

A disadvantage is that subjects within a cluster are often quite similar. Sometimes we don't learn much more by interviewing several subjects in a cluster than by interviewing a few of them. As a consequence, we usually need a larger sample size with a cluster sample than with a simple random sample in order to achieve a particular margin of error. (It's beyond the scope of this book to see how to get this margin of error.)

Stratified sampling

You're conducting a small survey of students in your school to estimate the mean number of hours a week that students work on outside employment. Using the student directory, you plan to take a simple random sample of $n = 40$ students. You plan to analyze how the mean compares for Freshmen, Sophomores, Juniors, and Seniors. Merely by chance, you may get only a few observations from one of the classes, making it hard to estimate the mean well for that class. If the student directory also identifies students by their class, you could amend the sampling scheme to take a simple random sample of 10 students from each class, still having an overall $n = 40$. The four classes are called **strata** of the population. This type of sample is called a **stratified random sample**.

Stratified Random Sample

A **stratified random sample** divides the population into separate groups, called **strata,** and then selects a simple random sample from each stratum.

Stratified random sampling is useful when you want to compare certain groups. The sampling uses the groups as the strata. During election years, polls measure voting intentions of registered voters. Suppose that a polling company wants to evaluate the percentages of people who plan to vote for the Democratic and the Republican candidates in each state and make comparisons of different states. Then it can stratify the sample by state.

Stratified sampling is called **proportional** if the proportion of the sample chosen from each stratum is the same as its proportion in the population. In your student survey about work, suppose you want to compare females and males in a student body that is 70% female and 30% male. Then, for a proportional stratified sample of size $n = 40$ with gender groups as the strata, you would take a simple random sample of $0.70(40) = 28$ females and a simple random sample of $0.30(40) = 12$ males.

Proportional sampling sometimes yields too few subjects from some group. You can then instead use **disproportional** sampling. For instance, if only 10% of the students are male, with an overall sample size of $n = 40$, a sample of $0.10(40) = 4$ males may be too small to be useful. Instead, you could take a simple random sample of 20 females and a simple random sample of 20 males, to ensure that your sample has enough of each gender to make useful comparisons.

A limitation of stratification is that you must know the stratum into which each subject in the sampling frame belongs. You might want to stratify students in your school by whether the student has a job, to make sure you get enough students who work and who don't work. However, this information may not be available for each student.

In summary, stratified random sampling has the
- Advantage that you can include in your sample enough subjects in each group (stratum) you want to evaluate.
- Disadvantage that you must have a sampling frame and know the stratum into which each subject belongs.

What's the difference between a stratified sample and a cluster sample? A stratified sample *uses every stratum*. By contrast, a cluster sample *uses a sample of the clusters*, rather than all of them. Figure 4.5 illustrates the distinction among sampling subjects (simple random sample), sampling clusters of subjects (cluster random sample), and sampling subjects from within strata (stratified random sample).

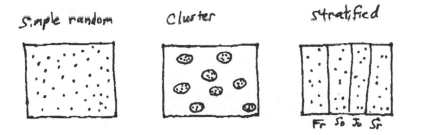

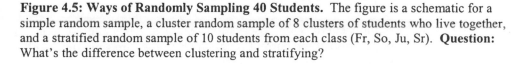

Figure 4.5: Ways of Randomly Sampling 40 Students. The figure is a schematic for a simple random sample, a cluster random sample of 8 clusters of students who live together, and a stratified random sample of 10 students from each class (Fr, So, Ju, Sr). **Question:** What's the difference between clustering and stratifying?

Multistage sampling

In practice, sampling designs often have two or more stages, incorporating some type of random sampling in each stage. When carrying out a large survey for predicting national elections, the Gallup Organization often (1) identifies election districts as clusters and takes a simple random sample of them, and (2) takes a simple random sample of households within each selected election district, rather than sampling every household in a district, which might be infeasible. This is called **two-stage cluster sampling**. Marketing companies also commonly use two-stage cluster sampling, first randomly selecting test market cities (the clusters) and then randomly selecting consumers within each test market city.

Some sampling designs have more than two stages. Most major surveys, such as the General Social Survey, Gallup polls, and the Current Population Survey incorporate both stratification and clustering.

Comparison of Different Random Sampling Methods

A good sampling design ensures that each subject in a population has an opportunity to be selected. The design should incorporate randomness. Table 4.3 summarizes the random sampling methods we've presented.

Table 4.3: Summary of Random Sampling Methods

Method	Description	Advantages
Simple random sample	Each possible sample is equally likely	Sample tends to be a good reflection of the population
Cluster random sample	Identify clusters of subjects, take simple random sample of the clusters	Do not need a sampling frame of subjects, less expensive to implement
Stratified random sample	Divide population into groups (strata), take simple random sample from each	Ensures enough subjects in each group that you want to compare
Multistage random sample	Can sample subjects in clusters rather than using everyone in the clusters (two-stage cluster sample), can use both clustering and stratification	Can combine advantages of clustering and stratification

In future chapters, when we use the term "random sampling," we'll mean *simple random sampling*. The formulas for most statistical methods assume simple random sampling. Similar formulas exist for other types of random sampling, but they are complex and beyond the scope of this text.

Types of Observational Studies

Sample surveys are observational studies. They attempt to take a cross-section of a population at the current time. Such studies are sometimes called **cross-sectional** studies.

Rather than taking a cross-section at some time, some observational studies are *backward looking* (**retrospective**) or *forward looking* (**prospective**). Observational studies in medicine are often retrospective. How can we study whether there is an association between cell phone use and brain cancer if we cannot perform an experiment? We can form a sample of subjects who have brain cancer and a similar sample of subjects who do not and then compare the past use of cell phones for the two groups. This approach was first applied to study smoking and lung cancer.

EXAMPLE 11: RETROSPECTIVE LINKING OF LUNG CANCER WITH SMOKING

Picture the Scenario
When was the link first shown between smoking and lung cancer? In 1950, the medical statisticians Austin Bradford Hill and Richard Doll conducted, one of the first studies on the relationship between smoking and lung cancer in London, England. In 20 hospitals, they matched 709 patients admitted with lung cancer in the preceding year with 709 non-cancer patients at the same hospital of the same gender and within the same 5-year grouping on age. All patients were queried about their smoking behavior. A smoker was defined as a person who had smoked at least one cigarette a day for at least a year. The study used a retrospective design to "look into the past" in measuring the patients' smoking behavior.

Table 4.4 shows the results. The 709 *cases* in the first column of the table were the patients with lung cancer. The 709 *controls* in the second column were the matched patients wiyhout lung cancer.

----- ---

Table 4.4: Results of Retrospective Study of Smoking and Lung Cancer. The cases had lung cancer and the controls did not. The *retrospective* aspect refers to studying whether subjects had been smokers in the past.

Smoker	Lung Cancer	
	Cases	Controls
Yes	688	650
No	21	59
Total	709	709

Question to Explore
Compare the proportions of smokers for the lung cancer cases and the controls. Interpret.

Think It Through
For the lung cancer cases, the proportion who were smokers was $688/709 = 0.970$, or 97%. For the controls (not having lung cancer), the proportion who were smokers was $650/709 = 0.917$, or about 92%. The lung cancer cases were more likely than the controls to have been smokers.

Insight
An inferential analysis showed that these results were statistically significant. This suggested that an association exists between smoking and lung cancer. Because this

was an observational study, cigarette companies argued vociferously that a lurking variable could have caused this association. Later studies, both retrospective and prospective, continually added more and more evidence that smoking causes lung cancer. For example, a prospective study began in 1951 with 35,000 doctors, and at its completion in 2001 the researchers[12] estimated that cigarettes took an average of 10 years off the lives of smokers who never quit. This study estimated that at least half the people who smoke from youth are eventually killed by their habit. Richard Doll and the biostatistician Richard Peto recently estimated that 30% of all cancer cases can be attributed to smoking.

To practice this concept, try Exercise 4.50.

Case-control studies

The retrospective smoking study of Example 11 is an example of what's called a **case-control study**.

IN WORDS	**Case-Control Study**
In Example 11, the response outcome of interest was having lung cancer. The cases had lung cancer, the controls did not have lung cancer, and they were compared on the explanatory variable-- whether the subject had been a smoker (yes or no).	A **case-control study** is an observational study in which subjects who have a response outcome of interest (the cases) and subjects who have the other response outcome (the controls) are compared on an explanatory variable.

This is a popular design for medical studies in which it is not practical or ethical to perform an experiment. We can't randomly assign subjects into a smoking group and a nonsmoking group – this would involve asking some subjects to start smoking. Since we can't use randomization to balance effects of potential lurking variables, usually the cases and controls are matched on such variables.

EXAMPLE 12: CASE-CONTROL STUDIES ABOUT CELL PHONE USE

Picture the Scenario
The U.S. and German studies about cell phone use mentioned in Example 1 were both case-control studies. In the U.S. study, the cases were brain cancer patients. The controls were hospitalized for benign conditions but were matched with the cases on age, gender, race, and month of admission. The German study used cases who had a type of eye cancer. In forming a set of controls, that study did not attempt to match subjects from the two groups.

Question to Explore
Why might researchers decide to match cases and controls on characteristics such as age?

Think It Through
We've seen that one way to balance groups on potential lurking variables such as age is to randomize in assigning subjects to groups. However, these were observational

---

[12] See article by R. Doll et al., *British Med. J.*, June 26, 2004, p. 1519.

studies, not experiments. So it was not possible for the researchers to use randomization to balance treatments on potential lurking variables. Matching is an attempt to achieve the sort of balance that randomization provides.

When researchers fail to use relevant variables to match cases with controls, those variables could influence the results. They could mask the true relationship. The results could suggest an association when actually there is not one, or the reverse. Without matching, results are more susceptible to effects of lurking variables.

Insight
The lack of matching in the German study may be one reason that the results from the two studies differed, the U.S. study not finding an association and the German study finding one.

◆

To practice this concept, try Exercise 4.54.

In a case-control study, the number of cases and the number of controls is fixed. The random part is observing the outcome for the explanatory variable. For instance, in the smoking study of Example 11, for the 709 patients of each type we looked back in time to see whether they smoked. We used the data to find percentages for the categories of the explanatory variable (smoker, non-smoker). It is not meaningful to form a percentage for a category of the response variable. For example, we cannot estimate the population percentage of subjects who have lung cancer, for smokers or non-smokers. By the study design, half the subjects had lung cancer. This does not mean that about half the population had lung cancer. Not being able to estimate this is one sacrifice we make from having to use a case-control study instead of an experiment to study this association. If an experiment were feasible, we *could* estimate the percentage of people who get lung cancer, for smokers or nonsmokers.

Case-control studies use convenience samples. Researchers are most easily able to find the needed cases and controls among patients in a hospital. The cases are in the hospital for the disease under study, and the controls are there for other reasons. Because such studies use convenience samples, we must be cautious about generalizing results to other populations.

Prospective studies

A *retrospective* study, such as a case-control study, looks into the past. By contrast, a *prospective* study follows its subjects into the future.

PHOTO to Come
Picture of nurses in scrubs – need to be sure and have a male nurse included.

EXAMPLE 13: THE PROSPECTIVE NURSES' HEALTH STUDY

Picture the Scenario
The Nurses' Health Study, conducted by researchers at Harvard University and funded by the National Institutes of Health,[13] began in 1976 with 121,700 female

---

[13] More information on this study can be found at clinicaltrials.gov/show/NCT00005152 and at www.channing.harvard.edu/nhs.

nurses age 30 to 55. The purpose of the study was to explore relationships among diet, hormonal factors, smoking habits, and exercise habits and the risk of coronary heart disease, pulmonary disease, and stroke. Since the initial survey in 1976, the nurses have been asked to fill out a questionnaire every two years. Response rates have been about 90% in each two-year cycle.

Question to Explore
What does it mean for this observational study to be called *prospective*?

Think It Through
The retrospective smoking study already knew whether subjects had lung cancer, and it looked into the past to see whether they had been smokers. By contrast, when the Nurses' Health Study began, it was not known whether a particular nurse would eventually have an outcome such as lung cancer. The study followed her into the future to see whether a nurse developed that outcome and to analyze whether certain explanatory variables (such as smoking) were associated with it.

Insight
Over the years, several analyses have been conducted using data from the Nurses' Health Study. A recent finding (reported in *The New York Times*, Feb. 11, 2003) was that highly overweight 18-year-olds were five times as likely as young women of normal weight to need hip replacement later in life.

♦

To practice this concept, try Exercise 4.49.

Confounding

When two explanatory variables are both associated with a response variable but are also associated with each other, there is said to be **confounding**. It is difficult to determine whether either of them truly causes the response, because a variable's effect could be at least partly due to its association with the other variable. For instance, age and height are confounded as predictors of a child's weight. They are both associated with weight, but they are also associated with each other.

In investigating the effect of an explanatory variable, good observational studies try to adjust for variables that are confounded with that explanatory variable. One way to do this is to split the sample into groups within which a confounding variable is constant--that is, "control" for the confounding variable by not letting it vary as you study the association between the variables of interest. This is analogous to blocking in an experimental design.

For instance, suppose you plan to study the association between whether a child is from a single-parent family and whether the child is a delinquent. You might expect family income to be a confounding variable. Specifically, you might expect a negative association between family income and being a juvenile delinquent, and a negative association between family income and being in a single-parent family. To control for family income, you could look at the association between single-parent families and delinquency while keeping family income roughly constant: You could study this association separately for low-income, medium-income, and high-income families. Doing so attempts to do with an observational study what randomization and blocking do for controlled experiments--keep the groups being compared balanced with respect to a variable that may affect the results.

In Chapter 3, Example 16 ("Is Smoking Actually Beneficial to Your Health") illustrates a study with a confounding variable. Over the 20-year study period, smokers had a greater survival rate than non-smokers. However, age was a confounding variable. Older subjects were less likely to be smokers, and older subjects were more likely to die. Within each age group, smokers had a *lower* survival rate than nonsmokers. Age had a dramatic influence on the association between smoking and survival status.

What's the difference between a confounding variable and a lurking variable?

A lurking variable is not measured in the study. It has the *potential* for confounding. If it were included in the study and if it were associated both with the response variable and the explanatory variable, it would become a confounding variable. It would affect the relationship between the response and explanatory variables.

The potential for lurking variables to affect associations is the main reason it is difficult to study many issues of importance, whether it be medical issues such as the effect of cell phone use on cancer, or social issues such as what causes crime or what causes the economy to improve or what causes students to succeed in school. It's not impossible--statistical methods can analyze the effect of an explanatory variable after controlling for confounding variables – but there's always the chance that an important variable was not included in the study.

The previous examples suggest the question, "Can we *ever* establish causation with an observational study?" No, we can't. But consider the issue of whether smoking has a causal effect on lung cancer. Ethically, we cannot design an experiment in which a researcher randomizes a group of subjects to determine who smokes. Researchers have had to use observational studies, such as case-control studies. So why are doctors so confident in declaring that smoking causes lung cancer? For a combination of reasons: Experiments conducted using animals have shown an association. In many countries, over time female smoking has increased relative to male smoking (It was once rare for women to smoke), and the incidence of lung cancer has increased in women compared to men. Most importantly, studies carried out on different populations of people have *consistently* concluded that smoking is associated with lung cancer, even after controlling for all potentially confounding variables that researchers have suggested.

As more studies are done that control confounding variables, this reduces the chance that a lurking variable remains that can explain the association. As a consequence, although we cannot definitively conclude that smoking causes lung cancer, physicians will not hesitate to tell you that they believe this to be true.

On the Shoulders of . . . Austin Bradford Hill and Richard Doll

"How did statisticians become pioneers in examining the effects of smoking on lung cancer?"

In the mid-twentieth century, Austin Bradford Hill was the leading medical statistician in Britain. In 1946, to assess the effect of streptomycin in treating pulmonary tuberculosis, he designed what is now regarded as the first controlled randomized medical experiment. About that time, doctors noticed the alarming increase in the number of lung cancer cases. The cause was unknown but suspected to be atmospheric pollution, particularly coal smoke. In 1950, Bradford Hill and the medical doctor and epidemiologist Richard Doll published results of the case-control study (Example 11) that identified smoking as a potentially important culprit. They were also pioneers in developing a sound theoretical basis for the case-control method. They followed their initial study with other case-control studies and with a long-term prospective study that showed that smoking is not only the predominant risk factor for lung cancer but also correlated with other diseases.

At first, Bradford Hill was cautious about claiming that smoking causes lung cancer. As the evidence mounted from various studies, he accepted the causal link as being overwhelmingly the most likely explanation. An influential article of his in 1965 proposed criteria that should be satisfied before you conclude that an association reflects a causal link. Over the past 50 years, the statistical work of Doll and Hill has resulted in the saving of millions of lives, as more and more people become aware of the dangers of smoking. For his work, in 1961 Austin Bradford Hill was knighted by Queen Elizabeth. Richard Doll was knighted in 1969. The world owes grateful thanks to Sir Austin (1897-1991) and Sir Richard for their statistical contributions to determining answers to important medical issues.

SECTION 4.4: PRACTICING THE BASICS

4.45 Health insurance: A researcher is planning a survey to estimate the proportion of employers in the U.S. that do not offer health insurance. She wants to study how the results depend upon the type of company for which a person works.
a. Suppose she considers three types of companies (individual-owned with less than 50 employees, individual-owned with more than 50 employees, public company) and can take a simple random sample of 100 companies from each type. Is this an example of a stratified sample, or a cluster sample? Explain.
b. Suppose instead she decides to randomly sample 300 companies and then cross classify them on their type and on whether they offer health insurance. Would this be a stratified sample, a cluster sample, or neither? Explain.

4.46 Multistage health survey: A researcher wants to study regional differences in dental care. He takes a multistage sample by dividing the U.S. into four regions, taking a simple random sample of ten schools in each region, randomly sampling three classrooms in each school, and interviewing all students in those classrooms about whether they've been to a dentist in the previous year. Identify each stage of this sampling design, indicating whether it involves stratification or clustering.

4.47 Club officers again: In Example 4, "Sampling from a Population of Club Officers," two officers were selected to attend a conference in New Orleans. Three of the officers are female and two are male. It is decided to send one female and one male to the convention.
a. Labeling the officers as 1, 2, 3, 4, 5, where 4 and 5 are male, draw a stratified random sample using random numbers. Explain how you did this.
b. Is this sampling proportional, or nonproportional? Explain.
c. Explain why this sampling design is not a simple random sample.

4.48 Clusters vs. strata:
a. With a (one-stage) cluster random sample, do you take a sample of (i) the clusters? (ii) the subjects within every cluster?
b. With a stratified random sample, do you take a sample of (i) the strata? (ii) the subjects within every stratum?
c. Summarize the main differences between cluster sampling and stratified sampling in terms of whether one samples the groups or samples from within the groups that form the clusters or strata.

4.49 More school accounts to audit: Example 5 discussed how to sample 10 accounts to audit in a school district that has 60 accounts.
a. Accounts 1-30 are large and accounts 31-60 are small. Explain why simple random sampling cannot guarantee sampling both types of accounts. (*Hint*: Is it possible that *all* the accounts sampled could be small?)
b. To ensure sampling the correct proportion of large and small accounts, use a proportional stratified random sample. Use random numbers to select 5 accounts of each type. Show the steps of your procedure.

4.50 Stratify citizens by region: A study is planned to compare various aspects of political beliefs for citizens living along the coast of Maine with citizens living in the interior of the state.
a. Although most citizens live on the coast, the study takes half the observations in each area, to have a large enough sample size to make comparisons. The study randomly samples 500 citizens from each area. What type of sample design is this?
b. If 70% live on the coast and 30% inland, and if the study randomly samples 700 citizens along the coast and 300 citizens inland, what type of sample is this?
c. If the study instead randomly samples 20 census tracts, interviewing citizens in each and making comparisons of responses in census tracts on the coast to responses in census tracts inland, what type of sample is this?

4.51 German cell phone study: The contingency table shows results from the German study about whether there was an association between mobile phone use and eye cancer (Stang, 2001).
a. The study was "retrospective". Explain what this means.
b. Explain what is meant by "cases" and "controls" in the heading of the table.
c. What proportion had used mobile phones, of those in the study who (i) had eye cancer, (ii) did not have eye cancer?

Eye Cancer and Use of Mobile Phones

Mobile phones	Cases	Controls
Yes	16	46
No	102	429
Total	118	475

⌨**4.52 Physicians' health study:** Read about the first Physicians' Health study at phs.bwh.harvard.edu.
a. Explain whether it was (i) an experiment or an observational study, (ii) a retrospective or prospective or cross-sectional study.
b. Identify the response variable and the explanatory variable(s), and summarize results.

4.53 Smoking and death: Example 1 in Chapter 3 described a survey of 1314 women during 1972-1974, in which each woman was asked whether she was a smoker. Twenty years later, a follow-up survey observed whether each woman was deceased or still alive. Was this study a cross-sectional study, or a retrospective study, or a prospective study? Explain.

4.54 Smoking and lung cancer: Refer to the smoking case-control study in Example 11. Since subjects were not matched according to *all* possible lurking variables, a cigarette company can argue that this study does not prove a causal link between smoking and lung cancer. Explain this logic, using diet as the lurking variable.

4.55 Age a confounder?: A study observes that the subjects in the study who say they exercise regularly reported only half as many serious illnesses per year, on the average, as those who say they do not exercise regularly. One paragraph in the results section of an article about the study starts out, "We next analyzed whether age was a confounding variable in studying this association."
a. Explain what this sentence means, and how age could potentially explain the association between exercising and illnesses.
b. If age was not actually measured in the study and the researchers failed to consider its effects, could it be a confounding variable, or a lurking variable? Explain.

4.56 Confounding vs. lurking:
a. Explain the difference between a lurking variable and a confounding variable.
b. Give an example of a study, identifying a potential confounding variable for analyzing the association between an explanatory variable and a response variable.

4.57 Death penalty and race: For murder trials in Florida between 1976 and 1987, the death penalty was given in 53 out of 467 cases in which a white defendant had a white victim, in 0 out of 16 cases in which a white defendant had a black victim, in 11 out of 48 cases in which a black defendant had a white victim, and in 4 out of 143 cases in which a black defendant had a black victim (M. Radelet and G. Pierce, *Florida Law Review*, Vol. 43, pp. 1-34, 1991).
a. Explain why this is an observational study.
b. In determining the effect of defendant's race on the death penalty verdict, form the contingency table relating these two variables for cases in which the victims were (i) white, (ii) black. Summarize results.

c. Now combine the two tables in (b) to form a table relating defendant's race and the death penalty verdict while ignoring information about victims' race. How does the association differ from that found in (b)? This shows what can happen when victim's race is a confounding variable.

CHAPTER SUMMARY

This chapter introduced methods for gathering data.

- An **experiment** assigns subjects to experimental conditions (such as drug or placebo) called **treatments**. These are categories of the explanatory variable. The outcome on the response variable is then observed for each subject.

- An **observational study** observes subjects on the response and explanatory variables. The study samples the population of interest, but merely observes rather than applies treatments to those subjects.

Since association does not imply causation, with observational studies we must be aware of potential **lurking variables** that influence the association. In an experiment, a researcher uses **randomization** in assigning experimental units (the subjects) to the treatments. This helps to balance the treatments on lurking variables. A randomized experiment can provide support for causation. To reduce bias, experiments should be **double-blind**, with neither the subject nor the data collector knowing to which treatment a subject was assigned.

A **sample survey** is an observational study that samples a population. Good methods for selecting samples incorporate random sampling.

- A **simple random sample** of n subjects from a population is one in which each possible sample of size n has the same chance of being selected.

- A **cluster random sample** takes a simple random sample of clusters of subjects (such as city blocks) and uses subjects in those clusters as the sample.

- A **stratified random sample** divides the population into separate groups, called **strata,** and then selects a simple random sample from each stratum.

Be cautious of results from studies that use a convenience sample, such as a **volunteer sample**, which Internet polls use. Even with random sampling, **biases** can occur due to sample undercoverage, nonresponse by many subjects in the sample, and responses that are biased because of question wording or subjects' lying.

Most sample surveys take a cross-section of subjects at a particular time. A **census** is a complete enumeration of an entire population. **Prospective** studies follow subjects into the future, as is true with many experiments. Medical studies often use **retrospective** observational studies, which look at subjects' behavior in the past.

- A **case-control study** is an example of a retrospective study. Subjects who have a response outcome of interest, such as cancer, serve as cases. Other subjects not having that outcome serve as controls. The cases and controls are compared on an explanatory variable, such as whether they had been smokers.

A categorical explanatory variable in an experiment is also called a **factor**. **Multi-factor experiments** have at least two explanatory variables. With a **completely randomized design**, subjects are assigned randomly to categories of each explanatory variable. Some designs instead use **blocks**--subgroups of experimental units having equal (or similar) values on a potential lurking variable. With a **randomized block design**, within each block the experimental units are assigned randomly to the treatments. **Matched-pairs designs** have two observations in each block. Often this is the same subject observed for each of two treatments, such as in a crossover study.

Data are only as good as the method used to obtain them. Unless we use a good study design, conclusions may be faulty.

ANSWERS TO FIGURE QUESTIONS

Figure 4.3: The design would have $2 \times 2 \times 2 = 8$ treatments.

Figure 4.4: The block design attempts to give the researcher more control over the influence of a third variable (whether the subject lives with a smoker) on the response variable. If the treatment Zyban is more effective only when a subjects lives with a smoker, we can learn this from the study with blocking but not from a completely randomized design.

Figure 4.5: In a stratified sample, every stratum is used. A simple random sample is taken within each stratum. A cluster sample takes a simple random sample of the clusters. All the clusters are <u>not</u> used.

CHAPTER PROBLEMS: PRACTICING THE BASICS

4.58 Observational vs. experimental study: Without using technical language, explain the difference between observational and experimental studies to someone who has not studied statistics. Illustrate with an example, using it also to explain the possible weaknesses of an observational study.

4.59 Unethical experimentation: Give an example of a scientific question of interest for which it would be unethical to conduct an experiment. Explain how you could instead conduct an observational study.

4.60 Breaking up is hard on your health: Married people seem to have better health than their single peers. If a marriage ends, the healthy edge tends to disappear, with divorced or separated people reporting higher rates of illness. This conclusion was based on a nationwide survey conducted by Statistics Canada of 9775 people, aged 20 to 64, about physical and mental health and relationship status, at 2-year intervals beginning in 1990.
a. Identify the explanatory variable and the response variable.

b. Explain why this is an example of an observational study.

c. Is this an example of a sample survey? Explain.

d. Would it be practically possible to design this study as an experiment? Explain why or why not.

Source: Journal of Marriage and Family 2002; 64:420-432

4.61 Dogs vs. cats: What is the preferred American pet? According to *Statistical Abstract of the United States 2002* and the American Veterinary Medical Association, of pet-owning households, 36% have a dog and 32% have a cat. In 2002, vet care was sought by 85% of households owning a dog and by 67% of households owning a cat.

a. What is the population being considered?

b. Do you think these statistics were based on an (i) observational study? (ii) experiment? (iii) sample survey? (iv) census?

4.62 The fear of asbestos: Your friend reads about a study that estimates that the chance is 15 out of 1 million that a teacher who works for 30 years in a school with typical asbestos levels gets cancer from asbestos. However, she also knows about a teacher who died recently who may have had asbestos exposure. In deciding whether to leave teaching, should she give more weight to the study or to the story she heard about the teacher who died? Why?

4.63 Dress in bathing suit?: In a recent Miss America beauty pageant, television viewers could cast their vote on whether to cancel the swimwear parade by phoning a number the network provided. Of the 1 million viewers who called and registered their opinion, 79% said they wanted to see the contestants dressed as bathing beauties. Since everyone had a chance to call, was this a simple random sample of all the viewers of this program? Why or why not?

4.64 Sampling your fellow students: You are assigned to direct a study on your campus to discover factors that are associated with strong academic performance. You decide to identify 20 students who have perfect GPAs of 4.0, and then measure explanatory variables you think may be important, such as high school GPA and average amount of time spent studying per day.

a. Explain what is wrong with this study design.

b. Describe a study design that you think would yield more useful information.

4.65 Beware of Internet polling: An Internet survey at a newspaper Web site reports that only 14% of respondents believe in gun control. Mention a lurking variable that could bias the results of such an online survey, and explain how it could affect the results.

4.66 Comparing female and male students: You plan to sample from the 3500 undergraduate students who are enrolled at the University of Rochester in order to compare the proportions of female and male students who would like to see the U.S. have a female President.

a. Describe the steps for how you would proceed, if you plan a simple random sample of 80 students. Illustrate, by picking the first three students for the sample.

b. Suppose that you use random numbers to select students, but you stop selecting females as soon as you have 40, and you stop selecting males as soon as you have 40. Is the resulting sample a simple random sample? Why or why not?

c. What type of sample is the sample in (b)? What advantage might it have over a simple random sample?

4.67 Football fans' opinions: A large southern university had problems with 17 football players being disciplined for team rule violations, arrest charges, and possible NCAA violations. The online *Atlanta Journal Constitution* ran a poll with the question, "Has the football coach lost control over his players?", with possible responses, "Yes, he's been too lenient," and "No, he can't control everything teenagers do."
a. Was there potential for bias in this study? If so, what types of bias?
b. The poll results after two days were:

    YES   6012   93%
    NO   487   7%

Does this large sample size guarantee that the results are unbiased? Explain.

4.68 Helping kids to sleep: The *New York Times (*May 13, 2003), reporting on an article appearing in the journal, *Pediatrics*, stated: "More than half of primary care pediatricians have prescribed drugs for sleep problems and nearly three-quarters have recommended over-the-counter medicines, a new study has found. The study surveyed 671 pediatricians for two months in 2001 about how they treated sleep problems. The survey covered doctors, participating anonymously, in the metropolitan areas of Atlanta, Cleveland, Dallas, Philadelphia, Providence and San Diego."
a. What is the population of interest?
b. Describe the sample selected to study this population.
c. What response variables were measured in the survey?
d. Do you have any concerns about the statement made in the article, based on how the survey was conducted?

4.69 Smallpox fears: A Gallup poll indicated that concerns over the possibility of terrorists using smallpox were greater in 2003 than in the weeks following the terrorist attacks on September 11, 2001. The poll, conducted January 23-25, 2003, found that 17% were very worried, 46% were somewhat worried, 28% were not too worried, 9% were not worried at all. A similar poll conducted November 9-13, 2001 found that 11% were very worried, 42% were somewhat worried, 26% were not too worried, and 21% were not worried at all. (*Source:*
www.usatody.com/news/nation/2003-01-27-smallpox-usat_x.htm)
a. Graph the results of both polls using the same bar graph. Explain how the graph supports the conclusion stated in the article.
b. The 2003 poll used telephone interviews with 1000 adults. What is the population of interest? What are possible sources of bias for this survey?
c. The article states that the margin of error was ±3 percentage points. Explain what this means and show how an approximate margin of error can be calculated.

4.70 What's your favorite poem?: In fall 1995, the BBC in Britain requested viewers to call the network and indicate their favorite poem. Of 7500 callers, more than twice as many voted for Rudyard Kipling's *If* than for any other poem. The BBC then reported that this was the clear favorite.

a.   Since any person could call, was this sample a simple random sample? Explain.
b.   Was this a volunteer sample? Explain.
c.   If the BBC truly wanted to determine Brits' favorite poem, how could they more reliably do so?

4.71 Video games mindless?: "Playing video games not so mindless." This was the headline in a *CNN* news report[14] about a study that concluded that young adults who play video games regularly demonstrated better visual skills than young adults who do not play regularly. Sixteen young men took a series of tests that measured their visual skills. The men who had played video games in the previous six months performed better on the tests than those who hadn't played.
a. What are the explanatory and response variables?
b. Was this an observational study or an experiment? Explain.
c. Specify a potential lurking variable. Explain your reasoning.

4.72 Reducing high blood pressure: A pharmaceutical company has developed a new drug for treating high blood pressure. They would like to compare its effects to those of the most popular drug currently on the market. Two hundred volunteers with a history of high blood pressure and who are currently not on medication are recruited to participate in a study.
a. Diagram how the researchers could conduct a randomized experiment. In your diagram, indicate the experimental units, the response variable, the explanatory variable, and the treatments.
b. Explain what would have to be done to make this study double-blind.

4.73 Aspirin prevents heart attacks?: During the 1980s approximately 22,000 physicians over the age of 40 agreed to participate in a long-term study called the Physicians' Health Study. One question investigated was whether aspirin helps to lower the rate of heart attacks. The physicians were randomly assigned to take aspirin or take placebo.
a. Identify the response variable and the explanatory variable.
b. Explain why this is an experiment, and identify the treatments.
c. There are other explanatory variables, such as age, that we would expect to be associated with the response variable. State one such variable, and explain how it is dealt with by the randomized nature of the experiment.

4.74 Exercise and heart attacks: Refer to the previous exercise. One potential confounding variable was the amount of exercise the physicians got. The randomization should have balanced the treatment groups on exercise. The contingency table shows the relationship between whether the physician exercised vigorously and the treatments. Find the percentage of physicians who exercised vigorously (a) in the aspirin group, (b) in the placebo group. Based on these results, did the randomization process do a good job of achieving balanced treatment groups in terms of exercise? Explain.

---

[14] See *http://www.cnn.com/2003/TECH/fun.games/05/28/action.video.ap/index.html*

Treatment	Exercise Vigorously?'		
	Yes	No	Total
Aspirin	7910	2997	10907
Placebo	7861	3060	10921

4.75 Smoking and heart attacks: Refer to the two previous exercises. Another potential confounding variable was whether the physicians smoked. The contingency table cross classifies treatment group by smoking status.

Treatment	Smoking Status			
	Never	Past	Current	Total
Aspirin	5431	4373	1213	11017
Placebo	5488	4301	1225	11014

a. Find the conditional proportions (Recall Section 3.1) on smoking status for each treatment group. Are they similar?

b. Do you think that the randomization process did a good job of achieving balanced treatment groups in terms of smoking behavior? Explain.

4.76 Aspirin, beta-carotene, and heart attacks: Refer to the previous three exercises. This completely randomized study actually used two factors: Whether received aspirin and whether received beta-carotene. Draw a table to portray the four treatments for this study.

4.77 Agricultural field trials: One of the first applications of a block design was in agricultural experiments that compared the yield of some crop for different fertilizers. Such experiments would use several fields. In each field, plots were marked, and a different fertilizer was used in each plot.

a. Is each field a block, or an experimental unit? Explain.

b. Is each plot in a field a block, or an experimental unit? Explain.

c. Explain an advantage of using each fertilizer in each field, with this block design, compared to randomly assigning the fields to the fertilizer treatments with a completely randomized design. (*Hint*: Two plots in the same field are likely to be much more similar than two plots from different fields.)

4.78 Conditioning program: A study was conducted at the University of Georgia to study the effects on work performance of introducing a new physical conditioning program into the high school physical education class of special needs students. Only 14 students were available for the study. The researchers wanted to compare the new conditioning program to the standard one. The researchers formed 7 pairs of students, attempting to match the students based on gender, age, and weight. Within each of the seven pairs, one student was randomly assigned to the new program and the other student was assigned to the standard program. Then results were compared for the two programs.

a. Identify the blocks in this randomized block design.

b. Suppose that instead the researchers randomly assigned the 14 students to the two programs, 7 to each. Is this design more properly identified as a randomized block design or a completely randomized design? Explain. (This was not done because the researchers were concerned that with such a small sample, this design might not sufficiently balance the two groups with respect to lurking variables.)

4.79 Best waterproofing method for jackets?: A nationwide chain of sporting-goods stores has developed a new waterproofing method for its jackets and wants to compare customer satisfaction with the new jackets compared to the old style. The researchers plan to recruit 100 volunteers in the Pacific Northwest (known for its rain) to participate, asking them at the end of a month to indicate their satisfaction with the jacket on a scale from 1 to 10. Before assigning the volunteers to a particular jacket, they express concern that some volunteers may be outdoors more and that they may be 'rougher' on the jackets.
a. Diagram how the researchers could set up an experiment using a completely randomized design. In your diagram, indicate the experimental units, the response variable, the explanatory variable, and the treatments. (*Hint*: Completely randomized designs do not use a blocking variable.)
b. Diagram an improved experimental design that would incorporate blocking. Explain why you chose this blocking variable.
c. Describe how the design in (a) or (b) could use double blinding.

4.80 Best waterproofing method for boots: Refer to the previous exercise. The store has also developed a new waterproofing method for its boots. The market researchers want to compare customer satisfaction for these new boots to the ones with the currently used waterproofing method. They recruit 100 volunteers to participate.
a. Diagram how the researchers could set up an experiment using a completely randomized design. In your diagram, indicate the experimental units, the response variable, the explanatory variable, and the treatments. (*Hint*: Completely randomized designs do not use a blocking variable.)
b. Diagram an improved experimental design that would incorporate blocking to take into account that some volunteers may be outdoors more and may be 'rougher' on the boots.
d. Explain how to design a single experiment to test the boots and the waterproof jacket simultaneously. What is an advantage of this over doing two separate experiments?

4.81 Samples not equally likely in a cluster sample?: In a cluster random sample with equal-sized clusters, every subject has the same chance of selection. However, the sample is not a simple random sample. Explain why not. *(Hint*: Is *every* possible sample of size *n* equally likely?)

4.82 Multistage sampling: You plan to sample residents of registered nursing homes in your county. You obtain a list of all 97 nursing homes in the county, which you number from 01 to 97. Using random numbers, you choose 5 of the nursing homes. You obtain lists of residents from those 5 homes and select a simple random sample of 10 residents from each home.
a. Are the nursing homes clusters, or strata?
b. Explain why the sample chosen is not a simple random sample of the population of interest to you.

4.83 Hazing: Hazing within college fraternities is a continuing concern. Before a national meeting of college presidents, a polling organization is retained to conduct a survey among fraternities on college campuses, gathering information on hazing for the meeting. The investigators from the polling organization realize that it is not possible to find a reliable sampling frame of all fraternities.

a. Using a list of all college institutions, they randomly sample 30 of them, and then interview the officers of each fraternity at each of these institutions that has fraternities. Would you describe this as a simple random sample, cluster random sample, or stratified random sample of the fraternities? Explain.

b. Suppose the college institutions were randomly sampled, and then three fraternities were randomly selected at each institution. What term would you use to describe this sampling design?

4.84 Benefits of quitting smoking: A study followed for five years 5300 middle-aged smokers who had chronic obstructive pulmonary disease (COPD) *(Amer. Jour. Epidemiology 2003;157:973-979)*. Smoking is claimed to be the number one cause of COPD, which itself is the fourth leading cause of death in the United States. During the study, some quit smoking. In comparing subjects who quit to those who did not, the average change in lung function was better for those who quit. In comparing male and female smokers who quit, the average change in lung function was better for women than for men.

a. Identify the response variable.

b. Two explanatory variables were mentioned above. Identify them.

c. Is this an example of (i) an experiment, (ii) a prospective study, (iii) a crosssectional study, or (iv) a retrospective study? Explain.

4.85 Religion's role: The Pew Research Center (http://people-press.org) conducted a nationwide poll of 2000 Americans in 2002 about religion's role at home and abroad. It found that 80% of Americans believe that the influence of religion is good, yet 65% believe that religion plays a significant role in most wars and conflicts. The participants for the phone interviews were sampled using a random digit sample of telephone numbers in the U.S. As described in the survey report, "The first eight digits of the sampled telephone numbers (area code, telephone exchange, bank number) were selected to be proportionally stratified by county and by telephone exchange within the county. The number of telephone numbers randomly sampled from within a given county is proportional to that county's share of telephone numbers in the U.S."

a. What sampling method best matches what they did? (i) simple random sample, (ii) two-stage cluster sampling, (iii) two-stage stratified sampling, (iv) volunteer sampling, (v) convenience sampling?

b. Which data collection method did the Pew researchers use? Describe a possible bias that could result from it.

c. Pew reported that the margin of error for their results was plus or minus 2.5 percentage points. Explain what this means.

4.86 TV linked to obesity in young children: A study of 2761 white, black, and Hispanic low-income adults who had preschool children reported that the prevalence of overweight children was significantly related to the number of hours per day reported that the children spend watching TV or videos. *(Source: Pediatrics 2002:109(6): 1028-1035)*

a. Identify the response and explanatory variables.

b. Was this observational study cross-sectional, prospective, or retrospective?

c. State a possible lurking variable, and describe its potential effect.

4.87 Commercial fleets reduce big fish: A study published in *Nature* (May 15, 2003) concluded that in the past half century, the spread of commercial fishing has cut by 90 percent the population of large predatory fishes. The study used data going back 47 years from nine oceanic and four continental shelf systems.
a. Describe some practical difficulties in obtaining appropriate samples. Is random sampling feasible?
b. Describe the practical difficulty in using a sample to determine the population size. (See Exercise 4.121 for one possible solution.)

4.88 Breast feeding and IQ: A Danish study of individuals born at a Copenhagen hospital between 1959 and 1961 reported higher mean IQ scores for adults who were breast-fed for longer lengths of time as babies (*J. Amer. Med. Assoc. 2002; 287: 2365-2371*). The subjects were enrolled in the study as babies and placed in categories based on duration of breastfeeding. The subjects were then given IQ tests as adults at age 19 and age 27.
a. Is this an example of a prospective or retrospective study? Explain.
b. Identify the response variable and the explanatory variable.
c. Explain how socioeconomic status (SES, a composite index summarizing educational attainment and wealth) could be a possible lurking variable, if SES is associated both with IQ and with the use of breastfeeding.

4.89 Cell phone study revisited: The American study reported in Example 1 about cell phones used 891 subjects with age ranging from 18 to 80 years. Of them, 469 people had brain cancer, and 422 people did not have brain cancer.
a. Explain why this was a retrospective study. Identify the cases and the controls.
b. How could researchers compare the groups, once data were collected?
c. Describe a potential source of bias for this study.

4.90 Twins and breast cancer: Excessive cumulative exposure to ovarian hormones is believed to cause breast cancer. A study (*New England J. Medic.* 2003;348: 2313-2322) used information from 1811 pairs of female twins, one or both of whom had breast cancer. Paired twins were compared with respect to age at puberty, when breast cancer was first diagnosed, and other factors. Their survey did not show an association between hereditary breast cancer and hormone exposure.
a. What type of observational study was used by the researchers?
b. Why do you think the researchers used this design instead of a randomized experiment?

CHAPTER PROBLEMS: CONCEPTS AND INVESTIGATIONS

▣4.91 Cell phone use: Using the Internet, find a study about cell phone use and its potential risk when used by drivers of automobiles.
a. Was the study an experiment, or was it an observational study?
b. Identify the response and explanatory variables.
c. Describe any randomization or control conducted in the study as well as attempts to take into account lurking variables.
d. Summarize conclusions of the study. Do you see any limitations of the study?

⌨**4.92 Read a medical journal:** Go to a Web site for an on-line medical journal, such as the *British Medical Journal* (www.bmj.com). Pick an article in the most recent issue.
a. Was the study an experiment, or was it an observational study?
b. Identify the response variable, and the primary explanatory variable(s).
c. Describe any randomization or control conducted in the study as well as attempts to take into account lurking variables.
d. Summarize conclusions of the study. Do you see any limitations of the study?

⌨**4.93 Internet poll:** Find an example of results of an Internet poll. Do you trust the results of the poll? If not, explain why not.

4.94 Search for a sample survey: Find an example of results from a sample survey reported by a newspaper, journal, the Internet, or some other media.
a. What was the population of interest?
b. Identify the response variable and the explanatory variable(s).
c. What type of sampling design was used? How many subjects were sampled?
d. What method was used to obtain data from the participants?
e. Describe potential sources of bias for this survey.

4.95 Search for an experimental study: Find an example of a randomized experiment from a newspaper, journal, the Internet, or some other media.
a. Identify the explanatory and response variables.
b. What were the treatments? What were the experimental units?
c. How were the experimental units assigned to the treatments?
d. Can you identify any potential sources of bias? Explain.

4.96 Search for an observational study: Find an example of an observational study from a newspaper, journal, the Internet, or some other medium.
a. Identify the explanatory and response variables.
b. What type of observational study was conducted? Describe how the data were gathered.
c. Were lurking variables considered? If so, discuss how. If not, can you think of potential lurking variables? Explain how they could affect the association.
d. Can you identify any potential sources of bias? Explain.

4.97 Judging sampling design: In each of the following situations, summarize negative aspects of the sample design.
a. A newspaper wants to determine whether its readers believe that government expenditures should be reduced by cutting social programs. They ask for readers to vote at their Internet site. Based on 1434 votes, they report that 93% of the city's residents believe that social programs should be reduced.
b. A congresswoman reports that letters to her office are running 3 to 1 in opposition to the passage of stricter gun control laws. She concludes that approximately 75% of her constituents oppose stricter gun control laws.
c. An anthropology professor wants to compare attitudes toward premarital sex of physical science majors and social science majors. She administers a questionnaire to her class of Anthropology 437, Comparative Human Sexuality. She finds no appreciable difference in attitudes between the two majors, so she concludes that the two student groups are about the same in their views about premarital sex.

d. A questionnaire is mailed to a simple random sample of 500 household addresses in a city. Ten are returned as bad addresses, 63 are returned completed, and the rest are not returned. The researcher analyzes the 63 cases and reports that they represent a "simple random sample of city households."

4.98 More poor sampling designs: Repeat the previous exercise for the following scenarios:
a. A principal in a large high school wants to sample student attitudes toward a proposal that seniors must pass a general achievement test in order to graduate. She lists all of the first-period classes. Then, using a random number table, she chooses a class at random and interviews every student in that class about the proposed test.
b. A new restaurant opened in January. In June, after six months of operation, the owner applied for a loan to improve the building. The loan application asked for the annual gross income of the business. The owner's record book contains receipts for each day of operation since opening. She decides to calculate the average daily receipt based on a sample of the daily records and multiply that by the number of days of operation in a year. She samples every Friday's record. The average daily receipt for this sample was then used to estimate the yearly receipts.

4.99 Age for legal alcohol: You want to investigate the opinions students at your school have about whether the age for legal drinking of alcohol should be 18.
a. Write a sentence to ask about this in a sample survey in such a way that results would be biased. Explain why it would be biased.
b. Now write an alternative sentence that should result in unbiased responses.

4.100 Quota sampling: An interviewer stands at a street corner and conducts interviews until obtaining a quota in various groups representing the relative sizes of the groups in the population. For instance, the quota might be 50 factory workers, 100 housewives, 60 elderly people, 30 Hispanics, and so forth. This is called **quota sampling**. Is this a random sampling method? Explain, and discuss potential advantages or disadvantages of this method. (The Gallup Organization used quota sampling until it predicted, incorrectly, that Dewey would easily defeat Truman in the 1948 Presidential election.)

4.101 Smoking and heart attacks: A Reuters story (4/2/2003) reported that "The number of heart attack victims fell by almost 60 percent at one hospital six months after a smoke-free ordinance went into effect in the area (Helena, Montana), a study showed, reinforcing concerns about second-hand smoke." The number of hospital admissions for heart attack dropped from just under seven per month to four a month during the six months after the smoking ban.
a. Did this story describe an experiment, or an observational study?
b. In the context of this study, describe how you could explain to someone who has never studied statistics that association does not imply causation. For instance, give a potential reason that could explain this association.

4.102 Discrimination in bargaining for a new car: A study investigated whether car dealers bargain differently with customers according to the customers' gender and race *(The American Economic Review: 1995;85: 304-321)*. One part of the study used pairs of customers, one female with one male in each pair. The customers were trained to deal for a car in a similar bargaining manner. Car dealerships were randomly selected. Pairs of customers were randomly selected to be sent to a particular dealership. The order that the customer within a pair went to the dealership was randomized. The initial price quoted by the car dealer to the customer was recorded. Write a brief description of the study design used by the researchers. Include the type of study, the explanatory and response variables, and the comparison groups. Why do you believe the researchers selected this design?

4.103 Issues in clinical trials: A randomized clinical trial is planned for AIDS patients to investigate whether a new treatment provides improved survival over the current standard treatment. It is not known whether it will be better, or worse.
a. Why do researchers use randomization in such experiments, rather than letting the subjects choose which treatment they will receive?
b. When patients enrolling in the study are told the purpose of the study, explain why they may be reluctant to be randomly assigned to one of the treatments.
c. If a researcher planning the study thinks the new treatment is likely to be better, explain why he or she may have an ethical dilemma in proceeding with the study.

4.104 Teenagers and alcohol: A study by Columbia University's National Center on Addiction and Substance Abuse took a nonproportional stratified random sample to make sure it sampled enough young people age 12-20 to make valid comparisons of them with adults. Their study reported that 25% of all alcohol consumption was by under-age drinkers. However, *The New York Times* reported (2/27/2002) that this was a serious over-estimate, because in their statistical analysis the researchers failed to account for the fact that although almost 40% of those surveyed were of age 12-20, that group makes up less than 20% of the population. A better estimate was that teenagers consumed 11% of all alcohol.
a. For this sample design, explain why if 40% of the alcohol consumed by the sample was by teenagers, it is not proper to conclude that 40% of the alcohol consumed by the population was by teenagers.
b. If the sample had been a simple random sample, would this sort of extrapolation be more valid? Explain.

4.105 Prospective vs. retrospective: Prospective studies are more difficult to implement than retrospective studies. Explain why you think this is.

4.106 Compare smokers with non-smokers?: Example 11 and Table 4.4 described a case-control study on smoking and lung cancer. Explain carefully why it is not sensible to use the study's proportion of smokers who had lung cancer (that is, $688/(688+650)$) and proportion of non-smokers who had lung cancer ($21/(21+59)$) to estimate corresponding proportions of smokers and non-smokers who have lung cancer in the overall population.

4.107 Is a vaccine effective?: A vaccine is claimed to be effective in preventing a rare disease that occurs only in about one of every 100,000 people. Explain why a randomized clinical trial comparing 100 people who get the vaccine to 100 people who do not get it is unlikely to be worth doing. Explain how you could use a case-control study to investigate the efficacy of the vaccine.

For multiple-choice questions 4.108 - 4.115, select the best response.

4.108 What's a simple random sample?: A simple random sample of size n is one in which:

a. Every *n*th member is selected from the population.

b. Each possible sample of size *n* has the same chance of being selected.

c. There is *exactly* the same proportion of women in the sample as is in the population.

d. You keep sampling until you have a fixed number of people having various characteristics (e.g., males, females).

4.109 Getting a random sample: When we use random numbers to take a simple random sample of 50 students from the 20,000 students at a university,

a. It is impossible to get the random number 00000 or 99999, since they are not random sequences.

b. If we get 20001 for the first random number, for the second random number that number is less likely to occur than the other possible five-digit random numbers.

c. The draw 12345 is no more or less likely than the draw 11111.

d. Since the sample is random, it is impossible that it will be nonrepresentative, such as having only females in the sample.

4.110 Be skeptical of medical studies?: An analysis of published medical studies about heart attacks (Crossen, 1994, p. 168) noted that in the studies having randomization and strong controls for bias, the new therapy provided improved treatment 9% of the time. In studies without randomization or other controls for bias, the new therapy provided improved treatment 58% of the time.

a. This result suggests it is better not to use randomization in medical studies, because it is harder to show that new ideas are beneficial.

b. Some newspaper articles that suggest a particular food, drug, or environmental agent is harmful or beneficial should be viewed skeptically, unless we learn more about the statistical design and analysis for the study.

c. This result shows the value of case-control studies over randomized studies.

d. The randomized studies were poorly conducted, or they would have found the new treatment to be better much more than 9% of the time.

4.111 Opinion and question wording: A recent General Social Survey asked subjects if they supported legalizing abortion in each of seven different circumstances. The percentage who supported legalization varied between 45% (if the woman wants it for any reason) to 92% (if the woman's health is seriously endangered by the pregnancy). This indicates that

a. Responses can depend greatly on the question wording.

b. Observational studies can never be trusted.

c. The sample must not have been randomly selected.

d. The sample must have had problems with response bias.

4.112 Campaign funding: When the Yankelovich polling organization asked, "Should laws be passed to eliminate all possibilities of special interests giving huge sums of money to candidates?" 80% of the sample answered *yes*. When they posed the question, "Should laws be passed to prohibit interest groups from contributing to campaigns, or do groups have a right to contribute to the candidate they support?" only 40% said *yes* (Source: *A Mathematician Reads the Newspaper*, by J. A. Paulos, New York: Basic Books, 1995, p. 15). This example illustrates problems that can be caused by

a. Nonresponse.

b. Bias in the way a question is worded.

c. Sampling bias.
d. Undercoverage.

4.113 Emotional health survey: An Internet poll conducted in the U.K. by Netdoctor.co.uk asked individuals to respond to an "Emotional health survey" (Fienberg, 2001). There were 400 volunteer respondents. Based on the results, the British Broadcasting Corporation (BBC) reported that "Britons are miserable – it's official." This conclusion reflected the poll responses, of which one-quarter feared a "hopeless future,"one in three felt "downright miserable," and nearly one in ten thought "their death would make things better for others". Which of the following is *not* correct about why these results may be misleading?
a. Many people who access a medical website and are willing to take the time to answer this questionnaire may be having emotional health problems.
b. Some respondents may not have been truthful or may have been Internet surfers who take pleasure in filling out a questionnaire multiple times with extreme answers.
c. The sample is a volunteer sample rather than a random sample.
d. It's impossible to learn useful results about a population from a sample of only 400 people.

> Include Calvin and Hobbes cartoon showing Calvin filling out a chewing gum survey showing how he loves to mess with data!

4.114 Sexual harassment: In 1995 in the United Kingdom, the Equality Code used by the legal profession added a section to make members more aware of the dangers of sexual harassment. It states that "research for the Bar found that over 40 percent of female junior tenants said they had encountered sexual harassment during their time at the Bar." This was based on a study conducted at the University of Sheffield that sent a questionnaire to 334 junior tenants at the Bar, of whom 159 responded. Of the 159, 67 were female. Of those females, 3 said they had experienced sexual harassment as a major problem, and 24 had experienced it as a slight problem.
a. The quoted statement might be misleading because the nonresponse was large.
b. No one was forced to respond, so everyone had a chance to be in the sample, which implies it was a simple random sample.
c. This was an example of a completely randomized experiment, with whether a female junior tenant experienced sexual harassment as the response variable.
d. This was a retrospective case-control study, with those who received sexual harassment as the cases.

4.115 Effect of response categories: A study (N. Schwarz et al., *Public Opinion Quarterly*, vol. 49, p. 388, 1985) asked German adults how many hours a day they spend watching TV on a typical day. When the possible responses were the six categories (up to ½ hour, ½ to 1 hour, 1 to 1½ hours, … , more than 2½ hours), 16% of respondents said they watched more than 2 ½ hours per day. When the six categories were (up to 2½ hours, 2½ to 3 hours, …, more than 4 hours), 38% said they watched more than 2½ hours per day.
a. The samples could not have been random, or this would not have happened.
b. This shows the importance of question design, especially when people may be uncertain what the answer to the question really is.
c. This study was an experiment, not an observational study.

♦♦4.116 Blocking and stratifying: Explain the similarity between blocking in an experiment and stratifying in a sample survey.

♦♦4.117 Sample size and margin of error:
a. Find the approximate margin of error when $n = 1$.
b. Show the two possible percentage outcomes you can get with a single observation. Explain why the result in (a) means that with only a single observation, you have essentially no information about the population percentage.
c. How large a sample size is needed to have a margin of error of about 5% in estimating a population percentage? (*Hint*: Take the formula for the approximate margin of error, and solve for the sample size.)

♦♦4.118 Systematic sampling: A researcher wants to select 1% of the 10,000 subjects from the sampling frame. She selects subjects by picking one of the first 100 on the list at random, and then skipping 100 names to get the next subject, skipping another 100 names to get the next subject, and so on.   This is called a **systematic random sample**.
a. With simple random sampling, (i) every subject is equally likely to be chosen, and (ii) every possible sample of size n is equally likely. Indicate which, if any, of (i) and (ii) are true for systematic random samples. Explain.
b. An assembly-line process in a manufacturing company is checked by using systematic random sampling to inspect 2% of the items. Explain how this sampling process would be implemented.

♦♦4.119 Complex multi-stage GSS sample: Go to the Web site for the GSS (www.icpsr.umich.edu/GSS), click on *Appendix*, and then click on *Sampling Design and Weighting*. There you will see described the complex multi-stage design of the GSS. Explain how the GSS uses (a) clustering, (b) stratification, (c) simple random sampling.

♦♦4.120 Mean family size: You'd like to estimate the mean size of families in your community. Explain why you'll tend to get a smaller sample mean if you sample n families than if you sample n individuals (asking them to report their family size). (*Hint*: When you sample individuals, explain why you are more likely to sample a large family than a small family. To think of this, it may help to consider the case $n = 1$ with a population of two families, one with 10 people and one with only 2 people.)

♦♦4.121 Capture-recapture: Biologists and naturalists often use sampling to estimate sizes of populations, such as deer or fish, for which a census is impossible. Capture-recapture is one method for doing this. A biologist wants to count the deer population in a certain region. She captures 50 deer, tags each, and then releases them. Several weeks later, she captures 125 deer and finds that 12 of them were tagged. Let N = population size, M = size of first sample, n = size of second sample, R = number tagged in second sample. The table shows how results can be summarized.

		In first sample?		
		Yes (tagged)	No (not tagged)	Total
In second	**Yes**	R		n
sample?	**No**			
	Total	M		N

a.   Identify the values of M, n, and R for the biologist's experiment.

b. One way to estimate N lets the sample proportion of tagged deer equal the population proportion of tagged deer. Explain why this means that

$$\frac{R}{n} = \frac{M}{N},$$

and hence that the estimated population size is $N = (M \times n)/R$.

c. Estimate the number of deer in the deer population using the numbers above.

♦♦**4.122 Capture-recapture and the census:** The U.S. Census Bureau uses capture-recapture to make adjustments to the census by estimating the undercount. The capture phase is the census itself (persons are 'tagged' by having returned their census form and being recorded as counted) and the recapture phase (the second sample) is the Post Enumerative Survey (PES) conducted after the census.

a. Label the table in the previous exercise in terms of the census application.

b. Make up some plausible values for M, n, and R, and find the population size estimate that they imply.

CHAPTER PROBLEMS: CLASS EXPLORATIONS

♦♦**4.123 Munchie capture-recapture:** Your class can use the capture-recapture method described in the previous exercise to estimate the number of goldfish in a bag of Cheddar Goldfish. Pour the Cheddar Goldfish into a paper bag, which represents the pond. Sample 10 of them. For this initial sample, use Pretzel Goldfish to replace them, to represent the tagged fish. Then select a second sample and derive an estimate for N, the number of Cheddar Goldfish in the original bag. See how close your estimate comes to the actual number of fish in the bag. (Your teacher will count the population of cheddar goldfish in the bag before beginning the sampling). If the estimate is not close, what could be responsible, and what would this reflect as difficulties in a real life application such as sampling a wildlife population?

♦♦⌨**4.124 Margin of error:** Activity 2 in Chapter 1 used the 'Sampling' applet to simulate randomness and variability. We'll use that applet again here, but with a much larger sample size, 1000.

a. For a population proportion of 0.50 for outcome 1, simulate a random sample of size 1000. What is the sample proportion of outcome 1? Do this 10 separate times, keeping track of the 10 sample proportions.

b. Find the approximate margin of error for a sample proportion based on 1000 observations.

c. Using the margin of error found in part (b) and the 10 sample proportions found in part (a), form ten intervals of believable values for the true proportion. How many of these intervals captured the actual population proportion, 0.50?

d. Collect the 10 intervals from each member of the class. (If there are 20 students, 200 intervals will be considered.) How many of these intervals captured the actual population proportion, 0.50?

♦♦**4.125 Activity: Sampling the states**
This activity illustrates how sampling bias can result when you use a non-random sample, even if you attempt to make it representative: You are in a geography class, discussing center and spread for various characteristics of the states in the contiguous United States. A particular value of center is the mean area of the states. A map and

a list of the states with their areas (in square miles) are shown in the figure and table below. Area for a state includes dry land and permanent inland water surface. Although we could use these data to calculate the actual mean area, let's explore *how well sampling performs in estimating the mean area* by sampling 5 states and finding the sample mean. The most convenient sampling design is to use our eyes to pick 5 states from the map that we think have areas representative of all the states. Another possible sampling design is simple random sampling.

Which sampling method, using your eyes or simple random sampling, do you think would tend to estimate more accurately the actual mean area for the 48 states in the continental United States?

a.   Using your eyes, pick five states from the U.S. map that you believe have areas representative of the actual mean area of the states. Compute their sample mean area.

b.   Collect the sample means for all class members. Construct a dot plot of these means. Describe the distribution of sample means. Note the shape, center, and spread.

c.   Using random numbers (Table 4.1, the "Random Number" applet, or software) select five random numbers between 01 and 48. Go to the table below and obtain a simple random sample of five states. Compute the sample mean area of these five states.

d.   Collect the sample means of all class members. Construct a dot plot of the sample means using the same horizontal scale as in (b). Describe this distribution of sample means. Note the shape, center, and spread.

e.   The true mean total land area for the 48 states can be calculated from the table below, by dividing the total at the bottom of the table by 48. Which sampling method, using your eyes or using random selection, tended to be better at estimating the true population mean? Which method seems to be less biased? Explain.

f.   Write a short summary comparing the two distributions of sample means.

Map of the Continental United States

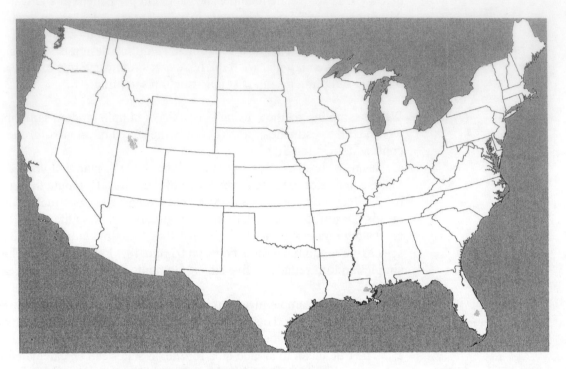

Table: Areas of the 48 Continental United States

State	Area
Alabama	52,419
Arizona	113,998
Arkansas	53,179
California	163,696
Colorado	104,094
Connecticut	5,543
Delaware	2,489
Florida	65,755
Georgia	59,425
Idaho	83,570
Illinois	57,914
Indiana	36,418
Iowa	56,272
Kansas	82,277
Kentucky	40,409
Louisiana	51,840
Maine	35,385
Maryland	12,407
Massachusetts	10,555
Michigan	96,716

Minnesota	86,939
Mississippi	48,430
Missouri	69,704
Montana	147,042
Nebraska	77,354
Nevada	110,561
New Hampshire	9,350
New Jersey	8,721
New Mexico	121,589
New York	54,556
North Carolina	53,819
North Carolina	70,700
Ohio	44,825
Oklahoma	69,898
Oregon	98,381
Pennsylvania	46,055
Rhode Island	1,545
South Carolina	32,020
South Dakota	77,116
Tennessee	42,143
Texas	268,581
Utah	84,899
Vermont	9,614
Virginia	42,774
Washington	71,300
West Virginia	24,230
Wisconsin	65,498
Wyoming	97,814
US TOTAL	**3,119,819**

BIBLIOGRAPHY

Armitage, P. (2001), "Austin Bradford Hill." In *Statisticians of the Centuries*, edited by C. C. Heyde and E. Seneta. New York: Springer.

Best, J. (2001), *Damned Lies and Statistics*. Berkeley: Univ. of California Press.

Bryson, M.C. (1976), "The *Literary Digest* poll: Making of a statistical myth," *American Statistician*, Vol. 30, pp.184-185.

Burrill, G., C. Franklin, L. Godbold, and L. Young (2003). *Navigating through Data Analysis in Grades 9-12*, NCTM

Crossen, C. (1994). *Tainted Truth: The Manipulation of Fact in America*. New York: Simon & Schuster.

Fienberg, H. (2001). "Internet Polls: Why Size Doesn't Matter," www.hfienberg.com/clips/pollspiked.htm

Fist, S. (1997), "Cell Phones Cancer Connection," *Australian* (Sydney), April 29. www.rense.com/health/cancer.htm

Jorenby, D. et al. (1999), "A Controlled Trial of Sustained-Release Bupropion, a Nicotine Patch, or Both for Smoking Cessation." *New England Journal of Medicine*, 340(9): 685-691.

Leake, J. (2001) "Scientist Link Eye Cancer to Mobile Phones," *The London Times*, Jan. 14. www.rfsafe.com/pages/articles/the_times_011401.htm

Muscat, J., et al. (2000). "Handheld Cellular Telephone Use and Risk of Brain Cancer." *Journal of the American Medical Association* 284: 3001-3007.

Repacholi, M.H. (1997). Radiofrequency field exposure and cancer. *Environ. Health Prospect* 105: 1565-1568.

Senn, S. (2003). *Dicing with Death*. Cambridge University Press.

Stang A, et al. (2001). "The possible role of radio frequency radiation in the development of uveal melanoma." *Epidemiology* 12(1): 7-12

Chapter 5
Probability in Our Daily Lives

EXAMPLE 1: WHAT'S THE CHANCE YOUR TAX RETURN IS AUDITED?

Picture the Scenario
For many Americans, April 15th--"Tax Day-- is a day filled with dread as income taxes from the previous year are due. Tax preparation, collection, and defense represent a huge business in the United States as CPAs, tax attorneys, and software programs crunch through earnings and deductions to determine how much is owed. Each year, a small percentage of Americans are subjected to an audit of their tax return to determine whether they did, indeed, pay their fair share of taxes. In fact, in 2004, audits performed by the IRS raised an additional $43 billion dollars. According to IRS Commissioner, Mark Everson, audits increased 40 percent in 2004 for taxpayers with adjusted gross incomes in excess of $100,000 and the total number of audits of taxpayers exceeded a million for the first time since 1999

Questions to Explore
- What is the likelihood of an American taxpayer being audited?
- What is the likelihood of being audited if your annual income is (i) $100,000 or higher, (ii) below $25,000?

Thinking Ahead
In Example 5 in Section 5.2 we'll look at some recent data on numbers of people who are audited. We'll see that the chance of an audit is quite small. But does that chance depend on your income level? At the beginning of Section 5.3 we'll see how to use data from the IRS to answer this.

The answers to these questions are found using **probability** methods. In this chapter, we'll learn how to find probabilities and how to interpret them. This knowledge will help you, as a consumer of information, understand how to assess probabilities in many aspects of your life that involve uncertainty.

In everyday life, you often must make decisions when you are uncertain about the outcome. Should you invest money in the stock market? Should you get dental insurance, or should you get extra collision insurance on your car? Should you start a new business, such as opening a pizza place across from campus? Should you take an advanced course that may be hard to pass? Daily, you face even mundane decisions, such as whether or not to buy a lottery ticket, to place a bet on a sporting event, or to carry an umbrella with you in case it rains.

This chapter introduces **probability**--the way we quantify uncertainty. You'll learn how to measure the chances of the possible outcomes for **random phenomena**--the many things in your life for which the outcome is uncertain. Using probability, for instance, you can find the chance you'll win a lottery. You can find the likelihood that an employer's drug test correctly detects whether or not you've used drugs. You can measure the uncertainty that comes with randomized experiments and with random sampling in surveys. The ideas in this chapter set the foundation for how we'll use probability to make inferences based on data in the rest of the book.

5.1   HOW CAN PROBABILITY QUANTIFY RANDOMNESS?

As we discovered in Chapter 4, there's an essential component that statisticians rely on to try to avoid bias in designing experiments and in sampling in observational studies. This is **randomness**--randomly assigning subjects to treatments or randomly selecting people for a sample. Randomness also applies to the outcomes of a response variable. The possible outcomes are known, but it is uncertain which will occur for any given observation.

Since we were children, we've all employed randomization in games we've played. Some popular randomizers are rolling dice, spinning a wheel, flipping a coin, and drawing cards. Randomization helps to make a game fair. Flips of coins, rolls of dice, and outcomes of random numbers are also simple ways to represent the randomness that occurs with randomized experiments and with sample surveys. For instance, the "head" and "tail" outcomes of a coin flip can represent "drug" and "placebo" when a medical study randomly assigns a subject to receive one of two treatments.

With a *small* number of observations, outcomes of random phenomena may look quite different from what you expect. For instance, you may expect to see a random pattern with different outcomes; but instead, exactly the same outcome may happen a few times in a row. That's not necessarily a surprise, as unpredictability for any given observation is the essence of randomness. We'll discover, however, that with a *large* number of observations, summary statistics become surprisingly *non-random*. For instance, as we make more and more observations, the proportion of times that a given outcome occurs gets closer and closer to a particular number. This long-run proportion provides the basis for the definition of *probability*.

EXAMPLE 2: IS A DIE FAIR?

Picture the Scenario
The board game you've been playing has a die that determines the number of spaces moved on the board. After you've rolled the die 100 times, the number 6 has appeared 21 times, more frequently than each of the other 5 numbers, 1 through 5. At one stage, it turns up three times in a row, resulting in your winning a game. Your opponent then complains that the die favors the number 6 and is not a fair die.

Questions to Explore
a. If a fair die is rolled 100 times, how many 6s do you expect?
b. Would it be unusual to get 21 6s out of 100 rolls? Would it be surprising to roll three 6s in a row at some point?

DID YOU KNOW?

The singular of *dice* is *die*. For a proper die, numbers on opposite sides add to 7, and when 4 faces up the die can be turned so that 2 faces the player, 1 is at the player's right, and 6 is at the player's left (Ainslie, 2003).

Think It Through

a. With many rolls of a fair die, you would expect the six numbers to appear about equally often. A 6s should occur about one-sixth of the time. In 100 rolls, we expect a 6s to come up about $(1/6)100 = 16.7 \approx 17$ times.

b. How can we determine whether or not it is unusual for 6 to come up 21 times out of 100 rolls, or three times in a row at some point? One way is to use simulation. We could roll the die 100 times and see what happens, roll it another 100 times and see what happens that time, and so on. Does a 6 appear 21 (or more) times in many of the simulations? Does a 6 occur three times in a row in many simulations?

This simulation using a die would be tedious. Fortunately, we can use an applet or other software to simulate rolling a fair die. Each simulated roll of a die is called a **trial**. After each trial we record whether a 6 occurred. We will also keep a running record of the proportion of times that a 6 has occurred. At each value for the number of trials, this is called a **cumulative proportion**. Table 5.1 shows partial results of the first simulation of 100 rolls. To find the cumulative proportion after a certain number of trials, divide the number of 6s at that stage by the number of trials. For example, by the eighth roll (trial), there had been one 6 in eight trials, so the cumulative proportion is $1/8 = 0.125$. Figure 5.1 plots the cumulative proportions against the trial number.

APPLET

You can try this yourself at www.prenhall.com/TBA using the *long run probability* applet. This is designed to generate "binary" data, which means that each trial has only *two* possible outcomes.

Table 5.1: Simulation Results of Rolling a Fair Die 100 Times. Each trial is a simulated roll of the die, with chance 1/6 of a 6. At each trial, we record whether a 6 occurred as well as the cumulative proportion of 6s by that trial.

Trial	6 Occurs?	Cumulative Proportion of 6s	
1	no	0/1	= 0.0
2	no	0/2	= 0.0
3	yes	1/3	= 0.333
4	no	1/4	= 0.250
5	no	1/5	= 0.200
6	no	1/6	= 0.167
7	no	1/7	= 0.143
8	no	1/8	= 0.125
...	...	...	...
80	no	14/80	= 0.175
81	yes	15/81	= 0.185
82	yes	16/82	= 0.195
83	yes	17/83	= 0.205
84	no	17/84	= 0.203
85	yes	18/85	= 0.212
...			
100	no	19/100	= 0.190

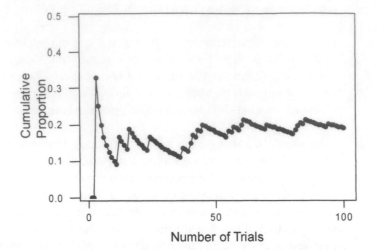

Figure 5.1: The Cumulative Proportion of Times a 6 Occurs, for a Simulation of 100 Rolls of a Fair Die. The horizontal axis reports the number of the trial, and the vertical axis reports the cumulative proportion of 6s observed by then. **Question:** The first four rolls of the die were 5, 2, 6, 2. How can you find the cumulative proportion of 6s after each of the first four trials?

Insert - illustration of rolling dice

In this simulation of 100 rolls of a die, a 6 occurred 19 times, near but not equal to the expected value of about 17. From Table 5.1, a 6 appeared three times in a row for trials 81 through 83.

Insight

One simulation does not prove anything. It suggests, however, that rolling three 6s in a row sometime out of 100 rolls may not be highly unusual. To find out whether 21 6s is unusual, we need to repeat this simulation many times.

♦

To practice this concept, try Exercise 5.10.

ACTIVITY 1: SIMULATE MANY ROLLS OF A DIE

Since one simulation is insufficient to tell us what is typical, go to the *long run probability* applet at www.prenhall.com/ TBA, which can conduct simulations of this type in which each trial has two possible outcomes. At the menu, set the "success probability" equal to 1/6 and the sample size equal to 100. How many 6s did you observe out of the 100 simulated rolls? After simulating 100 rolls, how close was the cumulative proportion of 6s to the expected value of 1/6?

Do the same simulation 25 times to get a feeling for how the sample cumulative proportion at 100 simulated rolls compares to the expected value of 1/6 (that is, 16.7%). You've probably seen that it's not terribly unusual for 21 or more (that is, at least 21%) of the 100 rolls to result in 6s. If you keep doing this simulation over and over again, about 15% of the time you will get 21 or more 6s out of the 100 rolls. Also, about 30% of the time you will see at least three 6s in a row somewhere out of the 100 rolls.

Now, change the sample size for each simulation to 1000, simulating rolling the die 1000 times instead of 100 times. Here's a prediction: The sample cumulative proportion of 6s at 1000 simulated rolls will tend to fall *closer* to its expected value of 1/6. In fact, we'll now be very surprised if at least 21% of the 1000 rolls result in 6s. In theory, this would happen only about once in every 5000 simulations of 1000 rolls of the die.

Long-Run Behavior of Random Outcomes

We've seen that the number of 6s in 100 rolls of a die can vary. One time we might get 19 6s, another time we might get 22, another time 13, and so on. So, what do we mean when we say that there's a one-in-six chance of a 6 on any given roll?

Let's continue the simulation from Example 2 and Table 5.1. In that example we stopped after only 100 rolls, but now let's continue simulating for a very large number of rolls. Figure 5.2 shows the cumulative proportion of 6s plotted against the trial number from trial 100, up to a total of 10,000 rolls. The cumulative proportions fall in the range of 0.16 to 0.185. As the trial number increases, the cumulative proportion of 6s gradually settles down. After 10,000 simulated rolls, Figure 5.2 shows that the cumulative proportion is *very* close to the value of 1/6.

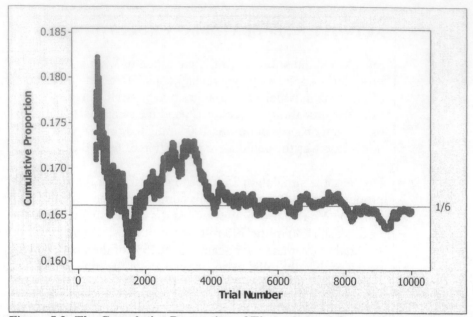

Figure 5.2: The Cumulative Proportion of Times that a 6 Occurs for a Simulation of 10,000 Rolls of a Fair Die. Figure 5.1 showed the first 100 trials, and this figure shows results of trials 100 through 10,000. As the trial number gets larger, the cumulative proportion gets closer to 1/6. The probability of a 6 on any single roll is defined to be this *long-run* value. **Question**: What would you expect for the cumulative proportion after you flipped a balanced coin 10,000 times?

With a relatively short run, such as 10 rolls of a die, the cumulative proportion of 6s can fluctuate a lot. It need not be close to 1/6 after the final trial. However, as the number of trials keeps increasing, the proportion of time the number 6 occurs becomes more predictable and less random: It gets closer and closer to 1/6. *With random phenomena, the proportion of times that something happens is highly random and variable in the short run but very predictable in the long run.*

Probability Quantifies Long-Run Randomness

In 1689, the Swiss mathematician Jacob Bernoulli proved mathematically that as the number of trials increases, the proportion of occurrences of any given outcome approaches a particular number (such as 1/6) "in the long run." He assumed that the outcome of any one trial does not depend on the outcome of any other trial. Bernoulli's result is known as the **law of large numbers.**

We will interpret the **probability** of an outcome to represent "long-run" results. Imagine a randomized experiment or a random sampling of subjects that provides a very long sequence of observations. Each observation does or does not have that outcome. Its probability is the proportion of times that the outcome occurs, in the long run.

Probability

With a randomized experiment or a random sample or other random phenomenon (such as a simulation), the **probability** of a particular outcome is the proportion of times that the outcome would occur in a long run of observations.

When we say that a roll of a die has outcome 6 with probability 1/6, this means that the proportion of times that a 6 would occur in a long run of observations is 1/6. A weather forecaster might say that the probability of rain today is 70%. This means that in a large number of days with atmospheric conditions like those today, the proportion of days in which rain occurs is 0.70. Since a probability is a *proportion*, it takes a value between 0 and 1. Sometimes probabilities are expressed as percentages, as in the weather forecast just mentioned. They then fall between 0 and 100, but we will mainly use the proportion scale, expressing the probability of rain as 0.70.

Why does probability refer to the *long run*? Because we can't accurately assess a probability with a small number of trials. If you sample ten people and they are all right-handed, you can't conclude that the probability of being right handed equals 1.0. It takes a much larger number of people to predict accurately the proportion of people in the population that are right-handed.

Independent Trials

With random phenomena, many believe that when some outcome has not happened in awhile, it is *due* to happen: Its probability goes up until it happens. In many rolls of a fair die, if a 6 has not occurred in a long time, some think it's due and that the chance of a 6 on the next roll is greater than 1/6. If a family has four girls in a row and is expecting another child, are they due to get a boy? Does the next child have more than a ½ chance of being a boy?

Example 2 showed that over a short run, observations may well deviate from what is expected (remember three 6s in a row?). But with many random phenomena, such as rolling a die, what happens on previous trials does not affect the trial that's about to occur. The trials are **independent** of each other.

Independent Trials

Different trials of a random phenomenon are **independent** if the outcome of any one trial is not affected by the outcome of any other trial.

With independent trials, whether you get a 6 on one roll of a fair die does not affect whether you get a 6 on the following roll. It doesn't matter if you had 20 rolls in a row that are not 6s, the next roll still has probability 1/6 of being a 6. The die has no memory. If you have lost many bets in a row, don't assume that you are due to win if you continue to gamble. Many gamblers have gone broke that way. The *law of large numbers*, which gamblers invoke as the *law of averages*, only guarantees *long run* performance. Over the short amount of time in which a gambler's money can disappear, the variability may well exceed what you expect.

In fact, Bernoulli's law of large numbers assumes independent trials. If the outcome of one trial *can* affect the next trial, then the proportion of times that a given outcome occurs need not approach a single value in the long run.

How Can We Find Probabilities?

In practice, how can we find probabilities? Sometimes we can find them by making assumptions about the nature of the random phenomenon. For instance, by symmetry, it may be reasonable to assume that the possible outcomes are *equally likely*. In rolling a die, we might assume that based on the physical makeup of the die, each of the six sides has an equal chance. Thus, we take the probability of rolling any particular number to equal 1/6. If we flip a coin and assume that the coin is balanced, we take the probability of flipping a tail (or a head) to equal 1/2. In Section 5.2 we'll learn how to find probabilities involving results from several trials.

IN PRACTICE

In theory, we could observe several trials and use the proportion of times an outcome occurs as its probability. In practice, this is imperfect, because only for a *very large* number of trials is the cumulative sample proportion necessarily close to the probability. When we take a random sample or perform a randomized experiment, a sample proportion merely estimates an actual probability. In Chapters 6 and 7, we'll see how the sample size determines just how good that estimate is.

Types of Probability: Relative Frequency and Subjective

We've defined the probability of an outcome as a long-run proportion (relative frequency) of times that the outcome occurs in a very large number of trials. However, this definition is not always helpful. Before the launch of the first space shuttle, how could NASA scientists assess the probability of success? No data were available about long-run observations of past flights. If you decide to start a new type of business when you graduate, you won't have a long run of trials with which to estimate the probability that the business is successful. Likewise, no long run of trials is available if you want to assess the probability of life after death, the probability that a defendant being judged in a jury trial is truly guilty, or the probability that a marriage you are contemplating will be a success and will not end in divorce.

In such situations, you must rely on your own judgment based on all the information you have. You rely on **subjective** information rather than solely on **objective** information such as data. You assess the probability of an outcome by taking into account all available information. Such probabilities are not based on a long run of trials. In this **subjective definition of probability**, the probability of an outcome is defined to be your degree of belief that the outcome will occur, based on all the available information. A branch of statistics uses subjective probability as its foundation. It is called **Bayesian statistics**, in honor of Thomas Bayes, a British clergyman who discovered a probability rule on which it is based. The subjective approach is less common than the approach we discuss in this text and is beyond our scope. We'll merely warn you to be wary of anyone who gives a subjective probability of 1 (certain occurrence) or of 0 (certain nonoccurrence) for some outcome. As Benjamin Franklin said, nothing is certain but death and taxes!

SECTION 5.1: PRACTICING THE BASICS

5.1 Probability: Explain what is meant by the long-run relative frequency definition of probability.

5.2 Testing a coin: Your friend decides to flip a coin repeatedly to analyze whether the probability of a head on each flip is 1/2. He flips the coin 10 times and observes a head 7 times. He concludes that the probability of a head for this coin is 7/10 = 0.70.
a. Your friend claims that the coin is not balanced, since the probability is not 0.50. What's wrong with your friend's claim?
b. If the probability of flipping a head is actually 1/2, what would you have to do to ensure that the cumulative proportion of heads falls very close to 1/2?

5.3 Vegetarianism: You randomly sample 10 people in your school, and none of them is a vegetarian. Does this mean that the probability of being a vegetarian for students at your school equals 0? Explain.

5.4 Fear of flying?: As of the year 2000, the Boeing 737 airplane had had 10 fatal accidents in 31 million flights.
a. Can you consider this a long run or short run of trials? Explain.
b. Estimate the probability of a fatal accident on a particular flight. (By contrast, the probability of death in a 1000-mile auto trip in a Western country is on the order of 50 times this probability.)

5.5 Due for a basket: True or false? If a pro basketball player misses 10 shots in a row, his chance of making a shot keeps going up until he makes one. Explain your answer.

5.6 Random digits: Consider a random number generator. Which of the following is *not* correct, and why?
a. For each random digit generated, each integer between 0 and 9 has probability 0.10 of being selected.
b. If you generate 10 random digits, each integer between 0 and 9 must occur exactly once.
c. If you generated a very large number of random digits, then each integer between 0 and 9 would occur close to 10% of the time.
d. The cumulative proportion of times that a 0 is generated tends to get closer to 0.10 as the number of random digits generated gets larger and larger.

5.7 Polls and sample size: A pollster wants to estimate the proportion of Canadian adults who support the prime minister's performance on the job. He comments that by the law of large numbers, to ensure a sample survey's accuracy, he does not need to worry about how to select the sample, only that the sample has a very large sample size. Do you agree with the pollster's comment? Explain.

5.8 Heart transplant: Before the first human heart transplant, Dr. Christiaan Barnard of South Africa was asked to assess the probability that the operation would be successful. Did he need to rely on the relative frequency definition or the subjective definition of probability? Explain.

5.9 Life on other planets?: Is there intelligent life on other planets in the universe? If you are asked to state the probability that there is, would you need to rely on the relative frequency or the subjective definition of probability? Explain.

⌨ **5.10 Applet for coin flipping:** Use the *long run probability* applet at www.prenhall.com/TBA or other software to illustrate the long-run definition of probability by simulating short-term and long-term results of flipping a balanced coin. Let the "success" outcome for the applet refer to getting a head in a flip of a balanced coin.
a.   Set the probability equal to 0.50 and the sample size to $n = 10$. Run the applet ten times, and record the cumulative proportion of heads for each of the 10 simulations.
b.   Now set the sample size $n = 100$. Run the applet 10 times, and record the ten cumulative proportions of heads for the separate sets. Do they vary much?
c.   Now set $n = 1000$. Run the applet 10 times, and record the ten cumulative proportions of heads. Do they vary more, or less, than the proportions in (b) based on $n = 100$?
d.   Summarize the effect of the number of trials n on the variability of the proportion. How does this reflect what's implied by the law of large numbers?

5.11 Unannounced Pop Quiz: A student's worst fear is going to class and having the teacher announce a pop quiz for which the student is not prepared. The quiz consists of 100 'true-false' questions. The student has no choice but to guess the answer randomly for all 100 questions.
a.   Simulate taking this quiz by random guessing. Number a sheet of paper 1 to 100 to represent the 100 questions. Write either a T (true) or F (false) for each question, by predicting what you think would happen if you repeatedly flipped a coin and let a tail represent a T guess and a head represent a F guess. (Don't actually flip a coin, but merely write down what you think a random series of guesses would look like.)
b.   How many questions would you expect to answer correctly simply by guessing?
c.   The table shows the 100 correct answers. The answers should be read across rows. How many questions did you answer correctly?

Pop Quiz Correct Answers

T F T T F F T T T T T F T F F T T T F T F
F F F F F F F T F F T F T F F T F T T F
T F F F F F T F T T F T T T F F F F F T
T F F T F F T T T T F F F F F F F T F F
F F T F F T T F F T F T F T T T T F F F

---

d. The above "answers" were actually randomly generated by the *long run probability* applet at www.prenhall.com/ TBA, with probability set at 0.50. What percentage were 'true', and what percentage would you expect? Why are they not necessarily identical?

e. Are there groups of numbers within the sequence of 100 answers that appear nonrandom? For instance, what is the longest run of Ts or Fs? By comparison, which is the longest run of T's or F's within your sequence of 100 answers? (There is a tendency in guessing what randomness looks like to identify too few long runs in which the same outcome occurs several times in a row.)

■**5.12 Stock market randomness:** An interview in an investment magazine (*In the Vanguard*, Autumn 2003) asked mathematician John Allen Paulos "What common errors do investors make?" He answered, "People tend not to believe that markets move in random ways. Randomness is difficult to recognize. If you have people write down 100 Hs and Ts to simulate 100 flips of a coin, you will always be able to tell a sequence generated by a human from one generated by real coin flips. When humans make up the sequence, they don't put in enough consecutive Hs and Ts, and they don't make the lengths of those runs long enough or frequent enough. And that is one of the reasons people look at patterns in the stock market and ascribe significance to them."

a. Suppose that on each of the next 100 business days the stock market has a ½ chance of going up and a ½ chance of going down, and its behavior one day is independent of its behavior on another day. Use the *lon- run probability* applet at www.prenhall.com/TBA or other software to simulate whether the market goes up or goes down for each of the next 100 days. What is the longest sequence of consecutive Hs (represented in the applet with a 1) or Ts (represented in the applet with a 0) that you observe?

b. Run the applet 9 more times, with 100 observations for each run, and each time record the longest sequence of consecutive Hs or Ts that you observe. For the 10 runs, summarize the proportion of times that the longest sequence was 1, 2, 3, 4, 5, 6, 7, 8 or more. (Your class may combine results to estimate this more precisely.)

c. Based on your findings, explain why if you are a serious investor you should not get too excited if sometime in the next few months you see the stock market go up for five days in a row or go down for five days in a row.

5.2   HOW CAN WE FIND PROBABILITIES?

We've learned that probability enables us to quantify uncertainty and randomness. Now, let's explore some basic rules that can help us find probabilities.

Sample Spaces

The first step is to list all the possible outcomes. The set of possible outcomes for a random phenomenon is called the **sample space.**

Definition: Sample Space

For a random phenomenon, the **sample space** is the set of all possible outcomes.

For example, when you roll a die once, the sample space consists of the six possible outcomes, {1, 2, 3, 4, 5, 6}. When you flip a coin twice, the sample space consists of the four possible outcomes, {(H, H), (H, T), (T, H), (T, T)}, where, for instance, (T, H) represents a tail on the first flip and a head on the second flip.

EXAMPLE 3: SAMPLE SPACE FOR A MULTIPLE-CHOICE POP QUIZ

Picture the Scenario

The instructor in your statistics course decides to give a short unannounced pop quiz (not a favorite class activity for students) with three multiple-choice questions. Each question has five options, and the student's answer is either correct (C) or incorrect (I). If a student answered the first two questions correctly and the last question incorrectly, the student's outcome on the quiz can be symbolized by CCI.

Question to Explore

What is the sample space for the correctness of a student's answers on this pop quiz?

Think It Through

One technique for visualizing and listing the outcomes in a sample space is to draw a **tree diagram**. This has branches showing what can happen on different trials. For the possible student performance on three questions, the tree has three sets of branches, as Figure 5.3 shows. From the tree diagram, a student's performance has 8 possible outcomes. They are

$$\{CCC, CCI, CIC, CII, ICC, ICI, IIC, III\}.$$

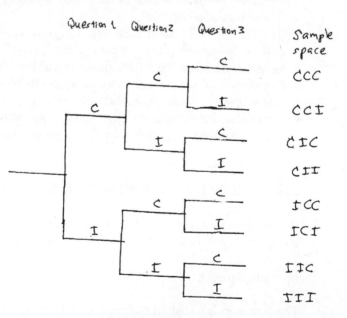

Figure 5.3: Tree Diagram for Student Performance on a Three-Question Pop Quiz.
Each path from the first set of 2 branches to the third set of 8 branches determines an outcome in the sample space. **Questions**: How many possible outcomes would there be if the quiz had four questions?

Insight

The number of branches doubles at each stage. There are 2 branches for question 1, $2 \times 2 = 4$ branches at question 2, and $2 \times 2 \times 2 = 8$ branches at question 3.

♦

To practice this concept, try Exercise 5.13.

How many possible outcomes are in a sample space when there are many trials? To determine this, multiply together the number of possible outcomes for each trial. For example, a pop quiz with 4 questions has $2 \times 2 \times 2 \times 2 = 16$ possible outcomes denoting whether each answer is correct or incorrect. Can you identify them with a tree diagram?

What if we want to consider the actual sequence of responses made? For a quiz with four multiple-choice questions and five possible responses on each question, the number of possible response sequences is $5 \times 5 \times 5 \times 5 = 625$. The tree diagram is ideal for visualizing sample spaces with a small number of outcomes. However, as the number of trials or the number of possible outcomes on each trial increases, it becomes impractical to construct a tree diagram. We will learn about methods for finding probabilities so that listing entire sample spaces is not always necessary.

Events

We are often interested in some *subset* of the outcomes in a sample space. An example is the subset of outcomes for which a student passes the pop quiz, by answering at least two of the three questions correctly. A subset of a sample space is called an **event**.

Definition: Event

An **event** is a subset of the sample space.

IN WORDS

An **event** corresponds to a particular outcome or a collection of possible outcomes. An example is the event of passing the pop quiz.

Events are often denoted by letters from the beginning of the alphabet, such as A and B, or by a letter or string of letters that describes the event. For a student taking the 3-question pop quiz, some possible events are:

A = student answers all 3 questions correctly = {CCC}
B = student passes (at least 2 correct) = {CCI, CIC, ICC, CCC}.

Finding Probabilities of Events

Each outcome in a sample space has a probability. So does each event. To find such probabilities, we list the sample space and specify assumptions that are plausible about its outcomes. For instance, sometimes we can assume that the outcomes are equally likely. The probabilities for the outcomes in a sample space must follow two basic rules:

Probabilities for a Sample Space

- The probability of each individual outcome is between 0 and 1.

- The total of all the individual probabilities equals 1.

EXAMPLE 4: ASSIGNING SUBJECTS TO TREATMENTS IN AN EXPERIMENT

Picture the Scenario

In several medical clinics, subjects volunteer to be part of a multi-center randomized experiment to compare an herbal remedy to a placebo for treating depression. In each center, half of the volunteers are randomly chosen to receive the herbal remedy, and the other half will receive the placebo. The center at UCLA has four volunteers, of whom two are men (Jamal and Ken) and two are women (Linda and Mary).

Questions to Explore

a. Identify the possible samples to receive the herbal remedy. For each possible sample, what is the probability that it is the one chosen?
b. What's the probability of the event that the sample chosen to receive the herbal remedy consists of one man and one woman?

Think It Through

a. The six possible samples to assign to the herbal remedy are {(Jamal, Ken), (Jamal, Linda), (Jamal, Mary), (Ken, Linda), (Ken, Mary), (Linda, Mary)}. This is the sample space for randomly choosing two of the four people. For a simple random sample, every sample is equally likely. Since there are 6 possible samples, each one has probability 1/6. These probabilities fall between 0 and 1, and their total equals 1, as is necessary for probabilities for a sample space.

b. The event in which the sample chosen has one man and one woman consists of the outcomes {(Jamal, Linda), (Jamal, Mary), (Ken, Linda), (Ken, Mary)}. These are the possible pairings of one man with one woman. Each outcome has probability 1/6, so the probability of the event is 4(1/6) = 4/6 = 2/3.

Insight

When each outcome is equally likely, the probability of a single outcome is simply 1/(number of possible outcomes), such as 1/6 in (a) above. The probability of an event is then (number of outcomes in the event)/(number of possible outcomes), such as 4/6 in (b) above.

♦

To practice this concept, try Exercise 5.19.

This example shows that to find the probability for an event, we can (1) find the probability for each outcome in the sample space, and (2) add the probabilities of each outcome that the event contains.

> **Probability of an Event**
> - The probability of an event A, denoted by P(A), is obtained by adding the probabilities of the individual outcomes in the event.
> - When all the possible outcomes are equally likely,
>
> $$P(A) = \frac{\text{number of outcomes in event A}}{\text{number of outcomes in the sample space}}.$$

For example, to find the probability of choosing one man and one woman in Example 4, we first determined the probability of each of the six possible outcomes. Because the probability is the same for each, 1/6, and because the event contains four of those outcomes, the probability is 1/6 added 4 times. This probability equals 4/6, the number of outcomes in the event divided by the number of outcomes in the sample space.

Except for simplistic situations such as simple random sampling or flipping balanced coins or rolling fair dice, different outcomes are not usually equally likely. Then, it is not so obvious how to obtain probabilities. Often, they are estimated, using sample proportions from simulations or from large samples of data.

> **RECALL**
>
> Section 3.1 introduced **contingency tables** to summarize the relationship between two categorical variables.

EXAMPLE 5: WHAT ARE THE CHANCES OF A TAXPAYER BEING AUDITED?

Picture the Scenario

April 15th is tax day in the U.S. – the deadline for filing federal income tax forms. The main factor in the amount of tax owed is a taxpayer's income level. It is assumed that taxpayers will honestly report all income and take only legitimate deductions. Each year, the Internal Revenue Service (IRS) audits a sample of tax forms to verify their accuracy. Table 5.2 is a contingency table that cross-tabulates the 80.2 million long-form federal returns received in 2002 by the taxpayer's income level and whether the tax form was audited.

Table 5.2: Contingency Table Cross-Tabulating Tax Forms by Income Level and Whether Audited. There were 80.2 million forms, and for simplicity the frequencies are reported in thousands and rounded. For example, 90 represents 90,000 tax forms that reported income under $25,000 and were audited.

	WHETHER AUDITED		
INCOME LEVEL	**Yes**	**No**	**Total**
Under $25,000	90	14010	14100
$25,000-$49,999	71	30629	30700
$50,000-$99,999	69	24631	24700
$100,000 or more	80	10620	10700
Total	310	79890	80200

Source: Statistical Abstract of the U.S.: 2003

Questions to Explore
a. What is the sample space?
b. For a randomly selected taxpayer in 2002, what is the probability of (i) an audit, (ii) an income of $100,000 or more?

Think It Through
a. The sample space is the set of possible outcomes. These are the 8 cells for the income and whether audited combinations in this table, such as (Under $25,000, Yes), (Under $25,000, No) , ($25,000-$49,999, Yes), and so forth.
b. If a taxpayer was randomly selected, from Table 5.2:
(i) The probability of an audit was 310/80200 = 0.004.
(ii) The probability of an income of $100,000 or more was 10700/80200 = 0.133.

Insight
Even though audits are not welcome news to a taxpayer, in 2002 only about 0.4% of taxpayers were audited. Is this typical? Inspecting similar data for earlier years shows that this percentage has decreased considerably in recent years. For instance, in 1996 the percentage audited was 1.2%, three times as high as in 2002.

♦

To practice this concept, try Exercise 5.24a-b.

Basic Rules for Finding Probabilities about a Pair of Events

Some events are expressed as the outcomes that (a) are *not* in some other event, or (b) are in one event *or* in another event, or (c) are in one event *and* in another event. We'll next see how to calculate probabilities for these three cases.

The complement of an event

For an event A, the rest of the sample space that is *not* in that event is called the **complement** of A.

IN WORDS

A^c reads as "A-complement." The c in the superscript denotes the term, *complement*. You can think of A^c as meaning "*not A*."

Definition: Complement of an Event

The **complement** of an event A consists of all outcomes in the sample space that are *not* in A. It is denoted by A^c. The probabilities of A and of A^c add to 1, so

$$P(A^c) = 1 - P(A).$$

In Example 5, for instance, the event of having an income *less than* $100,000 is the complement of the event that of having an income of $100,000 or more. Because the probability that a randomly selected taxpayer had an income of $100,000 or more is 0.13, the probability of income less than $100,000 is $1 - 0.13 = 0.87$.

Figure 5.4 illustrates the complement of an event. The box represents the entire sample space. The event A is the circle in the box. The complement of A, which is shaded, is everything else in the box that is not in A. Together, A and A^c cover the sample space. A diagram like this that uses areas inside a box to represent events is called a **Venn diagram**.

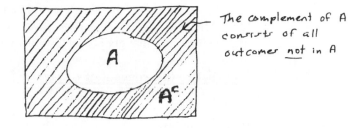

The complement of A consists of all outcomes *not* in A

Figure 5.4: Venn Diagram Illustrating an Event A and its Complement A^c. **Question**: Can you sketch a Venn diagram of two events A and B such that they share some common outcomes, but some outcomes are not in A or in B?

 Suppose that an event contains nearly all the outcomes of a sample space. To find its probability, it is usually simpler to evaluate the probability of its complement (which has a small number of outcomes) and then subtract that probability from 1.

EXAMPLE 6: SHOULD WE BE SURPRISED BY FINDING NO WOMEN ON A JURY?

Picture the Scenario
A jury of 12 people is chosen for a trial. The defense attorney claims it must have been chosen in a biased manner, because 50% of the city's adult residents are female yet the jury contains no women.

Questions to Explore
If the jury were randomly chosen from the population, what is the probability the jury would have (a) no females, (b) at least one female?

Think It Through
Let's use a symbol with 12 letters, with F for female and M for male, to represent a possible jury selection. For instance, MFMMMMMMMMMM denotes the jury in which only the second person selected is female. The number of possible outcomes is $2 \times 2 \times 2 \times \ldots \times 2$, that is, 2 multiplied 12 times, which is $2^{12} = 4096$. This is a case in which listing the entire sample space is not practical. Since the population is 50% male and 50% female, these 4096 possible outcomes are equally likely.

a. Only 1 of the 4096 possible outcomes corresponds to a no-female jury, namely, MMMMMMMMMMMM. So, the probability of this outcome is 1/4096, or 0.0002 to four decimal places. This is extremely unlikely, if a jury is truly chosen by random sampling.

b. As noted above, it would be tedious to list all possible outcomes in which at least one female is on the jury. But this is not necessary. The event that the jury contains *at least one* female is the complement of the event that it contains *no* females. Thus,

P(at least one female) = 1 – P(no females) = 1 – 0.0002 = 0.9998.

Insight

You might instead let the sample space be the possible values for the *number* of females on the jury, namely 0, 1, 2, ..., 12. But these outcomes are not equally likely. For instance, only one of the 4096 possible samples has 0 females, but 12 of them have 1 female: The female could be the first person chosen, or the second (as in MFMMMMMMMMMM), or the third, and so on. Chapter 6 will show a formula (the "binomial") that gives probabilities for this alternative sample space.

◆

To practice this concept, try Exercise 5.16.

Disjoint events

Events that do not share any outcomes in common are said to be **disjoint**.

> **Definition: Disjoint Events**
>
> Two events, A and B, are **disjoint** if they do not have any common outcomes.

Example 3 discussed a pop quiz with three questions. The event that the student answers exactly one question correctly is {CII, ICI, IIC}. The event that the student answers exactly two questions correctly is {CCI, CIC, ICC}. These two events have no outcomes in common, so they are disjoint. In a Venn diagram, they have no overlap. See Figure 5.5. By contrast, neither is disjoint from the event that the student answers the first question correctly, which is {CCC, CCI, CIC, CII}, because this event has outcomes in common with each of the other two events.

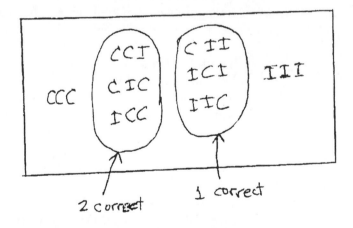

Figure 5.5: Venn Diagram Illustrating Disjoint Events. The event of a student answering one question correctly is disjoint from the event of answering two questions correctly. **Question**: Identify on this figure the event that the student answers the first question correctly. Is this event disjoint from either of the two events identified in the Venn diagram?

Consider an event A and its complement, A^c. They share no common outcomes, so they are disjoint events.

Intersection and union of events

Compound events are events composed from two or more other events. For instance, for two events A and B, the event that *both* occur is a compound event. Called the **intersection** of A and B, it consists of the outcomes that are in both events. By contrast, the compound event that the outcome is in A *or* B or both is the **union** of A and B. It is a larger set, containing the intersection as well as outcomes that are in A but not in B and outcomes that are in B but not in A. Figure 5.6 shows Venn diagrams illustrating the intersection and union of two events.

Definition: Intersection and Union of Two Events

The **intersection** of A and B consists of outcomes that are in both A *and* B.

The **union** of A and B consists of outcomes that are in A *or* B. In probability, "A *or* B" denotes that A occurs or B occurs or both occur.

IN WORDS
Intersection means A **and** B (the "overlap" of the events).
Union means A **or** B or both.

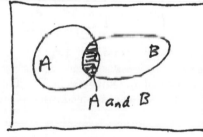
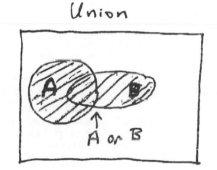

Figure 5.6: The Intersection and the Union of Two Events. Intersection means A occurs *and* B occurs, denoted "A and B." It consists of the shaded "overlap" part in the first diagram. Union means A occurs *or* B occurs *or* both occur, denoted "A or B." It consists of all the shaded parts in the second diagram. **Question**: How could you find P(A or B) if you know P(A), P(B), and P(A and B)?

INCOME	AUDITED	
	Yes	No
Under $25,000	90	14010
$25,000-$49,999	71	30629
$50,000-$99,999	69	24631
≥$100,000	80	10620

For instance, for the three-question pop quiz, consider the events:
A = student answers first question correctly = {CCC, CCI, CIC, CII}
B = student answers two questions correctly = {CCI, CIC, ICC}.

Then the intersection, A *and* B, is {CCI, CIC}, the two outcomes common to A and B. The union, A *or* B, is {CCC, CCI, CIC, CII, ICC}, the outcomes that are in A or in B or in both A and B.

How do we find probabilities of intersections and unions of events? Once we identify the possible outcomes, we can use their probabilities. For instance, for Table 5.2 (reproduced in the margin) for 80.2 million tax forms, let

A denote {audited = yes}
B denote {income ≥ $100,000}

The intersection A and B is the event that a taxpayer is audited and has income ≥ $100,000. This probability is simply the proportion for the cell in which these two events occurred, namely P(A and B) = 80/80200 = 0.001. Another way to find probabilities of intersections and unions uses the probabilities of the events themselves, as we'll see in the next two subsections.

How Can We Find the Probability that A or B Occurs?

Since the union A or B contains outcomes from A and from B, we can add P(A) to P(B). However, this sum counts the outcomes that are in *both* A and B (their intersection) twice. See Figure 5.7. We need to subtract the probability of the intersection from P(A) + P(B) so that it is only counted once. If there is no overlap, that is, if the events are disjoint, no outcomes are common to A and B. Then we can simply add the probabilities.

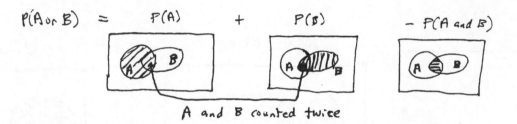

Figure 5.7: The Probability of the Union, Outcomes in A or B or Both. Add P(A) to P(B) and subtract P(A and B) to adjust for outcomes counted twice. **Question:** When does P(A or B) = P(A) + P(B)?

Addition Rule: Probability of the Union of Two Events

For the **union** of two events, P(A or B) = P(A) + P(B) − P(A and B).

If the events are **disjoint,** then P(A and B) = 0, so P(A or B) = P(A) + P(B).

For instance, for the events A (audited = yes) and B (income ≥ $100,000), in Example 5 and above we found P(A) = 0.004, P(B) = 0.133, and P(A and B) = 0.001. Thus,

P(A or B) = P(A) + P(B) − P(A and B) = 0.004 + 0.133 − 0.001 = 0.136

is the probability of being audited or having the highest income level, or both.

How Can We Find the Probability that A and B Occur?

The probability of the intersection of events A and B has a formula to be introduced in Section 5.3. In the special case discussed next, it equals $P(A) \times P(B)$.

Consider a basketball player who shoots two free throws. Let M1 denote making free throw 1, and let M2 denote making free throw 2. For any given free throw, he has an 80% chance of making it. So, $P(M1) = P(M2) = 0.80$. What is the probability of M1 and M2, making free throw 1 *and* free throw 2? In the long-run of many pairs of free throws, suppose that for 80% of the cases in which he made the first three throw, he also made the second. Then the percentage of times he made both is the 80% of the 80% of times he made the first one, for a probability of $0.80 \times 0.80 = 0.64$ (that is, 64%).

This multiplication calculation is valid only under an assumption, **independent trials**: Whether a player makes the second free throw is independent of whether he makes the first. The chance of making the second is 80%, regardless of whether or not he made the first. (For pro basketball players, independence is approximately true: Whether a player makes his first shot has almost no influence on whether he makes the second one. See Exercise 5.34 for some data.)

To find the probability of the intersection of two events, we can multiply probabilities whenever the events are independent. We'll see a formal definition of independent events in the next section, but it essentially means that whether one event occurs does not affect the probability that the other event occurs.

Multiplication Rule: Probability of the Intersection of Independent Events

For the **intersection** of two **independent** events, A and B,

$$P(A \text{ and } B) = P(A) \times P(B).$$

The paradigm for independent events is repeatedly flipping a coin or rolling a die, where what happens on one trial does not affect what happens on another. For instance, for two rolls of a die,

$$P(6 \text{ on roll 1 } and \text{ 6 on roll 2}) = P(6 \text{ on roll 1}) \times P(6 \text{ on roll 2}) = \frac{1}{6} \times \frac{1}{6} = \frac{1}{36}.$$

This multiplication rule extends to more than two independent events.

EXAMPLE 7: GUESSING ON A POP QUIZ

Picture the Scenario
For a 3-question multiple-choice pop quiz, suppose a student is totally unprepared and randomly guesses the answer to each question. If each question has 5 options, then the probability of selecting the correct answer for any given question is 1/5, or 0.20. With guessing, the response on one question is not influenced by the response

on another question. Thus, whether one question is answered correctly is independent of whether or not another question is answered correctly.

Questions to Explore
a. Find the probabilities of the possible student outcomes for the quiz, in terms of whether each response is correct (C) or incorrect (I).
b. Find the probability the student passes, answering at least two questions correctly.

Think It Through
a. For each question $P(C) = 0.20$ and $P(I) = 1-0.20 = 0.80$. The probability that the student answers all three questions correctly is

$$P(CCC) = P(C) \times P(C) \times P(C) = 0.20 \times 0.20 \times 0.20 = 0.008.$$

This would be very unusual. Similarly, the probability of answering the first two questions correctly and the third question incorrectly is

$$P(CCI) = P(C) \times P(C) \times P(I) = 0.20 \times 0.20 \times 0.80 = 0.032.$$

This is the same as $P(CIC)$ and $P(ICC)$, the other possible ways of getting two correct. Figure 5.8 is a tree diagram showing how to multiply probabilities to find the probabilities for all 8 possible outcomes.

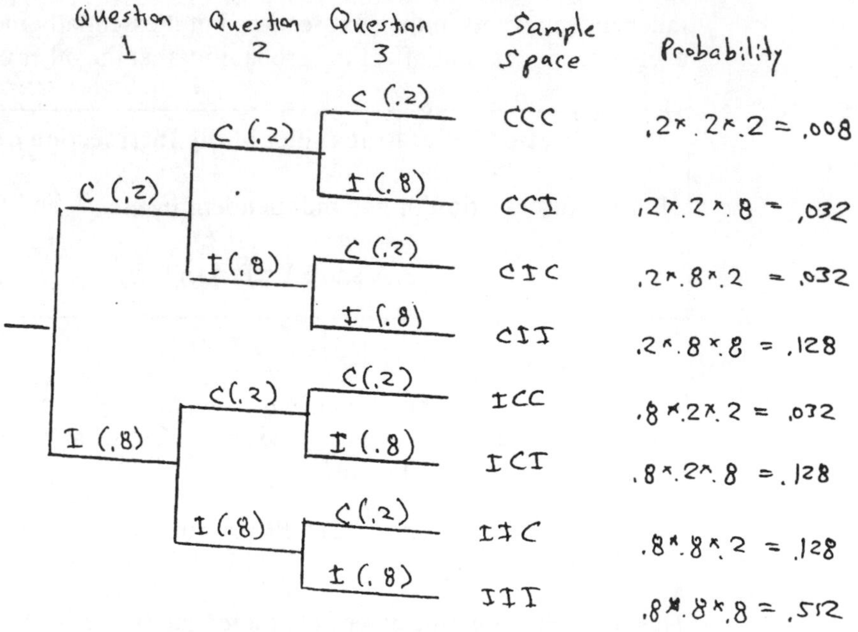

Figure 5.8: Tree Diagram for a Student Guessing on a Three-Question Pop Quiz. Each set of branches represents one question. Each path from the first set of branches to the third set determines one sample space outcome. Multiplication of the probabilities along that path gives its probability, when trials are independent. **Question**: Would you expect trials to be independent if a student is *not* merely guessing on every question? Why or why not?

b. The probability of *at least* 2 correct responses is

$$P(CCC) + P(CCI) + P(CIC) + P(ICC) = 0.008 + 3(.032) = 0.104.$$

There is only about a 10% chance of passing when a student randomly guesses the answers.

Insight
As a check, you can see that the probabilities of the 8 possible outcomes sum to 1.0. The probabilities indicate that it is in a student's best interests not to have to rely on random guessing.

◆

To practice this concept, try Exercise 5.15.

Events Often Are Not Independent

In practice, events need not be independent. For instance, on the previous quiz that had only two questions, the instructor found the following proportions for the actual responses of her students (again, I – incorrect, C – correct).

Outcome:	II	IC	CI	CC
Probability:	0.26	0.11	0.05	0.58

Let A denote {first question correct} and let B denote {second question correct}. Based on these probabilities,

$$P(A) = P(\{CI, CC\}) = 0.05 + 0.58 \quad 0.63$$
$$P(B) = P(\{IC, CC\}) = 0.11 + 0.58 = 0.69$$
and
$$P(A \text{ and } B) = P(\{CC\}) = 0.58.$$

If A and B were independent, then

$$P(A \text{ and } B) = P(A) \times P(B) = 0.63 \times 0.69 = 0.43$$

Since P(A and B) actually equaled 0.58, A and B were not independent.

Responses to different questions on a quiz are typically not independent. Most students do not guess randomly, and students who get the first question correct are usually more likely to get the second question correct than students who do not get the first question correct.

IN PRACTICE

Don't assume that events are independent unless you have given this assumption careful thought and it seems plausible. In Section 5.3, you will learn more about how to find probabilities when events are not independent.

Using the Basics of Probability in Probability Models

We have now developed several rules for finding probabilities. Let's summarize them.

Rules for Finding Probabilities

- The probability of each individual outcome is between 0 and 1, and the total of all the individual probabilities equals 1. The **probability of an event** is the sum of the probabilities of the individual outcomes in that event.

- For an event A and its **complement**, A^c (not in A), $P(A^c) = 1 - P(A)$.

- The **union** of two events has

$$P(A \text{ or } B) = P(A) + P(B) - P(A \text{ and } B),$$

which is $P(A) + P(B)$ when A and B are **disjoint** (no common elements).

- In the special case when A and B are **independent**, the **intersection** of two events has

$$P(A \text{ and } B) = P(A) \times P(B).$$

We've used these rules to find probabilities in several rather idealized situations. In practice, it's sometimes not obvious when different outcomes are equally likely or different events are independent. Then, when calculating probabilities, it is advisable to specify a **probability model** that spells out all the assumptions made.

Probability Model

A **probability model** specifies the possible outcomes for a sample space and provides assumptions on which are based the calculation of probabilities for events composed of those outcomes.

The next example illustrates a probability model used together with the rules for finding probabilities.

EXAMPLE 8: ASSESSING SAFETY OF THE SPACE SHUTTLE

Picture the Scenario
Since two space shuttle missions have resulted in disasters, considerable attention has focused on estimating probabilities of success or failure for a mission. But before the first flight, there were no trials to provide data for estimating probabilities.

Question to Explore
Based on all the information available, a scientist is willing to predict the probability of success for any particular mission. How can you use this to find the probability of *at least one* disaster in a total of 100 missions?

Think It Through
Let S1 denote the event that the first mission is successful, S2 the event that the second mission is successful, and so on up to S100, the event that mission 100 is successful. If all these events occur, there is no disaster in the 100 flights. The event of *at least one* disaster in 100 flights is the complement of the event that they are all successful (0 disasters). By the probability for complementary events,

$$P(\text{at least 1 disaster}) = 1 - P(0 \text{ disasters})$$
$$= 1 - P(S1 \text{ and } S2 \text{ and } S3 \ldots \text{ and } S100).$$

The intersection of events S1 through S100 is the event that *all* 100 missions are successful.

We now need a probability model in order to evaluate P(S1 and S2 and S3 ... and S100). If our probability model assumes that the 100 shuttle flights are *independent*, then the multiplication rule for independent events implies that

$$P(S1 \text{ and } S2 \text{ and } S3 \ldots \text{ and } S100) = P(S1) \times P(S2) \times P(S3) \times \ldots \times P(S100).$$

If our probability model assumes that each flight has the *same probability* of success, say P(S), then this product equals $[P(S)]^{100}$. To proceed further, we need a value for P(S). According to a discussion of this issue in the excellent PBS video series *Against All Odds: Inside Statistics*, a risk assessment study by the Air Force in 1983 used the estimate P(S) = 0.971. With it,

$$P(\text{at least 1 disaster}) = 1 - [P(S)]^{100} = 1 - [0.971]^{100} = 1 - 0.053 = 0.947.$$

Different estimates of P(S) can result in very different answers. For instance, the *Against All Odds* episode mentioned that a 1985 NASA study estimated that P(S) = 0.9999833. Using this,

$$P(\text{at least 1 disaster}) = 1 - [P(S)]^{100} = 1 - [0.9999833]^{100} = 1 - 0.998 = 0.002.$$

This was undoubtedly overly optimistic. We see that different probability models can result in drastically different probability assessments.

Insight
In fact, out of the first 113 flights, there were two disasters, one of which was the Columbia disaster on flight 113. After that flight, an estimate of P(S) was 111/113 – 0.982. The answer above depended strongly on the assumed probability of success for each mission. But, the assumptions of independence and of the same probability for each flight may also be suspect. For instance, other variables (e.g., temperature at launch, experience of crew, age of craft used, quality of O-ring seals) could affect that probability.

◆

To practice this concept, try Exercise 5.21.

In practice, probability models merely *approximate* reality. They are rarely *exactly* satisfied. For instance, with simple random sampling we assume that every possible sample is equally likely. Because of practical difficulties, such as finding a sampling frame that contains all the relevant subjects and then contacting them all and getting them all to respond, this is not exactly true. Whether probability calculations using a particular probability model are accurate depends on whether assumptions in that model are close to the truth or quite unrealistic.

SECTION 5.2: PRACTICING THE BASICS

5.13 Election study: An election study records for each sampled subject either D = voted for Democratic candidate, R = voted for Republican candidate, I = voted for an independent candidate, or N = did not vote.
a) Show how to summarize possible outcomes for two elections with a tree diagram. For example, one possible outcome is (D, R), voting for the Democrat in the first election and for the Republican in the second election.
b) If the study considers three elections, how many possible outcomes are in the sample space?

5.14 Random digit: A single random digit is selected using software or a random number table.
a.   State the sample space for the possible outcomes.
b.   State the probability for each possible outcome, based on what you know about the way random numbers are generated.
c.   Each outcome in a sample space must have probability between 0 and 1, and the total of the probabilities must equal 1. Show that your assignment of probabilities in (b) satisfies this.

5.15 Pop quiz: A teacher gives a four-question unannounced true-false pop quiz, with two possible answers to each question.
a.   Use a tree diagram to show the possible response patterns, in terms of whether any given response is correct or incorrect. How many outcomes are in the sample space?
b.   An unprepared student guesses all the answers randomly. Find the probabilities of the possible outcomes on the tree diagram.
c.   Refer to (b). Using the tree diagram, evaluate the probability of passing the quiz, which the teacher defines as answering *at least* three questions correctly.

5.16 More true-false questions: Your teacher gives a true-false pop quiz with 10 questions.
a.   Show that the number of possible outcomes for the sample space of possible sequences of 10 answers is 1,024.
b.   With random guessing, show that the probability of getting *at least* one question wrong is 0.999.

5.17 Rain tomorrow?: Tomorrow, it might rain sometime, or it might not rain at all. Since the sample space has two possible outcomes, each must have probability ½. True or false? Explain.

5.18 Two girls: A family has two children. If each child is equally likely to be a girl or boy,
a. Construct a sample space for the genders of the two children.
b. Find the probability that the family has two girls.
c. Answer (b) if in reality the chance of a girl is 0.49.

5.19 All girls in a family: A family has four children, all girls. Is this unusual?
a. Construct a sample space for the possible genders of four children. (Hint: The outcome (G, B,G,G) represents the case where only the second child born is a boy.)
b. If each child is equally likely to be a girl or a boy, find the probability that all four children are female.
c. What other assumption is necessary for the calculation in (b) to be valid?

5.20 Wrong sample space: Refer to the previous exercise. What's wrong with the following logic for part (b)? A family with 4 children can have 0, 1, 2, 3, or 4 girls. Since there are five possibilities, the probability of 4 girls is 1/5.

5.21 Rosencrantz and Guildenstern: In the opening scene of Tom Stoppard's play *Rosencrantz and Guildenstern are Dead*, about two Elizabethan contemporaries of Hamlet, Guildenstern flips a coin 91 times and gets a head each time. Suppose the coin was balanced.
a. Specify the sample space, such that each outcome would be equally likely.
b. What's the probability of the event of getting a head 91 times in a row?
c. What's the probability of at least one tail, in the 91 flips?
d. What assumption do your solutions in (b) and (c) make?

5.22 Insurance: Every year the insurance industry spends considerable resources assessing risk probabilities. To accumulate a risk of about one in a million of death, you can drive 100 miles, take a cross country plane flight, work as a police officer for 10 hours, work in a coal mine for 12 hours, smoke two cigarettes, be a nonsmoker but live with a smoker for two weeks, or drink 70 pints of beer in a year (Wilson and Crouch, 2001, pp. 208-209). Show that a risk of about one in a million of death is also approximately the probability of flipping 20 heads in a row with a balanced coin.

5.23 True-blooded Republicans: Of those who voted in the 2000 and 2004 U. S. Presidential elections, about 48% voted Republican in 1996 and about 51% voted Republican in 2000. True or false: Since $0.48 \times 0.51 = 0.24$, about 24% voted for the Republican (George W. Bush) in both elections. If true, explain why. If false, explain the flaw in the logic.

5.24 Seat belt use and auto accidents: Based on records of automobile accidents in a recent year, the Department of Highway Safety and Motor Vehicles in Florida reported the counts who survived (S) and died (D), according to whether they wore a seat belt (Y = yes, N = no). The data are presented in the contingency table shown.

Outcome of auto accident by whether subject wore seat belt

Wore Seat Belt	Survived (S)	Died (D)	Total
Yes (Y)	412,368	510	412,878
No (N)	162,527	1601	164,128
Total	574,895	2111	577,006

---

a. What is the sample space of possible outcomes for a randomly selected individual involved in an auto accident? Use a tree diagram to illustrate the possible outcomes. (Hint: One possible outcome is (Y, S)).

b. Using these data, estimate (i) P(D), (ii) P(N).

c. Estimate the probability that an individual did not wear a seat belt and died.

d. Based on (b), what would the answer to (c) have been if the events N and D were independent? So, are N and D independent, and if not, what does that mean in the context of these data?

5.25 Catalog sales: You are the marketing director for a museum that raises money by selling gift items from a mail-order catalog. For each catalog sent to a potential customer, the customer's entry in the data file is Y if they ordered something and N if they did not (Y = yes, N = no). After you have mailed the fall and the winter catalogs, you estimate the probabilities of the buying patterns based on those who received the catalog by:

Outcome (fall, winter):	YY	YN	NY	NN
Probability:	0.30	0.10	0.05	0.55

a. Let F denote buying from the fall catalog and W denote buying from the winter catalog. Find P(F) and P(W).

b. Explain what the event "F and W" means, and find P(F and W).

c. Are F and W independent events? Explain why you would not normally expect

d. customer choices to be independent.

5.26 Arts and crafts sales: A local downtown arts and crafts shop found from past observation that 20% of the people who enter the shop actually buy something. Three potential customers enter the shop.

a. How many outcomes are possible for whether the clerk makes a sale to each customer? Construct a tree diagram to show the possible outcomes. (Let Y = sale, N = no sale)

b. Find the probability of at least one sale to the three customers.

c. What did your calculations assume in (b)? Describe a situation in which that assumption would be unrealistic.

5.27 Polygraph testing in pre-employment screening: The Employee Polygraph Protection Act of 1988 (EPPA) prohibits private employers from using polygraph testing for pre-employment screening, but it doesn't prohibit public employers such as police agencies. The APA Research Center at Michigan State University conducted a survey of approximately 700 of the largest police agencies in the U.S. to estimate how many police agencies use polygraphs for pre-employment screening. The survey revealed the following.[1]

Polygraph screen program	Actively use (U)	Do not use (N)	Discontinued (D)
Proportion	0.62	0.31	0.07

A recent graduate of a police academy decides to choose two large police agencies to apply for employment.

a. How many outcomes are possible for the polygraph status of the two sampled agencies? Sketch a tree diagram to show the sample space.

b. Let A = event that both sampled police agencies have the same polygraph status. Which outcomes are in A?

c. The graduate decides to apply to the two agencies she read about in a research study that had picked two police agencies at random and studied the way they were run. Find P(A) for random selection of two agencies.

5.3 CONDITIONAL PROBABILITY: WHAT'S THE PROBABILITY OF A, GIVEN B?

Many employers require potential employees to take a diagnostic test for drug use. The diagnostic test has two categorical variables of interest: (1) whether the person has recently used drugs (yes or no) and (2) whether the diagnostic test says the person has used them (yes or no). Suppose the diagnostic test says the person has recently used drugs. What's the probability that the person truly did use drugs?

This section introduces **conditional probability**. This deals with finding the probability of an event when you know that the outcome was in some particular part of the sample space. Most commonly, it is used to find a probability about a category for one variable (for instance, a person being a drug user), when we know the outcome on another variable (for instance, a test result that indicates drug use).

Finding the Conditional Probability of an Event

	AUDITED		
INCOME	Yes	No	Total
< $25,000	90	14010	14100
$25-50,000	71	30629	30700
$50-100,000	69	24631	24700
≥ $100,000	80	10620	10700
Total	310	79890	80200

Example 5 showed a contingency table on income level and whether audited by the Internal Revenue Service for taxpayers in 2002. The table is shown again in the margin. Using this table, we found that the probability a randomly selected taxpayer was audited equaled 310/80200 = 0.004. Were the chances higher if a taxpayer was at the highest income level? What was the probability of being audited, given that income was ≥ $100,000? From the margin table, the number having income ≥ $100,000 was 80 + 10620 = 10700. Of them, 80 were audited, for a probability of 80/10700 = 0.007. This is higher than 0.004, but still less than 1%.

In practice, we often know probabilities rather than counts for the outcomes in the sample space. Table 5.3 shows *joint probabilities*, based on the frequencies in the

---

[1] "The Polygraph Place-- Frequently Asked Questions" at www.polygraphplace.com/docs/information.shtml

margin table, for the eight possible outcomes. The probability of income $\geq$ \$100,000 was 0.1334, whereas the probability of both income $\geq$ \$100,000 and being audited was 0.0010. So, of the proportion 0.1334 with highest income, the proportion, 0.0010 were audited, for a fraction 0.0010/0.1334 = 0.007, the same answer we obtained above using the frequencies.

Table 5.3: Joint Probabilities for Taxpayers at the Eight Possible Combinations of Categories of Income and Whether Audited. Each frequency in Table 5.2 was divided by 80200 to obtain the joint probabilities shown here, such as 90/80200 = 0.0011.

	WHETHER AUDITED		
INCOME LEVEL	**Yes**	**No**	**Total**
Under \$25,000	0.0011	0.1747	0.1758
\$25,000-\$49,999	0.0009	0.3819	0.3828
\$50,000-\$99,999	0.0009	0.3071	0.3080
$\geq$ **\$100,000**	0.0010	0.1324	0.1334
Total	0.0039	0.9961	1.0000

Let event A denote {audited = yes} and event B denote {income $\geq$ \$100,000}. Of the highest income group, also being audited is the intersection event A and B. So, given B, the probability of A is the proportion of A and B cases out of the B cases. This is P(A and B)/P(B). This ratio is the **conditional probability** of the event A, given the event B.

IN WORDS	**Definition: Conditional Probability**
P(A | B) is read as "the probability of event A, *given* event B." The vertical slash represents the word "given." Of the times that B occurs, P(A | B) is the proportion of times that A also occurs.	For events A and B, the **conditional probability** of event A, given that event B has occurred, is $$P(A \mid B) = \frac{P(A \text{ and } B)}{P(B)}.$$

From Table 5.3,

$$P(A \mid B) = \frac{P(A \text{ and } B)}{P(B)} = \frac{0.0010}{0.1334} = 0.007.$$

Given that a taxpayer has income $\geq$ \$100,000, the chances of being audited are 0.007. Given that a taxpayer has income $\geq$ \$100,000, the chances of *not* being audited are 0.1324/0.1334 = 0.993. Since not being audited is the complement of being audited, conditional on income $\geq$ \$100,000, P(A^c \| B) = 1 – P(A \| B) = 1 - 0.007 = 0.993.

To practice this concept, try Exercise 5.30 .

$\blacklozenge$

In Section 3.1, in learning about contingency tables we saw that we could find **conditional proportions** for a categorical variable, at any particular category of a second categorical variable. These enable us to study how the outcome on a response variable depends on the outcome on the second categorical variable. The conditional probabilities just found are conditional proportions found for the population of all taxpayers, treating audit status as a response variable and income level as an explanatory variable. We could find similar conditional probabilities on audit status at each level of income. We'd then get the results shown in Table 5.4. We get each conditional probability by dividing the joint probability for that particular cell (which refers to an intersection of an income event and an audit event, as shown in the margin table) by the row total that is the probability of income at that level. In each row of Table 5.4, the conditional probabilities sum to 1.0.

RECALL: Table 5.3 was

INCOME	AUDITED	
	Yes	No
Under $25,000	.0011	.1747
$25,000-$49,999	.0009	.3819
$50,000-$99,999	.0009	.3071
≥ $100,000	.0010	.1324

The total of these *joint probabilities* over the 8 cells is 1.0.

Table 5.4: Conditional Probabilities on Whether Audited, Given the Income Level. Each probability in Table 5.3 was divided by the row marginal total probability to obtain the conditional probabilities shown here, such as 0.0011/0.1758 = 0.0064.

	WHETHER AUDITED		
INCOME LEVEL	**Yes**	**No**	**Total**
Under $25,000	0.0063	0.9937	1.000
$25,000-$49,999	0.0024	0.9976	1.000
$50,000-$99,999	0.0029	0.9971	1.000
≥ $100,000	0.0075	0.9925	1.000

← 0.0075 = 0.0010/0.1334 from Table 5.3

Figure 5.9 is a graphical illustration of the definition of conditional probability. "Given event B" means that we restrict our attention to the outcomes in that event. This is the set of outcomes in the denominator. The proportion of those cases in which A occurred are those outcomes that are in event A as well as B. So, the intersection of A and B is in the numerator.

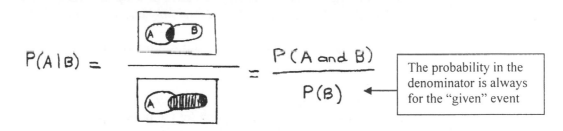

$$P(A \mid B) = \frac{}{} = \frac{P(A \text{ and } B)}{P(B)}$$

The probability in the denominator is always for the "given" event

Figure 5.9: Venn Diagram of Conditional Probability of Event A, Given Event B. Of the cases in which B occurred, it is the proportion in which A also occurred. **Questions:** Sketch a representation of $P(B \mid A)$. Is $P(A \mid B)$ necessarily equal to $P(B \mid A)$?

EXAMPLE 9: EVALUATING THE TRIPLE BLOOD TEST FOR DOWN SYNDROME

Picture the Scenario
Diagnostic testing is used to detect many medical conditions. Some of these tests use blood samples, such as the PSA test for prostate cancer, the ELISA screening test for HIV, and selected tests for pregnancy. Some tests use X-rays or other imaging devices such as the mammogram for diagnosing breast cancer and the MRI body scan. Diagnostic testing is also used for drug detection of employees in a workplace or athletes in a competition, often with urine samples.

A diagnostic test for some condition is said to be **positive** if it states that the condition is present and **negative** if it states that the condition is absent. How accurate are diagnostic tests? One way to assess accuracy is to measure the probabilities of the two types of possible error:

False positive: Test states the condition is present, but it is actually absent.
False negative: Test states the condition is absent, but it is actually present.

The Triple Blood Test screens pregnant women for the genetic disorder, Down syndrome. This syndrome, which occurs in about 1 in 800 live births, arises from an error in cell division that results in a fetus having an extra copy of chromosome 21. It is the most common genetic cause of mental impairment. The chance of having a baby with Down syndrome increases after a woman is 35 years old.

A study of 5282 women aged 35 or over analyzed the Triple Blood Test[2] to test its accuracy. It was reported that of the 5282 women, "48 of the 54 cases of Down syndrome would have been identified using the test and 25 percent of the unaffected pregnancies would have been identified as being at high risk for Down syndrome (these are false positives)."

Questions to Explore
a. Construct the contingency table that shows the counts for the possible outcomes of the blood test and whether the fetus has Down syndrome.
b. Assuming the sample is representative of the population, find the estimated probability of a positive test for a randomly chosen pregnant woman 35 years or older.
c. Given that the diagnostic test result is positive, find the estimated probability that Down syndrome truly is present.

Think It Through
a. We'll use the following notation for the possible outcomes of the two variables:

Down syndrome status: D = Down syndrome present, D^c = unaffected
Blood test result: POS = positive, NEG = negative.

Table 5.5 shows the four possible combinations of outcomes. From the article quote, there were 54 cases of Down syndrome. This is the first row total. Of them, 48 tested positive and 54 − 48 = 6 tested negative. These are the counts in the first row. Since $n = 5282$, if there were 54 Down cases, there were 5282 − 54 = 5228

---

[2] J. Haddow et al., *New England Journal of Medicine,* vol. 330, pp. 1114 – 1118, 1994

unaffected cases, event D^c. That's the second row total. Now, 25% of those 5228, or $0.25 \times 5228 = 1307$, would have a positive test. The remaining $5228 - 1307 = 3921$ would have a negative test. These are the counts for the two cells in the second row.

Table 5.5: Contingency Table for Triple Blood Test of Down Syndrome

	Blood Test Result		
Down Syndrome Status	**POS**	**NEG**	**Total**
D (Down)	48	6	54
D^c (unaffected)	1307	3921	5228
Total	1355	3927	5282

b. From Table 5.5, the estimated probability of a positive test is $P(POS) = 1355/5282 = 0.257$.

c. The probability of Down syndrome, given that the test is positive, is the conditional probability, $P(D \mid POS)$. Conditioning on a positive test means we consider only the cases in the first column of Table 5.5. Of the 1355 who tested positive, 48 cases had Down syndrome, so the estimated probability is $48/1355 = 0.035$. Let's see how to get this from the definition of conditional probability,

$$P(D \mid POS) = \frac{P(D \text{ and } POS)}{P(POS)}.$$

Since $P(POS) = 0.257$ and $P(D \text{ and } POS) = 48/5282 = 0.009$, we estimate $P(D \mid POS) = 0.009/0.257 = 0.035$. In summary, of the women who tested positive, fewer than 4% actually had fetuses with Down syndrome. This is somewhat comforting news to a woman who has a positive test result.

Insight
So, why should a woman undergo this test, since most positives are false positives? From Table 5.5, notice that $P(D) = 54/5282 = 0.010$, so we estimate a 1% chance of Down syndrome for women aged 35 or over. Also from Table 5.5, $P(D \mid NEG) = 6/3927 = 0.0015$, a bit more than 1 in a thousand. A woman can have much less worry about Down syndrome if she has a negative test result, because the chance of Down is then a bit more than 1 in a thousand, compared to 1 in a hundred overall.

♦

To practice this concept, try Exercises 5.34 and 5.35.

IN PRACTICE

When you read or hear a news report that uses a probability statement, be careful to distinguish whether it is reporting a conditional probability. Most statements are conditional on some event, and must be interpreted in that context. For instance, probabilities reported by opinion polls are often conditional on a person's gender, or race, or age group.

Multiplication Rule for Finding P(A and B)

From Section 5.2, when A and B are independent events, $P(A \text{ and } B) = P(A) \times P(B)$. The definition of conditional probability provides a more general formula for $P(A \text{ and } B)$ that holds regardless of whether A and B are independent. We can rewrite the definition $P(A \mid B) = P(A \text{ and } B)/P(B)$, multiplying both sides of the formula by $P(B)$, as

$$P(A \text{ and } B) = P(A \mid B) \times P(B).$$

Multiplication Rule for Evaluating P(A and B)

For events A and B, the probability that A and B both occur equals

$$P(A \text{ and } B) = P(A \mid B) \times P(B).$$

Applying the conditional probability formula to $P(B \mid A)$, we also see that

$$P(A \text{ and } B) = P(B \mid A) \times P(A).$$

EXAMPLE 10: HOW LIKELY IS A DOUBLE FAULT IN TENNIS?

Picture the scenario:
In a tennis match, on a given point the player who is serving has two chances to hit the ball in play. The ball must fall in the correct marked box area on the opposite side of the net. A serve that misses that box is called a *fault*. Most players hit the first serve very hard, resulting in a fair chance of making a fault. If they do make a fault, they hit the second serve less hard and with some spin, making it more likely to be successful. Otherwise, with two misses - a "*double fault*," they lose the point.

Question to Explore
The 2004 men's champion in the Wimbledon tennis tournament was Roger Federer of Switzerland. During that tournament he made 64% of his first serves. So, he faulted on the first serve 36% of the time $(100 - 64 = 36)$. Given that he made a fault with his first serve, he made a fault on his second serve only 6% of the time. Assuming these are typical of his serving performance, when he serves, what is the probability that he makes a double fault?

Think It Through
Let F1 be the event that Federer makes a fault with the first serve, and let F2 be the event that he makes a fault with the second serve. From the information above, we know $P(F1) = 0.36$ and $P(F2 \mid F1) = 0.06$, as shown in the margin figure. The event that Federer makes a double fault is F1 and F2. From the multiplication rule, its probability is

$$P(F1 \text{ and } F2) = P(F2 \mid F1) \times P(F1) = 0.06 \times 0.36 = 0.02.$$

PHOTO of Roger Federer serving

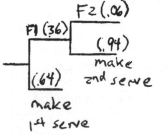

Insight

Federer makes a fault on his first serve 36% of the time, and in 6% of those cases he makes a fault on his second serve also. So, he makes a double fault in 6% of the 36% of points in which he faults on the first serve, or in the proportion $0.06 \times 0.36 = 0.02$, 2% of the points.

◆

To practice this concept, try Exercise 5.38.

Sampling with or without Replacement

In many sampling processes, once subjects are selected from a population, they are not eligible to be selected again. This is called *sampling without replacement*. At any stage of such a sampling process, probabilities of potential outcomes depend on the previous outcomes. Conditional probabilities are then used in finding probabilities of the possible samples.

EXAMPLE 11: HOW LIKELY ARE YOU TO WIN THE LOTTO?

Picture the Scenario

Many states have lotteries. The biggest jackpot, typically millions of dollars, usually comes from the Lotto game. In Georgia's Lotto, 6 numbers are randomly sampled without replacement from the integers 1 to 46. For example, a possible sample is (4, 9, 23, 26, 40, 46). Their order of selection is not important.

Question to Explore

You buy a Lotto ticket. What is the probability that it is a winning ticket?

Think It Through

The probability of winning is the probability that the six numbers chosen are the six that you have on your ticket. For the first number chosen, 6 of the 46 numbers that can be selected are ones you hold, so P(you have 1st number) = 6/46. Given that you have the first number, for the second trial there are 5 numbers left that you hold out of 45 possible, so P(have 2nd number| have 1st number) = 5/45. Given that you have the first two, for the third trial there are 4 numbers left that you hold out of 44 possible, so P(have 3rd number | have 1st and 2nd numbers) = 4/44. Continuing with this logic, using an extension of the multiplication rule with conditional probabilities,

P(have all 6 numbers) = P(have 1st and 2nd and 3rd and 4th and 5th and 6th)

= P(have 1st)P(have 2nd | have 1st)P(have 3rd | have 1st and 2nd) ...
... P(have 6th | have 1st and 2nd and 3rd and 4th and 5th)

= $(6/46) \times (5/45) \times (4/44) \times (3/43) \times (2/42) \times (1/41) = 720/6,744,109,680 = 0.0000001$.

This is about 1 chance in ten million.

DID YOU KNOW

Comparable to the probability of 0.0000001 of winning Lotto are the annual probabilities of about 0.0000004 of being hit by a meteorite, 0.0000002 of dying in a tornado, and 0.00000016 of dying by a lightning strike (Wilson and Crouch 2001, p. 200). It's also roughly the probability that a person of average mortality will die in the next 4 minutes.

Insight

Let's give this small number some perspective. The probability that a given person will die during the next year in a car crash is about 0.00015, 1500 times the chance of winning this Lotto game. You would be extremely lucky to win. The chance of this is less than your chance of being hit by a meteorite in the next year. If you have money to spare, go ahead and play the lottery, but understand why many call it "sport for the mathematically challenged." You are very unlikely to see a big return on your money. This is a regressive way for states to collect money. Many contributors are people who cannot afford it. As Mark Twain said,[3] "There are two times in a man's life when he should not speculate: when he can't afford it, and when he can."

♦

To practice this concept, try Exercise 5.42.

Since Lotto uses *sampling without replacement*, once a number is chosen, it cannot be chosen again. If, by contrast, Lotto allowed numbers to be used more than once, the sampling scheme would be *sampling with replacement*.

As we saw in the Lotto example, each time a subject is sampled from a small population without replacement, the population remaining is reduced by one and the conditional probability of a particular outcome changes, possibly considerably. With large population sizes, reducing the population by one does not much affect the probability from one trial to the next. In practice, when selecting random samples, we normally sample without replacement. With populations that are large compared to the sample size, the probability at any given observation depends little on the previous observations.

Independent Events Defined Using Conditional Probability

Two events A and B are **independent** if the probability that one occurs is not affected by whether or not the other event occurs. This is expressed more formally using conditional probabilities.

Definition: Independent Events, in Terms of Conditional Probabilities

Events A and B are **independent** if $P(A \mid B) = P(A)$. If this holds, then also necessarily $P(B \mid A) = P(B)$ and $P(A \text{ and } B) = P(A) \times P(B)$.

For instance, let's consider two flips of a balanced coin, for which the sample space is {HH, HT, TH, TT}. With a balanced coin, these four outcomes are equally likely. Let A denote {first flip is a head} and let B denote {second flip is a head}. Then $P(A) = \frac{1}{2}$, since two of the four possible outcomes have a head on the first flip. Likewise, $P(B) = \frac{1}{2}$. Also, $P(A \text{ and } B) = \frac{1}{4}$, since this corresponds to the single outcome, HH. So, from the definition of conditional probability,

$$P(A \mid B) = P(A \text{ and } B)/P(B) = (\tfrac{1}{4})/(\tfrac{1}{2}) = \tfrac{1}{2}.$$

---

[3] In *Following the Equator*

Thus, $P(A \mid B) = \frac{1}{2} = P(A)$, so A and B are independent events. Intuitively, given B (that is, that the second flip is a head), the probability that the first flip was a head is $\frac{1}{2}$, since one outcome (HH) has this, out of the two possibilities (HH, TH).

Let A denote {first flip is a head} and C denote {both flips are heads}. Then $P(C) = \frac{1}{4}$, since HH is one of the four equally-likely outcomes. Also, $P(C \mid A) = \frac{1}{2}$, since one outcome has two heads out of the two outcomes (HH, HT) for which the first flip is a head. Since $P(C)$ and $P(C \mid A)$ are different, A and C are not independent. They are *dependent* events.

In sampling without replacement, outcomes of different trials are dependent. For instance, in the Lotto game described in Example 11, let A denote {your first number is chosen} and B denote {your second number is chosen}. Then $P(A) = 6/46$, but $P(A \mid B) = 5/45$ (since there are 5 possibilities out of the 45 numbers left), which differs slightly from $P(A)$.

EXAMPLE 12: HOW TO CHECK WHETHER TWO EVENTS ARE INDEPENDENT

Picture the Scenario

Table 5.5 showed a contingency table relating the result of a diagnostic blood test (POS = positive, NEG = negative) to whether a pregnant woman has Down syndrome (D = Down syndrome, D^c = unaffected). The estimated joint probabilities using the frequencies in that table are shown in the margin.

	Blood Test		Total
Status	POS	NEG	
D	0.009	0.001	0.010
D^c	0.247	0.742	0.990
Total	0.256	0.743	1.00

Questions to Consider

a. Are the events POS and D independent, or dependent?

b. Are the events POS and D^c independent, or dependent?

Think It Through

a. The probability of a positive test result is 0.256. However, the probability of a positive result, given Down syndrome, is

$$P(POS \mid D) = P(POS \text{ and } D)/P(D) = 0.009/0.010 = 0.90.$$

Since $P(POS \mid D) = 0.90$ differs from $P(POS) = 0.256$, the events POS and D are dependent.

b. Likewise,

$$P(POS \mid D^c) = P(POS \text{ and } D^c)/P(D^c) = 0.247/0.990 = 0.250.$$

This differs slightly from $P(POS) = 0.256$, so POS and D^c are also dependent events.

Insight

As we'd hope, the probability of a positive result depends on whether the fetus has Down syndrome, and it's much higher if the fetus does. The diagnostic test would be worthless if the disease status and the test result were independent.

If A and B are dependent events, then so are A and B^c and so are A^c and B and so are A^c and B^c. For instance, if A depends on whether B occurs, then so does A depend on whether B does not occur. So once we find that POS and D are dependent events, we know that POS and D^c are dependent events also.

◆

To practice this concept, try Exercise 5.43.

We can now justify the formula given in Section 5.2 for the probability of the intersection of two independent events, P(A and B) = P(A)$\times$P(B). This is a special case of the multiplication rule for finding P(A and B),

$$P(A \text{ and } B) = P(A \mid B) \times P(B).$$

If A and B are independent, then P(A | B) = P(A), so the multiplication rule simplifies to

$$P(A \text{ and } B) = P(A) \times P(B).$$

Checking for Independence

Here are three ways to check whether events A and B are independent:

- Is P(A | B) = P(A)?

- Is P(B | A) = P(B)?

- Is P(A and B) = P(A)$\times$P(B)?

If any of these is true, the others are also true and the events A and B are independent.

EXAMPLE 13: ARE DISJOINT EVENTS INDEPENDENT?

Picture the Scenario
Suppose A and B are disjoint events, each with positive probability.

Question to Explore
Are they independent events also?

Think It Through
Since A and B are disjoint, they have no outcomes in common. So, P(A and B) = 0. Thus P(A)$\times$P(B) cannot equal P(A and B), so they are dependent events. In fact, P(B | A) = P(A and B)/P(A) = 0, so given that A occurs, B *cannot* occur.

Insight
Consider events POS = positive diagnostic test and NEG = negative diagnostic test. They are disjoint, since they have no common outcomes. Knowing that a test result

was negative, it cannot have been positive, so P(POS | NEG) = 0. Thus, these events are very strongly dependent.

♦

To practice this concept, try Exercise 5.43.

SECTION 5.3: PRACTICING THE BASICS

5.28 Alcohol and college students: For the results of the 2001 Harvard School of Public Health College Alcohol Study Survey about binge drinking, the estimated probability of being a binge drinker was 0.49 for males and 0.41 for females. Introducing notation, express each of these as a conditional probability.

5.29 Spam: Because of the increasing nuisance of spam e-mail messages, many start-up companies have emerged to develop e-mail filters. One such filter was recently advertised as being 95% accurate. This could mean that (a) 95% of spam is blocked, (b) 95% of valid e-mail is allowed through, (c) 95% of the e-mail allowed through is valid, or (d) 95% of the blocked e-mail is spam. Let S denote {message is spam}, and let B denote {filter blocks message}. Using these events and their complements, identify each of these four possibilities as a conditional probability.

Income	Audited Yes	Audited No
< $25,000	0.0011	0.1747
$25-50,000	0.0009	0.3819
$50-100,000	0.0009	0.3071
≥ $100,000	0.0010	0.1324

5.30 Audit and low income: Table 5.3 on audit status and income is shown in the margin. Show how to find the probability of
a. Being audited, given that the taxpayer is in the lowest income category.
b. Being in the lowest income category, given that the taxpayer is audited.

5.31 Commuting to work: The 2000 U.S. census reported the frequencies in the following table for how workers 16 years and over commute to work.

Method Used to Commute to Work	Frequency
Car, truck, or van--drove alone	97,102,050
Car, truck, or van--carpooled	15,634,051
Public transportation (including taxicab)	6,067,703
Walked	3,758,982
Other means	1,532,219
Worked at home	4,184,223
Total	128,279,228

a. What is the probability that a randomly selected worker commuted to work in a car, truck, or van?
b. Given that a person commuted to work in a car, truck, or van, what is the probability of car pooling?

5.32 Cancer deaths: Current estimates are that about 25% of all deaths are due to cancer, and of the deaths that are due to cancer, 30% are attributed to tobacco, 40% to diet, and 30% to other causes.
a. Define events, and identify which of these four probabilities refer to conditional probabilities.
b. Find the probability that a death is due to cancer and tobacco.

309

Outcome

Belt	S	D	Total
Yes	412,368	510	412,878
No	162,527	1601	164,128
Total	574,895	2111	577,006

5.33 Revisiting seat belts and auto accidents: The table in the margin is from Exercise 5.24 classifying auto accidents by survival status (S = survived, D = died) and seat belt status of the individual involved in the accident.
a. Estimate the probability that the individual died (D) in the auto accident.
b. Estimate the probability that the individual died, given that they (i) wore, (ii) did not wear a seat belt. Interpret results.
c. Are the events of dieing and wearing a seat belt independent? Justify your answer.

5.34 Go Celtics!: Larry Bird, who played pro basketball for the Boston Celtics, was known for being a good shooter. In games during 1980-1982, when he missed his first free throw, 48 out of 53 times he made the second one, and when he made his first free throw, 251 out of 285 times he made the second one.
a. Form a contingency table that cross tabulates the outcome of the first free throw (made or missed) in the rows and the outcome of the second free throw (made or missed) in the columns.
b. For a given pair of free throws, estimate the probability that Bird (i) made the first free throw, (ii) made the second free throw.
c. Estimate the probability that Bird made the second free throw, given that he made the first one. Does it seem as if his success on the second shot depends strongly, or hardly at all, on whether he made the first?

Blood Test

Down	POS	NEG	Total
D	48	6	54
D^c	1307	3921	5228
Total	1355	3927	5282

5.35 Down syndrome again: Example 9 discussed the triple blood test for Down syndrome, using data summarized in a table shown again in the margin.
a. Given that a test result is negative, show that the probability the fetus actually has Down syndrome is $P(D \mid NEG) = 0.0015$.
b. Is $P(D \mid NEG)$ equal to $P(NEG \mid D)$? If so, explain why. If not, find $P(NEG \mid D)$.

5.36 America the fast food nation: A Gallup Organization poll conducted in July 2003 reported the results of a survey asking approximately 1000 randomly chosen Americans (18 years or older) questions about fast food consumption. It reported that 59% of the male respondents and 46% of the female respondents eat fast food at least once weekly.
a. What type of percentages (probabilities) were reported by Gallup: Ordinary, or conditional? Explain, by specifying events to which the probabilities refer.
b. Suppose that half those polled were female and half were male. Construct a contingency table showing counts for gender and fast food status.

5.37 Happiness in marriage: Are people happy in their marriages? The table shows results from the 2002 General Social Survey for married adults classified by gender and level of happiness.

	Level of Happiness			
Gender	Very Happy	Pretty Happy	Not too Happy	Total
Male	221	95	9	325
Female	149	120	9	278
Total	370	215	18	603

a. Estimate the probability that a married adult is very happy.

b. Estimate the probability that a married adult is very happy, (i) given that their gender is male, (ii) given that their gender is female, (iii) not given any information about their gender.

c. For these subjects, are the events being very happy and being a male independent? (Your answer will be merely for this sample. Chapter 10 will show how to answer this for the population.)

5.38 Serena Williams serves: In the 2004 Wimbledon tennis championship, Serena Williams made 63% of her first serves. When she faulted on her first serve, she made 93% of her second serves. Assuming these are typical of her serving performance, when she serves, what is the probability that she makes a double fault?

5.39 Shooting free throws: Pro basketball player Shaquille O'Neal is a poor free-throw shooter. Consider situations in which he shoots a pair of free throws. The probability that he makes the first free throw is 0.50. Given that he makes the first, suppose the probability that he makes the second is 0.60. Given that he misses the first, suppose the probability that he makes the second one is 0.40.

a. What is the probability that he makes both free throws?

b. Find the probability that he makes *one* of the two free throws (i) using the multiplicative rule with the two possible ways he can do this, (ii) by defining this as the complement of making neither or both of the free throws.

c. Are the results of the free throws independent? Explain.

d. If the probability that O'Neal makes any given free throw is 0.50 and if the results of different free throws *are* independent, what is the probability that of the two free throws he makes (i) 0, (ii) 1, (iii) 2?

5.40 Drawing cards: A standard card deck has 52 cards consisting of 26 black and 26 red cards. Three cards are dealt from a shuffled deck, *without replacement*.

a. True or false: The probability of being dealt 3 black cards is $(1/2) \times (1/2) \times (1/2) = 1/8$. If true, explain why. If false, show how to get the correct probability.

b. Let $A =$ first card red and $B =$ second card red. Are A and B independent? Explain why or why not.

c. Answer (a) and (b) if each card is replaced in the deck after being dealt.

5.41 Screening smokers for lung cancer: An article about using a diagnostic test (helical computed tomography) to screen adult smokers for lung cancer warned that a negative test may cause harm by providing smokers with false reassurance, and a false positive test results in an unnecessary operation opening the smoker's chest (Mahadevia et al., *JAMA*, vol. 289, 313-322, 2003). Explain what false negatives and false positives mean in the context of this diagnostic test.

5.42 Big loser in Lotto: Example 11 showed that the probability of having the winning ticket in Georgia's Lotto was 0.0000001. Find the probability of holding a ticket that has zero winning numbers out of the 6 numbers selected (without replacement) for the winning ticket out of the 46 possible numbers.

5.43 Checking independence: In three independent flips of a balanced coin, let A denote {first flip is a head}, B denote {second flip is a head}, C denote {first two flips are heads}, and D denote {three heads on the three flips}.

a. Find the probabilities of A, B, C, and D.

b. Which, if any, pairs of these events are independent? Explain.

5.44 Find P(A | B): For two events A and B, P(A) = 0.60, P(B) = 0.40, and P(B | A) = 0.60. Find P(A | B).

5.45 Stay in school: Suppose 80% of students finish high school. Of them, 50% finish college. Of them, 20% get a masters' degree. Of them, 30% get a PhD.
a. What percentage of students get a PhD?
b. Explain how your reasoning in (a) used a multiplication rule with conditional probabilities.
c. Conditional on finishing college, find the probability of getting a PhD.

5.4   APPLYING THE PROBABILITY RULES

You may not have thought about it, but probability relates to many aspects of your daily life--for instance, when you make decisions that affect your financial well being, and when you evaluate risks due to lifestyle decisions (flying somewhere, wearing a seat belt while in a vehicle, smoking, dieting, exercising). Objectively or subjectively you need to consider questions such as: What's the chance that the business you're thinking of starting will succeed? What's the chance that you will get cancer if you continue to smoke? What's the chance that the extra collision insurance you're thinking of getting for your car will be needed?

To finish this chapter, we'll apply the basics of probability to a few more examples to practice key concepts from this chapter. We'll start by seeing what probability tells us about coincidence in our lives.

Is a "Coincidence" Truly an Unusual Event?

Some events in our lives seem to be highly coincidental. One of the authors, who lives in Georgia, once spent a summer vacation in Newfoundland, Canada. One day during that trip, she made an unplanned stop at a rest area and observed a camper with a Georgia license tag from a neighboring county to her hometown. She discovered that the camper's owner was a patient of her physician husband. Was this meeting as highly coincidental as it seemed?

Events that seem highly coincidental are often not so unusual when viewed in the context of *all* the possible random occurrences at all times. Lots of topics can trigger an apparent "coincidence--the person you meet having the same last name as yours, the same birth place, the same high school or college alma mater, the same profession, the same birthday, the same make, year and color of car, a common friend, and so on. Think of the huge number of possible coincidences that can potentially occur. It's really not so surprising that in your travels you will sometime have such a coincidence, such as seeing a friend or meeting someone who knows someone you know. This sort of thing happens to all of us. The other author has had several such incidences, such as encountering people he knows in various airports, on the underground in London, England, and at a restaurant in Naples, Italy.

With a large enough sample of people or times or topics, seemingly surprising things are actually quite sure to happen. Events that are rare per person occur rather commonly with large numbers of people. If a particular event happens to one person in a million each day, then in the U.S. we expect about 300 such events a day and more than 100,000 every year. The one in a million chance regularly occurs, however surprised we may be if it should happen to us. The **law of very large numbers** states that if something has a very large number of opportunities to happen, occasionally it will happen, even if it seems highly unusual.

If you take a coin now and flip it 10 times, would you be surprised to get 10 heads? Probably you would. But if you flipped the coin for a long time, would you be surprised to get 10 heads in a row at some stage? Perhaps you would, but you should not be. For instance, if you flip a balanced coin 2000 times, then you can expect the longest run of heads during those flips to be about 10. When a seemingly unusual event happens to you, think about whether it is like seeing 10 heads on the next ten flips, or more like seeing 10 heads in a row sometime in a long series of flips. If the latter, it's really not such an unusual occurrence.

Coincidence and seemingly unusual patterns

Once we have data, it's easy to find patterns: ten heads in a row, ten tails in a row, five tails followed by five heads, and so forth for lots of other possibilities. Our minds are programmed to look for patterns. Out of the huge number of things that can happen to us and to others over time, it's not surprising to see patterns that seem unusual occasionally.

Sometimes a cluster of occurrences of some disease in a neighborhood will cause worry in residents that there is some environmental cause. But, *some* disease clusters will appear around a nation just by chance. If we look at a large number of places and times, by the law of very large numbers we should expect some disease clusters. By themselves, they seem unusual, but viewed in a broader context they may not be. Epidemiologists are statistically trained scientists who face the difficult task of determining which such events can be explained by ordinary random variation and which cannot.

To illustrate that an event you may perceive as a coincidence is actually not that surprising, let's answer the question, "What is the chance that at least two people in your class have the same birthday?"

ACTIVITY 2: MATCHING BIRTHDAYS

If your class is small to moderate in size (say, fewer than about 60 students), the instructor may have all your classmates state their birth dates. Does a pair of students in the class share the same birthday?

EXAMPLE 14: IS A MATCHING BIRTHDAY SURPRISING?

Picture the Scenario

With a small class, since there are 365 possible birth dates (without counting February 29), our intuition tells us that the probability is small that there will be any birthday matches. Suppose a class has 25 students. Assume that the birth date of a student is equally likely to be any one of the 365 days in a year and that students' birth dates are independent (for example, there are no twins in the class). Let's evaluate the chance of having a birthday match using the probability rules of this chapter.

Question to Explore

What is the probability that *at least* two of the 25 students have the same birthday? Is our intuition correct that the probability of a match is small?

Think It Through

The event of *at least one* birthday match includes 1 match, 2 matches, or more. To find the probability of *at least one* outcome, it is simpler to find the complement probability of *no* outcomes. Then

$$P(\text{at least one match}) = 1 - P(\text{no matches}).$$

To begin, suppose a class has only 2 students. The first student's birthday could be any of 365 days. Given that student's birthday, the chance that the second student's birthday is different is 364/365, since 364 of the possible 365 birthdays are different. The probability is $1 - 364/365 = 1/365$ that the two students share the same birthday.

Now, suppose a class has 3 students. The probability that all three have different birthdays is

P(no matches) = P(students 1 and 2 and 3 have different birthdays) =
P(students 1 and 2 different) $\times$ P(student 3 different | students 1 and 2 different)
$$= (364/365) \times (363/365).$$

The second probability in this product equals 363/365 because there are 363 days left for the third student that differ from the different birthdays of the first two students.

For 25 students, similar logic applies. By the time we get to the 25th student, for that student's birthday to differ from the other 24, there are 341 choices left out of 365 possible birthdays. So

P(no matches) = P(students 1 and 2 and 3 ... and 25 have different birthdays)

$$= (364/365) \times (363/365) \times (362/365) \times \ldots\ldots\ldots\ldots \times (341/365).$$
  student 2,   student 3,   student 4,   next 20 students   student 25,
    given 1   given 1,2   given 1,2,3   given 1,...,24

This product equals 0.43. Using the probability for the complement of an event,

$$P(\text{at least one match}) = 1 - P(\text{no matches}) = 1 - 0.43 = 0.57.$$

The probability exceeds ½ of at least one birthday match in a class of 25 students.

Insight
Is this higher than you expected? It should not be once you realize that with 25 students, there are 300 *pairs* of students who can share the same birthday (See Exercise 5.48). Remember that with lots of opportunities for something to happen, coincidences are really not so surprising.

Did you have a birthday match in your class? Was it coincidence? If the number of students in your class is at least 23, the probability of at least one match is greater than ½. For a class of 50 students, the probability of at least one match is 0.97. For 100 students, it is 0.9999997 (there are then 4950 different *pairs* of students). Here are some other facts about matching birthdays that may surprise you:

- With 88 people, there's a ½ chance that at least 3 people have the same birthday.

- With 18 people, there's a ½ chance that at least 3 people were born on the same numbered day of the month (i.e., all 3 on the 1st, or all 3 on the 2nd).

- With 14 people, there's a ½ chance that at least two people have a birthday within a day of each other in the calendar year.

◆

To practice this concept, try Exercise 5.46.

Probabilities and Diagnostic Testing

We've seen the important role that probability plays in diagnostic testing, illustrated in Section 5.3 (Example 9) with the pregnancy test for Down syndrome. Table 5.6 summarizes conditional probabilities about the result of a diagnostic test, given whether some condition or state (such as Down syndrome) is present. We let S denote that the state is present and S^c denote that it is not present.

Table 5.6: Probabilities of Correct and Incorrect Results in Diagnostic Testing. The probabilities in the body of the table refer to the test result, conditional on whether the state (S) is truly present. The sensitivity and specificity are the probabilities of the two types of correct diagnoses.

State present?	Diagnostic Test Result Positive (POS)	Negative (NEG)	Total Probability
Yes (S)	Sensitivity $P(POS\|S)$	False negative rate $P(NEG\|S)$	1.0
No (S^c)	False positive rate $P(POS\|S^c)$	Specificity $P(NEG\|S^c)$	1.0

> **RECALL**
>
> The vertical slash means "given." $P(POS \mid S)$ denotes the conditional probability of a positive test result, given that the state is present.

Sensitivity and **specificity** refer to correct test results, given the actual state. For instance, given that the state tested for is present, sensitivity is the probability the test detects it by giving a positive result, that is, $P(POS \mid S)$.

Sensitivity and specificity refer to the probability of the test result, given whether the state is present. Other conditional probabilities refer to the probability the state is present, given the test result. Table 5.7 lists these and introduces terminology by which these probabilities are often reported. These conditional probabilities reverse the direction of what is given and what is not given compared to the sensitivity and specificity. Positive and negative **predictive values** refer to correct diagnoses, given the test result. **Prevalence** measures how often the state tested for is truly present.

Table 5.7: Diagnostic Testing Probabilities of a Correct Result. Event notation is (S, S^c) for state (present, absent), (POS, NEG) for diagnostic test result.

Prevalence: P(state present) = P(S)

Test, given state:
  Sensitivity: P(test positive | state present) = P(POS | S)
  Specificity: P(test negative | state not present) = P(NEG | S^c)

State, given test:
  Positive predictive value: P(state present | test positive) = P(S | POS)
  Negative predictive value: P(state not present | test negative) = P(S^c | NEG)

Medical journal articles that discuss diagnostic tests commonly report the sensitivity and specificity. However, what's often more relevant to you once you take a diagnostic test are the conditional probabilities that condition on the test result. If a diagnostic test for cancer is positive, you want to know the probability that cancer is truly present. If you know the sensitivity, specificity, and prevalence, can you find P(S | POS) using the rules of probability?

The easiest way to do this is with a tree diagram, as shown in the margin. The first branches show the probabilities of the two possible states, P(S) and P(S^c). The next set of branches show the known conditional probabilities, such as P(POS | S), for which we are given the true state. Then the products P(POS | S)P(S) and P(POS | S^c)P(S^c) give intersection probabilities P(S and POS) and P(S^c and POS), which can be used to get probabilities such as

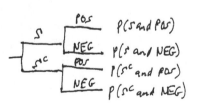

$$P(POS) = P(S \text{ and } POS) + P(S^c \text{ and } POS)$$
$$\text{and } P(S \mid POS) = P(S \text{ and } POS)/P(POS).$$

EXAMPLE 15: RANDOM DRUG TESTING OF AIR TRAFFIC CONTROLLERS

Picture the Scenario
Air traffic controllers have a job that affects the lives of millions of people each day-- monitoring the flights of aircraft and helping to ensure safe takeoffs and landings. In the U.S., because the Federal Aviation Administration (FAA) considers it important to promote a drug-free environment, air traffic controllers are required to undergo periodic random drug testing. A urine test is used as an initial screening test due to its low cost and ease of implementation. One such urine test is the *Triage Panel for*

Drugs of Abuse plus TCA.[4] This urine test detects the presence of drugs such as amphetamines, opiates, and barbiturates. Its sensitivity and specificity have been reported as 0.96 and 0.93 (Peace, 2000). Based on past drug testing of air traffic controllers, the FAA reports that the probability of drug use at a given time is approximately 0.007 (that is, less than 1%).

Questions to Explore

a. A positive test result puts the air traffic controller's job in jeopardy. What is the probability of a positive test result?

b. Find the probability a person truly uses drugs, given that the test is positive.

c. How does the probability in (b) change if the prevalence rate increases, say to 15%?

Think It Through

a. We've been given the probability of drug use, which is the prevalence rate for that state, $P(S) = 0.007$. We also know that the sensitivity is $P(POS \mid S) = 0.96$ and the specificity is $P(NEG \mid S^c) - 0.93$. A tree diagram is useful for visualizing these probabilities and for finding $P(POS)$. Figure 5.10 shows this diagram.

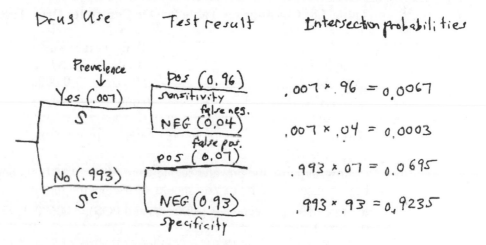

Figure 5.10: Tree Diagram for Random Drug Testing of Air Traffic Controllers. The first set of branches show the probabilities for drug use. The second set of branches shows the conditional probabilities for the test result, given whether used drugs. Multiplication of the probabilities along each path gives the probabilities of intersections of the events.

From the tree diagram and the multiplicative rule for intersection probabilities,

$$P(S \text{ and } POS) = P(POS \mid S)P(S) = 0.96 \times 0.007 = 0.0067.$$

The other route with a positive test result has probability

$$P(S^c \text{ and } POS) = P(POS \mid S^c)P(S^c) = 0.07 \times 0.993 = 0.0695.$$

---

[4] Screening assay from Biosite Diagnostics, San Diego, CA.

To find P(POS), we add the probabilities of these two possible positive test routes. Thus, P(POS) = P(S and POS) + P(S^c and POS) = 0.0067 + 0.0695 = 0.0762.

b. The probability of the drug use state, given a positive test, is P(S | POS). From the definition of conditional probability,

P(S | POS) = P(S and POS)/P(POS) = 0.0067/0.0762 = 0.09.

When the test is positive, only 9% of the time had the person actually used drugs.

If you're a bit unsure about how to answer this and part (a) with a tree diagram, you can construct a contingency table, as shown in Table 5.8. The table shows the summary value of P(S) = 0.007 in the right margin, the other right-margin value P(S^c) = 1 – 0.007 = 0.993 determined by it, and the intersection probabilities P(S and POS) = 0.0067 and P(S^c and POS) = 0.0695 found in (a). From that table, of the proportion 0.0762 of positive cases, 0.0067 truly had used drugs, so the conditional probability is 0.0067/0.0762 = 0.09.

.

Table 5.8: Contingency Table for Air Controller Drug Test

State Present	Drug Test Result		
	POS	NEG	Total
Yes (S)	0.0067	0.0003	0.007
No (S^c)	0.0695	0.9235	0.993
Total	0.0762	0.9238	1.000

c. Now, suppose the prevalence rate were P(S) = 0.15 (let's hope it's not really this high!) instead of 0.007. You should verify that the four intersection probabilities in the tree diagram become 0.144 (for S and POS), 0.006, 0.0595 (for S^c and POS), and 0.7905. Thus,

P(S | POS) = P(S and POS)/P(POS) = 0.144 /(0.144 + 0.0595) = 0.71.

The probability that the person truly used drugs is then much higher, 0.71 compared to 0.09.

Insight

This test has a sensitivity and specificity that are both above 0.90. However, the chance that a positive test result is truly correct depends very much on the prevalence rate. The lower the prevalence rate, the lower the chance. The more drug-free the population, the less likely an individual who tests positive for drugs truly used them.

If the prevalence rate is near 0.007, the chances are very high that an individual who tests positive is actually not a drug user. Does this mean the individual will automatically lose their job? No, if the urine test comes back positive, the individual is given a second test that is more accurate but more expensive than the urine test.

♦

To practice this concept, try Exercises 5.53 and 5.54.

In the previous example, we started with probabilities of the form P(POS | S) and used them to find a conditional probability P(S | POS) that reverses what is given and what is to be found. The method used to find this reverse conditional probability can be summarized in a single formula (see Exercise 5.126). It is known as **Bayes' rule**, named after the person who originally discovered it. We have not shown that formula here because it is easier to understand the logic behind evaluating this conditional probability using tree diagrams or contingency tables.

To help with understanding the logic, it's a good idea to show on a tree diagram what would be expected to happen with a typical group of people. For instance, let's investigate how P(S | POS) can be so small for the air controllers. Figure 5.11 shows a tree diagram for what we'd expect to happen for 1000 air controllers.

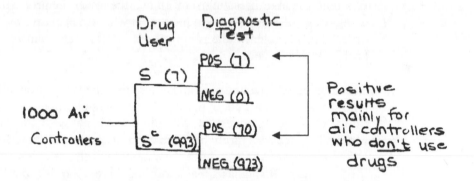

Figure 5.11: Expected Results of Drug Tests for 1000 Air Controllers. This shows typical results when the proportion 0.007 (7 in 1000) are drug users, a positive result has probability 0.96 for those who use drugs, and a negative result has probability 0.93 for those who do not use drugs. Most positive results occur with individuals who are *not* using drugs, since there are so many such individuals (more than 99% of the population). **Question**: How would results change if a higher percentage of the population were using drugs?

Since the proportion 0.007 (which is 7 in 1000) of the target population uses drugs, the tree diagram shows 7 drug users and 993 non-drug users out of the 1000 air controllers. For the 7 drug users, there's a 0.96 chance the test detects this. So, we'd expect all 7 or perhaps 6 of the 7 to be detected with the test (Figure 5.11 shows 7). For the 993 non-drug users, there's a 0.93 chance the test is negative. So, we'd expect about $0.93 \times 993 = 923$ individuals to have a negative result and the other 993 − 923 = 70 non-drug users to have a positive result, as shown on Figure 5.11.

In summary, Figure 5.11 shows 7 + 70 = 77 individuals having a positive test result, but only 7 actually were drug users. Of those with a positive test result, the proportion who truly were drug users is 7/77 = 0.09. What's happening is that the 7% of errors for the large majority of individuals who do not use drugs is much smaller than the 96% of correct decisions for the small number of individuals who do use drugs.

Probability Answers to Questions Can Be Surprising

The results in Example 15 on drug testing of air controllers may seem surprising, but actually they are not uncommon. For instance, consider breast cancer, which is the most common form of cancer in women, affecting about 10% of women at some time in their lives. The chance of breast cancer increases as a woman ages, and the American Cancer Society recommends an annual mammogram after age 40 to test for its presence. The likelihood of a false test result varies according to the breast density and the radiologist's level of experience. One recent study (by W. Barlow et al., *J. Natl. Cancer Inst.*, 2002, vol. 94, p. 1151) estimated sensitivity = 0.86 and specificity = 0.88. These correspond to chances of 0.14 for false negatives and 0.12 for false positives. Of the women who undergo mammograms at any given time, about 1% are typically estimated to actually have breast cancer.

Of the women who receive a positive mammogram result, what proportion actually have breast cancer? In Exercise 5.53 you can work out that it is only about 0.07. Example 15 on drug testing showed that the lower the proportion who were actually using drugs, the less likely an individual who tests positive will truly have used drugs. The prevalence rate is only about 1% for breast cancer, one reason this probability is so poor.

These diagnostic testing examples point out that when probability is involved, answers to questions about our daily lives are sometimes not as obvious as you may think. Consider the question, "Should a woman have an annual mammogram?" Medical choices are rarely between certainty and risk, but rather between different risks. If a woman fails to have a mammogram, this is risky--She may truly have breast cancer. If she has a mammogram, there's the risk of a false positive or false negative, such as the risk of needless worry and additional invasive procedures (often biopsy and other treatments) due to a false positive. There's also the risk from the Xray itself and from traveling to the appointment, small as these may be.

Likewise, with the Triple Blood Test for Down syndrome, most positive results are false, yet the recommendation after receiving a positive test may be to follow up the test with an amniocentesis, a procedure that gives a more definitive diagnosis but has the risk of causing a miscarriage. With some diagnostic tests (such as the PSA test for prostate cancer), testing can detect the disease, but the evidence is unclear about whether or not early detection has much, if any, effect on life expectancy.

Simulation to Estimate a Probability

One way to explore what to expect with randomness is to simulate. We have carried out simulations in previous chapters, in chapter exercises, and in Activity 1 in this chapter. Simulations are also useful for estimating probabilities that are difficult to find with ordinary reasoning. The steps for a simulation are as follows:

Carrying Out a Simulation

- Identify the random phenomenon to be simulated.
- Describe how to simulate observations of the random phenomenon.
- Carry out the simulation many times (ideally, at least 1000 times).
- Summarize results and state the conclusion.

EXAMPLE 16: BE BOLD OR BE CAUTIOUS TO SAVE YOUR BUSINESS?

Picture the Scenario
The business you started last year has only $5000 left in capital. A week from now you need to repay a $10,000 loan, or you will go bankrupt, so you are desperate. You see two possible ways to raise the money you need. You can approach a large company and see if they will invest $10,000 in your company. You would spend the $5000 you have to woo them, and your guess is that the probability of success is 0.20. Alternatively, you can approach in sequence several small companies and see if each of them will invest $2000. You would spend $1000 each to woo them, and again you guess that the probability of success is 0.20 for each appeal.

Questions to Explore
What's the probability you will raise enough money to re-pay the loan, with each strategy? Which is the better strategy, to be bold with one large bet or to be cautious and use several smaller bets?

Think It Through
With the bold strategy, the probability of success is 0.20, the chance that the large company you approach agrees to back you. With the cautious strategy, it's not easy to find the probability, so you can use simulation. For a randomizer, you can use a random number table or generator on a computer or calculator, or an applet such as the *random number* applet at www.prenhall.com/TBA.

Carrying Out the Simulation
- *Identify the random phenomenon to be simulated*: Observing the success or failure of each appeal until you have $10,000 or have gone broke.
- *Describe how to simulate observations of the random phenomenon:* You can use random numbers. Consider the digits 0 to 9. The probability of a successful appeal is 0.20, so we let digits 1 and 2 (20% of the digits) represent success and 3, 4, 5, 6, 7, 8, 9, 0 represent failure. With a sequence of randomly generated digits, you keep track of the cash total until you have $10,000 or are broke. For example, a possible sequence of random digits is

Random digit:	9	8	3	5	1	7	9
Cash flow (thousands):	4	3	2	1	2	1	0

  The first random digit is 9, not 1 or 2, so the first company did not agree to back you. You lose the $1000 spent to woo that company, leaving you with $4000 (shown on the next line). Likewise, the second random digit is 8, not 1 or 2, so now you are down to $3000. Only the fifth appeal was successful (a random digit of 1, and your capital increased at that stage from $1000 to $2000). You went broke after the seventh appeal.
- *Carry out the simulation many times:* Let's carry out this simulation 10 times, with the cash flow in parentheses after each random digit. The first case repeats the results shown above.

SIMULATION #	SEQUENCE OF OUTCOMES	RESULT
1	9 (4) 8 (3) 3 (2) 5 (1) 1 (2) 7 (1) 9 (0)	broke
2	8 (4) 3 (2) 3 (1) 7 (0)	broke
3	7 (4) 5 (3) 2 (4) 1 (5) 1 (6) 4 (5) 2 (6)	
	1 (7) 4 (6) 8 (5) 9 (4) 4 (3) 5 (2) 7 (1) 3 (0)	broke
4	8 (4) 6 (3) 9 (2) 6 (1) 4 (0)	broke
5	6 (4) 5 (3) 1 (4) 5 (3) 6 (2) 8 (1) 9 (0)	broke
6	3 (4) 7 (3) 7 (2) 8 (1) 0 (0)	broke
7	3 (4) 1 (5) 1 (6) 7 (5) 8 (4) 0 (3) 2 (4)	broke
	7 (3) 0 (2) 3 (1) 5 (0)	
8	2 (6) 8 (5) 9 (4) 3 (3) 3 (2) 4 (1) 2 (2) 5 (1)	
	6 (0)	broke
9	1 (6) 1 (7) 8 (6) 4 (5) 2 (6) 5 (5) 0 (4) 6 (3)	broke
	1 (4) 1 (5) 5 (4) 5 (3) 0 (2) 1 (3) 4 (2) 4 (1)	
	1 (2) 1 (3) 1 (4) 5 (3) 9 (2) 2 (3) 2 (4) 3 (3)	
	6 (2) 3 (1) 6 (0)	
10	8 (4) 7 (3) 8 (2) 7 (1) 2 (2) 1 (3) 7 (2) 5 (1)	broke
	1 (2) 5 (1) 6 (0)	

- *Summarize results and state the conclusion*: You went broke in all 10 simulations. With this strategy, the estimated probability of success is $0/10 = 0$. The bolder strategy seems to be the better one.

Insight

You can't estimate a probability well with only 10 simulations. In practice, it is common to carry out at least 1000 simulations, to get a better sense of what happens in the long run. Carry out 10 simulations on your own, and (if possible) combine them with results from your entire class. How would you now estimate this probability?

If many, many simulations were performed, which can be easily done with a computer, you would discover that the probability of success with the more cautious strategy is only 0.001. The bolder strategy is better.

                                                  ♦

To practice this concept, try Exercise 5.60.

Probability Is the Key to Statistical Inference

The concepts of probability hold the key to methods for conducting statistical inference--making conclusions about populations using sample data. To help preview this connection, let's consider an opinion poll. Suppose a poll indicates that 45% of those sampled favor legalized gambling. A basic statistical question is, "What's the probability that this sample percentage falls within plus or minus 3% (or some other value) of the true population percentage?" The next chapter will build on the probability foundation of this chapter and enable us to answer such a question. We will study how to evaluate the probabilities of all the possible outcomes in a survey sample or an experiment. This will then be the basis of the statistical inference methods we'll learn about in Chapters 7 and 8.

SECTION 5.4: PRACTICING THE BASICS

5.46 Birthdays of Presidents: Of the first 43 presidents of the United States (George Washington through George W. Bush), two had the same birthday (Polk and Harding). Is this highly coincidental? Answer by finding the probability of at least one birthday match among 43 people.

5.47 Matching your birthday: You consider your birth date to be special since it falls on January 1. Suppose your class has 25 students.
a. Is the probability of finding at least one student with birthday that matches yours greater, the same, or less than the probability found in Example 14 of a match for at least two students? Explain.
b. Find that probability.

5.48 Lots of pairs: Show that with 25 students, there are 300 *pairs* of students who can have the same birthday. So it's really not so surprising if at least two students have the same birthday. *(Hint:* You can pair 24 other students with each student, but how can you make sure you don't count each pair twice?)

5.49 Horrible 11 on 9/11: The digits in 9/11 add up to 11 (9 + 1 + 1), American Airlines flight 11 was the first to hit the World Trade Towers (which took the form of the number 11), there were 92 people on board (9 + 2 = 11), September 11 is the 254th day of the year (2 + 5 + 4 = 11), and there are 11 letters in Afghanistan, New York City, the Pentagon, and George W. Bush (see article by L. Belkin, *New York Times*, August 11, 2002). How could you explain to someone who has not studied probability that, because of the way we look for patterns out of the huge number of things that happen, this is not necessarily an amazing coincidence?

5.50 Coincidence in your life: State an event that has happened to you or to someone you know that seems highly coincidental (such as seeing a friend while on vacation). Explain why that event may not be especially surprising, once you think of all the similar types of events that could have happened to you or someone that you know, over the course of several years.

5.51 Monkeys typing Shakespeare: Since events of low probability eventually happen if you observe enough trials, a monkey randomly pecking on a typewriter could eventually write a Shakespeare play just by chance. Let's see how hard it would be to type the title of *Macbeth* properly. Assume 50 keys for letters and numbers and punctuation. Find the probability that the first seven letters that a monkey types are macbeth. (Even if the monkey can type 60 strokes a minute and never sleeps, if we consider each sequence of 7 keystrokes as a trial, we would wait on the average over 100,000 years before seeing this happen!)

5.52 A *true* coincidence at Disneyworld: Wisconsin has 5.4 million residents. On any given day, the probability is 1/5000 that a randomly selected Wisconsin resident decides to visit Disneyworld in Florida.
a. Find the probability that they all will decide to go tomorrow, in which case Disneyworld has more than 5.4 million people in line when it opens in the morning.
b. What assumptions did your solution in (a) make? Are they realistic? Explain.

5.53 Mammogram diagnostics: For use of the mammogram to detect breast cancer, typical values reported are sensitivity = 0.86 and specificity = 0.88. Of the women who undergo mammograms at any given time, about 1% are typically estimated to actually have breast cancer.
a. Construct a tree diagram in which the first set of branches shows whether a woman has breast cancer and the second set of branches shows the mammogram result. At the end of the final set of branches, show that P(S and POS) = 0.01×0.86 = 0.0086, and report the other intersection probabilities also.
b. Restricting your attention to the two routes that have a positive test result, show that P(S) = 0.1274.
c. Of the women who receive a positive mammogram result, what proportion actually have breast cancer?
d. The tree diagram shown here illustrates how P(S | POS) can be so small, using a typical group of 100 women who have a mammogram.   Explain how to get the frequencies shown on the branches, and explain why this suggests that P(S | POS) is only about 0.08.

Typical results of mammograms for 100 women

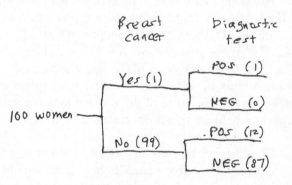

5.54 More screening for breast cancer: Refer to the previous exercise. The prevalence rate for all women undergoing a mammogram was given as 0.01. For young women, the prevalence of breast cancer is lower. Suppose the sensitivity is 0.86 and the specificity is 0.88, but the chance of breast cancer for a randomly selected young woman is only 0.001.
a. Suppose that a test comes out positive. Find the probability that the woman truly has breast cancer.
b. Show how to use a tree diagram with frequencies for a typical sample of 1000 women to explain to someone who has not studied statistics why the probability found in (a) is so low.
c. Of the cases that are positive, explain why the proportion in error is likely to be larger for a young population than for an older population.

5.55 Convicted by mistake: In a jury trial, suppose the probability the defendant is convicted, given guilt, is 0.95, and the probability the defendant is acquitted, given innocence, is 0.95. Suppose that 90% of all defendants truly are guilty.

a. Given that the defendant is convicted, find the probability he or she was actually innocent. Draw a tree diagram or construct a contingency table to help you answer.
b. Repeat (a), but under the assumption that 50% of all defendants truly are guilty.

5.56 Was OJ actually guilty?: Former pro football star O. J. Simpson was accused of murdering his wife. In the trial, a defense attorney pointed out that although Simpson had been guilty of earlier spousal abuse, annually less than 40 women are murdered per 100,000 incidents of partner abuse. This means that P(murdered by partner | partner abuse) = 40/100,000, but more relevant is P(murdered by partner | partner abuse and woman murdered). Every year it is estimated that 5 of every 100,000 women in the U.S. who are murdered are killed by someone other than their partner (Gigerenzer, 2002, p. 144). Part of a tree diagram is shown starting with 100,000 women.

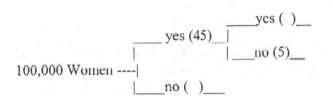

a. Based on the results stated, explain why the numbers 45 and 5 are entered as shown on two of the branches.
b. Fill in the two blanks shown in the tree diagram.
c. Conditional on partner abuse and the woman being murdered (by someone), explain why the probability the woman was murdered by her partner is 40/45. Why is this so dramatically different from P(murdered by partner | partner abuse) = 40/100,000?

5.57 Air traffic controllers: In the diagnostic test for drugs with air controllers discussed in Example 15, the sensitivity and specificity were reported as 0.96 and 0.93, and the probability of drug use at a given time was given as 0.007.
a. Use a tree diagram or contingency table to show that the conditional probability that the individual is *not* a drug user, given that the test result is negative, is 0.9997.
b. Suppose the prevalence rate were 0.15. Given that a test result is positive, show all steps to verify that the probability the individual actually is a drug user is 0.71.

5.58 DNA evidence compelling?: DNA evidence can be extracted from biological traces such as blood, hair, and saliva. "DNA fingerprinting" is increasingly used in the courtroom as well as in paternity testing. Given that a person is innocent, suppose that the probability of their DNA matching that found at the crime scene is only 0.000001, one in a million. Further, given that a person is guilty, suppose that the probability of their DNA matching that found at the crime scene is 0.99. Jane Doe's DNA matches that found at the crime scene.
a. Find the probability that Jane Doe is actually innocent, if unconditionally her probability of innocence is 0.50. Interpret. Show your solution by introducing notation for events, specifying probabilities that are given, and using a tree diagram to find your answer.
b. Repeat (a) if the unconditional probability of innocence is 0.99. Compare results.

c. Explain why it is very important for a defense lawyer to explain the difference between P(DNA match | person innocent) and P(person innocent | DNA match).

Blood Test			
Down	POS	NEG	Total
Yes	48	6	54
No	1307	3921	5228
Total	1355	3927	5282

5.59 Triple blood test: Example 9 about the Triple Blood Test for Down syndrome found the results shown in the table in the margin.
a. Find the prevalence rate.
b. Find the (i) sensitivity, (ii) specificity.
c. Find the (i) positive predictive value, (ii) negative predictive value.
d. Explain how the probabilities in (b) and (c) give four ways of describing the probability that a diagnostic test makes a correct decision.

5.60 Simulating Donations to Local Blood Bank: The director of a local blood bank is in need of a donor with type AB blood. The distribution of blood types for Americans is estimated by the American Red Cross to be:

Type	A	B	O	AB
Probability	40%	10%	45%	5%

The director decides that if more than 20 donors are required before the first donor with AB blood appears, she will need to issue a special appeal for AB blood.
a. Conduct a simulation 10 times, using *the random number* applet at www.prenhall.com/TBA or a calculator or software, to estimate the probability this will happen. Show all steps of the simulation, including any assumptions that you make. Refer to Example 16 as a model for carrying out this simulation.
b. In practice, you would do at least 1000 simulations to estimate this probability well. You'd then find that the probability of exceeding 20 donors before finding a type AB donor is 0.36. Actually, simulation is not needed. Show how to find this probability using the methods of this chapter.

5.61 Revisiting the blood bank: Refer to the previous exercise. The director is also concerned about getting a donation of Type B blood within the next 20 donors. Carry out a simulation to help the director estimate the probability this will happen.
a. Clearly describe the steps of the simulation.
b. Based on ten simulations, estimate the probability that more than 20 donors will be required.
c. If you were to conduct a large number of simulations, what would you find for the probability in (b)? Use results of this chapter to answer exactly.

CHAPTER SUMMARY

In this chapter, we've seen how to quantify uncertainty and randomness. With many independent trials of random phenomena, outcomes do show certain regularities. The proportion of times an outcome occurs, in the long run, is its **probability**.

The **sample space** is the set of all possible outcomes of a random phenomenon. An **event** is a subset of the sample space. Two events A and B are **disjoint** if they have no common elements.

We find probabilities using basic rules:

- The **probability** of each individual outcome falls between 0 and 1, and the total of all the individual probabilities equals 1.

- The **probability of an event** is the sum of the probabilities of individual outcomes in that event.

- For an event A and its **complement** A^c (the outcomes not in A),
  $$P(A^c) = 1 - P(A).$$

- $P(A \text{ or } B) = P(A) + P(B) - P(A \text{ and } B)$.

  This simplifies to $P(A \text{ or } B) = P(A) + P(B)$ when the events are disjoint.

- $P(A \text{ and } B) = P(A \mid B) \times P(B)$, where $P(A \mid B)$ denotes the **conditional probability** of event A, given that event B occurs. Equivalently, the conditional probability satisfies

  $$P(A \mid B) = \frac{P(A \text{ and } B)}{P(B)}.$$

  Likewise, $P(A \text{ and } B) = P(B \mid A)P(A)$, and $P(B \mid A) = P(A \text{ and } B)/P(A)$.

- When A and B are **independent**, $P(A \mid B) = P(A)$ and $P(B \mid A) = P(B)$. Then, also $P(A \text{ and } B) = P(A) \times P(B)$.

A **probability model** states certain assumptions and, based on them, finds the probabilities of events of interest.

Understanding probability helps us to make informed decisions in our lives. In the next three chapters, we'll see how an understanding of probability gives us the basis for making statistical inferences about a population.

SUMMARY OF NEW NOTATION IN CHAPTER 5

A, B, C   Events

A^c Complement of event A (the outcomes *not* in A)

$P(A)$ Probability of event A

$P(A \mid B)$ Conditional probability of event A, given event B (| denotes "given")

ANSWERS TO THE CHAPTER FIGURE QUESTIONS

Figure 5.1:   *The cumulative proportion after each roll is evaluated as the frequency of six's rolled through that trial (roll) number divided by the total number of trials (rolls). The* cumulative proportions are 0/1=0 for roll 1, 0/2 = 0 for roll 2, 1/3=0.33 for roll 3, and 1/4 = 0.25 for roll 4.

Figure 5.2:   It should be very, very close to 1/6.

Figure 5.3:   2 x 2 x 2 x 2 = 16 branches.

Figure 5.4:   Venn diagrams will vary. One possible Venn diagram is shown in the margin.

Figure 5.5:   The favorable outcomes to the event that the student answers the first question correctly are {CCC, CCI, CIC, CII}. This event is not disjoint from either of the two labeled events in the Venn diagram.

Figure 5.6: $P(A \text{ or } B) = P(A) + P(B) - P(A \text{ and } B)$

Figure 5.7: $P(A \text{ or } B) = P(A) + P(B)$ when the events A and B are disjoint.

$P(B/A) =$

Figure 5.8:   We would not expect the trials to be independent if the student is not guessing. If the student is using prior knowledge, the probability of getting a correct answer on a question, such as question 3, may be more likely if the student answers questions 1 and 2 correctly.

Figure 5.9: $P(A|B)$ is not necessarily equal to $P(B|A)$, as seen in the margin.

Figure 5.11:   As the prevalence rate increases, of the individuals that test positive, a higher percentage of positive test results will occur with individuals that actually use drugs. For example, if 5% of the population uses drugs, of the 115 individuals out of 1000 that would be expected to test positive, 48 would be actual drug users while 67 would be non-drug users.

CHAPTER PROBLEMS: PRACTICING THE BASICS

5.62 Barry Bonds home runs: In his prime, the probability that major league baseball player Barry Bonds hit a home run when he came to bat was about 1/10.
a. Does this mean that if we watched him at bat 10 times in two games, he would almost certainly get exactly one home run? Explain.
b. Explain what this probability means in terms of watching him at bat over a longer period, say 1000 times over two seasons of his prime.

5.63 Due for a boy?: A couple has five children, all girls. They are expecting a sixth child. The father tells a friend that by the law of large numbers the chance of a boy is now much greater than 1/2. Comment on the father's statement.

5.64 ESP: The General Social Survey has occasionally asked, "How often have you felt as though you were in touch with someone when they were far away from you?" Of 3887 sampled subjects, 1407 said 'never' and 2480 said 'at least once'.
a. Using the sample data, estimate the probability that a subject from the sampled population has had this ESP experience.
b. Can you expect the proportion found in part (a) to be *exactly* equal to the probability for the population? Explain.

5.65 P(life after death): Explain the difference between the relative frequency and subjective definitions of probability. Illustrate by explaining how it is possible to give a value for (a) the probability of life after death, (b) the probability that in the morning you remember at least one dream that you had in the previous night's sleep, based on what you observe every morning for the next year.

5.66 Wheel of Fortune: The Wheel of Fortune is a popular game show that lets three contestants attempt to solve a word puzzle by revealing an unknown phrase one letter at a time, with the option to spin the wheel to win money for each letter they uncover. Suppose the wheel has 30 sections, all of equal size. With a fair spin of a wheel, what is the probability of landing on any one section?

5.67 Choices for lunch: For the set lunch at Amelia's Restaurant, customers can select one meat dish, one vegetable, one beverage, and one dessert. The menu offers two meats (beef and chicken), three vegetables (corn, green beans, or potatoes), three beverages (Coke, ice tea, or coffee), and one dessert (Amelia's apple pie).
a. How many possible meals are there?
b. Use a tree diagram to list the possible outcomes.

5.68 Randomly selecting a family: A local school has a popular pre-kindergarten program. Four families applied for the final two slots. The director decides to randomly select two families. Families Breiner and Vince have a boy while families Edwards and Wilson have a girl.
a. Show the sample space for the two families selected.
b. Assign probabilities to the possible outcomes in the sample space.
c. Find the probability that at least one boy is selected.

5.69 Caught doctoring the books: After the major accounting scandals with Enron, a large energy company, the question may be posed, "Was there any way to examine Enron's accounting books to determine if they had been 'doctored'?" One way uses *Benford's Law.* This states that in a variety of circumstances, numbers as varied as populations of small towns, figures in a newspaper or magazine, and tax returns and other business records begin with the digit "1" more often than other digits. This law states that the probabilities for the digits 1 through 9 are approximately:

Digit	1	2	3	4	5	6	7	8	9
Probability	0.30	0.18	0.12	0.10	0.08	0.07	0.06	0.05	0.04

a. If we were to randomly pick one of the digits between 1 and 9 using a random number table or software, what is the probability for each digit?
b. When people attempt to fake numbers, there's a tendency to use 5 or 6 as the initial digit more often than predicted by Benford's law. What is the probability of 5 or 6 as the first digit by (i) Benford's law, (ii) random selection?

5.70 Life after death: In a General Social Survey, in response to the question "Do you believe in life after death?", 907 answered yes and 220 answered no.
a. Based on this survey, estimate the probability that a randomly selected adult in the U.S. believes in life after death.
b. A married couple is randomly selected. Estimate the probability that both subjects believe in life after death.
c. What assumption did you make in answer (b)? Explain why that assumption

is probably unrealistic, making this estimate unreliable.

5.71 Death penalty jury: In arguing against the death penalty, Amnesty International has pointed out supposed inequities, such as the many times a black person has been given the death penalty by an all-white jury. If jurors are selected randomly from an adult population, find the probability that all 12 jurors are white when the population is (a) 90% white, (b) 50% white.

5.72 Driver's exam: Three fifteen-year-old friends with no particular background in driver's education decide to take the written part of the Georgia Driver's Exam. Each exam was graded as a pass (P) or a failure (F).
a. How many outcomes are possible for the grades received by the three friends together? Using a tree diagram, list the sample space.
b. If the outcomes in the sample space in (a) are equally likely, find the probability that all three pass the exam.
c. In practice, the outcomes in (a) are *not* equally likely. Suppose that statewide 70% of fifteen-year-olds pass the exam. If these three friends were a random sample of their age group, find the probability that all three pass.
d. In practice, explain why probabilities that apply to a random sample are not likely to be valid for a sample of three friends.

5.73 Visit France and Italy: Students in a geography class are asked whether they've ever visited France, and whether they've ever visited Italy.
a. For a randomly selected student, would you expect these events to be independent, or dependent? Explain.
b. How would you explain to someone who has never studied statistics what it means for these events to be independent?

5.74 Insurance coverage: The Gallup Organization conducted a national poll in June 2003. The adults in the poll were asked the type of health insurance coverage they have, given they had used an emergency room at least once in the last 12 months. The table shows results. For the population who had used an emergency room at least once in the past 12 months:
a. Estimate the probability that a patient has health insurance.
b. For the next two independent patients admitted, estimate the probability that they both use private insurance.
c. Given that a subject has health insurance, estimate the probability it is private.

Type of Insurance Coverage	Proportion
Private insurance	0.22
Medicare/Medicaid	0.36
No health insurance	0.27
Other	0.15

5.75 Using the emergency room: In the Gallup poll described in the previous exercise, the sampled adults were asked the reason for the visit to the emergency room. The table shows results. For the population who had used an emergency room at least once in the past 12 months
a. Estimate the probability that an admission is *not* due to a doctor's recommendation.

b. Given that an admission was not due to an emergency or accident, estimate the probability that it was because the subject did not have a doctor.

Reason for using emergency room	Proportion
Emergency or accident	0.50
Office closed/after hours	0.31
Doctor recommended	0.06
Easier/faster to be seen	0.04
Do not have a doctor	0.01
Other	0.06
No opinion	0.02

5.76 Teens and drugs: In 2000 the Center on Addiction and Substance Abuse (CASA) at Columbia University conducted a survey asking 1000 randomly chosen teenagers questions about their parents and the teenagers' views on the use of illegal substances. Twenty-six percent of the sampled teenagers cited drugs as their biggest concern. Of them, 31 % said "drugs can ruin your life and cause harm." Which of these two percentages refers to a conditional probability? For it, identify the event conditioned on and the event to which the probability refers.

5.77 Teens and parents: In the CASA teen survey described in the previous exercise, 27% of the teens said they lived with "hands-on" parents – parents who have established a household culture of rules and expectations for their teen's behavior. An excellent relationship with their fathers was reported by 47% of teens living in "hands-on" households but only 13% of teens living in "hands off" households.
a. Which of these percentages estimates a conditional probability? For each that does, identify the event conditioned on and the event to which the probability refers.
b. For the variables "type of parent" and "whether the teen has an excellent relationship with father," construct a contingency table showing counts for the 1000 subjects.
c. Using the contingency table, estimate the probability that a randomly selected teenager has an excellent relationship with his/her father.
d. Using the contingency table, given that a teen has an excellent relationship with the father, estimate the probability that the teenager has "hands off" parents.

5.78 Laundry detergent: A manufacturer of laundry detergent has introduced a new product that it claims to be more environmentally sound. An extensive survey gives the percentages shown in the table.

Result of Advertising for New Laundry Product

	Tried the New Product	
Advertising Status	Yes	No
Seen the ad	10%	25%
Have not seen ad	5%	60%

....

a. Estimate the probability that a randomly chosen consumer would have seen advertising for the new product and tried the product.

b. Given that a randomly chosen consumer has seen the product advertised, estimate the probability that the person has tried the product.

c. Let A be the event that a consumer has tried the product. Let B be the event that a consumer has seen the product advertised. Express the probabilities found in (a) and (b) in terms of A and B.

d. Are A and B independent? Justify your answer.

5.79 Shoppers at the local mall: As part of a class project, two marketing majors conducted a survey at the local mall, using a cluster random sample. The following table shows some survey demographic results.

		Age	
Gender	Under 25	25-40	Over 40
Male	26%	12%	14%
Female	?	13%	18%

a. What percentage of shoppers in the survey were female and under 25?

b. Is the event that a shopper is male disjoint from the event that a shopper is

c. over 40? Why or why not?

d. Estimate the probability that a randomly selected shopper is under 25, given that their gender is (i) male, (ii) female.

5.80 Working women: A researcher in the business school at a major university studied the relationship between the salary and the number of children for a working woman. Based on a statewide survey, she reported the joint probabilities in the table. Let A denote the event that a working woman has a low salary. Let B denote the event that a working woman has more than 2 children.

Salary by Number of Children

	Children	
Salary	2 or fewer	more than 2
High	0.15	0.05
Medium	0.30	0.15
Low	0.15	0.20

a. Find P(A) and P(B).

b. Find P(A and B).

c. Find P(A | B), and interpret.

d. Are A and B independent? Justify using the previous parts of this exercise.

5.81 Board games and dice: A board game requires players to roll a pair of balanced dice for each player's turn. Denote the outcomes by (die 1, die2), such as (5, 3) for a 5 on die 1 and a 3 on die 2.

a. List the sample space of the 36 possible outcomes for the two dice.

b. Let A be the event that you roll doubles (that is, each die has the same outcome). List the outcomes in A, and find its probability.

c. Let B be the event that the sum on the pair of dice is seven. Find P(B).

d. Find the probability of (i) A and B, (ii) A or B, (iii) B given A.

5.82 Rolling two dice: In rolling a pair of dice, let event A be rolling doubles, event B be rolling a sum of seven, and event C be rolling a 6 on the first die.

a. Are events A and B independent, or disjoint, or neither? Explain.

b. Are events A and C independent? Explain.

5.83 Roll two more dice: Refer to the previous exercise. Define event D as rolling an odd sum with two dice and define event E as rolling an even sum.

a. Find P(D) and P(E).

b. B denotes a sum of seven on the two dice. Find P(B and D). When an event B is contained within an event D, as here, explain why P(B and D) = P(B).

c. Find P(B or D). When an event B is contained within an event D, explain why P(B or D) = P(D).

5.84 Conference dinner: Of the participants at a conference, 50% attended breakfast, 90% attended dinner, and 40% attended both breakfast and dinner. Given that a participant attended breakfast, find the probability that she also attended dinner.

5.85 Snack preferences: A student survey found that for snacking, the probability of eating chips was 0.45, the probability of drinking a soft drink was 0.80, and the probability of eating chips and drinking a soft drink was 0.36. Let A be the event of eating chips and B be the event of drinking a soft drink.

a. Are events A and B independent? Justify your answer.

b. Find the probability that a student eats chips or drinks a soft drink.

c. Find the probability that a randomly selected student (i) eats chips, given that

d. he or she drinks a soft drink, (ii) drinks a soft drink, given that he or she eats chips. Why are these not equal?

5.86 Waste dump sites: A federal agency is deciding which of two waste dump projects to investigate. A top administrator estimates that the probability of federal law violations is 0.30 at the first project and 0.25 at the second project. Also, he believes the occurrences of violations in these two projects are disjoint.

a. What is the probability of federal law violations in the first project or in the

b. second project?

c. Given that there is not a federal law violation in the first project, find the probability that there is a federal law violation in the second project.

d. In reality, the administrator confused disjoint and independent, and the events are actually independent. Answer (a) and (b) with this correct information.

5.87 Dice and craps: In the game of craps, you roll two dice, and you win if the total is 7 or 11 and you lose if the total is 2, 3, or 12. You keep rolling until one of these totals occurs. Using conditional probability, find the probability that you win.

5.88 No coincidences: Over time, you have many conversations with a friend about your favorite actress, favorite musician, favorite book, favorite TV show, and so forth for 100 separate topics. On any given topic, there's only a 0.02 probability that you agree. If you did agree on a topic, you would consider it to be coincidental.
a. If whether you agree on any two different topics are independent events, find the probability that you and your friend *never* have a coincidence on these 100 topics.
b. Find the probability that you have a coincidence on at least one topic.

5.89 Amazing roulette run?: A roulette wheel in Monte Carlo has 18 even-numbered slots, 18 odd-numbered slots, and a slot numbered zero. On August 18, 1913, it came up even 26 times in a row.[5] As more and more evens occurred, the proportion of people betting on an odd outcome increased, as they figured it was "due."
a. Comment on this strategy.
b. Find the probability of 18 evens in a row if each slot is equally likely.
c. Suppose that over the past 100 years there have been 1000 roulette wheels, each being used hundreds of times a day. Is it surprising if sometime in the previous 100 years one of these wheels had 18 evens in a row? Explain.

5.90 Death penalty and false positives: For the decision about whether to convict someone charged with murder and give them the death penalty, consider the variables "reality" (defendant innocent, defendant guilty) and "decision" (convict, acquit).
a. Explain what the two types of errors are in this context.
b. Jurors are asked to convict a defendant if they feel the defendant is guilty "beyond a reasonable doubt." Suppose this means that given the defendant is executed, the probability that he or she truly was guilty is 0.99. For the 873 people put to death from the time the death penalty was reinstated in 1977 until September 2003, find the probability that (i) they were all truly guilty, (ii) at least one of them was actually innocent.

5.91 Beyond a reasonable doubt: Refer to the previous exercise. How do the answers in (b) change if the probability of true guilt is actually (a) 0.999, (b) 0.95?

5.92 Screening for heart attacks: Biochemical markers are used by emergency room physicians to aid in diagnosing patients who have suffered acute myocardial infarction (AMI), or what's commonly referred to as a heart attack. One type of biochemical marker used is creatine kinase (CK). Based on a review of published studies on the effectiveness of these markers (by E. M. Balk et al., *Annals of Emergency Medicine*, vol. 37, pp. 478-494, 2001), CK had an estimated sensitivity of 37% and specificity of 87%. Consider a population having a prevalence rate of 25%.
a. Explain in context what is meant by the prevalence rate being 25%.
b. Explain in context what is meant by the sensitivity equaling 37%.
c. Explain in context what is meant by the specificity equaling 87%.
d. Construct a tree diagram for this diagnostic test. Label the branches with the appropriate probabilities.

---

[5] *What Are the Chances?*, by B. K. Holland (Johns Hopkins University Press, 2002, p. 10)

5.93 Screening for colorectal cancer: Gigerenzer (2002, p. 105) reported that on the average, "Thirty out of every 10,000 people have colorectal cancer. Of these 30 people with colorectal cancer, 15 will have a positive hemoccult test. Of the remaining 9,970 people without colorectal cancer, 300 will still have a positive hemoccult test."

a. Sketch a tree diagram or construct a contingency table to display the counts.
b. Estimate the sensitivity.
c. Estimate the specificity.
d. Estimate the prevalence.
e. Of the 315 people mentioned above who have a positive hemoccult test, what proportion actually have colorectal cancer? Interpret.

5.94 Late again: If Maura remembers to set her alarm clock, 90% of the time she gets to work on time. If she forgets to set it, 50% of the time she gets to work on time. She remembers to set her alarm clock 80% of the time. What percentage of work days does she get to work on time? Use a tree diagram to answer.

5.95 Color blindness: For genetic reasons, color blindness is more common in men than women: 5 in 100 men and 25 in 10,000 women suffer from color blindness.

a. Defining events, identify these proportions as conditional probabilities.
b. If the population is half male and half female, what proportion of the population is color blind? Use a tree diagram or contingency table with frequencies to portray your solution, showing what you would expect to happen with 20,000 people.
c. Given that a randomly chosen person is color blind, what's the probability that person is female? Use the tree diagram or table from (b) to find the answer.

5.96 HIV testing: For a combined ELISA-Western blot blood test for HIV positive status, the sensitivity is about 0.999 and the specificity is about 0.9999 (Gigerenzer 2002, pp. 124, 126).

a. Consider a high-risk group in which 10% are truly HIV positive. Construct a tree diagram to summarize this diagnostic test.
b. Display the intersection probabilities from (a) in a contingency table.
c. A person from this high-risk group has a positive test result. Use the tree diagram or the contingency table, find the probability that this person is truly HIV positive.

5.97 Low risk HIV: Refer to the previous exercise. Now consider a lower-risk group in which only 1% are truly HIV positive.

a. Using a tree diagram and/or a contingency table, find the probability that a person from this group who has a positive test result is truly HIV positive.
b. Explain why a positive test result is more likely to be in error when the prevalence is lower. Use tree diagrams or contingency tables with frequencies for 10,000 people with the 10% and 1% prevalence cases to illustrate your arguments.

5.98 Prostate cancer: A study of the PSA blood test for diagnosing prostate cancer in men (by R. M. Hoffman et al., *BMC Family Practice*, vol. 3: p. 19, 2002) used a sample of 2620 men in the Albuquerque metropolitan area who were 40 years and older. When a positive diagnostic test result was defined as a PSA reading of at least 4, the sensitivity was estimated to be 0.86 but the specificity only 0.33.

a. Given that a man did not have prostate cancer, what is the estimated probability that the test suggested that he did have it?

b. Suppose that 10% of those who took the PSA test truly had prostate cancer. Given that the PSA was positive, use a tree diagram and/or contingency table to estimate the probability that the man truly had prostate cancer.

c. Illustrate your answer in (b) by using a tree diagram or contingency table with frequencies showing what you would expect for a typical sample of 1000 men.

5.99 More about the PSA: Refer to the previous exercise. Lowering the PSA boundary to 2 for a positive result changed the sensitivity to 0.95 and the specificity to 0.20. Explain why, intuitively, if the cases increase for which a test is positive, the sensitivity will go up but the specificity will go down.

CHAPTER PROBLEMS: CONCEPTS AND INVESTIGATIONS

5.100 Simulate law of large numbers: Using the *long run probability* applet at www.prenhall.com/TBA or other software, simulate the flipping of a balanced coin.

a. Report the cumulative proportion of heads after (i) 10 flips, (ii) 100 flips, (iii) 1000 flips, (iv) 10,000 flips. Explain how the results illustrate the law of large numbers and the long-run relative frequency definition of probability.

b. Repeat the simulation, but with an unbalanced coin, by setting the probability of a "success" equal to 0.1. Explain how the results illustrate the law of large numbers and the long-run relative frequency definition of probability.

c. Using the *lon- run probability* applet or other software, simulate the roll of a die with a success defined as rolling a 3 or 4, using (i) 10 rolls, (ii) 100 rolls, (iii) 1000 rolls, (iv) 10,000 rolls. Summarize results.

5.101 Illustrate probability terms with scenarios:

a. What is a sample space? Give an example of a sample space for a scenario involving (i) a designed experiment, (ii) an observational study.

b. What are disjoint events? Give an example of two events that are disjoint.

c. What is a conditional probability? Give an example of two events in your everyday life that you would expect to be (i) independent, (ii) dependent.

5.102 Short term vs. long run: Short-term aberrations do not affect the long run. To illustrate, suppose that you flip a coin 10 times and you get 10 heads. Find the cumulative proportion of heads, including these first ten flips, if (a) in the 100 flips after those 10 you get 50 heads (so there are now 60 heads in 110 flips), (b) in the 1000 flips after those 10 you get 500 heads, (c) in the 10,000 flips after those 10 you get 5000 heads. What number is the cumulative proportion tending toward as *n* increases?

5.103 Risk of space shuttle: After the *Columbia* space shuttle disaster, a former NASA official who faulted the way the agency deals with safety risk warned (in an AP story, March 7, 2003) that NASA workers believed, "If I've flown 20 times, the risk is less than if I've flown just once."

a. Explain why it would be reasonable for someone to form this belief, from the way we use empirical evidence and margins of error to estimate an unknown probability.

b. Explain a criticism of this belief, using coin flipping as an analogy.

5.104 Mrs. Test: "Mrs. Test" (see www.mrstest.com) sells diagnostic tests for various conditions. Their Web site gives only imprecise information about the accuracy of the tests. The test for pregnancy is said to be "over 99% accurate." Describe at least four different probabilities to which this could refer.

5.105 Marijuana leads to heroin?: Nearly all heroin addicts have used marijuana sometime in their lives. So, some argue that marijuana should be illegal because marijuana users are likely to become heroin addicts. Use a Venn diagram to illustrate the fallacy of this argument, by sketching sets for M = marijuana use and H = heroin use such that $P(M \mid H)$ is close to 1 but $P(H \mid M)$ is close to 0.

5.106 Fewer false positives with mammograms: The Netherlands has tried to reduce the high rate of false positive mammograms by introducing more restrictive criteria for defining a positive test. "This policy has reduced the rate of false positives, but at the cost of increasing the rate of false negatives. In other words, more cancers are now missed during screening" (Gigerenzer 2002, p. 63). Explain why the rate of false negatives would increase with this policy.

5.107 Is this test good?: A diagnostic test has sensitivity = specificity = 0.95, quite high. Does this imply that this is a good test? Discuss some of the issues involved that determine pros and cons of using this test.

5.108 How good is a probability estimate?: In Example 9 about Down syndrome, we estimated the probability of a positive test result (predicting that Down syndrome is present) to be $P(POS) = 0.257$, based on observing 1355 positive results in 5228 observations. How good is such an estimate? From Section 4.2, $1/\sqrt{n}$ is an approximate margin of error in estimating a proportion with n observations.
a. Find the approximate margin of error to describe how well this proportion estimates the true probability, $P(POS)$.
b. The *long-run* in the definition of probability refers to letting n get very large. What happens to this margin of error formula as n keeps growing, eventually toward infinity? What's the implication of this?

5.109 Probability for bell-shaped distributions: By the Empirical Rule in Section 2.4, about 95% of a bell-shaped distribution falls within two standard deviations of the mean. Heights of women have a bell-shaped distribution.
a. What is the probability that two randomly selected women both have height within two standard deviations of the mean.
b. Would the calculation in (a) apply to a randomly selected pair of sisters? Why or why not? (Hint: Are events concerning their heights likely to be independent?)

5.110 Protective bomb: Before the days of high security at airports, there was a legendary person who was afraid of traveling by plane because someone on the plane might have a bomb. He always brought a bomb himself on any plane flight he took, believing that the chance would be astronomically small that two people on the same flight would both have a bomb. Explain the fallacy in his logic, using ideas of independent events and conditional probability.

5.111 Streak shooter: Sportscaster Maria Coselli claims that players on the New York Knicks professional basketball team are streak shooters. To make her case, she looks at the statistics for all the team's players over the past three games and points out that one of them (Joe Smith) made 6 shots in a row at one stage. Coselli argues, "Over a season, Smith makes only 45% of his shots. The probability that he would make 6 in a row if the shots were independent is $(0.45)^6 = 0.008$, less than 1 in a hundred. This would be such an unusual occurrence that we can conclude that Smith is a streak shooter."

a. Explain the flaws in Coselli's logic.

b. Use this example to explain how some things that look highly coincidental may not be so when viewed in a wider context.

Select the appropriate answer(s) to multiple-choice Exercises 5.112-5.118. Some exercises may have more than one correct answer.

5.112 MC1: For two events A and B, P(A) = 0.5 and P(B) = 0.2. Then P(A or B) equals:

a. 0.10, if A and B are independent

b. 0.70, if A and B are independent

c. 0.60, if A and B are independent

d. 0.70, if A and B are disjoint

5.113 MC2: Which of the following is always true?

a.   If A and B are independent, then they are also disjoint.

b. $P(A \mid B) + P(A \mid B^c) = 1$

c.   If $P(A \mid B) = P(B \mid A)$, then A and B are independent.

d.   If A and B are disjoint, then A and B cannot occur at the same time.

5.114 MC3: The first baseman in a baseball game is thrown two balls during an inning. Define A = {catches both balls}, B = {catches at least one ball}, C = {misses both balls}.

a. $P(A) + P(B) + P(C) = 1$

b. A and B are independent

c. A and C are disjoint

d. $P(B \mid A) = 1$

5.115 MC4: For two events A and B, P(A) = 0.4, P(B) = 0.3, and P(A and B) = 0. It follows that A and B are:

a. independent but not disjoint

b. disjoint but not independent

c. neither disjoint nor independent

d. both disjoint and independent

5.116 Coin flip: A balanced coin is flipped 100 times. By the law of large numbers:

a. There will almost certainly be *exactly* 50 heads and 50 tails.

b. If we got 100 heads in a row, almost certainly the next flip will be a tail.

c. For the 100 flips, the probability of getting 100 heads equals the probability of getting 50 heads.

d. It is absolutely impossible to get a head every time.

e. None of the above.

5.117 Dreams come true: You have a dream in which you see your favorite movie star in person. The very next day, you are visiting Manhattan and you see her walking down Fifth Avenue.

a. This is such an incredibly unlikely coincidence that you should report it to your local newspaper.

b. This is somewhat unusual, but given the very large number of dreams you could have during your lifetime about people you know or know of, it is not an incredibly unlikely event.

c. If you had not had the dream, you would definitely not have seen the film star the next day.

d. This proves the existence of ESP.

5.118 Comparable risks: Mammography is estimated to save about 1 life in every 1000 women. "Participating in annual mammography screening ... has roughly the same effect on life expectancy as reducing the distance one drives each year by 300 miles" (Gigerenzer 2002, pp. 60, 73). Which of the following do you think has the closest effect on life expectancy as that of smoking throughout your entire life?

a. Taking a commercial airline flight once a year.

b. Driving about ten times as far every year as the average motorist.

c. Drinking a cup of coffee every day.

d. Eating a fast-food hamburger once a week.

e. Never having a mammogram (if you are a woman)

Select true or false for Exercises 5.119 -5.121

5.119 TF1: If $P(A) > 0$, $P(B) > 0$, and $P(A \text{ and } B) = 0$, the events A and B are independent.

5.120 TF2: When you flip a coin ten times, you are more likely to get the sequence HTHTHTHTHT than the sequence HHHHHHTTTT.

5.121 TF3: When you flip a balanced coin twice, there are three things that can happen: 0 heads, 1 head, or 2 heads. Since there are three possible outcomes, they each have probability 1/3. (This was claimed by the French mathematician, Jean le Rond d'Alembert, in 1754. To reason this, write down the sample space for the possible sequence of results for the two flips.)

5.122 Driving vs. flying: In the U.S. in 2002, about 43,000 people died in auto crashes and 0 people died in commercial airline accidents. G. Gigerenzer (2002, p. 31) states, "The terrorist attack on September 11, 2001, cost the lives of some 3,000 people. The subsequent decision of millions to drive rather than fly may have cost the lives of many more." Explain the reasoning behind this statement.

5.123 Prosecutor's fallacy: An eyewitness to the crime says that the person who committed it was male, between 15 and 20 years old, Hispanic, and drove a blue Honda. The prosecutor points out that a proportion of only 0.001 people living in that city match all those characteristics, and one of them is the defendant. Thus, the prosecutor argues that the probability that the defendant is not guilty is only 0.001. Explain what is wrong with this logic. (Hint: is $P(\text{match}) = P(\text{not guilty} \mid \text{match})$?)

♦♦ **5.124 Generalizing the addition rule:** For events A, B, and C such that each pair

339

of events is disjoint, use a Venn diagram to explain why

$$P(A \text{ or } B \text{ or } C) = P(A) + P(B) + P(C).$$

♦♦ **5.125 Generalizing the multiplication rule:** For events A, B, and C, explain why

$$P(A \text{ and } B \text{ and } C) = P(A) \times P(B \mid A) \times P(C \mid A \text{ and } B).$$

♦♦ **5.126 Bayes's rule**: Suppose we know $P(A)$, $P(B \mid A)$, and $P(B^c \mid A^c)$, but we want to find $P(A \mid B)$, such as when we know the prevalence, sensitivity, and specificity of a diagnostic test but want to find the positive predictive value.
a. Using the definition of conditional probability for $P(A \mid B)$ and for $P(B \mid A)$, explain why $P(A \mid B) = P(A \text{ and } B)/P(B) = [P(B \mid A)P(A)]/P(B)$.
b. Splitting the event that B occurs into two parts, according to whether or not A occurs, explain why $P(B) = P(B \text{ and } A) + P(B \text{ and } A^c)$.
c. Using (b) and the definition of conditional probability, explain why
$$P(B) = P(B \mid A)P(A) + P(B \mid A^c)P(A^c).$$
d. Combining what you have shown in (a)-(c), reason that

$$P(A \mid B) = \frac{P(B \mid A)P(A)}{P(B \mid A)P(A) + P(B \mid A^c)P(A^c)}.$$

This formula is called **Bayes's rule**. It is named after a British clergyman who discovered this formula in 1763.

CHAPTER PROBLEMS: CLASS EXPLORATIONS

🖳**5.127 Cards in board games:** Many board games have special cards that either help or hinder players. A particular board game has 30 such cards, 25 of which help and 5 of which hinder.
a. If the 30 cards are well shuffled, what is the probability of drawing an unfavorable card?
b. Simulate 20 times the drawing of a card from these 30 special cards. To do this, you may use the *random number* applet (using numbers from 1 to 30, designating 1 – 25 as favorable numbers and 26-30 as unfavorable numbers) at www.prenhall.com/TBA or a software random number generator. Or, you may use a deck of regular playing cards (let 25 black cards be the favorable and 5 red cards be the unfavorable). In that case, shuffle the cards well and draw a card. Note if the card is unfavorable. Put the card back in the deck. Repeat this 19 more times, shuffling the cards each time. As you do this, construct a graph similar to Figure 5.1 showing the cumulative proportion of draws of an unfavorable card. What happens to the cumulative proportion as you increase the number of draws?
c. Collect your classmates' results and create one graph, similar to Figure 5.2. Describe what happens to the cumulative proportion for drawing an unfavorable card. Is it close to the answer in part (a)?

🖳**5.128 Simulating matching birthdays:** Do you find it hard to believe that the probability of at least one birthday match in a class of 25 students is 0.57? Let's

simulate the answer. Using the *random number* applet at www.prenhall.com/TBA, each student in the class should simulate 25 integers between 1 and 365. Observe whether there is at least one match of two integers out of the 25 simulated integers.
a. Combine class results. What proportion of the simulations had at least one birthday match?
b. Repeat (a) for a birthday match in a class of 50 students. Again, combine class results. What is the simulated probability of at least one match?

5.129 M&Ms: How many peanut M&Ms of a particular color can you expect in a package?
a. As a class exercise, each student should take a bag of peanut M&Ms and count the number of M&Ms for each color. Combine the class results and fill in the table.

Color	Frequency	Proportion
Brown		
Red		
Blue		
Tan		
Orange		
Green		

b. Mars, Inc. states on their website (www.mms.com) that the colors for peanut M&M's in a bag should be distributed as follows:

COLOR	Brown	Red	Blue	Tan	Orange	Green
PROPORTION	0.20	0.20	0.20	0.20	0.10	0.10

The proportions reported by Mars can be regarded as long-run proportions. Compare the class results to them. Describe a factor that might cause your class results to differ from the proportions reported by Mars, Inc.

🖳5.130 Simulate table tennis: In table tennis, the first person to get at least 21 points while being ahead of the opponent by at least two points wins the game. In games between you and an opponent, suppose successive points are independent, and suppose the probability of your winning any given point is 0.40.
a. Do you have any reasonable chance to win a game? Showing all steps, simulate two games, using the the *random number* applet at www.prenhall.com/???, or other software. (*Hint*: Let the integers 0-3 represent winning a point and 4-9 represent losing a point.)
b. Combining results for all students in the class, estimate the probability of winning.

🖳5.131 Which tennis strategy is better?: A tennis match can consist of the best of three sets (that is, the winner is the first to win two sets) or the best of five sets (the winner is the first to win three sets). Which would you be better off playing if you are the weaker player and have probability 0.40 of winning any particular set?
a. Simulate 10 matches of each type, using *the random number* applet at www.prenhall.com/TBA or other software. Show the steps of your simulation, and specify any assumptions.

b. Combining results for all students in the class, estimate the probability of the weaker player winning the match under each type of match.

⌨**5.132 Sample representation**: Your community is 50% white, 40% black, and 10% Hispanic. If you take a simple random sample of 10 people, what is the probability that your sample contains at least one of each?
a. Show all the steps of a simulation using 10 runs, using the *random number* applet at www.prenhall.com/TBA or other software.
b. Combining results for all students in the class, estimate the probability.

BIBLIOGRAPHY

Against All Odds: Inside Statistics (1989). Produced by Consortium for Mathematics and its Applications, for PBS. See www.pbs.org/als/against_odds

Ainslie, T. (2003). *Ainslie's Complete Hoyle*, New York: Barnes and Noble Books.

Diaconis, P., and F. Mosteller. (1989) "Methods for studying coincidences," *J. Amer. Statist. Assoc.*, vol. 84, 853-861.

Everitt, B. (1999). *Chance Rules*. New York: Springer-Verlag.

Gigerenzer, G. (2002). *Calculated Risks*. New York: Simon and Schuster.

Osborne, J.A. (2003) "Markov chains for the RISK board game revisited," *Mathematics Magazine,* vol.76, 129-135.

Peace, M., L.Tarnai, and A. Poklis. (2000). "Performance Evaluation of Four On-Site Drug-Testing Devices for Detection of Drugs of Abuse in Urine," *Journal of Analytical Toxicology,* vol. 24.

Wilson, R., and E. A. C. Crouch. (2001). *Risk-Benefit Analysis*. Cambridge, MA: Harvard Univ. Press.

Chapter 6   Probability Distributions

EXAMPLE 1: PREDICTING ELECTION RESULTS USING EXIT POLLS

Picture the Scenario
In the United States, Canada, the United Kingdom, Australia, and many other countries, news organizations (often through television networks) declare the outcomes of many election contests well before all the votes have been counted. They develop this projection with a technique known as "exit polling," an opinion poll in which voters are randomly sampled after leaving the voting booth. Using an exit poll, a polling organization predicts a winner after learning how a small number of people voted, often only a few thousand out of possibly millions of voters.   What amazes many people is that these predictions almost always turn out to be correct. (A famous exception will be discussed at the end of the chapter.)

In California in October 2003, a special election was held to determine whether or not Governor Gray Davis should be recalled from office. In the same election, Californians voted for the candidate who would replace Davis as governor if he were recalled. The four candidates included the Hollywood actor Arnold Schwarzenegger. If a majority voted yes for the recall, the candidate who received the most votes would be the new governor. The exit poll on which TV networks relied for their projections found, after sampling 3160 voters, that 54% of the sample voted to recall Gray Davis and 46% voted not to recall him (www.cnn.com/ELECTION/2003/recall). At the time of the exit poll, the percentage of the entire voting population that voted in favor of the recall was unknown. In determining whether they could predict a winner, the TV networks had to study whether the exit poll data gave enough evidence to predict that the population percentage voting in favor of recalling Davis was above 50%.

Questions to Explore
- How close can we expect a sample percentage to be to the population percentage? For instance, if 54% of 3160 sampled voters supported the recall, how close to 54% is the population percentage who voted for it?

- How does the sample size influence our analyses? For instance, could we sample 100 voters instead of 3160 voters and make an accurate inference about the population percentage favoring recall of Davis?

Thinking Ahead
In this chapter, we'll apply the probability tools of the previous chapter to analyze how likely it is that sample results will be "close" to population values. For example, in Examples 16 and 17, we'll see why these results for 3160 voters allow us to make a prediction about the outcome for the entire population of nearly 8 million voters. This will be our first example of the use of *inferential statistics*.

The answers to the questions above will help us answer many other questions that have nothing to do with election data but to which the same principles apply. For instance, here's a question we'll investigate for those optimists who hope to make their fortune with a $1 lottery ticket: "Why is it usually worse to play a lottery many times rather than a few times, or not at all?"

Having learned the basics of probability in Chapter 5, we'll next study how probability provides the basis for making statistical inferences. As Chapter 1 explained, inferential statistical methods use sample data to make decisions and predictions about a population. The population summaries are called **parameters**. Inferential methods are the main focus of the rest of this book.

Before addressing inferential statistics, we need to learn a few more fundamentals about probability. We'll learn how to summarize possible outcomes and their probabilities with a **probability distribution**. We'll then study two very commonly used probability distributions--the **binomial** and the **normal**. We'll see in the rest of the book that the normal distribution, which has a bell-shaped graph, plays a key role in statistical inference.

6.1 HOW CAN WE SUMMARIZE POSSIBLE OUTCOMES AND THEIR PROBABILITIES?

With the proper methods of data collection presented in Chapter 4, the numerical values that a variable assumes are the result of some random phenomenon. For example, the randomness may involve selecting a random sample from a population or performing a randomized experiment. In such cases, to emphasize that the outcome of the variable is *random*, the variable is called a **random variable**.

> **Random Variable**
>
> A **random variable** is a numerical measurement of the outcome of a random phenomenon. Often, the randomness results from the use of random sampling or a randomized experiment to gather the data.

We've used letters near the end of the alphabet, such as x, to symbolize variables. We'll also use letters such as x for the possible value of a random variable. When we want to refer to the random variable itself, rather than a particular value, we'll use a capital letter, such as X. For instance, X = number of heads in three flips of a coin is a random variable, whereas $x = 2$ is one of its possible values.

Because a random variable refers to the outcome of a random phenomenon, each possible outcome has a specific probability of occurring. The **probability distribution** of a random variable specifies its possible values and their probabilities. An advantage of a variable being also a "random variable" is that it's possible to specify such probabilities. Without randomness, it's just not possible to know the probabilities of the possible outcomes.

Probability Distributions for Discrete Random Variables

When a random variable has separate possible values, such as 0, 1, 2, 3 for the number of heads in three flips of a coin, it is called **discrete**. The **probability distribution** of a discrete random variable assigns a probability to each possible value. Each probability falls between 0 and 1, and the sum of the probabilities of all possible values equals 1. We let *p(x)* denote the probability of a possible value *x*, such as *p(2)* for the probability the random variable takes value 2.

Definition: Probability Distribution of a Discrete Random Variable

A **discrete** random variable *X* takes a set of separate values (such as 0, 1, 2, …). Its **probability distribution** assigns a probability *p(x)* to each possible value *x*.

- For each *x,* the probability *p(x)* falls between 0 and 1.

- The sum of the probabilities for all the possible *x* values equals 1.

IN WORDS

From Chapter 5, a **probability** is a long-run **proportion**. So, it can take any value in the interval from 0 up to 1. The probabilities for all the possible outcomes add up to 1.0, so that's the sum of the probabilities in a probability distribution.

For instance, let *X* be a random digit selected using software for random numbers as part of the process of identifying subjects from a sampling frame to include in a random sample. The possible values for *X* are *x* = 0, 1, 2, ..., 8, 9. Each digit is equally likely, so the probability distribution is

$$p(0) = p(1) = p(2) = \ldots = p(9) = 0.10.$$

Each probability falls between 0 and 1, and the probabilities add up to 1.0.

Random variables can also be **continuous**, having possible values that are an interval rather than a set of separate values. We'll learn about continuous random variables and their probability distributions in the third section of this chapter.

EXAMPLE 2: WHO GETS TO MOVE FIRST IN *MONOPOLY*?

Picture the Scenario
Board game enthusiasts know that rolling dice is often an integral part of the game. Whether it is rolling one die in *Clue*, two dice in *Monopoly*, or five dice in *Yahtzee*, the outcome of the rolls dictate moves and options the players can have. Let's consider *Monopoly*. Dice are first used to determine which player gets to move first. Suppose there are two players in the game. Each player rolls a die and the player with the higher number gets to move first. If the numbers are the same, the players roll again. Throughout the game, each player rolls two dice, and their sum determines the number of spaces the player moves on the board.

Questions to Explore
To determine who moves first, each of two players rolls a die that is equally likely to have any of the 6 possible outcomes.
a. What is the probability distribution for the higher number?
b. What is the probability that the higher number is 3 or less?

RECALL

Section 5.2 introduced the **sample space**, which lists all the possible outcomes.

Think It Through
a. Let's first construct the sample space--the set of all possible outcomes when two dice are rolled. Table 6.1 shows this sample space. There are 36 equally likely outcomes, each having probability 1/36.

Table 6.1: The 36 Possible Outcomes from Rolling Two Dice. In a cell of this table, the first number shows the outcome of the first die and the second number shows the outcome of the second die.

DIE ONE	\multicolumn{6}{c}{DIE TWO}					
	1	2	3	4	5	6
1	1, 1	1, 2	1, 3	1, 4	1, 5	1, 6
2	2, 1	2, 2	2, 3	2, 4	2, 5	2, 6
3	3, 1	3, 2	3, 3	3, 4	3, 5	3, 6
4	4, 1	4, 2	4, 3	4, 4	4, 5	4, 6
5	5, 1	5, 2	5, 3	5, 4	5, 5	5, 6
6	6, 1	6, 2	6, 3	6, 4	6, 5	6, 6

RECALL

$p(2)$ denotes the probability that the random variable takes the value 2, that is, the probability that the larger of the numbers on the two dice equals 2.

Let the random variable $X =$ the higher number for the two rolls. From Table 6.1, the higher number equals 2 for the outcomes (1, 2), (2, 1), and (2, 2). These are 3 of the 36 possible outcomes, so $p(2) = 3/36$. Table 6.2 shows the entire probability distribution. Check to make sure you can verify these entries, such as $p(6) = 11/36$.

Table 6.2: Probability Distribution of $X =$ Higher Number for Rolls of Two Dice. The x column contains the possible outcomes and the $p(x)$ column gives their probabilities, such as $p(2) = 3/36$ giving the probability that the higher number is 2.

x	$p(x)$
1	1/36
2	3/36
3	5/36
4	7/36
5	9/36
6	11/36

b. The probability that the higher number is 3 or less is

$$p(1) + p(2) + p(3) = 1/36 + 3/36 + 5/36 = 9/36 = 0.25.$$

Insight

You shouldn't be optimistic about being the first player to move if you roll a 3 or less. The chance that it is the higher number is only 0.25, and in a third of the 9 cases in Table 6.1 in which this happens, the two numbers are the same, so the dice would have to be rolled again. We'll find the probability distribution for the sum of the numbers from rolling two dice in Exercise 6.1.

Notice in Table 6.2 that the two conditions in the definition of a probability distribution are satisfied:

- For each x, the probability $p(x)$ falls between 0 and 1.

- The sum of the probabilities for all the possible x values equals 1. ♦

To practice this concept, try Exercise 6.1.

The probability distribution of a discrete random variable can be displayed in a *table*. Table 6.2 above is an example. Alternatively, it can be displayed with a *graph* and sometimes with a *formula*. As in describing sample data, a graphical display is useful for revealing the three key components of a distribution: shape, center, and spread. Figure 6.1 displays a graph for the probability distribution in Table 6.2. It is skewed to the left. How do we find summary measures of the center and spread of such a probability distribution?

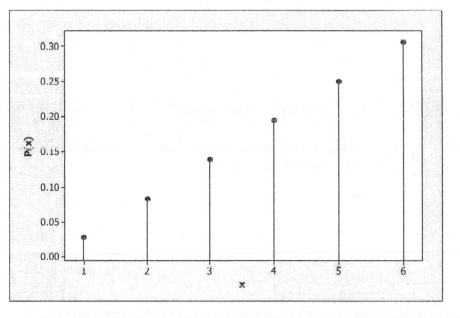

Figure 6.1: MINITAB Histogram for Probability Distribution of *X* = **Higher Number from Rolling Two Dice.** The stem over a possible value has height equal to its probability. **Question:** How would you describe the shape of the distribution?

The Mean of a Probability Distribution

IN WORDS

μ is the Greek letter *mu*, pronounced "mew." σ is the lowercase Greek letter *sigma*. The corresponding Roman letter *s* is used for the standard deviation of *sample* data. Recall that the sample mean is denoted by $\bar{x}$.

To describe characteristics of a probability distribution, we can use any of the numerical summaries defined in Chapter 2. These include the mean, median, quartiles, and standard deviation. It is most common to use the *mean* to describe the center and the *standard deviation* to describe the spread.

Recall that numerical summaries of populations are called **parameters**. You can think of a **population distribution** as merely being a type of probability distribution – one that applies for selecting a subject at random from the population. Like numerical summaries of populations, numerical summaries of probability distributions are also referred to as **parameters**. Most parameters are denoted by Greek letters. The mean of a probability distribution is denoted by μ and the standard deviation is denoted by σ.

Suppose we repeatedly observed values of a random variable, such as repeatedly noting the higher number shown on their faces when we roll two dice. The mean μ of the probability distribution for that random variable is the value we would get, in the long run, for the average of those values observed in repeated tosses of the two dice. This "long-run" interpretation parallels the interpretation of probability itself as a summary of the long-run behavior of a random phenomenon (recall Section 5.1).

EXAMPLE 3: WHAT'S THE EXPECTED NUMBER OF HOME RUNS IN A BASEBALL GAME?

Photo – Red Sox Celebration

Picture the Scenario

In 2004, the Boston Red Sox baseball team won the World Series (the crowning achievement in major league baseball) for the first time since 1918. This ended many years of frustration, which many attributed to a curse on the team for trading Babe Ruth to the New York Yankees in 1920. For the 2004 regular season of 162 games, Table 6.3 shows an estimate of the probability distribution for X = number of home runs the Red Sox hit in a game.[1]

Table 6.3: Probability Distribution of Number of Home Runs in a Game for Boston Red Sox in 2004.

Number of Home Runs	Probability
0	0.23
1	0.38
2	0.22
3	0.13
4	0.03
5	0.01

Question to Explore
Find the mean of this probability distribution, and interpret it.

Think It Through
Since $p(0) = 0.23$, over the long run you expect $x = 0$ (that is, no home runs) in 23% of the games. Likewise, you expect $x = 1$ home run 38% of the time, and so forth. In 100 games, for example, you expect $x = 0$ about 23 times, $x = 1$ about 38 times, and so forth. Since the mean equals the total of the observations divided by the sample size, for 100 games you expect a mean of about

$$\mu = \frac{(0+0+...+0)+(1+1+...+1)+(2+...+2)+(3+...+3)+(4+4+4)+5}{100}$$

$$= \frac{0(23)+1(38)+2(22)+3(13)+4(3)+5(1)}{100} = \frac{138}{100} = 1.38$$

[1] Source: www.usatoday.com/sports/baseball/al/redsox/home.htm. For instance, they hit 0 home runs in 38 of the 162 games, so the probability of 0 is estimated to be 38/162 = 0.23.

Note that rather than adding a number 38 times, you can multiply the number by 38, such as 1(38). If this probability distribution applies in a large number of games, you'd expect the mean number of home runs hit by the Red Sox to be about 1.4, more than one a game on the average, but not much higher than one. Across all teams in major league baseball, in 2004 the mean number of home runs per game was 1.12.

Insight

Since 23/100 = 0.23, 38/100 = 0.38, and so forth, the above calculation has the form

$$\mu = \frac{0(23)+1(38)+2(22)+3(13)+4(3)+5(1)}{100}$$

$$= 0(0.23) + 1(0.38) + 2(0.22) + 3(0.13) + 4(0.03) + 5(0.01)$$

$$= 0 \times p(0) + 1 \times p(1) + 2 \times p(2) + 3 \times p(3) + 4 \times p(4) + 5 \times p(5).$$

Each possible value x is multiplied by its probability. In fact, for any discrete random variable, the mean of its probability distribution results from multiplying each possible value x by its probability $p(x)$, and then adding.
◆

To practice this concept, try Exercise 6.3.

IN WORDS	**Definition: Mean of a Discrete Probability Distribution**
To get the **mean** of a probability distribution, multiply each possible value of the random variable by its probability, and then add all these products.	The **mean of a probability distribution** for a discrete random variable is $$\mu = \Sigma\ x\,p(x),$$ where the sum is taken over all possible values of x.

IN WORDS

A **weighted average** is used when each x value is not equally likely. If a particular x value is more likely to occur, it has a larger influence on the mean, which is the balance point of the distribution.

For Table 6.3, for example,

$$\mu = \Sigma\ x\,p(x) = 0p(0) + 1p(1) + 2p(2) + 3p(3) + 4p(4) + 5p(5)$$
$$= 0(0.23) + 1(0.38) + 2(0.22) + 3(0.13) + 4(0.03) + 5(0.01) = 1.4.$$

The mean $\mu = \Sigma\ x\,p(x)$ is called a **weighted average**: Values x that are more likely receive greater weight $p(x)$. It does not make sense to take a simple average of the possible values of x, $(0 + 1 + 2 + 3 + 4 + 5)/6 = 2.5$, because some outcomes are much more likely than others.

Consider the special case in which the outcomes are equally likely. Suppose there are n such possible outcomes, each with probability $1/n$. Then

$$\mu = \Sigma\ x\,p(x) = \Sigma\ x\,(1/n) = (\Sigma\ x)/n.$$

349

This is the ordinary formula we used for a sample mean in Chapter 2. So the formula $\mu = \Sigma\, x\, p(x)$ generalizes the ordinary formula to allow different outcomes not to be equally likely and to apply to probabilities for random variables as well as to sample data.

The mean of the probability distribution of a random variable X is also called the **expected value of X**. The expected value reflects not what we'll observe in a *single* observation, but rather what we expect for the *average* in a long run of observations. In the above example, the expected value of the number of Boston Red Sox home runs in a game is $\mu = 1.4$. As with means of sample data, the mean of a probability distribution doesn't have to be one of the possible values for the random variable. We will not see 1.4 home runs in a game, but the long-run average of observing some games with 0 home runs, some with 1 home run, some with 2, and so forth, is 1.4.

EXAMPLE 4: ARE YOU A RISK TAKER OR RISK AVERSE?

Picture the Scenario
Are you "risk averse," a person who prefers the sure thing to a risky action that could give you a better or a worse outcome? Or are you "risk taking," willing to gamble in hopes of achieving the better outcome?

Questions to Explore
a. You are given $1000 to invest. You must choose between (i) a sure gain of $500 and (ii) a 0.50 chance of a gain of $1000 and a 0.50 chance to gain nothing. What is the expected gain with each strategy? Which do you prefer?
b. You are given $2000 to invest. You must choose between (i) a sure loss of $500 and (ii) a 0.50 chance of losing $1000 and a 0.50 chance to lose nothing. What is the expected loss with each strategy? Which do you prefer?

Think It Through
a. The expected gain is $500 with the sure strategy (i), since that strategy has gain $500 with probability 1.0. With the risk-taking strategy (ii), the expected gain is $\mu = \Sigma\, x\, p(x) = \$0(0.50) + \$1000(0.50) = \$500$.
b. The expected loss is $500 with the sure strategy (i). With the risk-taking strategy (ii), the expected loss is $\mu = \Sigma\, x\, p(x) = \$0(0.50) + \$1000(0.50) = \$500$.

Insight
In each case, the expected values are the same with each strategy. But most people prefer the sure gain strategy (i) in case (a) but the risk-taking strategy (ii) in case (b). They are risk averse in case (a) but risk taking in case (b). This was explored in research by Daniel Kahneman and Amos Tversky, for which Kahneman won the Nobel Prize in 2002.

♦

To practice this concept, try Exercise 6.8.

Summarizing the Spread of a Probability Distribution

As with distributions of sample data, it's useful to summarize both the *center* and the *spread* of a probability distribution. For instance, suppose two different investment strategies have the same expected payout. Does one strategy have more variability in its payoffs?

The **standard deviation** of a probability distribution, denoted by σ, measures its spread. Larger values for σ correspond to greater spread. Roughly, σ describes how far the random variable tends to fall, on the average, from the mean of its distribution. We won't deal with the formula for calculating the standard deviation until we study a particular type of probability distribution in the next section.

EXAMPLE 5: WHAT'S THE IDEAL NUMBER OF CHILDREN FOR A FAMILY?

Picture the Scenario
Let X denote the response of a randomly selected person to the question, "What is the ideal number of children for a family to have?" This is a discrete random variable, with possible values such as 0, 1, 2, 3, and so on. The probability distribution of X in the United States is approximately[2] as shown in Table 6.4, according to the gender of the person asked the question.

Table 6.4: Probability Distribution of X = Ideal Number of Children.

x	$p(x)$ females	$p(x)$ males
0	0.00	0.04
1	0.03	0.03
2	0.63	0.57
3	0.23	0.23
4	0.11	0.13

For instance, from the column of Table 6.4 for females, 0.63 is approximately the proportion of females in the population who think that 2 is the ideal number of children. So the probability is also 0.63 that a randomly selected female from that population makes that choice. (All potential x values higher than 4 have a negligible probability.) The means are similar for females and males,

$$\mu = \Sigma \ x\, p(x) = 0(0.04) + 1(0.03) + 2(0.57) + 3(0.23) + 4(0.13) = 2.38$$

for males and μ = 2.42 for females, more than 2 but fewer than 3 children.

Question to Explore
Which of the two probability distributions shown in Table 6.4 would you expect to have the larger standard deviation?

---

[2] These probability distributions are merely approximations, suggested by results from a recent General Social Survey.

Think It Through
The ideal family size values are similar, on the average, for females and male. But they are a bit more spread out for males, as the smallest value of 0 and largest value of 4 both have higher probabilities for males than females. We would expect the standard deviation to be larger for the males' probability distribution.

Insight
In fact, it turns out that $\sigma = 0.89$ for males but $\sigma = 0.72$ for females. What's the practical implication of this? Although both groups have about the same mean, females hold slightly more consistent views than males about the ideal family size.

◆

To practice this concept, try Exercise 6.7, part (c).

Sometimes we can use a sample space to help us find a probability distribution, as we did for the higher outcome of two dice in Example 2. Sometimes the probability distribution is available with a *formula* or a *table* or a *graph*, as we'll see in the next two sections. Sometimes, though, it's necessary to use simulation to approximate the distribution, as shown in Activity 2 at the end of the chapter.

To practice concepts of this section, try Activity 2 at the end of the chapter.

Section 6.1 Practicing the Basics

6.1 Rolling dice:
a. State the probability distribution for the outcome of rolling a balanced die. (This is called the **uniform distribution** on the integers 1, 2, … , 6.)
b. Two balanced die are rolled. Show that the probability distribution for X = total on the two dice is as shown in the figure below. (*Hint*: You can use results from Table 6.1 to set up a table similar in format to Table 6.2.)
c. Show that the probabilities in (b) satisfy the two conditions for a probability distribution.

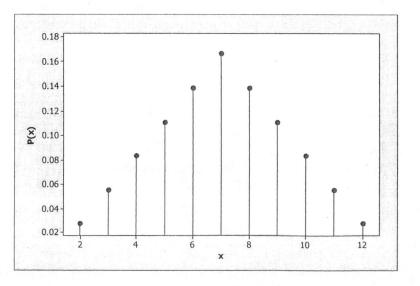

6.2 Means for dice games: Refer to the previous exercise.
a. Find the mean of the probability distribution for the outcome of the roll of one die.
b. From inspection of the graph in Exercise 6.1, what is the mean of the probability distribution of X = total on two dice? (In fact, when you add two random variables, the mean of the probability distribution of the sum is the sum of the two separate means.)

6.3 Boston Red Sox hitting: In 2004, the table shows the probability distribution of the number of bases for a randomly selected time at bat for a Boston Red Sox player (excluding times when the player got on base because of a "walk" or being "hit by pitch"). In 71.8% of the at-bats the player was out, 17.4% of the time the player got a "single" (one base), 6.5% of the time the player got a "double" (two bases), 0.4% of the time the player got a "triple," and 3.9% of the time the player got a home run.
a. Verify that the probabilities give a legitimate probability distribution.
b. Find the mean of this probability distribution. Interpret it.

Boston Red Sox hitting

Number of bases	Probability
0	0.718
1	0.174
2	0.065
3	0.004
4	0.039

6.4 Grade distribution: An instructor always assigns final grades such that 20% are A, 40% are B, 30% are C, and 10% are D. The grade-point scores are 4 for A, 3 for B, 2 for C, and 1 for D.
a. Specify the probability distribution for the grade point of a randomly selected student of this instructor.
b. Find the mean of this probability distribution.

a. **6.5 Bilingual Canadians?:** Let X = number of languages in which a person is fluent. According to Statistics Canada, for residents of Canada this has probability distribution $p(0) = 0.02$, $p(1) = 0.81$, $p(2) = 0.17$, with negligible probability for higher values of x
b. Explain why it does not make sense to compute the mean of this probability distribution as $(0 + 1 + 2)/3 = 1.0$.
c. Find the correct mean.

6.6 Playing the lottery: The state of Ohio has several statewide lottery options (www.ohiolottery.com). One of the simpler ones is a "Pick 3" game in which you pick one of the 1000 3-digit numbers between 000 and 999. The lottery selects a 3-digit number at random. With a bet of $1, you win $500 if your number is selected and nothing ($0) otherwise. The Florida lottery has the same game (called CASH 3, with the STRAIGHT play option) and the same winnings structure.
a. With a single $1 bet, what is the probability that you win $500?
b. Let X denote your winnings for a $1 bet, so x = $0 or x = $500. Construct the probability distribution for X.

c. Show that the mean of the distribution equals 0.50, corresponding to an expected return of 50 cents for the dollar paid.

6.7 Lottery profit: Refer to the previous exercise. The profit Y from buying a $1 ticket equals the winnings X minus the dollar paid for the ticket; that is, $Y = X - 1$.
a. Construct the probability distribution of Y.
b. Find the mean of the distribution of Y. Interpret. Note this can also be obtained by subtracting 1 from the mean of the distribution of X.
c. Would you expect the standard deviation of the distribution of Y to be equal to, larger than, or smaller than the standard deviation of the distribution of X? Explain.

6.8 Which wager do you prefer?: You are given $100 and told that you must pick one of two wagers, for an outcome based on flipping a fair coin:
>   1: You win $200 if it comes up heads and lose $50 if it comes up tails.
>   2: You win $350 if it comes up heads and lose your original $100 if it comes up tails.

a. Without doing any calculation, which wager would you prefer? (There is no "correct" answer. Just specify which you'd choose.)
b. Find the expected outcome for each wager. Which wager is better in this sense? (Most people pick strategy 1, because of the smaller possible loss.)

6.9 Law school admissions: Five students, Ann, Briana, Clinton, Douglas, and Eduardo, are rated equally qualified for admission to law school, ahead of other applicants. However, all but two positions have been filled for the entering class. Since the admissions committee can admit only two more students, it decides to randomly select two of these five candidates. For this strategy, let X summarize the number of females admitted. Using the first letter of the name to denote a student, the different combinations that could be admitted are (A,B), (A,C), (A,D), (A,E), (B,C), (B,D), (B,E), (C,D), (C,E), and (D,E). For instance, $x = 2$ for (A,B) and $x = 1$ for (A,C).
a. What is the probability of each of the ten possible samples?
b. Construct a table showing the probability distribution for X.
c. Find the mean of the probability distribution of X.

6.10 Ohio's irresistible lottery: In Ohio's "Pick 4"lottery, you pick one of the 10,000 4-digit numbers between 0000 and 9999 and (with a $1 bet) win $5000 if you get it correct. In terms of your expected winnings, with which game are you better off--playing Pick 4, or playing Pick 3 in which you win $500 for a correct choice of a 3-digit number? Justify your answer.

6.11 Streaks of makes or misses: Let $X =$ the longest number of free throws made in a row, *or* missed in a row, out of 20 shots in a basketball game by a player who has probability 0.50 of making any particular shot. Activity 2 at the end of the chapter investigates a related probability distribution. When successive shots are independent, the probability distribution of X is:

x	0	1	2	3	4	5	6	7	8	9	10	≥ 11
$p(x)$	0.00	0.00	0.02	0.21	0.31	0.22	0.12	0.06	0.03	0.01	0.01	0.01

a. Show that this satisfies the two conditions for a probability distribution.

b. The mean of this probability distribution equals 4.7. Explain how this relates to values of X observed for a large number of games with 20 shots in each.

6.2 HOW CAN WE FIND PROBABILITIES WHEN EACH OBSERVATION HAS TWO POSSIBLE OUTCOMES?

Some probability distributions merit special attention because they are useful for many applications. They have formulas that provide probabilities of the possible outcomes. We now study the most important probability distribution for discrete random variables. Learning about it helps us answer the questions we asked at the beginning of the chapter--or instance, whether results of an exit poll allow us to predict the outcome of an election.

The Binomial Distribution: Probabilities for Counts with Binary Data

In many applications, each observation is **binary**: It has one of two possible outcomes. For instance, a person may

  - accept, or decline, an offer from a bank for a credit card

  - have, or not have, health insurance

  - vote yes or no in a referendum, such as whether to recall a governor from office

With a sample, we summarize such variables by counting the *number* or the *proportion* of cases with an outcome of interest. For instance, with a sample of size n = 5, let the random variable X denote the number of people who vote "yes" about some issue in a referendum. The possible values for X are 0, 1, 2, 3, 4, and 5. Under certain conditions, a random variable X that counts the number of observations of a particular type has a probability distribution called the **binomial**.

Consider n cases, called **trials**, in which we observe a binary random variable. This is a *fixed number*, such as n = 5 for a sample of 5 voters. The number X of trials in which the outcome of interest occurs can take any one of the integer values 0, 1, 2, ..., n. The binomial distribution gives probabilities for these possible values of X when the following three conditions hold:

Conditions for Binomial Distribution

  - Each of n trials has two possible outcomes. It's common to refer to the outcome of interest as a "success" and the other outcome as a "failure."
  - Each trial has the same probability of a success. This is denoted by p.
  - The n trials are independent. That is, the result for one trial does not depend on the results of others.

The **binomial random variable** X is the number of successes in the n trials.

Flipping a coin is a prototype for when the conditions for the binomial apply:

- Each trial is a flip of the coin. There are two possible outcomes for each flip, head or tail. Let's identify (arbitrarily) head as "success."

- The probability p of a head equals 0.50 for each flip if head and tail are equally likely.

- The flips are independent, since the result for the next flip does not depend on the outcomes of previous flips.

The binomial random variable X counts the number of the n flips that are heads (the outcome of interest). With $n = 3$ coin flips, X = number of heads could equal 0, 1, 2, or 3.

EXAMPLE 6: FINDING BINOMIAL PROBABILITIES FOR AN ESP EXPERIMENT

Picture the Scenario
John Doe claims to possess extrasensory perception (ESP). An experiment is conducted in which a person in one room picks one of the integers 1, 2, 3, 4, 5 at random and concentrates on it for one minute. In another room, John Doe identifies the number he believes was picked. The experiment is done with three trials. After the third trial, the random numbers are compared with John Doe's predictions. Doe got the correct result twice.

Question to Explore
If John Doe does not actually have ESP and is actually guessing the number, what is the probability that he'd make a correct guess on two of the three trials?

Think It Through
Let X = number of correct guesses in $n = 3$ trials. Then $X = 0$, 1, 2, or 3. Let p denote the probability of a correct guess for a given trial. If Doe is guessing, $p = 0.2$ for Doe's prediction of one of the five possible integers. Also, $1 - p = 0.8$ is the probability of an incorrect prediction on a given trial. Denote the outcome on a given trial by S or F, representing "success" or "failure" for whether Doe's guess was correct or not. Table 6.5 shows the eight outcomes in the sample space for this experiment. For instance, FSS represents a correct guess on the second and third trials. It also shows their probabilities, by using the multiplication rule for independent events.

RECALL
From Section 5.3, for independent events, $P(A \text{ and } B) = P(A)P(B)$. Thus, $P(FSS) = P(F)P(S)P(S) = 0.8 \times 0.2 \times 0.2$.

Table 6.5: Sample Space and Probabilities for Three Guesses. The probability of a correct guess is 0.2 on each of the three trials, if John Doe does not have ESP.

Outcome	Probability	Outcome	Probability
SSS	$0.2 \times 0.2 \times 0.2 = (0.2)^3$	SFF	$0.2 \times 0.8 \times 0.8 = (0.2)^1(0.8)^2$
SSF	$0.2 \times 0.2 \times 0.8 = (0.2)^2(0.8)^1$	FSF	$0.8 \times 0.2 \times 0.8 = (0.2)^1(0.8)^2$
SFS	$0.2 \times 0.8 \times 0.2 = (0.2)^2(0.8)^1$	FFS	$0.8 \times 0.8 \times 0.2 = (0.2)^1(0.8)^2$
FSS	$0.8 \times 0.2 \times 0.2 = (0.2)^2(0.8)^1$	FFF	$0.8 \times 0.8 \times 0.8 = (0.8)^3$

The three ways John Doe could make two correct guesses in three trials are SSF, SFS, and FSS. Each of these has probability equal to $(0.2)^2(0.8) = 0.032$. The total probability of two correct guesses is

$$3(0.2)^2(0.8) = 3(0.032) = 0.096.$$

Insight

In terms of the probability $p = 0.2$ of a correct guess on a particular trial, the solution $3(0.2)^2(0.8)$ for $x = 2$ correct in $n = 3$ trials equals $3\,p^2(1-p)^1 = 3p^x(1-p)^{n-x}$. The multiple of 3 represents the number of ways that 2 successes can occur in 3 trials. You can use similar logic to evaluate the probability that $x = 0$, or 1, or 3. Try $x = 1$, for which you should get $p(1) = 0.384$.

♦

To practice this concept, try Exercises 6.12, part (a).

The formula for binomial probabilities

When the number of trials n is large, it's tedious to write out all the possible outcomes in the sample space. But there's a formula we can use to find binomial probabilities for *any n*.

Probabilities for a Binomial Distribution

Denote the probability of success on a trial by p. For n independent trials, the probability of x successes equals

$$p(x) = \frac{n!}{x!(n-x)!}p^x(1-p)^{n-x}, \quad x = 0, 1, 2, \dots, n.$$

The symbol $n!$ is called **n factorial**. It represents $n! = 1 \times 2 \times 3 \times \dots \times n$, the product of all integers from 1 to n. That is, $1! = 1$, $2! = 1 \times 2 = 2$, $3! = 1 \times 2 \times 3 = 6$, $4! = 1 \times 2 \times 3 \times 4 = 24$, and so forth. Also, 0! is defined to be 1. For given values for p and n, you can find the probabilities of the possible outcomes by substituting values for x into the binomial formula.

Let's use this formula to find the answer for Example 6 above about ESP:

- The random variable X represents the number of correct guesses (successes) in $n = 3$ trials of the ESP experiment.

- The probability of a correct guess in a particular trial is $p = 0.2$.

- The probability of exactly 2 correct guesses is the binomial probability with $n = 3$ trials, $x = 2$ correct guesses, and $p = 0.2$ probability of a correct guess for a given trial,

$$p(2) = \frac{n!}{x!(n-x)!}\, p^x\,(1\text{-}p)^{n-x} = \frac{3!}{2!1!}(0.2)^2\,(0.8)^1 = 3(0.04)(0.8) = 0.096.$$

What's the role of the different terms in this binomial formula?

- The factorials term tells us the number of possible outcomes that have $x = 2$ successes. Here, $[3!/(2!1!)] = (3\times2\times1)/(2\times1)(1) = 3$ tells us there were 3 possible outcomes with 2 successful guesses, namely SSF, SFS, and FSS.

- The term $(0.2)^2\,(0.8)^1$ with the power exponents gives the probability for each such sequence. Here, the probability is $(0.2)^2\,(0.8) = 0.032$ for each of the three sequences having $x = 2$ successful guesses, for a total probability of $3(0.032) = 0.096$.

Try to calculate $p(1)$ by letting $x = 1$ in the binomial formula with $n = 3$ and $p = 0.2$. You should get 0.384. You'll see that there are again 3 possible sequences, now each with probability $(0.2)^1\,(0.8)^2 = 0.128$. Table 6.6 summarizes the calculations for all four possible x values. You can also find binomial probabilities using statistical software, such as Minitab or a calculator with statistical functions.

Table 6.6: The Binomial Distribution for $n = 3$, $p = 0.20$. When $n = 3$, the binomial random variable X can take any integer value between 0 and 3. The total probability equals 1.0.

x	$p(x)$	$=$	$[n!/(x!(n\text{-}x)!)]p^x\,(1\text{-}p)^{n-x}$
0	0.512	=	$[3!/(0!3!)](0.2)^0\,(0.8)^3$
1	0.384	=	$[3!/(1!2!)](0.2)^1\,(0.8)^2$
2	0.096	=	$[3!/(2!1!)](0.2)^2\,(0.8)^1$
3	0.008	=	$[3!/(3!0!)](0.2)^3\,(0.8)^0$

EXAMPLE 7: ARE WOMEN PASSED OVER FOR MANAGERIAL TRAINING?

Picture the Scenario

While the binomial distribution is useful for finding probabilities for such trivial pursuits as flipping a coin or checking ESP claims, it's also an important tool in helping us understand more serious issues. An example is the presence of bias in hiring, promotion, and daily life. Take, for example, the case of a large supermarket chain in Florida. The chain occasionally selects employees to receive management training. A group of women there claimed that female employees were passed over for this training in favor of their male colleagues.[3] The company denied this claim. A similar claim of gender bias was recently made about promotions and pay for the 1.6 million women who work or who have worked for Wal-Mart.[4]

---

[3] *Tampa Tribune*, April 6, 1996
[4] *New York Times*, June 25, 2004.

Question to Explore

Suppose the large employee pool (more than a thousand people) that can be tapped for management training is half female and half male. Since this program began, none of the 10 employees chosen for management training were female. How can we investigate statistically the women's assertion of gender bias?

Think It Through

Other factors being equal, at each choice the probability of selecting a female equals 0.50 and the probability of selecting a male equals 0.50. If the employees are selected randomly in terms of gender, about half of the employees picked should be females and about half should be male. Due to ordinary sampling variation, however, it need not happen that *exactly* 50% of those selected are female. A simple analogy is in flipping a coin 10 times. We won't necessarily see exactly 5 heads. The number of heads is a random variable with a binomial distribution.

Since none of the 10 employees chosen for management training has been female, we might be inclined to support the women's claim. However, would these results be all that unlikely, if there were no gender bias? Equivalently, if we flip a coin 10 times, would it be very surprising if we got 0 heads?

Let X denote the number of females selected for management training in a random sample of 10 employees. Then, the possible values for X are 0, 1, ... , 10, and X has the binomial distribution with $n = 10$ and $p = 0.50$. For each x between 0 and 10, we can find the probability that x of the 10 people selected are female. We use the binomial formula for that x value with $n = 10$ and $p = 0.50$, namely

$$p(x) = \frac{n!}{x!(n-x)!} p^x (1-p)^{n-x} = \frac{10!}{x!(10-x)!} (0.50)^x (0.50)^{10-x}, \quad x = 0, 1, 2, \ldots, 10.$$

The probability that no females are chosen ($x = 0$) equals

$$p(0) = \frac{10!}{0!10!} (0.50)^0 (0.50)^{10} = (0.50)^{10} = 0.001.$$

Any number raised to the power of 0 equals 1. Also, $0! = 1$, and the $10!$ terms in the numerator and denominator cancelled, leaving $p(0) = (0.50)^{10}$. If the employees were chosen randomly, it is very unlikely (one chance in a thousand) that none of the 10 selected for management training would have been female.

Insight

In summary, because this probability is so small, seeing no women chosen would make us highly skeptical that the choices were random with respect to gender.

◆

To practice this concept, try Exercises 6.24 and 6.25.

In this example, we found the probability that $x = 0$ female employees would be chosen for management training. To get a more complete understanding of just

```
binomPdf(10,.5,6
)
            .205
```

which outcomes are likely, you can find the probabilities of all the other possible x values. Table 6.7 lists the entire binomial distribution for $n = 10$, $p = 0.50$. If your statistical software provides binomial probabilities, see if you can verify that $p(0) = 0.001$ and $P(1) = 0.010$. The margin shows a screen shot of the TI-83+ calculator providing a binomial probability.

Table 6.7: The Binomial Probability Distribution for $n = 10$ and $p = 0.5$.

x	$p(x)$	x	$p(x)$
0	0.001	6	0.205
1	0.010	7	0.117
2	0.044	8	0.044
3	0.117	9	0.010
4	0.205	10	0.001
5	0.246		

In Table 6.7, the least likely values for x are 0 and 10, followed by 1 and 9. The combined probability of these four outcomes is only 0.022. The probability is $1 - 0.022 = 0.978$ that X falls between 2 and 8, inclusive. If the sample were randomly selected, somewhere between about two and eight females would probably be selected. It is highly unlikely that none or ten would be selected.

Figure 6.2 is a graph of the binomial distribution with $n = 10$ and $p = 0.50$. It has a symmetric appearance around $x = 5$. For instance, $x = 10$ has the same probability as $x = 0$. The binomial distribution is perfectly symmetric only when $p = 0.50$.

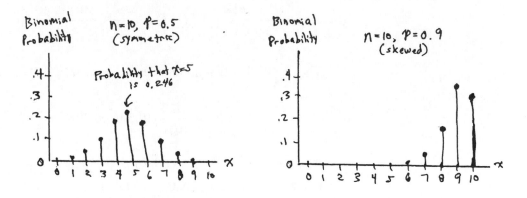

Figure 6.2: Binomial Distributions when $n = 10$ for $p = 0.5$ and for $p = 0.9$. The binomial probability distribution is symmetric when $p = 0.5$, but it can be quite skewed for p near 0 or near 1. **Question**: How do you think the distribution would look if $p = 0.1$?

When $p \neq 0.50$, the binomial distribution has a skewed appearance. The degree of skew increases as p gets closer to 0 or 1. To illustrate, Figure 6.2 also shows the binomial distribution for $n = 10$ when $p = 0.9$. If 90% of the potential trainees were female, it would not be especially surprising to observe 10, 9, 8, or even 7 females in the sample, but the probabilities drop sharply for smaller x-values.

Do the Binomial Conditions Apply?

Before you use the binomial distribution, check that its three conditions apply. These are (1) binary data (success or failure), (2) the same probability of success for each trial (denoted by p), and (3) independent trials. To judge this, ask yourself whether the observations resemble coin flipping, the simple prototype for the binomial. For instance, for Example 7 on gender bias in selecting employees for management training, this does seem plausible. In this instance:

- The data are binary because (female, male) plays the role of (head, tail).

- If employees are randomly selected, the probability p of selecting a female on any given trial is 0.50.

- With random sampling, the outcome for one trial is separate and independent from the outcome of another trial.

EXAMPLE 8: BINOMIAL SAMPLING FOR MANAGERIAL TRAINING?

Picture the Scenario
Consider the gender bias investigation in Example 7. Suppose the population of potential trainees contained only four people, two men and two women (instead of more than 1000 people, as in Example 7), and the sample size was $n = 2$.

Question to Explore
Do the binomial conditions apply for calculating the probability, under random sampling, of selecting 0 women trainees?

Think It Through
For the first trainee selected, the probability of a woman is $2/4 = 0.50$. The usual sampling is "sampling without replacement," in which the first person selected is no longer in the pool for future selections. So, given that the first person selected was male, the conditional probability that the second person selected is male equals $1/3$, since the pool of potential trainees now has 1 man and 2 women. So, the outcome of the second selection *depends* on that of the first. The trials are not independent, which the binomial requires.

Insight
This example suggests a caution with applying the binomial to a random sample from a population: For trials to be sufficiently "close" to independent with common probability p of success, the population size must be large relative to the sample size.

                                                                    ♦

To practice this concept, try Exercise 6.28.

> **RECALL**
>
> As Section 5.3 discussed, if it is the case that once subjects are selected from a population they are not eligible to be selected again, this is called *sampling without replacement*. Example 11 there showed the effect on sampling from a small population versus a large population.

Here's a guideline about the relative sizes of the sample and population for which the binomial formula works well:

> **IN PRACTICE: Population Size Needed to Use the Binomial**
>
> For sampling n separate subjects from a population (that is, sampling without replacement), the exact probability distribution of the number of successes is too complex to discuss in this text, but the binomial distribution approximates it well when n is less than 10% of the population size. In practice, sample sizes are usually small compared to population sizes, and this guideline is satisfied.

Suppose your school has 4000 students. Then the binomial formula is adequate as long as the sample size is less than about 10% of 4000, which is 400. Why? Because with a sample of 400 or fewer out of 4000, the probabilities of the two possible outcomes for any one selection will be practically the same regardless of what happens on other selections. Likewise, Example 7 dealt with the selection of 10 employees for management training when the employee pool for such training was more than 1000 people. Again, the sample was less than 10% of the population size, so using the binomial is valid.

Mean and Standard Deviation of the Binomial Distribution

Example 7 applied the binomial distribution for the number of women selected for management training when $n = 10$ and $p = 0.50$. If p truly equals 0.50, out of 10 selections, what do you expect for the number of women selected for management training?

As with any discrete probability distribution, we can use the formula $\mu = \Sigma\, x\, p(x)$ to find the mean. However, finding the mean μ and standard deviation σ is actually simpler for the binomial distribution, because they have special formulas based on the number of trials n and the probability p of success on each trial.

> **Binomial Mean and Standard Deviation**
>
> The binomial probability distribution for n trials with probability p of success on each trial has mean μ and standard deviation σ given by
>
> $$\mu = np, \quad \sigma = \sqrt{np(1-p)}.$$

The formula for the mean makes sense. If the probability of success is p for a given trial, then we expect about a proportion p of the n trials to be successes, or about np total. If we sample $n = 10$ people from a population in which half are female, then we expect that about $np = 10(0.50) = 5$ in the sample will be female.

The formula $\sigma = \sqrt{np(1-p)}$ for the standard deviation shows that σ increases as n does. This means that more variability tends to occur as the number of trials increases. In flipping a coin n times, the distribution of $X =$ number of heads is more spread out with $n = 1000$ coin flips than with $n = 10$ coin flips.

When the number of trials n is large, it is tedious to calculate binomial probabilities of all the possible outcomes. Often, it's adequate merely to use the mean and standard deviation to describe where most of the probability falls. When the binomial distribution has a bell shape, we can use the Empirical Rule to do this.

EXAMPLE 9: HOW CAN WE CHECK FOR RACIAL PROFILING?

Picture the Scenario
In the 1990s, the U.S. Justice Department and other groups studied possible abuse by Philadelphia police officers in their treatment of minorities. One study, conducted by the American Civil Liberties Union,[5] analyzed whether African-American drivers were more likely than others in the population to be targeted by police for traffic stops. They studied the results of 262 police car stops during one week in 1997. Of those, 207 of the drivers were African-American, or 79% of the total.

Question to Explore
At that time, Philadelphia's population was 42.2% African-American. Does the number of African-Americans stopped suggest possible bias, being higher than we would expect (other things being equal, such as the rate of violating traffic laws)?

Think It Through
We'll treat the 262 police car stops as $n = 262$ trials. Suppose that the percentage of drivers in Philadelphia during that week who were African-American was the same as their representation in the population (42.2%). Then, if there is no bias, the chance for any given police car stop that the driver is African-American is $p = 0.422$. Suppose also that successive police car stops are independent. (They would not be, for example, if once a car was stopped, the police followed that car and stopped it repeatedly.) Under these assumptions, for the 262 police car stops, the number of African Americans stopped has a binomial distribution with $n = 262$ and $p = 0.422$.

The binomial distribution with $n = 262$ and $p = 0.422$ has

$$\mu = np = 262(0.422) = 111, \quad \sigma = \sqrt{np(1-p)} = \sqrt{262(0.422)(0.578)} = 8.$$

RECALL

By the Empirical Rule (Section 2.4), when a distribution is bell-shaped, about 68%, 95%, and 100% of it falls within 1, 2, and 3 standard deviations of the mean.

Since $p = 0.422$ is close to 0.50 and since n is large, this binomial distribution is approximately symmetric and bell-shaped. By the Empirical Rule, the probability within 3 standard deviations of the mean is close to 1.0. This is the interval between

$$\mu - 3\sigma = 111 - 3(8) = 87 \quad \text{and} \quad \mu + 3\sigma = 111 + 3(8) = 135.$$

If no racial profiling is happening, we would not be surprised if between about 87 and 135 of the 262 people stopped were African-Americans. See the smooth curve approximation for the binomial in the margin. However, the actual number stopped (207) is well above these values. This suggests that the number of African-Americans stopped may be too high, even taking into account random variation.

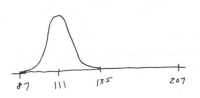

87   111   135   207

---

[5] www.archive.aclu.org/profiling/report

Insight

This approximate analysis based on the Empirical Rule told us that we would not expect to see 207 African-Americans stopped if there were truly only a 0.422 chance of finding an African-American behind the wheel on each stop. In fact, if we were to use software to do a more precise analysis by calculating the binomial probabilities of *all* possible values 0, 1, 2, …, 262 when $n = 262$ and $p = 0.422$, we'd find that the probability of getting 207 or a larger value out in the right tail of the distribution is 0 to the precision of many decimal places (that is, 0.00000000…). However, a limitation of both this analysis and the one based on the Empirical Rule is that different people do different amounts of driving, so we don't really know that 42.2% of the *potential* stops were African-American. For instance, many suburban residents (who are predominantly white) drive into Philadelphia for work, so that percentage may be less than 42.2%.

♦

 To practice this concept, try Exercise 6.23.

The solution in Example 9 treated the binomial distribution as being approximately bell-shaped. For sufficiently large n, this is true even if p is not close to 0.50. Here's a guideline:[6]

IN PRACTICE: When Is the Binomial Distribution Bell Shaped?

The binomial distribution has close to a symmetric, bell shape when the expected number of successes, np, and the expected number of failures, $n(1 - p)$, are both at least 15.

In Example 9, of those stopped, the expected number who were African-Americans was $np = 262(0.422) = 111$. The expected number who were *not* African-American was $n(1 - p) = 262(0.578) = 151$. Both exceed 15, so this binomial distribution has a bell shape. Section 6.3 presents a bell-shaped probability distribution, called the *normal distribution*, which approximates the binomial distribution well when this guideline is satisfied.

One of the conditions for the binomial distribution is binary data. However, the binomial is useful even when each trial has *more than two* possible outcomes. For instance, consider Example 1 about the recall of Gray Davis as governor of California in 2003. If the recall were successful, there were four candidates to be the next governor--the Republican candidate Arnold Schwarzenegger, the Democratic candidate Cruz Bustamante, and two independent candidates. The winner did not need more than 50 percent of the votes, only more votes than anyone else. Since relatively few votes were not for Schwarzenegger or Bustamante, to predict a winner it was sufficient to consider only the votes for those two candidates and predict which was larger. If we wanted to let p refer to the population proportion voting for Schwarzenegger out of *all* votes, we could still apply the binomial by collapsing the

---

[6] A lower bound of 15 is actually a bit conservative. The shape is reasonably close to a bell even when both np and $n(1-p)$ are about 10. We use 15 here because it ties in better with a guideline in the next chapter for using the bell shape for inference about proportions.

four outcomes into the binary form (Schwarzenegger, other candidate). If we prefer to use all outcomes at once, there is an extension of the binomial distribution to deal with multiple outcomes (called the **multinomial distribution**).

Section 6.2: Practicing the Basics

6.12 ESP: Jane Doe claims to possess extrasensory perception (ESP). She says she can guess more often than not the outcome of a flip of a balanced coin in another room. In an experiment, a coin is flipped 3 times. If she does not actually have ESP, find the probability distribution of the number of her correct guesses.
a. Do this by constructing a sample space, finding the probability for each point, and using them to construct the probability distribution.
b. Do this using the formula for the binomial distribution.

6.13 More ESP: In Example 6 on ESP, John Doe had to predict which of five numbers was chosen in each of 3 trials. Doe did not actually have ESP. Explain why this experiment satisfies the three conditions for the binomial distribution.
a. For the analogy with coin flipping, what plays the role of (head, tail)?
b. Explain why it is sensible to assume the same probability of a correct guess on each trial.
c. Explain why it is sensible to assume independent trials.
d. If Doe had to guess one of 3 numbers on the first trial, one of 5 numbers on the second trial, and one of 10 numbers on the third trial, would X = number of correct guesses have the binomial distribution? Why or why not?

6.14 Symmetric binomial: Construct a table with the possible values and their probabilities for a binomial distribution with $n = 3$ and $p = 0.5$. Construct a graph, and note that it is symmetric, which is always true when $p = 0.5$.

6.15 Nonsymmetric binomial: Construct a table with the possible values and their probabilities for a binomial distribution with $n = 3$ and $p = 0.2$, using either the binomial formula or software. Is its graph symmetric?

6.16 Number of girls in family: Each newborn baby has a probability of approximately 0.49 of being female and 0.51 of being male. For a family with 4 children, let X = number of children who are girls.
a. Explain why the three conditions are satisfied for X to have the binomial distribution.
b. Identify n and p for the binomial distribution.
c. Find the probability that the family has 2 girls and 2 boys.

6.17 Number of boys: Refer to the previous exercise about a family with 4 children.
a. If Y = number of boys, and X = number of girls, explain why $Y = 4 - X$.
b. Identify n and p for the binomial distribution of Y = number of boys.
c. Explain how the binomial distribution of X can be used to figure out the binomial distribution of Y. Illustrate, by showing how to find the probability of $y = 2$ by using one of the probabilities from the distribution of X.

6.18 NBA shooting: In the National Basketball Association, the top free-throw shooters usually have probability about 0.90 of making any given free throw.
a. During a game, one such player (Dolph Schayes) shot 10 free throws. Let X = number of free throws made. What must you assume in order for X to have a binomial distribution? (Studies have shown that such assumptions are well satisfied for this sport.)
b. Specify the values of n and p for the binomial distribution of X in (a).
c. Find the probability that he made (i) all ten free throws, (ii) 9 free throws.

6.19 Two free throws: Refer to the previous exercise. The player is fouled and takes two free throws. Whether he makes the second one is independent of whether he makes the first one. Let X be the number of free throws that he makes.
a. Identify n and p for the binomial distribution.
b. Find all the probabilities in the probability distribution of X.

6.20 Season performance: Refer to the previous two exercises. Over the course of a season, this player shoots 400 free throws.
a. Find the mean and standard deviation of the probability distribution of the number of free throws he makes.
b. By the Empirical Rule, within what range would you expect the number made to almost certainly fall? Why?
c. Within what range would you expect the *proportion* made to fall?

6.21 Is the die balanced?: A balanced die with 6 sides is rolled 60 times.
a. For the binomial distribution of X = number of 6s, what is n and what is p?
b. Find the mean and the standard deviation of the distribution of X. Interpret.
c. If you observe $x = 0$, would you be skeptical that the die is balanced? Explain why, based on the mean and standard deviation of X.
d. Show that the probability that $x = 0$ is 0.0000177.

6.22 Exit poll: Example 1 discussed an exit poll of 3160 voters for the special election about the recall of California governor Gray Davis. Let X denote the number who voted for the recall. Suppose that in the population, exactly 50% voted for it.
a.   Explain why this scenario would seem to satisfy the three conditions needed to use the binomial distribution. Identify n and p.
b.   Find the mean and standard deviation of the probability distribution of X.

6.23 More on exit poll: Refer to the previous exercise.
a. Using the Empirical rule, give an interval in which you would expect X almost certainly to fall, if truly $p = 0.50$. *(Hint:* You can follow the reasoning of Example 9 on racial profiling.)
b. Now, suppose that the exit poll had $x = 1706$. What would this suggest to you about the actual value of p?

6.24 Jury duty: The juror pool for the upcoming murder trial of a celebrity actor contains the names of 100,000 individuals in the population who may be called for jury duty. The proportion of the available jurors on the population list who are Hispanic is 0.40. A jury of size 12 is selected at random from the population list of available jurors. Let X = the number of Hispanics selected to be jurors for this jury.
a. State the possible values that X can assume.

b. Is it reasonable to assume that X has a binomial distribution? If so, identify the values of n and p. If not, explain why not.

6.25 Jury probabilities: Refer to the previous exercise.
a. Find the probability that no Hispanic is selected.
b. Find the probability that at least one Hispanic is selected.
c. If no Hispanic is selected out of a sample of size 12, does this cast doubt on whether the sampling was truly random? Explain.

6.26 Euro: *Eurobarometer* has tracked opinions of Europeans about the common currency (the euro) that is now used in many European countries. When it was introduced in January, 2002, 67% of adult residents of the affected countries indicated they were happy that the euro had arrived. Suppose a poll of size 1000 is planned next year to estimate the percentage of people who now approve of the common currency. Suppose the population proportion still equals 0.67.
a. With a random sample, explain why it is reasonable to use the binomial for the probability distribution of the number of the 1000 who will indicate approval of the euro. For the binomial, what is n and what is p?
b. Find the mean and standard deviation of the probability distribution in (a).
c. Suppose the poll is taken and that 800 people indicate approval. Does this give strong evidence that the percentage approving of the euro has gone up? Why or why not?

6.27 Checking guidelines: For Example 7 on the gender distribution of management trainees, the population size was more than one thousand, half of whom were female. The sample size was 10.
a. Check that the guideline was satisfied about the relative sizes of the population and the sample, thus allowing you to use the binomial for the probability distribution for the number of females selected.
b. Check that the guideline was satisfied for this binomial distribution to have approximately a bell shape.

6.28 Class sample: Four of the 20 students in a class (20%) are fraternity or sorority members. Five students are picked at random. Does X = the number of students in the sample who are fraternity or sorority members have the binomial distribution with $n = 5$ and $p = 0.20$? Explain why or why not.

6.29 Binomial needs fixed n: For the binomial distribution, the number of trials n is a fixed number. Let X denote the number of girls in a randomly selected family in Canada that has three children. Let Y denote the number of girls in a randomly selected family in Canada (that is, the number of children could be any number). A binomial distribution approximates well the probability distribution for one of X and Y, but not for the other.
a. Explain why.
b. Identify the case for which the binomial applies, and identify n and p.

6.30 Binomial assumptions: For the following random variables, explain why at least one condition needed to use the binomial distribution is unlikely to be satisfied.
a. X = number of people in a family of size 4 who go to church on a given Sunday, when any one of them goes 50% of the time in the long run (binomial, $n = 4$, $p = 0.5$). (*Hint:* Is the independence assumption plausible?)

b. *X* = number voting for the Democratic candidate out of the 100 votes in the first precinct that reports results, when 60% of the population voted for the Democrat (binomial, *n* = 100, *p* = 0.60). (*Hint:* Is the first precinct a random sample?)

c. *X* = number of females in a random sample of 4 students from a class of size 20, when half the class is female (binomial, *n* = 4, *p* = 0.5).

SECTION 6.3 HOW CAN WE FIND PROBABILITIES FOR BELL-SHAPED DISTRIBUTIONS?

> **RECALL**
>
> Section 2.1 defined and gave examples of **continuous** variables. They have an infinite continuum of possible values in an interval. Examples are time, age, and size measures such as height and weight.

A random variable is called **continuous** when its possible values form an interval. For instance, a recent study by the U.S. Census Bureau analyzed the time that people take to commute to work. Commuting time can be measured with real number values, such as between 0 minutes and 150 minutes.

Probability Distributions for Continuous Random Variables

Probability distributions of continuous random variables assign probabilities to any interval of the possible values. For instance, a probability distribution for commuting time provides the probability that the travel time is less than 15 minutes or the probability that the travel time is between 30 and 60 minutes. The probability that a random variable falls in any particular interval is between 0 and 1, and the probability of the entire interval containing all the possible values equals 1.

When a random variable is continuous, the intervals of values for the bars of a histogram for probabilities can be chosen as desired. For instance, one possibility for commuting time is {0 to 30, 30 to 60, 60 to 90, 90 to 120, 120 to 150}, quite wide intervals. By contrast, using {0 to 1, 1 to 2, 2 to 3, ..., 149 to 150} gives lots of very narrow intervals. As the number of intervals increases, with their width narrowing, the shape of the histogram gradually approaches a smooth curve. We'll use such curves to portray probability distributions of continuous random variables. See the graphs in the margin.

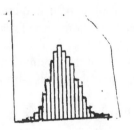

many intervals

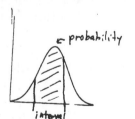

smooth curve approximation

Definition: Probability Distribution of a Continuous Random Variable

A **continuous** random variable has possible values that form an interval. Its **probability distribution** is specified by a curve that determines the probability that the random variable falls in any particular interval of values.

- Each interval has probability between 0 and 1. This is the area under the curve, above that interval.

- The interval containing all possible values has probability equal to 1, so the total area under the curve equals 1.

Figure 6.3 shows a graph for a probability distribution of *X* = commuting time for workers in the U.S. who commute to work. Historically, in surveys about travel to work, a commute of 45 minutes has been thought of as the maximum time that people would be willing to spend. However, a recent U.S. Census Bureau report[7] based on the 2000 census suggested that of the 97% of U.S. workers who do not work at home, 15% of the commuters have a commuting time above 45 minutes. The shaded area in Figure 6.3 refers to the probability of values higher than 45.0, for those who travel to work. This area equals 15% of the total area under the curve. So the probability is 0.15 of spending more than 45 minutes commuting to work. The report stated that the probability distribution is skewed to the right, with a mean of 25.5 minutes.

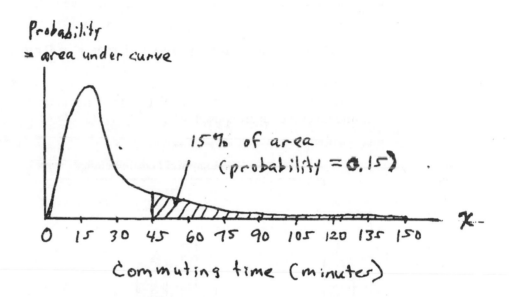

Figure 6.3: **Probability Distribution of Commuting Time.** The area under the curve for values higher than 45 is 0.15. **Questions**: Identify the area under the curve represented by the probability that commuting time is less than 15 minutes, which equals 0.29.

<table>
<tr><td>

DID YOU KNOW

For continuous random variables, the area above a single value equals 0. For example, the probability that the commuting time equals 23.7693611045... minutes is 0, but the probability that commuting time is between 23.5 and 24.5 minutes is positive.

</td><td>

For continuous random variables, in practice we need to round off with our measurements. Probabilities are given for *intervals* of values rather than individual values. In measuring commuting time, the U.S. Census Bureau asked, "How many minutes did it usually take to get from home to work last week?", and it gave as possible responses 0, 1, 2, 3, … . Then, for instance, a commuting time of 24 minutes actually means the interval of real numbers that round to the integer value 24. This is the area under the curve for the interval between 23.50 and 24.50. In practice, the probability that the commuting time falls in some given interval, such as above 45 minutes, or between 20 and 25 minutes, is of greater interest than the probability it equals some particular single value.

</td></tr>
</table>

---

[7] *Journey to Work: 2000*, issued March 2004 by U.S. Census Bureau

> **IN PRACTICE**
>
> In practice, **continuous** variables are measured in a **discrete** manner because of rounding. With rounding, a continuous random variable can take on a large number of separate values. A probability distribution for a continuous random variable is used to approximate the probability distribution for the possible rounded values.

EXAMPLE 10: PROBABILITY DISTRIBUTION FOR HEIGHT

Picture the Scenario
Figure 6.4 shows a histogram of heights of females at the University of Georgia, with a smooth curve superimposed.

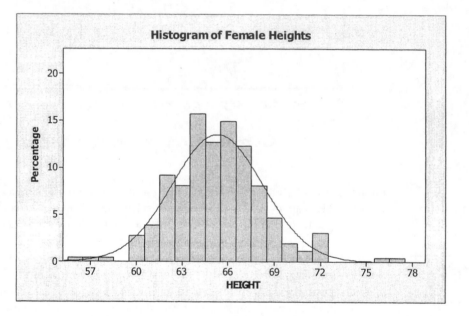

Figure 6.4: Histogram of Female Student Heights with Bell-Shaped Curve Superimposed. Height is continuous but is measured as discrete by rounding to the nearest inch. The smooth curve approximates the probability distribution for height, treating it as a continuous random variable. **Question:** How would you describe the shape, center, and spread of the distribution?

Question to Explore
What does the smooth curve represent?

Think It Through
In theory, height is a continuous random variable. In practice, it is measured here by rounding to the nearest inch. For instance, the bar of the histogram above 64 represents heights between 63.5 and 64.5 inches, which were rounded to 64 inches. The smooth curve approximates the probability distribution for the discrete way height is actually measured. We can also think of it as an approximation for the probability distribution that height would have if we could actually measure it

precisely as a continuous random variable. The area under this curve between two points approximates the probability that height falls between those points.

Insight
The smooth curve has a bell shape. We'll next study a probability distribution that has this shape and we'll learn how to find probabilities for it.

                                                                                    ♦

To practice this concept, try Exercise 6.32.

The Normal Distribution: A Probability Distribution with Bell-Shaped Curves

We now discuss a probability distribution commonly used for continuous random variables. The **normal distribution** is characterized by a particular[8] symmetric, bell-shaped curve with two parameters--the mean μ and the standard deviation σ.

Figure 6.4 showed that heights of female students at the University of Georgia have approximately a bell-shaped distribution. The approximating curve describes a probability distribution that has a mean of 65.0 and a standard deviation of 3.5. In fact, adult female heights in North America have approximately a normal distribution with $\mu = 65.0$ inches and $\sigma = 3.5$ inches. Adult male height is approximately normal with $\mu = 70.0$ and $\sigma = 4.0$ inches. The adult males tend to be a bit taller (since $70 > 65$) with heights a bit more spread out (since $4.0 > 3.5$). See Figure 6.5.

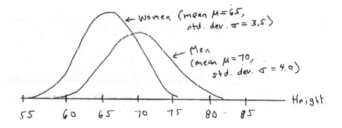

Figure 6.5: Normal Distributions for Women's Height and Men's Height. For each different combination of μ and σ values, there is a normal distribution with mean μ and standard deviation σ. **Question**: According to $\mu = 70$ and $\sigma = 4$, within what interval do almost all of the men's heights fall?

For any real number for the mean μ and any positive number for the standard deviation σ, there is a normal distribution with that mean and standard deviation.

---

[8] A mathematical formula specifies *which* bell-shaped curve is the normal distribution, but it is complex and we'll not use it in this text. In the next chapter, we'll learn about another bell-shaped distribution, one with "thicker tails" than the normal has.

Normal Distribution

The **normal distribution** is symmetric, bell shaped, and characterized by its mean μ and standard deviation σ. The probability falling within any particular number of standard deviations of μ is the same for all normal distributions. This probability equals 0.68 within 1 standard deviation, 0.95 within 2 standard deviations, and 0.997 within 3 standard deviations. Figure 6.6 illustrates.

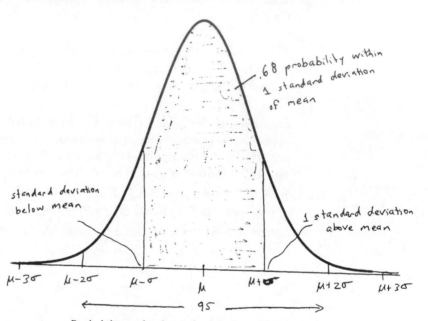

Probability within 2 standard deviations of the mean

Figure 6.6: The Normal Distribution. The probability equals 0.68 within one standard deviation of the mean, 0.95 within two standard deviations, and 0.997 within three standard deviations. **Question:** How do these probabilities relate to the Empirical Rule?

To illustrate, for adult female height, $\mu = 65.0$ and $\sigma = 3.5$ inches. Since

$$\mu - 2\sigma = 65.0 - 2(3.5) = 58.0 \text{ and } \mu + 2\sigma = 65.0 + 2(3.5) = 72.0,$$

about 95% of the female heights fall between 58 inches and 72.0 inches (6 feet). For adult male height, $\mu = 70.0$ and $\sigma = 4.0$ inches. About 95% fall between $\mu - 2\sigma = 70.0 - 2(4.0) = 62$ inches and $\mu + 2\sigma = 70.0 + 2(4.0) = 78$ inches (6 1/2 feet).

RECALL

From Section 2.4, the **z-score** for an observation is the number of standard deviations that it falls from the mean.

The property of the normal distribution in the box above tells us probabilities within 1, 2, and 3 standard deviations of the mean. The multiples 1, 2, and 3 of the number of standard deviations from the mean are denoted by the symbol z in general. For instance, $z = 2$ for two standard deviations. For each fixed number z, the probability within z standard deviations of the mean is the area under the normal curve between

$\mu - z\sigma$ and $\mu + z\sigma$. This is shown in Figure 6.7. For every normal distribution, this probability is 0.68 for $z = 1$, so 68% of the area (probability) of a normal distribution falls between $\mu - \sigma$ and $\mu + \sigma$. Similarly, this probability is 0.95 for $z = 2$, and nearly 1.0 for $z = 3$ (that is, between $\mu - 3\sigma$ and $\mu + 3\sigma$). The total probability for any normal distribution equals 1.0.

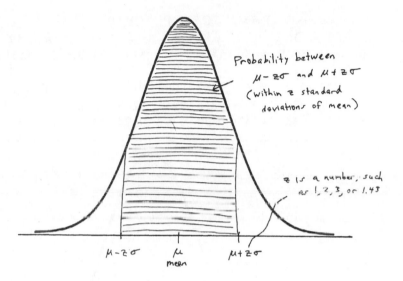

Figure 6.7: The Probability Between $\mu - z\sigma$ **and** $\mu + z\sigma$. This is the area highlighted under the curve. It is the same for every normal distribution and depends only on the value of z. Figure 6.6 showed this for $z = 1, 2,$ and 3, but z can be any other number also.

The normal distribution is the most important one in statistics. It's important partly because many variables have approximate normal distributions, as illustrated in Figure 6.4 with female heights. It's important also because it's used to approximate many discrete distributions when there are a large number of possible outcomes, as we saw for the binomial in Example 9 on racial profiling. The main reason for its prominence, however, is that many statistical methods use the normal distribution even when the data are not bell shaped. We'll see why later in this chapter.

Finding Probabilities for the Normal Distribution

As we'll discuss below, the probabilities 0.68, 0.95, and 0.997 within 1, 2, and 3 standard deviations of the mean are no surprise, because of the Empirical Rule. But what if we wanted to find the probability within, say, 1.43 standard deviations?

Table A at the end of the text enables us to find normal probabilities for *any* interval of values. It tabulates the normal **cumulative probability** falling *below* $\mu + z\sigma$.

The leftmost column of Table A lists the values for z to one decimal point, with the second decimal place listed above the columns. Table 6.8 shows a small excerpt from Table A. The tabulated probability for $z = 1.43$ falls in the row labeled 1.4 and in the column labeled 0.03. It equals 0.9236. For every normal distribution, the probability that falls below $\mu + 1.43\sigma$ equals 0.9236. Figure 6.8 illustrates.

Table 6.8: Part of Table A, Showing Normal Cumulative (Left-Tail) Probabilities. The top of the table gives the second digit for z. The table entry is the probability falling below $\mu + z\sigma$, for instance, 0.9236 below $\mu + 1.43\sigma$ for $z = 1.43$.

SECOND DECIMAL PLACE OF z

z	.00	.01	.02	**.03**	.04	.05	.06	.07	.08	.09
0.0	.5000	.5040	.5080	.5120	.5160	.5199	.5239	.5279	.5319	.5359
...										
1.3	.9032	.9049	.9066	.9082	.9099	.9115	.9139	.9147	.9162	.9177
1.4	.9192	.9207	.9222	**.9236**	.9251	.9265	.9278	.9292	.9306	.9319
1.5	.9332	.9345	.9357	.9370	.9382	.9394	.9406	.9418	.9429	.9441

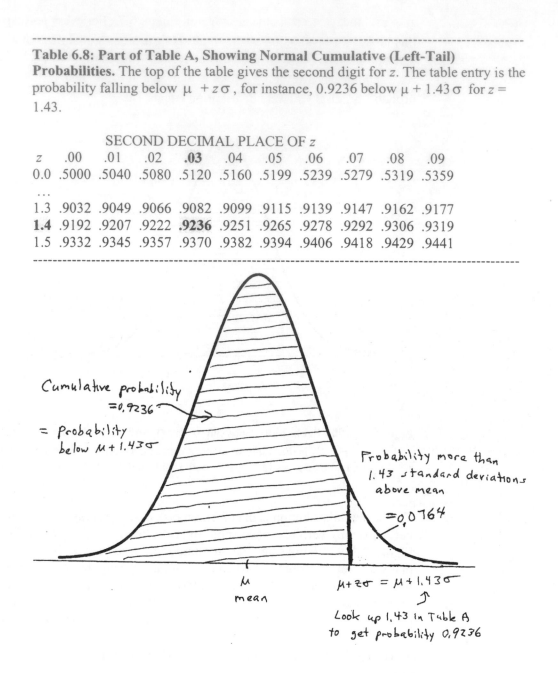

Figure 6.8: The Normal Cumulative Probability, Less than z Standard Deviations above the Mean. Table A lists a cumulative probability of 0.9236 for $z = 1.43$, so 0.9236 is the probability less than 1.43 standard deviations above the mean of any normal distribution (that is, below $\mu + 1.43\sigma$). The complement probability of 0.0764 is the probability *above* $\mu + 1.43\sigma$ in the right tail.

Since an entry in Table A is a probability *below* $\mu + z\sigma$, one minus that probability is the probability *above* $\mu + z\sigma$. For example, the right-tail probability above $\mu + 1.43\sigma$ equals $1 - 0.9236 = 0.0764$. By the symmetry of the normal curve, this

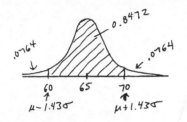

probability also refers to the left tail below $\mu - 1.43\sigma$, which you'll find in Table A by looking up $z = -1.43$. The negative z-scores in the table refer to cumulative probabilities for random variable values *below* the mean.

Since the probability is 0.0764 in each tail, the total probability *more than* 1.43 standard deviations from the mean equals $2(0.0764) = 0.1528$. The total probability equals 1, so the probability falling *within* 1.43 standard deviations of the mean equals $1 - 0.1528 = 0.8472$, about 85%. For instance, 85% of women in North America have height between $\mu - 1.43\sigma = 65.0 - 1.43(3.5) = 60$ inches and $\mu + 1.43\sigma = 65 + 1.43(3.5) = 70$ inches (that is, between 5 feet and 5 feet, 10 inches).

Table A lists cumulative probabilities, rather than central probabilities within z standard deviations of the mean. This is for convenience, because cumulative probabilities and their complements in the right tail are more useful in later chapters. Normal cumulative probabilities are also available in statistical software and on some calculators.

Normal Probabilities and the Empirical Rule

The Empirical Rule (Section 2.3) states that for an approximately bell-shaped distribution, approximately 68% of observations fall within one standard deviation of the mean, 95% within two standard deviations, and all or nearly all within three. In fact, those percentages came from probabilities calculated for the normal distribution.

For instance, a value that is two standard deviations below the mean has $z = -2.00$. The cumulative probability below $\mu - 2\sigma$ listed in Table A opposite $z = -2.00$ is 0.0228. The right-tail probability above $\mu + 2\sigma$ also equals 0.0228, by symmetry. See Figure 6.9. The probability falling more than two standard deviations from the mean in either tail is $2(0.0228) = 0.0456$. Thus, the probability that falls *within* two standard deviations of the mean equals $1 - 0.0456 = 0.9544$. When a variable has a normal distribution, 95.44% (about 95%) of the distribution falls within two standard deviations of the mean.

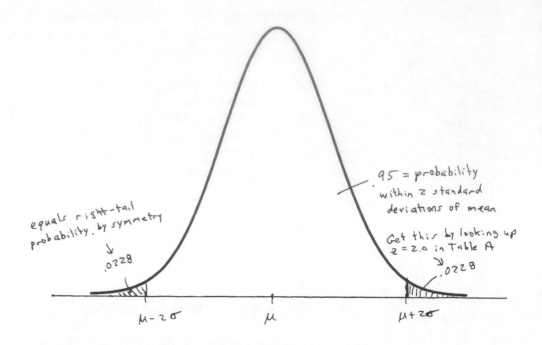

Figure 6.9: Normal Probability within Two Standard Deviations of the Mean.
Probabilities in one tail determine probabilities in the other tail by symmetry. Subtracting the total two-tail probability from 1.0 gives probabilities within a certain distance of the mean. **Question**: Can you do the analogous calculation for *three* standard deviations?

The approximate percentages that the Empirical Rule lists are the actual percentages for the normal distribution, rounded. For instance, you should verify that the probability within one standard deviation of the mean of a normal distribution equals 0.68. (*Hint:* Get the left-tail probability by looking up $z = -1.00$, double it for the two-tail probability, and subtract from 1.) The probability within three standard deviations of the mean equals 0.997, or 1.00 rounded off. The Empirical Rule stated *approximate* probabilities rather than *exact* probabilities because it referred to *all approximately* bell-shaped distributions, not just the normal.
To practice this concept, try Exercise 6.35.

How Can We Find the Value of z for a Certain Cumulative Probability?

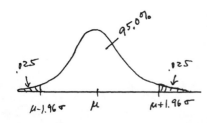

In practice, we'll sometimes need to find the value of z that corresponds to a certain normal cumulative probability. How can we do this? To illustrate, let's find the value of z for a cumulative probability of 0.025. We look up the cumulative probability of 0.025 in the body of Table A. It corresponds to $z = -1.96$, since it is in the row labeled -1.9 and in the column labeled 0.06. So a probability of 0.025 lies below $\mu - 1.96\sigma$. Likewise, a probability of 0.025 lies above $\mu + 1.96\sigma$. A total probability of 0.050 lies more than 1.96σ from μ. Precisely 95.0% of a normal distribution falls within 1.96 standard deviations of the mean. We've seen in the previous subsection that 95.4% falls within 2.00 standard deviations, and we now see that precisely 95.0% falls within 1.96 standard deviations.

EXAMPLE 11: WHAT IQ DO YOU NEED TO GET INTO MENSA?

Picture the Scenario

Mensa[9] is a society of high-IQ people whose members have a score on an IQ test at the 98th percentile or higher. The Stanford-Binet IQ scores that are used as the basis for admission into Mensa are approximately normally distributed with a mean of 100 and a standard deviation of 16.

Questions to Explore

a. How many standard deviations above the mean is the 98th percentile?
b. What is the IQ score for that percentile?

Think It Through

a. For a value to represent the 98th percentile, its cumulative probability must equal 0.98, by the definition of a percentile. See Figure 6.10.

> **RECALL**
>
> The **percentile** was defined in Section 2.3. For the 98th percentile, 98% of the distribution falls below that point.

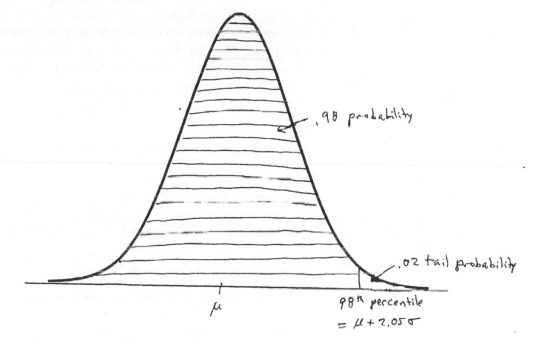

Figure 6.10: The 98th Percentile for a Normal Distribution. This is the value such that 98% of the distribution falls below it and 2% falls above. **Question**: Where is the 2nd percentile located?

The cumulative probability of 0.980 in the body of Table A corresponds to $z = 2.05$. The 98 th percentile is 2.05 standard deviations above the mean, at $\mu + 2.05\sigma$.

b. Since $\mu = 100$ and $\sigma = 16$, the 98th percentile of IQ equals

$$\mu + 2.05\sigma = 100 + 2.05(16) = 133.$$

---

[9] See www.mensa.org

In summary, 98% of the IQ scores fall below 133, and an IQ score of at least 133 is required to join Mensa.

Insight

About 2% of IQ scores are higher than 133. By symmetry, about 2% of IQ scores are lower than $\mu - 2.05\sigma = 100 - 2.05(16) = 67$. This is the 2nd percentile. The remaining 96% of the IQ scores fall between 67 and 133, which is the region within 2.05 standard deviations of the mean.

It's also possible to use software to find normal probabilities or z-scores. The margin shows a screen shot using the TI-83+/84 to find the 98th percentile of IQ.

To practice this concept, try Exercise 6.37 and 6.46.

```
invNorm(.98,100,
16)
            132.86
■
```

TI-83+/84 output

Using z = Number of Standard Deviations to Find Probabilities

We've used the symbol z to represent the *number of standard deviations* a value falls from the mean. If we have a value x of a random variable, how can we figure out the number of standard deviations it falls from the mean μ of its probability distribution? The difference between x and μ equals $x - \mu$. The **z-score** expresses this difference as a number of standard deviations, using $z = (x - \mu)/\sigma$.

RECALL

Section 2.4 showed that for *sample data*, the number of standard deviations that a value x falls from the mean of the sample is

$$z = \frac{\text{observed value - mean}}{\text{standard deviation}}.$$

Definition: *z*-Score for a Value of a Random Variable

The **z-score** for a value x of a random variable is the number of standard deviations that x falls from the mean μ. It is calculated as

$$z = \frac{x - \mu}{\sigma}.$$

The formula for the z-score is useful when we are given the value of x for some normal random variable and need to find a probability relating to that value. We convert x to a z-score and then use a normal table to find the appropriate probability. The next two examples illustrate.

EXAMPLE 12: FINDING YOUR RELATIVE STANDING ON THE SAT

Picture the Scenario
Scores on the verbal or math portion of the Scholastic Aptitude Test (SAT), a college entrance examination, are approximately normally distributed with mean $\mu = 500$ and standard deviation $\sigma = 100$. The scores range from 200 to 800.

Questions to Explore
a. If one of your SAT scores was $x = 650$, how many standard deviations from the mean was it?
b. What percentage of SAT scores was higher than yours?

Think It Through
a. The SAT score of 650 has a z-score of $z = 1.50$ because 650 is 1.50 standard deviations above the mean. In other words, $x = 650 = \mu + z\sigma = 500 + z(100)$, where $z = 1.50$. We can find this directly using the formula

$$z = \frac{x - \mu}{\sigma} = \frac{650 - 500}{100} = 1.50.$$

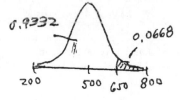

b. The percentage of SAT scores that were higher than 650 is the right-tail probability above 650, for a normal random variable with mean $\mu = 500$ and standard deviation $\sigma = 100$. From Table A, the z-score of 1.50 has cumulative probability 0.9332. That's the probability *below* 650, so the right-tail probability above it is $1 - 0.9332 = 0.0668$. Only about 7% of SAT test scores fall above 650. In summary, a score of 650 was well above average, in the sense that relatively few students scored higher.

Insight
Positive z-scores occur when the value x falls *above* the mean μ. Negative z-scores occur when x falls *below* the mean. For instance, SAT = 350 has a z-score of

$$z = \frac{x - \mu}{\sigma} = \frac{350 - 500}{100} = -1.50.$$

The SAT score of 350 is 1.50 standard deviations *below* the mean. The probability that an SAT score falls below 350 is also 0.0668. Figure 6.11 illustrates.

◆

To practice this concept, try Exercise 6.48.

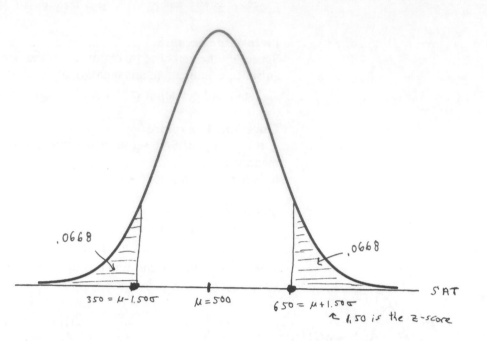

Figure 6.11: Normal Distribution for SAT. The SAT scores of 650 and 350 have *z*-scores of 1.50 and -1.50 because they fall 1.50 standard deviations above and below the mean. **Question**: Which SAT scores have *z* = 3.0 and *z* = -3.0?

EXAMPLE 13: WHAT PROPORTION OF STUDENTS GET A GRADE OF B?

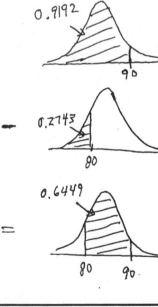

Picture the Scenario
On the midterm exam in introductory statistics, an instructor always gives a grade of B to students who score between 80 and 90.

Question to Explore
One year, the scores on the exam have approximately a normal distribution with mean 83 and standard deviation 5. About what proportion of students get a B?

Think It Through
A midterm exam score of 90 has a *z*-score of

$$z = \frac{x - \mu}{\sigma} = \frac{90 - 83}{5} = 1.40.$$

Its cumulative probability of 0.9192 (from Table A) means that about 92% of the exam scores were below 90. Similarly, an exam score of 80 has a *z*-score of

$$z = \frac{x - \mu}{\sigma} = \frac{80 - 83}{5} = -0.60.$$

Its cumulative probability of 0.2743 means that about 27% of the exam scores were below 80. See the margin figure. It follows that about 0.9192 − 0.2743 = 0.6449, or about 64 %, of the exam scores were in the B range.

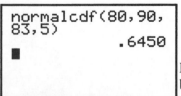

```
normalcdf(80,90,
83,5)
          .6450
■
```

TI-83+/84 output

380

Insight
Here, we used two separate cumulative probabilities to help us find a probability between two points.

◆

To practice this concept, try Exercise 6.49.

Here's a summary of how we've used z-scores:

Using z-Scores to Find Normal Probabilities or Random Variable x Values:

- If we're given a value x and need to find a probability, convert x to a z-score using $z = (x - \mu)/\sigma$, use a table of normal probabilities to get a cumulative probability, and then convert it to the probability of interest.

- If we're given a probability and need to find the value of x, convert the probability to the related cumulative probability, and find the z-score using a normal table and then evaluate $x = \mu + z\sigma$.

In Example 11, we used the equation $x = \mu + z\sigma$ to find a percentile score (namely, 98th percentile $= \mu + 2.05\sigma = 100 + 2.05(16) = 133$). In Examples 12 and 13, we used the equation $z = (x - \mu)/\sigma$ to determine how many standard deviations certain scores fell from the mean, which enabled us to find probabilities relating to those scores. Of course, in using a normal table, we need to keep in mind that it provides cumulative probabilities and make any needed adjustments, as in these examples.

Another use of z-scores is for comparing observations from different normal distributions in terms of their relative distances from the mean.

EXAMPLE 14: HOW CAN WE COMPARE TEST SCORES THAT USE DIFFERENT SCALES?

Picture the Scenario
There are two primary standardized tests used by college admissions, the SAT and the ACT.[10]

Question to Explore
When you applied to college, you scored 650 on an SAT exam, which had mean $\mu = 500$ and standard deviation $\sigma = 100$. Your friend took the comparable ACT in 2001, scoring 30. That year, the ACT had $\mu = 21.0$ and $\sigma = 4.7$. How can we compare these scores to tell who did better?

---

[10] See www.sat.org and www.act.org.

Think It Through

The test scores of 650 and 30 are not directly comparable because the SAT and ACT have different means and different standard deviations. But we can convert them to z-scores and analyze how many standard deviations each falls from the mean.

With $\mu = 500$ and $\sigma = 100$, we saw in Example 12 that a SAT test score of $x = 650$ converts to a z-score of $z = (x - \mu)/\sigma = (650 - 500)/100 = 1.50$. With $\mu = 21.0$ and $\sigma = 4.7$, an ACT score of 30 converts to a z-score of

$$z = \frac{x - \mu}{\sigma} = \frac{30 - 21}{4.7} = 1.91.$$

The ACT score of 30 is a bit higher than the SAT score of 650, since ACT = 30 falls 1.91 standard deviations above its mean, whereas SAT = 650 falls 1.50 standard deviations above its mean. In this sense, the ACT score is better even though its numerical value is smaller.

Insight

The SAT and ACT tests both have approximately normal distributions. From Table A, $z = 1.91$ (for the ACT score of 30) has a cumulative probability of 0.97. Of all students who took the ACT, only about 3% scored above 30. From Table A, $z = 1.50$ (for the SAT score of 650) has a cumulative probability of 0.93. Of all students who took the SAT, about 7% scored above 650.

◆

To practice this concept, try Exercise 6.50.

The Standard Normal Distribution has Mean 0 and Standard Deviation 1

Many statistical methods refer to a particular normal distribution called the **standard normal distribution**.

Standard Normal Distribution

The **standard normal distribution** is the normal distribution with mean $\mu = 0$ and standard deviation $\sigma = 1$. It is the distribution of normal z-scores.

For the standard normal distribution, the number falling z standard deviations above the mean is $\mu + z\sigma = 0 + z(1) = z$, simply the z-score itself. For instance, the value of 2.0 is two standard deviations above the mean, and the value of -1.3 is 1.3 standard deviations below the mean. As Figure 6.12 shows, the original values are the same as the z-scores, since

$$z = \frac{x - \mu}{\sigma} = \frac{x - 0}{1} = x.$$

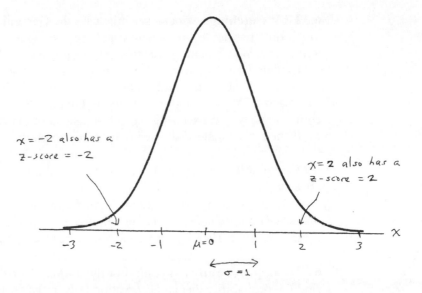

Figure 6.12: The Standard Normal Distribution. This has mean 0 and standard deviation 1. The random variable value *x* is the same as its *z*-score. **Question**: What are the limits within which almost all its values fall?

Examples 12 and 14 dealt with SAT scores, having μ = 500 and σ = 100. Suppose we convert each SAT score *x* to a *z*-score by using $z = (x - \mu)/\sigma = (x - 500)/100$. Then $x = 650$ converts to $z = 1.50$, and $x = 350$ converts to $z = -1.50$. When the values for a normal distribution are converted to *z*-scores, those *z*-scores have a mean of 0 and have a standard deviation of 1. That is, the entire set of *z*-scores has the standard normal distribution.

---

z-Scores and the Standard Normal Distribution

When a random variable has a normal distribution and its values are converted to *z*-scores by subtracting the mean and dividing by the standard deviation, the *z*-scores have the **standard normal** distribution (mean 0, standard deviation 1).

We'll find this result to be useful for statistical inference in coming chapters.

SECTION 6.3: PRACTICING THE BASICS

6.31 Uniform distribution: A random number generator is used to generate a real number at random between 0 and 1, equally likely to fall anywhere in this interval of values. (For instance, 0.3794259832… is a possible outcome.)
a. Sketch a curve of the probability distribution of this random variable, which is the continuous version of the **uniform distribution**.
b. What is the mean of this distribution?
c. What is the probability that this random variable falls between 0.25 and 0.75?

6.32 TV watching: A social scientist uses the General Social Survey to study how much time per day people spend watching TV. The variable denoted by TVHOURS at the GSS Web site measures this using the values 0, 1, 2, ... , 24.
a. Explain how, in theory, TV watching is a continuous random variable.
b. An article about the study shows two histograms, both skewed to the right, to summarize TV watching for females and males. Since TV watching is in theory continuous, why were histograms used instead of curves?
c. If the article instead showed two curves, explain what they would represent.

6.33 Probabilities in tails: For a normal distribution, find the probability that an observation is
a. at least one standard deviation above the mean,
b. at least one standard deviation below the mean.
c. In each case, sketch a curve and show the tail probability.

6.34 Tail probability in graph: For the normal distribution shown in the margin, find the probability that an observation falls in the shaded region.

6.35 Empirical Rule: Verify the Empirical Rule by showing that for a normally distributed random variable, the probability (rounded to two decimal places) within
a. 1 standard deviation of the mean equals 0.68.
b. 2 standard deviations of the mean equals 0.95.
c. 3 standard deviations of the mean is very close to 1.00.

6.36 Central probabilities: For a normally distributed random variable, verify that the probability within (a) 1.64 standard deviations of the mean equals 0.90, (b) 2.58 standard deviations of the mean equals 0.99.
a. Find the probability that falls within 0.67 standard deviations of the mean.
b. Sketch these three cases on a single graph.

6.37 z-score for given probability in tails:
a. Find the z-score for which a total probability of 0.02 falls more than z standard deviations (in either direction) from the mean of a normal distribution.
b. For this z, explain why the probability more than z standard deviations above the mean equals 0.01.
c. Explain why $\mu + 2.33\sigma$ is the 99th percentile.

6.38 Probability in tails for given z-score: For a normal distribution:
a. Show that a total probability of 0.01 falls more than $z = 2.58$ standard deviations from the mean.
b. Find the z-score for which the two-tail probability that falls more than z standard deviations from the mean in either direction equals (a) 0.05, (b) 0.10. Sketch the two cases on a single graph.

6.39 z-score for right-tail probability:
a. For the normal distribution shown in the margin, find the z-score.
b. Find the value of z (rounding to two decimal places) for right-tail probabilities of (i) 0.05, (ii) 0.01, and (iii) 0.005.

6.40 z-score and central probability: Find the z-score such that the interval within z standard deviations of the mean for a normal distribution contains
a. 50% of the probability.
b. 90% of the probability.
c. Sketch the two cases on a single graph.

6.41 z-score and percentile: Find the z-scores for a normal distribution for which
a. the 90th percentile is that many standard deviations above the mean.
b. the 95th percentile is that many standard deviations above the mean.
c. Sketch a graph showing where these percentiles are relative to the mean and to each other.

6.42 Female heights: The normal distribution for women's height in North America has μ = 65 inches, σ = 3.5 inches. Most major airlines have height requirements for flight attendants (www.cabincrewjobs.com). Although exceptions are made, the minimum height requirement is 62 inches. What proportion of adult females in North American would not be tall enough to be a flight attendant?

6.43 Blood pressure: A World Health Organization study (the MONICA project) of health in various countries reported that in Canada, systolic blood pressure readings have a mean of 121 and a standard deviation of 16. A reading above 140 is considered to be high blood pressure.
a) What is the z-score for a blood pressure reading of 140?
b) If systolic blood pressure in Canada has a normal distribution, what proportion of Canadians suffers from high blood pressure?
c) What proportion of Canadians has systolic blood pressures in the range from 100 to 140?

6.44 Working hours: According to *Current Population Reports*, self-employed individuals in the United States work an average of 44.6 hours per week, with a standard deviation of 14.5. If this variable is approximately normally distributed, find the proportion of the self-employed who work more than 40 hours per week. Sketch a graph, and mark off on it 44.6, 40, and the region to which the answer refers.

6.45 MDI: The Mental Development Index (MDI) of the Bayley Scales of Infant Development is a standardized measure used in observing infants over time. It is approximately normal with a mean of 100 and a standard deviation of 16.
a. What proportion of children has MDI of (i) at least 120? (ii) at least 80?
b. Find the MDI score that is the 99th percentile.
c. Find the MDI score such that only 1% of the population has an MDI score below it.

6.46 Quartiles and outliers: Refer to the previous exercise.
a. Find and interpret the lower quartile (Q1), median, and upper quartile (Q3) of the MDI.
b. Find the interquartile range (IQR) of MDI scores.
c. Section 2.3 defined an observation to be a potential outlier if it is more than $1.5 \times$ IQR below Q1 or above Q3. Find the intervals of MDI scores that would be considered potential outliers.

6.47 Murder rates: In 1999, the murder rates (per 100,000 residents) for the 50 states and the District of Columbia (D.C.) had a mean of 5.9 and a standard deviation of 6.3 (*Statistical Abstract of the United States, 2001*).
a. D.C. had a murder rate of 46.4. Find its z-score. If the distribution were roughly normal, would this be unusually high? Explain.
b. Find the murder rate of a state that has a z-score of 0.0.
c. Based on the mean and standard deviation, do you think that the distribution of murder rates is approximately normal. Why or why not?

6.48 Property taxes: Suppose that property taxes on homes in Columbus, Ohio, are approximately normal in distribution, with a mean of $3000 and a standard deviation of $1000. The property tax for one particular home is $3500.
a. Find the z-score for that property tax value.
b. What proportion of the property taxes exceeds $3500?

6.49 How many get a C?: Final exam scores have approximately a normal distribution with mean 76 and standard deviation 8. The instructor gives a C to students who score between 70 and 80. About what proportion of students get a C?

6.50 SAT versus ACT: For the SAT distribution ($\mu = 500$, $\sigma = 100$) and the ACT distribution ($\mu = 21$, $\sigma = 4.7$), which score is relatively higher, a SAT of 600 or an ACT of 25? Explain.

6.51 Relative height: Refer to the normal distributions for women's height ($\mu = 65$, $\sigma = 3.5$) and men's height ($\mu = 70$, $\sigma = 4.0$). A man's height of 75 inches and a woman's height of 70 inches are both 5 inches above their means. Which is relatively taller? Explain why.

6.4 HOW LIKELY ARE THE POSSIBLE VALUES OF A STATISTIC? THE SAMPLING DISTRIBUTION

RECALL

A **statistic** is a numerical summary of sample data, such as a sample proportion or a sample mean. A **parameter** is a numerical summary of a population, such as a population proportion or a population mean.

In this chapter, we've used probability distributions with known parameter values (such as $p = 0.50$ for the binomial) to find probabilities of possible outcomes of a random variable. In practice, we seldom know the values of parameters. They are estimated using sample data from observational studies or from experiments.

Let's consider the special 2003 California recall election mentioned in Example 1. Before all the votes were counted, the proportion of the population of voters who voted in favor of recalling Governor Gray Davis was an unknown parameter. An exit poll of 3160 voters reported that the sample proportion who voted in favor of the recall was 0.54. This statistic estimates the population proportion. How do we know it is a good estimate? After all, the total number of voters was nearly eight million, and the poll sampled a minuscule portion of them. To know how well the sample proportion estimates a population proportion, we need to be able to answer such questions as, "How likely is it that a sample proportion falls within 0.03 of the population proportion?" Within 0.05? Within any particular specified value? This section introduces a type of probability distribution that helps us determine how close to the population parameter a statistic is likely to fall.

Representing Sampling Variability by a Sampling Distribution

For an exit poll of voters, a person's vote is a variable because the outcome varies from voter to voter. A statistic such as a sample proportion can also be regarded as a variable, because its value varies from sample to sample. For example, before the exit-poll sample was selected, the value of the sample proportion who voted to recall the California governor was unknown. It was a random variable. The proportion of surveyed voters who voted for recall in one poll probably differed somewhat from the proportion for another sample of voters in another poll.

If many different polling agencies each selected a random sample of about 3000 voters, a certain predictable amount of variation would occur in the sample proportion values. Suppose the *population* proportion is 0.56. Would we see very similar sample proportions in the different polls, such as 0.54 in one, 0.57 in another, 0.55 in another? Or would they tend to be quite different, such as 0.40 in one, 0.59 in another, and 0.73 in another? To answer, we need to learn about a probability distribution that provides the possible values for the sample proportion and their probabilities. A probability distribution for a statistic (such as a sample proportion) is called a **sampling distribution**.

IN WORDS

Different polls take different random samples, which have different sample proportion values. The **sampling distribution** specifies the possible sample proportion values and their probabilities.

Definition: Sampling Distribution

The **sampling distribution** of a statistic is the probability distribution that specifies probabilities for the possible values the statistic can take.

For sampling 3000 voters in an exit poll, imagine *all* the distinct samples of 3000 voters you could possibly get. Each such sample has a value for the sample proportion voting a certain way (for instance, yes on a referendum). Suppose you could look at each possible sample, find the sample proportion for each one, and then construct the frequency distribution of those sample proportion values. That would be precisely the sampling distribution of the sample proportion.

A sampling distribution is merely a type of probability distribution. Rather than giving probabilities for an observation for an individual subject, it gives probabilities for the value of a statistic for a sample of subjects, such as a sample proportion. The next example illustrates, finding the sampling distribution of a sample proportion.

EXAMPLE 15: WHICH BRAND OF PIZZA DO YOU PREFER?

Picture the Scenario

The North End of Boston is a neighborhood that has, for many years, housed a large number of Italian immigrants to the U.S. The neighborhood supports a large number of Italian restaurants and cafés. Aunt Erma's Restaurant in the North End specializes in pizza baked in a wood-burning oven. The owners plan an advertising campaign with the claim that more people prefer the taste of their pizza (which we'll denote by A) than the current leading fast-food chain selling pizza (which we'll denote by D). To support their claim, they plan to randomly sample *n* people in Boston. Each person is asked to taste a slice of pizza A and a slice of pizza D. Subjects are blindfolded so they cannot see the pizza when they taste it, and the order of giving them the two slices is randomized. They are then asked which pizza they prefer.

Questions to Explore
a. In the entire Boston population, suppose that exactly half would prefer pizza A and half would prefer pizza D in such a taste experiment. Construct the sampling distribution of the sample proportion who prefer Aunt Erma's pizza, when $n = 3$.
b. In practice, we don't know the population proportion preferring each pizza type. Do the results in (a) suggest that we can gauge Boston's tastes well by sampling only 3 people?

Think It Through
a. Let's use a symbol with three entries to represent the responses for each possible sample of size $n = 3$. For instance, (A, D, D) represents a sample in which the first subject sampled preferred pizza A, and the second and third subjects preferred pizza D. The eight possible samples of size 3, and the number and the sample proportion of each sample that preferred A pizza, are:

Sample	Number Prefer A Pizza	Proportion
(A, A, A)	3	1
(A, A, D)	2	2/3
(A, D, A)	2	2/3
(D, A, A)	2	2/3
(A, D, D)	1	1/3
(D, A, D)	1	1/3
(D, D, A)	1	1/3
(D, D, D)	0	0

The sample proportion that prefers pizza A can be 0, 1/3, 2/3, or 1.0. When the population proportion preferring pizza A is $p = 0.50$, the 8 samples are equally likely, each occurring with probability 1/8. The sample proportion 0 occurs for only 1 of the 8 possible samples, namely (D, D, D), so its probability equals 1/8. The sample proportion 1/3 occurs for 3 samples, (A, D, D), (D, A, D), (D, D, A), so its probability equals 3/8. Similarly, one can find the probability for each possible sample proportion value. Table 6.9 shows the entire sampling distribution.

Table 6.9: Sampling Distribution of Sample Proportion. The random sample size is $n = 3$ from a population with population proportion $p = 0.50$. The 8 possible samples provide probabilities for the 4 possible sample proportion values. **Question**: How can we determine the probabilities when $p \neq 0.50$?

Sample Proportion	Probability
0	1/8
1/3	3/8
2/3	3/8
1	1/8

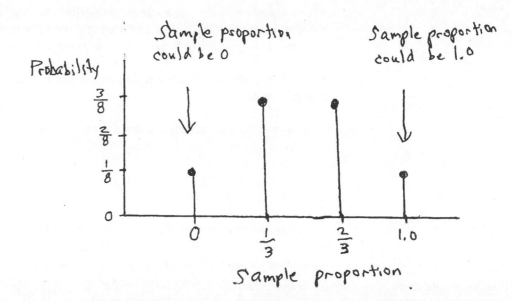

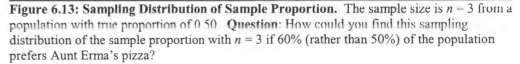

Figure 6.13: Sampling Distribution of Sample Proportion. The sample size is $n = 3$ from a population with true proportion of 0.50. **Question:** How could you find this sampling distribution of the sample proportion with $n = 3$ if 60% (rather than 50%) of the population prefers Aunt Erma's pizza?

b. Figure 6.13 portrays the sampling distribution of the sample proportion for $n = 3$, when the population proportion is $p = 0.50$. With a very small sample such as $n = 3$, the sample proportion need not fall near the population proportion. Aunt Erma's restaurant would not be able to gauge Boston's tastes well by sampling only 3 people.

Insight

Why would we care about finding a sampling distribution? Because it shows us whether a sample proportion is likely to be near the population proportion we're really interested in. In practice, of course, samples are almost always much larger than $n = 3$. We used a small n in this example merely so it was easy to find and show all the probabilities in the sampling distribution.

Since the population is quite large compared to the sample size, we can also find the probabilities for each possible sample proportion value using the binomial distribution. The *number* of people preferring pizza A in the sample is a binomial random variable with $n = 3$ and $p = 0.50$. For instance, the probability of a sample proportion of 1/3 equals the probability of 1 person out of 3 preferring pizza A. This equals

$$p(1) = \frac{n!}{x!(n-x)!}p^{x}(1-p)^{n-x} = (3!/(1!2!))(0.50)^{1}(0.50)^{2} = 3/8.$$

It's especially helpful to use the binomial formula when p differs from 0.50, since then the 8 possible samples listed above would not be equally likely.

♦

To practice this concept, try Exercise 6.52.

Every sample statistic has a sampling distribution. Besides the sample proportion, there is a sampling distribution of a sample mean, a sampling distribution of a sample median, a sampling distribution of a sample standard deviation, and so forth.

Sampling distribution of sample proportion is a sampling distribution of a sample mean

In Example 15 above, suppose that for each person we defined a binary random variable, X, to represent their preference, as:

> $x = 1$, subject prefers Aunt Erma's pizza A
> $x = 0$, subject prefers pizza D

For example, the sample (D, A, D) in which only the second subject sampled preferred pizza A is also represented by (0, 1, 0). When we denote the two possible outcomes for each observation by 0 and 1, the proportion of times that 1 occurs is the sample mean. For instance, for the sample (0, 1, 0) in which only the second person prefers pizza A, the sample mean equals $(0 + 1 + 0)/3 = 1/3$. This is the sample proportion of the 3 subjects that preferred pizza A. Using this alternative coding shows us that the sampling distribution of a sample proportion is actually a special case of a sampling distribution of a sample mean.

Describing the Sampling Distribution of a Sample Proportion

The binomial probability distribution, introduced in Section 6.2, is an example of a sampling distribution. It is the sampling distribution for the *number* of successes in n independent trials. For interpretation, it is simpler to use the sample *proportion* of successes, since the possible values fall between 0 and 1 regardless of the value of n. As the previous example showed, there is a close connection between the two.

Consider a binomial random variable X for $n = 3$ trials, such as the *number* of people who preferred Aunt Erma's pizza in the pizza taste comparison. The number who preferred it can equal $x = 0, 1, 2, 3$. These correspond to sample *proportions* of 0, 1/3, 2/3, 1. The proportion values are less spread out, by a factor of $1/n = 1/3$, going from 0 to 1 instead of from 0 to 3. This is because the proportion divides the count by 3. The smaller spread for the proportion values is illustrated by the distributions shown in the margin. In general, the formulas for the mean and the standard

> **RECALL**
>
> As seen in the Insight for Example 12 of Section 2.3, *the sample proportion is a special case of a sample mean* calculated for observations that equal 1 for the outcome of interest and 0 otherwise.

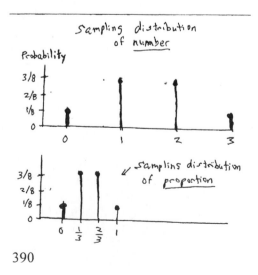

390

deviation of the sampling distribution of the *proportion* of successes are the formulas for the mean and standard deviation of the *number* of successes divided by *n*.

<table>
<tr><td>

RECALL

For a binomial random variable X = number of successes in n trials, Section 6.2 gave the formulas mean = np and standard deviation = $\sqrt{np(1-p)}$.

</td><td>

Mean and Standard Deviation of Sampling Distribution of a Proportion

For a binomial random variable with n trials and probability p of success for each, the sampling distribution of the *proportion* of successes has

$$\text{Mean} = p \text{ and standard deviation} = \sqrt{\frac{p(1-p)}{n}}.$$

To obtain these values, take the mean np and standard deviation $\sqrt{np(1-p)}$ for the binomial distribution of the *number* of successes and divide by n.

</td></tr>
</table>

For instance, consider a random sample of $n = 3$ people from a population in which the proportion who prefer A pizza equals $p = 0.50$. The sample proportion preferring A pizza can take values 0, 1/3, 2/3, 1. The sampling distribution has a mean of $p = 0.50$ and a standard deviation of $\sqrt{p(1-p)/n} = \sqrt{(0.50)(0.50)/3} = 0.29$.

EXAMPLE 16: EXIT POLL OF CALIFORNIA VOTERS REVISITED

Picture the Scenario
Example 1 at the beginning of the chapter discussed an exit poll of 3160 voters for the 2003 special California election to decide whether then-governor Gray Davis should be recalled from office. If the recall were successful, Davis would be removed from office and replaced as governor by the candidate who received the most votes in the election. The winning candidate was Arnold Schwarzenegger.

Questions to Explore
Suppose that exactly 50% of the population of all voters voted in favor of the recall. Describe the mean and standard deviation of the sampling distribution of:
a. the *number* in the sample who voted in favor of it.
b. the *proportion* in the sample who voted for it.

Think It Through
a. Each voter had two options (voting yes or voting no about the recall), so we have binary data. If the recall actually had 50% support in the population of voters, then the *number* of people in the exit-poll sample of 3160 voters who voted for it is a binomial random variable with $n = 3160$ and $p = 0.50$. That random variable has mean $\mu = np = 3160(0.50) = 1580$. We expect about 1580 to vote for the recall and about 1580 to vote against it. The standard deviation is $\sigma = \sqrt{np(1-p)} = \sqrt{3160(0.50)(0.50)} = 28.1$. There is variability around the mean, and we would probably not observe *exactly* 1580 voters supporting the recall. For example, we might observe 1550, or perhaps 1610, both values falling about a standard deviation away from the mean.

b. The *proportion* of people in the sample who voted for the recall is the binomial random variable divided by 3160. If the population proportion $p = 0.50$, then the sampling distribution of the sample proportion has

$$\text{Mean} = p = 0.50$$

$$\text{Standard deviation} = \sqrt{\frac{p(1-p)}{n}} = \sqrt{\frac{(0.50)(0.50)}{3160}} = \sqrt{0.000079} = 0.0089.$$

There is variability around the mean, and we would probably not observe *exactly* a sample proportion of 0.50 supporting the recall. For example, we might observe a sample proportion of 0.49, or 0.51, both values falling about a standard deviation from the mean.

Insight

The standard deviation of the sampling distribution of the sample *proportion* is very small (0.0089). This means that with $n = 3160$, the sample proportion will probably fall close to the population proportion of 0.50.

♦

To practice this concept, try Exercise 6. 57, parts (a)-(c).

The Standard Error

To distinguish the standard deviation of a *sampling* distribution from the standard deviation of an ordinary probability distribution, we refer to it as a **standard error**.

Definition: Standard Error
The standard deviation of a sampling distribution is called a **standard error**.

For instance, the standard deviation $\sqrt{p(1-p)/n}$ of the sampling distribution of the sample proportion is called the **standard error of the sample proportion**.

EXAMPLE 17: CAN YOU PREDICT THE OUTCOME OF THE SPECIAL ELECTION?

Picture the Scenario

Let's now conduct an analysis that uses the actual exit poll of 3160 voters for the special California election about recalling Governor Gray Davis. In that exit poll, the recall of Davis was supported by 54% of the 3160 voters sampled.

Question to Explore

a. If the population proportion supporting recall was 0.50, would it have been unlikely to observe the exit-poll sample proportion of 0.54 that was obtained?

b. Would the exit-poll result have been unlikely if the population proportion supporting recall were less than 0.50? Based on this answer, did this exit poll provide much evidence to predict whether the recall would pass?

RECALL

By the guideline in Section 6.2, the binomial distribution is bell shaped when the expected numbers of successes and failures, np and $n(1-p)$, are both at least 15. Here, $np = n(1-p) = 3160(0.50) = 1580$ are both much larger than 15.

Think It Through

a. We use the fact that the binomial distribution for the *number* voting for recall has a bell shape (See the Recall box). Likewise, so does the sampling distribution of the sample *proportion*, which we get by dividing the number by 3160. Figure 6.14 shows a normal distribution approximation for the sampling distribution of the sample proportion. When we assume $p = 0.50$, this distribution has a mean of $p = 0.50$ and standard error of $\sqrt{p(1-p)/n} = \sqrt{0.5(0.5)/3160} = 0.0089$. Because it's bell-shaped, nearly the entire distribution falls within three standard errors of the mean, or within about $3(0.0089) = 0.03$ of the mean of 0.50.

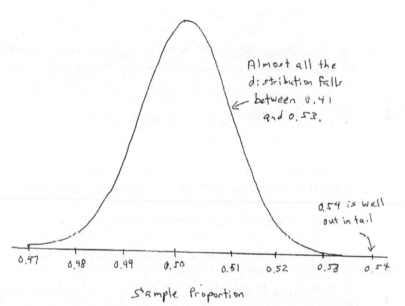

Figure 6.14: Sampling Distribution of Sample Proportion Voting for Recall in California Special Election. This shows where we expect sample proportions to fall for random samples of size $n = 3160$, when a population proportion $p = 0.50$ voted for the recall. In this case, nearly all the distribution falls between 0.47 and 0.53. The observed sample proportion of 0.54 is far out in the right tail, so it would be very unlikely if $p = 0.50$. Observing this would make us doubt that p actually equals 0.50.

In other words, if $p = 0.50$, it is very likely that the sample proportion falls between 0.47 and 0.53. So it would be surprising to observe a sample proportion of 0.54. Converting the sample proportion value of 0.54 to a z-score,

$$z = (0.54 - 0.50)/0.0089 = 4.5.$$

This tells us that the sample proportion of 0.54 is more than four standard errors from the expected value of 0.50. In summary, this analysis suggests that if the exit-poll sample of voters was random, then the sample proportion of 0.54 voting for recall would be very unlikely if the population support were $p = 0.50$.

b. A sample proportion of 0.54 would also be very unlikely if the population support were *less than* 0.50, because then 0.54 would be even more than 4.5 standard errors from 0.50. That is, it would be even more unusual. Because the sample proportion of 0.54 would be extremely unusual if $p = 0.50$ or less, we have strong evidence that actually p was larger than 0.50. In summary, the exit poll gives strong evidence to predict that Governor Gray Davis would be recalled.

Insight

In this special election, when all 7,974,834 votes were tallied, 55.4% voted for the recall and 44.6% voted against it. The prediction based on the exit poll that Governor Davis would be recalled was correct. He was removed from office, and Arnold Schwarzenegger became the next California governor.

If the sample proportion favoring the recall had been closer to 0.50, we would have been unwilling to make a prediction. For instance, a sample proportion of 0.51 for a sample of size 3160 would not have been especially unlikely if truly $p = 0.50$ or is slightly below 0.50. We've seen that the sample proportion might plausibly fall as much as about 0.03 from the true population proportion.

♦

To practice this concept, try Exercises 6.57 and 6.58.

In this example, it was sufficient merely to evaluate a region over which most of the sampling distribution concentrates under the stated condition and then check whether the sample proportion falls within that region. We did not need to use the formula for binomial probabilities. Indeed, calculating the probabilities would have been tedious for the 3160 possible values (0, 1, 2, … 3160) that the number voting for recall could equal. The probability of *any* single value can be extremely small when n is large.

A Sampling Distribution Shows How a Statistic Would Vary with Many Similar Studies

Sampling distributions describe the variability that occurs in collecting data and using statistics to estimate population parameters. If different polling organizations each take a sample and estimate the population proportion who voted a certain way, they will get different estimates, because their samples have different people. Figure 6.14 portrayed the variability for different samples of size $n = 3160$.

Sampling distributions are the foundation of inferential statistics. They help us to predict how close a statistic falls to the parameter it estimates. This accuracy of prediction depends on the sample size. For instance, Figure 6.14 suggests that for a random sample of size 3160, the probability is high that a sample proportion falls within 0.03 of the population proportion. The sampling distribution (shown in Figure 6.13) for a sample of size 3 is much more spread out, and such accuracy does not occur.

Here's a summary of what we've learned so far about the sampling distribution of a sample proportion:

Summary of Sampling Distribution of a Proportion

For a random sample of size n from a population with proportion p, the **sampling distribution of the sample proportion** has

$$\text{mean} = p, \text{ standard error} = \sqrt{\frac{p(1-p)}{n}}.$$

If n is sufficiently large that the expected *numbers* of outcomes of the two types, np and $n(1-p)$, are both at least 15, then this sampling distribution has a bell shape.

In fact, we'll see in the next section that, for large n, the sampling distribution of the sample proportion is approximately a *normal* distribution.

To get a feel for the important concept of a sampling distribution, it's helpful to simulate a sampling distribution by repeatedly taking samples of a particular size. That's the purpose of the following activity.

ACTIVITY 1: SIMULATING A SAMPLING DISTRIBUTION

Let's continue to explore how a sample proportion may vary from sample to sample. We'll do this using random numbers and then using the *sampling distribution* applet, under the assumption that the population proportion voting for recall of Governor Gray Davis in the California special election was $p = 0.50$. You can use random numbers and the *sampling distribution* applet with other values for p as well.

How can we simulate the sampling distribution of sample proportions for random samples of size $n = 1000$. How would you describe the shape, center, and variability?

We can simulate selecting a voter at random from a population by picking a random number. To simulate random sampling from a population in which exactly 50% of all voters voted for each candidate, we identify the 50 two-digit numbers between 00 and 49 as voting "yes" about the recall and the 50 two-digit numbers between 50 and 99 as voting "no." That is,

　　　00 01 02 03 04 ... 49　　　50 51 52 53 54 ... 99
　　　　　Vote "yes" on recall　　　　　Vote "no" on recall

Then each has a 50% chance of selection on each choice. To simulate random sampling when 45% of the population vote for recall, we would use only the digits 00 to 44 for a yes vote and 45 to 99 for a no vote.

The first two digits of the first column of the random number table in Chapter 4 (Table 4.1, part of the first column of which is shown in the margin) provide the random numbers 10, 22, 24, 42, 37, 77, and so forth. The first number, 10, falls between 00 and 49, so in this simulation the first person sampled voted for the recall. Of the first 6 people selected, 5 voted for the recall (they have numbers between 00 and 49). Selecting 1000 two-digit random numbers simulates the votes for randomly sampling 1000 residents of the much larger population.

Rather than laboriously using random numbers, you can use the *sampling distribution* applet (**www.prenhall.com/TBA**) to generate randomly a binomial random variable for a particular n and p. Select *binary* for the parent population, setting the population proportion as $p = 0.50$. Select for the sample size $n = 1000$. For the third graph, select the option *counts* from the menu. For the fourth graph, select the option *proportion*. For one of our simulations, we got $x = 523$ voting "yes" and 477 voting "no" on recall. The sample proportion of yes votes was $523/1000 = 0.523$, near the "true" proportion of 0.50. This particular estimate was good. Were we simply lucky? We repeated the process and used this applet to generate another binomial with $n = 1000$ and $p = 0.50$. This time the sample proportion of yes votes was $491/1000 = 0.491$, different but also quite good.

Now, you use the applet to take a random sample of size 1000. Did you get a sample proportion close to 0.50? Perform this simulation of a random sample of size 1000 ten times, each time observing from the graphs the counts and the corresponding sample proportion of yes votes.

We programmed software to perform this process of picking 1000 people a million separate times, so we could search for a pattern in the results. Figure 6.15 shows a histogram of the million outcomes for the sample proportion. The simulated sample proportions fell in a bell shape around the true population value of $p = 0.50$. Nearly all the sample proportions fell between 0.45 and 0.55, that is, within 0.05 of the true population value of 0.50. Did the ten sample proportion values in your simulation all fall between 0.45 and 0.55?

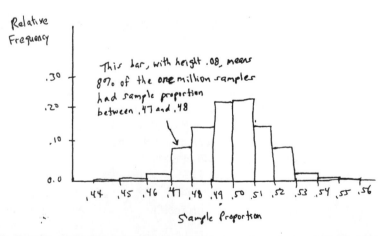

Figure 6.15: Results of Simulating the Sample Proportion Voting "Yes" about the Recall of Governor Gray Davis of California. One million random samples of 1000 subjects each were simulated from a population in which half favor each option. **Question**: For $n = 1000$, explain why it is very likely that the sample proportion falls within 0.05 of the population value of 0.50.

This result is not a surprise. We know that the standard error of this simulated sampling distribution is

$$\text{Standard error} = \sqrt{p(1-p)/n} = \sqrt{(0.50)(0.50)/1000} = \sqrt{0.00025} = 0.0158.$$

Because the simulated sampling distribution is approximately bell shaped, the probability is very close to 1.0 that the sample proportions would fall within 3 standard errors of 0.50, that is, within 3(0.0158) = 0.047.

Return to the *sampling distribution applet* and perform 10,000 simulations with a sample size of *n* = 1000 each, for *p* =0.50. Compare the sampling distribution in graph 4 of the 10,000 sample proportions to Figure 6.15. Are the results similar?

♦

To practice this activity, try Exercise 6.62.

SECTION 6.4: PRACTICING THE BASICS

6.52 Pizza: Refer to Example 15, "Which brand of pizza do you prefer?" Now use sample size *n* = 4, when the population proportion who preferred Aunt Erma's pizza is 0.50.
a. Enumerate the 16 possible samples and the possible sample proportion values.
b. Using (a), construct the sampling distribution of the sample proportion of people who preferred Aunt Erma's pizza. That is, set up a table (like Table 6.9) showing the sample proportion values and their probabilities.
c. In practice, you would not know the population proportion value. Based on (b), if you sample only 4 people, can you be confident that the sample proportion is close to the population proportion? Explain.

6.53 Pizza using the binomial: Refer to the previous exercise.
a. Explain how to use the binomial distribution to find the probability that two people preferred Aunt Erma's pizza.
b. Suppose now that the population proportion that prefers Aunt Erma's pizza is 0.60. Use the binomial distribution to find the probability that two people preferred Aunt Erma's pizza.
c. For (b), explain why the sample space approach would no longer have equally likely points.

6.54 Who gets into law school?: Exercise 6.9, "Law school admissions," referred to randomly selecting two of five students (A, B, C, D, E) for admission to law school, when A and B were female. The possible samples were (A,B), (A,C), (A,D), (A,E), (B,C), (B,D), (B,E), (C,D), (C,E), and (D,E).
a. Of the two students selected, what is the probability that the *proportion* of females selected equals 0?
b. Construct a table displaying the sampling distribution of the proportion of the students selected who are female.

6.55 Relative frequency of heads: Construct the sampling distribution of the proportion of heads for flipping a balanced coin

a. Once.
b. Twice. (*Hint*: The possible samples are (H, H), (H, T), (T, H), (T, T).)
c. Four times. (*Hint*: There are 16 possible samples.)
d. Refer to (a) - (c). Describe how the shape of the sampling distribution is changing as the number of flips *n* increases.

6.56 Migraine headaches: For the population of people who suffer occasionally from migraine headaches, suppose $p = 0.60$ is the proportion who get some relief from taking ibuprofen. For a random sample of 24 people who suffer from migraines:
a. State the mean of the sampling distribution of the sample proportion.
b. Find the standard error of the sampling distribution.
c. Explain what the standard error describes.
d. Describe the impact of the sample size by explaining how the sampling distribution changes if instead the sample size is 96 (four times as large).

6.57 Other scenario for exit poll: Refer to Examples 16 and 17 about the exit poll of California voters, for which the sample size was 3160. Suppose that 55% of all voters voted for the recall (which was actually the case).
a. Identify *n* and *p* for the binomial distribution that is the sampling distribution of *X* = the *number* in the sample who voted for the recall.
b. Find the mean and standard deviation of *X*.
c. Find the mean and standard error of the sampling distribution of the *proportion* in the sample who voted for the recall.
d. Based on (c) and the bell shape of the sampling distribution, give an interval of values within which the sample proportion will almost certainly fall.

6.58 Exit poll and *n*: Refer to the previous exercise.
a. Based on the result in (d), if you take an exit poll and observe a sample proportion of 0.60, would you be safe in concluding that the population proportion probably exceeds 0.55?
b. Repeat (c) and (d) of the previous exercise and (a) of this exercise when the sample size instead equals 100.
c. Explain the practical implication of the increase in the standard error when *n* is smaller.

6.59 Rolling a die: Two balanced dice are rolled. Example 2 showed that there are 36 equally likely pairs of numbers, (1,1), (1,2), …(1,6), (2,1), … (6,6).
a. Construct the sampling distribution for the sum of the two dice.
b. Construct the sampling distribution for the sample mean $\bar{x}$ of the two numbers rolled.
c. Show that the mean of the probability distribution of the outcome for each die is 3.5.
d. By the symmetric form of the sampling distribution in (b), show that its mean is 3.5. (We'll see in the next section that the mean of the sampling distribution of $\bar{x}$ equals the mean of the probability distribution for each observation.)

6.60 Effect of *n* on sample proportion: The figure illustrates two sampling distributions for sample proportions when the population proportion $p = 0.50$.

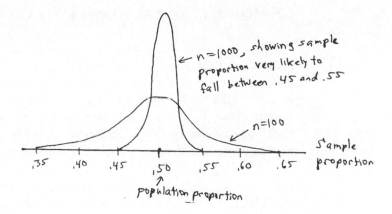

a. Find the standard error for the sampling distribution with (i) $n - 100$, (ii) $n - 1000$.

b. Explain why the sample proportion would be very likely (as the figure suggests) to fall (i) between 0.35 and 0.65 when $n = 100$, and (ii) between 0.45 and 0.55 when $n - 1000$.

c. Explain how the results in (b) indicate that the sample proportion tends to more precisely estimate the population proportion when the sample size is larger.

6.61 Purpose of sampling distribution: You'd like to estimate the proportion of all students in your school who are fluent in more than one language. You poll a random sample of students and get a sample proportion of 0.12. Explain why the standard error of the sampling distribution of the sample proportion gives you useful information for helping you gauge how close this sample proportion is to the unknown population proportion.

▣6.62 Simulating the exit poll: Simulate an exit poll of 100 voters, using the *sampling distribution* applet at www.prenhall.com/TBA, assuming that the population proportion is 0.50. Refer to Activity 1 for guidance on using the applet.

a. What sample proportion did you get? Why do you not expect to get exactly 0.50?

b. Simulate this exit poll 10 times. Keep the sample size at $n = 100$ and $p = 0.50$. Describe the 10 sample proportion values. If you took a huge number of samples of size 100 each, what would you predict for the value of the standard deviation of the sample proportions?

c. Now change the population proportion to 0.70, keeping the sample size $n = 100$. Simulate the exit poll ten times. How would you say the results differ from those in (b)?

d. Using the appropriate formulas from this section, for part (c) find the mean and the standard error of the sampling distribution of the sample proportion.

6.5 How Close Are Sample Means to Population Means?

The previous section introduced the *sampling distribution*. This is a probability distribution for the possible values of a statistic. For instance, we learned how much a sample proportion can vary among different random samples. As we've seen, when we denote the two possible outcomes by 0 and 1, the sample proportion is merely a sample mean $\bar{x}$. Because the sample mean is so commonly used, we'll now give its sampling distribution special attention. Using it, we'll see how to predict how close a particular sample mean $\bar{x}$ falls to the population mean μ.

There are two main results about the sampling distribution of the sample mean: One provides formulas for its mean and standard deviation (the "standard error"). The other indicates that its shape is often approximately a normal distribution, as we observed in the previous section for the sample proportion.

The Mean and Standard Error of the Sampling Distribution of $\bar{x}$

Before we select a sample, the sample mean $\bar{x}$ is a random variable--we don't know what value it will take. The value varies from sample to sample. By contrast, the *population* mean μ is a single fixed number. At a given time for a particular scenario, there can be many different samples but there's a single population. For random sampling, the sample mean $\bar{x}$ can fall either above or below the population mean μ. In fact, *the mean of the sampling distribution of the sample mean $\bar{x}$ equals the population mean μ.* The margin figure illustrates. It shows the sample mean $\bar{x}$ as a random variable with distribution concentrated around the value of the population mean μ.

$\bar{x}$ fluctuates around popul. μ

How can we describe the spread of the sampling distribution of $\bar{x}$? As we mentioned in the previous section, we use the standard deviation, and to distinguish the standard deviation of a sampling distribution from other standard deviations, we call it a **standard error**.

IN WORDS

The **standard error of** $\bar{x}$ describes how much the sample mean varies for all the possible samples.

> **Standard Error of the Sample Mean** $\bar{x}$
>
> The standard deviation of the sampling distribution of the sample mean $\bar{x}$ is called the **standard error of** $\bar{x}$.

Here's a way to think of the standard error: Suppose each of several researchers was studying some topic, independently of each other. Suppose that each researcher took a random sample of a particular size n and found a sample mean to estimate an unknown population mean μ. If we could collect all their $\bar{x}$−values, those values would tend to fluctuate around that unknown μ. Moreover, the standard deviation of all those $\bar{x}$-values would approximate the standard error. In fact, we'll see that the standard error does give us information about the variability in results we can expect from different studies about the same topic.

Do you think that sample mean values tend to vary more, or vary less, than individual observations? For instance, suppose you choose a random sample of 100 graduating seniors at your school and record the GPA for each, and you also find their mean GPA. Then you look at last year's seniors and do the same, and likewise for the past several years. Would the sample means vary more, or vary less, than the individual GPAs?

For random sampling, the standard error of $\bar{x}$ depends on the sample size n and the spread of the probability distribution from which the observations are sampled – the *population distribution*. It can be shown to equal

$$\text{Standard error} = \frac{\text{Population standard deviation}}{\text{Square root of sample size}} = \frac{\sigma}{\sqrt{n}}.$$

We'll shortly explain why and how the sample size n affects the standard error. But you can see from the formula that the standard error is less than the population standard deviation, σ: Means of samples tend to vary less than individual observations. The graduating senior GPAs might look like

     3.53, 2.18, 3.95, 2.68, 3.11, 3.80, 2.35, ... (standard deviation of about 0.40, say)

whereas the sample mean GPAs from year to year might look like

     3.10, 3.15, 3.03, 3.11, 3.09, 3.13, ... (standard deviation of about 0.04, say).

The individual GPAs vary much more than the sample mean GPAs do.

In summary, the following result describes the center and spread of the sampling distribution of the sample mean $\bar{x}$.

Mean and Standard Error of Sampling Distribution of Sample Mean $\bar{x}$

For a random sample of size n from a population having mean μ and standard deviation σ, the **sampling distribution of the sample mean $\bar{x}$** has:

     center described by the **mean** μ (the same as the mean of the population)

     spread described by the **standard error**, which equals $\sigma/\sqrt{n}$ (the population standard deviation divided by the square root of the sample size).

EXAMPLE 18: HOW MUCH DO MEAN SALES VARY FROM WEEK TO WEEK?

Picture the Scenario
The total daily sales of food and drink in Aunt Erma's pizza restaurant vary from day to day. The sales figures fluctuate around a mean of $\mu = \$900$, with a standard deviation of $\sigma = \$300$.

Questions to Explore

The mean sales for the seven days in a week are computed each week. The weekly means are plotted over time to help detect substantial increases or decreases in sales. Suppose the sales vary independently from day to day, according to a probability distribution with $\mu = \$900$ and $\sigma = \$300$.

a. What would the weekly mean sales figures fluctuate around?

b. How much variability would you expect in the sample weekly mean sales figures? Find the standard error, and interpret.

Think It Through

a. The sample means fluctuate around the mean of the probability distribution for each observation. This is $\mu = \$900$.

b. The sampling distribution of the sample mean for $n = 7$ has mean $900. Its standard error equals

$$\frac{\sigma}{\sqrt{n}} = \frac{300}{\sqrt{7}} = 113.$$

If we were to observe the sample mean sales for several weeks, the sample means would vary around $900, with a spread described by the standard error of 113.

Insight

Figure 6.16 portrays a possible population distribution for daily sales. It also shows the sampling distribution of the mean sales $\overline{x}$ for $n = 7$. There is less variability from week to week in the mean sales than there is from day to day in the daily sales.

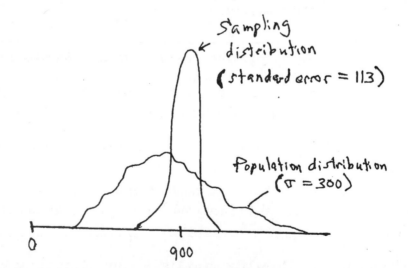

Figure 6.16: A Population Distribution for Daily Sales and the Sampling Distribution of Weekly Mean Sales $\overline{x}$ ($n = 7$). These show there is more variability day-to-day in the daily sales than week-to-week in the weekly mean sales.

In practice, the assumptions in this example are a bit unrealistic. Some days of the week (such as weekends) may be busier than others, and certain periods of the year may be busier (such as the summer, or New England's leaf-viewing season in the fall). Also, we would not know values of population parameters such as the population mean and standard deviation. However, this example shows that even under some simplifying assumptions, we'd still expect to see quite a bit of variability from week to week in the sample mean sales. We should not be surprised if the mean is $800 this week, $1000 next week, and so forth. Knowing how to find a standard error gives us a mechanism for understanding how much variability to expect in sample statistics "just by chance."

♦

To practice this concept, try Exercise 6.63, part (a) and Exercise 6.67, part (a).

Effect of *n* on the Standard Error

Let's consider again the formula $\sigma/\sqrt{n}$ for the standard error of the sample mean. Notice that as the sample size *n* increases, the denominator increases, so the standard error decreases. This has an important practical implication: With larger samples, the sample mean is more likely to fall close to the population mean.

EXAMPLE 19: WHAT'S THE LONG-RUN CONSEQUENCE TO YOUR FINANCES OF PLAYING A LOTTERY?

Picture the Scenario
Many U.S. states, Canadian provinces, and countries have lottery games. Most lotteries use "numbers games." They are usually several-digit generalizations of the following game: You bet a dollar on a number between 0 and 9 that you pick at random. If you are correct, you win $5, and otherwise, you win nothing.

X = winnings in one game	
x	$P(x)$
0	0.90
5	0.10

Let *X* denote your winnings for a single play of the lottery. Then the possible values for *X* are $0 and $5. The probability distribution of *X* is $p(0) = 0.90$ and $p(5) = 0.10$, since there is a 1 in 10 chance that you pick the correct number and win $5. This probability distribution has mean

$$\mu = \Sigma x\, p(x) = 0(0.90) + 5(0.10) = 0.50.$$

This is the expected value of the outcome *X* for a single play. The expected return on the dollar bet is only half a dollar. This does not seem attractive, but it is a typical return. (Some are even worse; see the New Jersey lottery in Exercise 6.82.) Naturally, the expected return is less than you pay to play, because the government wants to make money! It can be shown that the standard deviation of this probability distribution is $\sigma = 1.50$.

Question to Explore
Find the mean and standard error of the sampling distribution of your mean winnings if you play this game (a) once a *week* for the next year (that is, 52 times), (b) once a *day* for the next year (that is, 365 times).

Think It Through
The sampling distribution of the sample mean winnings $\bar{x}$ has the same mean as the probability distribution of X for a single play, namely, $\mu = 0.50$. For playing the lottery n times, its standard error is

$$\sigma / \sqrt{n} = 1.50 / \sqrt{n}.$$

a. If you play $n = 52$ times, the standard error equals $1.50 / \sqrt{52} = 0.21$.
b. If you're really serious and play 365 times over the next year, the standard error equals $\sigma / \sqrt{n} = 1.50 / \sqrt{365} = 0.08$.

Insight
Figure 6.17 portrays the dependence of the sampling distribution on the sample size. As n increases, the standard error gets smaller, so the sampling distribution gets narrower. This means that the sample mean winnings tend to get closer to the mean of the probability distribution for each game, 0.50, which is the expected value of X. The **law of large numbers** states that as n increases, the sample mean tends to get closer to the expected value of X. As you play the lottery more, the mean of your winnings tends to get closer to 0.50. This is bad news (for you, not the government) when you pay $1 each time to play.

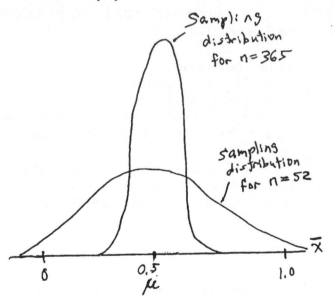

Figure 6.17: The Sampling Distribution of $\bar{x}$ when $n = 52$ and when $n = 365$. The standard error is smaller when n is larger. **Question**: What is the practical implication of the effect of n on the standard error?

◆

To practice this concept, try Exercise 6.65 and conduct the simulation study in Exercise 6.134 to get some feeling for how sampling distributions depend on n.

This example reinforces what we observed in the previous section about the effect of increased sample size on the precision of a sample proportion in estimating a population proportion. The larger the sample size, the smaller the standard error and, thus, less potential error in estimating a population parameter.

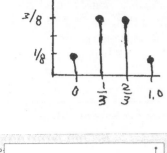

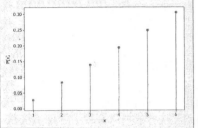

Sample Means Have Normal Sampling Distributions for Large *n*

Example 15 constructed the sampling distribution for the sample proportion preferring Aunt Erma's pizza in a taste test for a sample of size *n* = 3. The sampling distribution had somewhat of a mound shape. See the margin. A simulation in Activity 1 for the sample mean with *n* = 1000 showed even more of a bell shape. Is it typical for sampling distributions to have bell shapes?

Let's look at another example, this time one in which the probability distribution of *X* is skewed. Example 2 derived the probability distribution of *X* = maximum value on two rolls of a die. The figure in the margin displays the distribution. It is skewed to the left with a mean of $\mu = 4.5$ and a standard deviation of $\sigma = 1.4$. Suppose we repeatedly sample from this distribution with random sample sizes of *n* = 30. Figure 6.18 displays the histogram of the sample means for a huge number of simulated samples from this skewed distribution. (In theory, the sampling distribution refers to an infinite number of samples of size 30 each; using software, we took 10,000 samples of size 30 each, which gives similar results.)

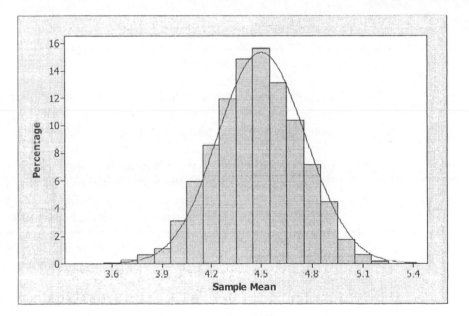

Figure 6.18: Histogram of Simulated Sample Means with *n* = 30. Each sample was simulated from a discrete probability distribution (shown in the margin above) that is skewed to the left with $\mu = 4.5$ and $\sigma = 1.4$. The superimposed normal distribution has mean 4.5 and standard deviation 0.26. **Question:** How does this sampling distribution relate with respect to shape, center, and spread to the probability distribution from which the samples were taken?

Even though the probability distribution from which we took the samples (in the margin above) is highly skewed, this sampling distribution has a bell shape. The histogram in Figure 6.18 has a normal distribution superimposed that has the same mean as the sampling distribution ($\mu = 4.5$) and the same standard deviation as the standard error of the sampling distribution, which is $\sigma/\sqrt{n} = 1.4/\sqrt{30} = 0.26$.

We've seen here a quite surprising result: Even when a probability distribution is not at all bell-shaped, the sampling distribution of the sample mean $\bar{x}$ can have a bell shape. This is a consequence of the second main result of this section. It states that the sampling distribution of $\bar{x}$ often has approximately a normal distribution. This result is called the **central limit theorem**. (It's called a "theorem" because it can be proved mathematically.)

IN WORDS
As the sample size increases, the sampling distribution of the sample mean has more of a bell shaped appearance. Figure 6.19 below shows that even if we sample from a population that is not bell-shaped, the sampling distribution becomes more bell-shaped as *n* increases.

Central Limit Theorem: Normal Sampling Distribution for Sample Mean $\bar{x}$

For random sampling with a large sample size n, the sampling distribution of the sample mean $\bar{x}$ is approximately a normal distribution.

This result applies *no matter what the shape* of the probability distribution from which the samples are taken. This probability distribution is called the *population distribution*, because in practice it describes a population from which we take a sample. This is remarkable. For relatively large sample sizes, the sampling distribution is approximately bell shaped even if the population distribution is highly discrete or highly skewed. We observed this in Figure 6.18 with a skewed, highly discrete population, using $n = 30$, which is not all that "large" but is the size of sample we sometimes see in practice.

Figure 6.19 displays sampling distributions of the sample mean $\bar{x}$ for four different shapes for the population distribution from which samples are taken. The population shapes are shown at the top of the figure. Below them are portrayed the sampling distributions of $\bar{x}$ for random sampling of sizes $n = 2$, 5, and 30. Even if the population distribution itself is uniform (column 1 of the figure) or U-shaped (column 2) or skewed (column 3), the sampling distribution has approximately a bell shape when n is at least about 30, and sometimes for n as small as 5. In addition, the spread of the sampling distribution noticeably decreases as n increases, because the standard error decreases.

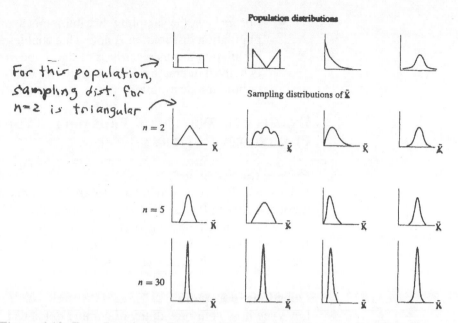

For this population, sampling dist. for n~2 is triangular

Figure 6.19: Four Population Distributions and the Corresponding Sampling Distributions of $\overline{x}$. Regardless of the shape of the population distribution, the sampling distribution becomes more bell shaped as the random sample size *n* increases. You can use the *sampling distribution* applet at www.prenhall.com/TBA to simulate how the first population distribution shown has a sampling distribution for the sample mean that becomes more nearly normal as *n* increases. See Exercise 6.134 for details.

IN PRACTICE

The sampling distribution of $\overline{x}$ takes more of a bell shape as the random sample size *n* increases. The more skewed the population distribution, the larger *n* must be before the shape is close to normal. In practice, the sampling distribution is usually close to normal when the sample size *n* is at least about 30.

If the population distribution is approximately normal, then the sampling distribution is approximately normal for *all* sample sizes. The rightmost column of Figure 6.19 shows this case.

How does the central limit theorem help us make inferences?

> **RECALL**
>
> From Section 6.3, about 95% of a normal distribution falls within 2 standard deviations of the mean, and 99.7% within 3 standard deviations.

The central limit theorem and the formula for the standard error of the sample mean $\overline{x}$ have many implications. These include the following:

- When the sampling distribution of the sample mean $\overline{x}$ is approximately normal, $\overline{x}$ falls within 2 standard errors of the population mean μ with probability close to 0.95, and $\overline{x}$ almost certainly falls within 3 standard errors of μ. Results of this type are vital to inferential methods that predict how close sample statistics fall to unknown population parameters.

- For *large n*, the sampling distribution is approximately normal even if the population distribution is not. This enables us to make inferences about population means regardless of the shape of the population distribution. This is helpful in practice, since usually we don't know the shape of the population distribution. Often, it is quite skewed.

EXAMPLE 20: WHAT'S THE PROBABILITY YOU COME OUT AHEAD IN PLAYING THE LOTTERY?

Picture the Scenario

In Example 19, we considered the following lottery game: You bet a dollar on a number between 0 and 9 that you pick at random. If you are correct, you win $5, and otherwise, you win nothing. Let X denote your winnings. Then, the possible values for X are $0 or $5. From Example 19, the probability distribution is $p(0) = 0.90$ and $p(5) = 0.10$, which has mean $\mu = 0.50$ and standard deviation $\sigma = 1.50$.

Questions to Explore

Let's see how your probability of coming out ahead (winning more than you pay to play) depends on how often you play. What's the probability that you come out ahead if you play this game (a) once, (b) once a week for the next year, (c) once a day for the next year?

Think It Through

You come out ahead if your mean winnings $\bar{x}$ exceed $1, the amount you pay to play each time.

<table>
<tr>
<td>

RECALL

The sampling distribution of $\bar{x}$ has mean equal to the population mean μ and standard error equal to $\sigma/\sqrt{n}$, where σ is the population standard deviation. For large *n*, by the central limit theorem it has approximately a normal distribution.

</td>
<td>

a. In playing $n = 1$ time, you come out ahead only if you win. The probability of this is the probability that you win in playing once, which is 0.10.

b. If you play once a week for the next year, then $n = 52$. We need to find the probability that the sample mean exceeds $1. By the central limit theorem, for large *n*, the sampling distribution of $\bar{x}$ is approximately normal. This is true even though the population distribution is very non-normal. In fact, the population distribution here is binary and highly skewed, consisting of a large probability at $x = 0$ and a small probability at $x = 5$. The sampling distribution has mean $\mu = 0.50$ and standard error $\sigma/\sqrt{n} = 1.50/\sqrt{n}$.

</td>
</tr>
</table>

If you play $n = 52$ times, the standard error equals $1.50/\sqrt{52} = 0.208$. Thus, by the central limit theorem, the sampling distribution of $\bar{x}$ is approximately a bell shaped curve with mean 0.50 and standard error 0.208. Figure 6.20 shows this curve.

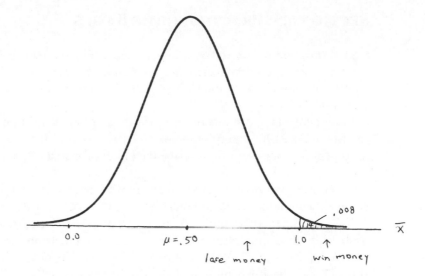

Figure 6.20: Approximate Sampling Distribution of the Sample Mean $\overline{x}$ **when** $\mu = 0.50$
and Standard Error = 0.208. We use this to approximate the probability that $\overline{x}$ falls above
$1 and you win more money than you pay to play the lottery 52 times. **Question**: Does this
have the same shape as the probability distribution for each game? Explain.

The probability that you come out ahead after playing 52 times (that is, the sample
mean exceeds $1) is approximately the probability above 1.0 in the right tail of this
figure. So we need to find the probability that a normal random variable with mean
0.50 and standard error 0.208 takes a value above 1. The value 1.0 has a z-score of

$$z = (\text{value} - \text{mean})/(\text{standard error}) = (1.0 - 0.50)/0.208 = 2.40.$$

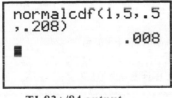

TI-83+/84 output

From Table A, this has a cumulative probability of 0.992. Thus, the right-tail
probability is $1 - 0.992 = 0.008$. There is roughly 1 chance in a hundred that you'll
win more than you paid to play, if you play 52 times (Exercise 6.142 gives a more
precise answer). It seems that you were better off playing only once, in which case
the probability was 0.10 of coming out ahead.

c. What happens if you're really serious and play once a day for the next year, for a
total of 365 times? Then, the standard error equals $\sigma / \sqrt{n} = 1.50/\sqrt{365} = 0.0785$.
A sample mean winnings of 1 then has a z-score of

$$z = (1.0 - 0.50)/0.0785 = 6.4.$$

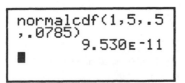

TI-83+/84 output

You won't find this in Table A, but software tells us that the probability in the right
tail is only 0.0000000001, which is 1 in ten billion. This chance of coming out ahead
is less than the chance that you would be chosen if one person of everyone in the
entire world were picked at random!

Insight
Over the course of a year, it's nearly impossible to win more than you paid to play.
With large *n,* it's hard for gamblers to come out ahead (unless they can cheat).

To practice this concept, try Exercises 6.66 and 6.68.

SECTION 6.5: PRACTICING THE BASICS

6.63 Education of the self-employed: According to a recent *Current Population Reports*, the population distribution of number of years of education for self-employed individuals in the United States has a mean of 13.6 and a standard deviation of 3.0.
a. Find the mean and standard error of the sampling distribution of $\bar{x}$ for a random sample of size 100. Interpret them.
b. Repeat (a) for $n = 400$. Describe the effect of increasing n.

6.64 Rolling two dice: Exercise 6.59 showed that the sampling distribution for the sample mean outcome when you roll two dice is $p(1) = 1/36$, $p(1.5) = 2/36$, $p(2) = 3/36, \ldots, p(3.5) = 6/36$, $p(4) = 5/36, \ldots, p(6) = 1/36$. Construct a graph of the (a) probability distribution for each roll, (b) sampling distribution of the sample mean outcome for $n = 2$. (The first column of Figure 6.19 portrays a case like this, and shows how it becomes more bell-shaped for $n = 5$ and 30.)

6.65 Playing roulette: A roulette wheel in Las Vegas has 38 slots. If you bet a dollar on a particular number, you'll win \$35 if the ball ends up in that slot and \$0 otherwise. Roulette wheels are calibrated so that each outcome is equally likely.
a. Let X denote your winnings when you play once. State the probability distribution of X.
b. Show that the mean of the distribution in (a) is $35/38 = 0.92$. The standard deviation of the distribution of X is 5.603.
c. You decide to play once a minute for 12 hours a day for the next week, a total of 5040 times. Find the mean and the standard error of the sampling distribution of your mean winnings. Given that you bet a dollar each time you play, what does this suggest about the wisdom of playing roulette a large number of times?

6.66 Great expectations: Refer to (c) in the previous exercise. Using the central limit theorem, find the probability that with this amount of roulette playing, your mean winnings is at least \$1, so that you have not lost money after this week of playing.

6.67 Skewed population: Figure 6.18 showed the sampling distribution of sample means from a highly skewed population with $\mu = 4.47$ and $\sigma = 1.40$. For repeated random samples of size 100 from this population:
a. Find the mean and standard error.
b. Explain why the sampling distribution of the sample mean is bell-shaped, even though the population was highly skewed.

6.68 Canada lottery: In one lottery option in Canada (www.lotterycanada.com), you bet on a six-digit number between 000000 and 999999. For a \$1 bet, you win \$100,000 if you are correct. The mean and standard deviation of the probability distribution for the lottery winnings are $\mu = 0.10$ and $\sigma = 31.623$. Joe figures that if he plays enough times every day, eventually he will strike it rich, by the law of large numbers. Over the course of several years, he plays one million times. Let $\bar{x}$ denote his average winnings.
a. Find the mean and standard error of the sampling distribution of $\bar{x}$.

b. About how likely is it that Joe's average winnings exceed $1, the amount he paid to play each time? Use the central limit theorem in answering.

6.69 Income of farm workers: For the population of farm workers in New Zealand, suppose that weekly income has a distribution that is skewed to the right with a mean of μ = $500 (N.Z.) and a standard deviation of σ = $160. A researcher, unaware of these values, plans to randomly sample 100 farm workers and use the sample mean annual income $\overline{x}$ to estimate μ.
a. Show that the standard error of $\overline{x}$ equals 16.0.
b. Explain why it is almost certain that the sample mean will fall within $48 of $500.

6.70 Unknown mean: Refer to the previous exercise. The sampling distribution of $\overline{x}$ provides the probability that $\overline{x}$ falls within a certain distance of μ, regardless of the value of μ. Show how to calculate the probability that $\overline{x}$ falls within $20 of μ for all such workers. (*Hint*: Convert the distance 20 to a *z*-score for the sampling distribution.)

6.71 Restaurant profit?: Jan's All You Can Eat Restaurant finds that their customer expense (based on how much the customer eats and the expense of labor) has a distribution that is skewed to the right with a mean of $8.20 and a standard deviation of $3. Jan's charges $8.95 per customer to eat at the restaurant.
a. If the 100 customers on a particular day have the characteristics of a random sample from their customer base, find the mean and standard error of the sampling distribution of their sample mean customer expense.
b. Using the central limit theorem, find the probability that the restaurant makes a profit that day, with the sample mean expense being less than $8.95.

6.72 Female heights: In North America, female adult heights are approximately normal with μ = 65 inches and σ = 3.5 inches. The heights of 50 females were measured at a national collegiate volleyball tournament. The sample mean height was found to be 70 inches.
a. Using the population parameters given above, what is the probability of obtaining a sample mean height of 70 inches or higher with a random sample of *n* = 50? (Hint: Apply the central limit theorem.)
b. Does this probability make you question the population mean stated for female heights? Justify why you believe this sample mean may not be representative of the population of female heights.

6.73 Survey accuracy: A study investigating the relationship between age and annual medical expenses randomly samples 100 individuals in Springfield, Illinois. It is hoped that the sample will have a similar mean age as the entire population.
a. If the standard deviation of the ages of all individuals in Springfield is σ = 15, find the probability that the mean age of the individuals sampled is within 2 years of the mean age for all individuals in Springfield. (*Hint*: Find the sampling distribution of the mean age and use the central limit theorem. You don't need to know the mean to answer this, but if it makes it easier for you, use a value such as μ = 30.)
b. Would the probability be larger, or smaller, if σ = 10? Why?

6.74 Blood pressure: Vincenzo De Cerce was diagnosed with high blood pressure. He was able to keep his blood pressure in control for several months by taking blood

411

pressure medicine (amlodipine besylate). De Cerce's blood pressure is monitored by taking 3 readings a day, in early morning, at mid-day, and in the evening.
a. During this period, suppose that the probability distribution of his systolic blood pressure reading had a mean of 130 and a standard deviation of 6. If the successive observations behave like a random sample from such a distribution, find the mean and standard error of the sampling distribution of the sample mean for each day.
b. Suppose that the probability distribution of his blood pressure reading is normal. What is the shape of the sampling distribution? Why?
c. Refer to (b). Find the probability that the sample mean exceeds 140, which is considered excessively high. (*Hint*: Use the sampling distribution, not the probability distribution for each observation.)

6.75 Blood pressure increasing?: Refer to the previous exercise. The two most recently calculated sample means both fall above 140. Perhaps the actual mean blood pressure is drifting upward. De Cerce plans to ask his doctor to evaluate whether to change the dose of the medicine or to change the medicine itself. Assuming that any one sample mean is independent of any other, find the probability that two sample means in a row both fall above 140, if the probability distribution is truly as stated above.   Explain your reasoning.

6.6 How Can We Make Inferences about a Population?

Sampling distributions are fundamental to statistical inference -- making decisions and predictions about populations based on sample data. We'll end the chapter with a preview of this connection, which we'll explore in detail in the next two chapters. But first, let's first review three quite distinct types of distributions: the **population distribution**, the **data distribution**, and the **sampling distribution of a statistic**. Here is a capsule description:

Review: Population Distribution, Data Distribution, Sampling Distribution

- **Population distribution**: This is the probability distribution from which we take the sample. Values of its *parameters*, such as the population proportion p for a categorical variable and the population mean μ for a quantitative variable, are usually unknown. They're what we'd like to learn about.

- **Data distribution**: This is the distribution of the sample data. It's the distribution we actually see in practice. It's described by sample *statistics,* such as a sample proportion or a sample mean. With random sampling, the larger the sample size n, the more closely it resembles the population distribution. With larger n, the higher the probability that a sample *statistic* (such as a sample proportion or the sample mean) falls close to the population parameter (such as the population proportion p or the population mean μ).

- **Sampling distribution**: This is the probability distribution of a sample statistic, such as a sample proportion or sample mean. With random sampling, it provides probabilities for all the possible values of the statistic. We'll see in the next two chapters that the sampling distribution provides the key for telling us how close a sample statistic falls to the unknown parameter we'd like to make an inference about. For large n, it's approximately a normal distribution, by the central limit theorem. Its standard deviation is called the *standard error*.

  For binary data, the sampling distribution of the sample proportion has mean equal to the population proportion p, and
  $$\text{standard error} = \sqrt{p(1-p)/n}\,.$$
  For a quantitative variable, the sampling distribution of the sample mean $\overline{x}$ has mean equal to the population mean μ, and
  $$\text{standard error} = \sigma/\sqrt{n}\,,$$
  where σ is the population standard deviation.

EXAMPLE 21: POPULATION, DATA, AND SAMPLING DISTRIBUTIONS FOR FAMILY SIZE

Picture the Scenario

The distribution of family size in a particular tribal society is skewed to the right, with population mean $\mu = 5.2$ and population standard deviation $\sigma = 3.0$. These values are not known to an anthropologist who samples families in this society to estimate mean family size. For a random sample of 36 families, she gets a mean of 4.6 and a standard deviation of 3.2.

Questions to Explore

a. Identify the *population distribution*, and state its mean and standard deviation.
b. Identify the *data distribution*, and state its mean and standard deviation.
c. Identify the *sampling distribution* of $\overline{x}$, and state its mean and standard error.

Think It Through

a. The *population distribution* is the set of family-size values for everyone in the population of this tribal society. It is skewed to the right and described by the population mean of 5.2 and the population standard deviation of 3.0.

b. The *data distribution* is the collection of family-size values for the 36 families in the sample. It is described by the sample mean of 4.6 and the sample standard deviation of 3.2. It would probably look much like the population distribution, being skewed to the right.

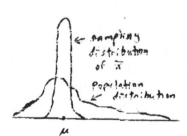

c. Consider all the possible samples of size 36 that could be collected from this population. If we could find the sample mean $\overline{x}$ for each, then the frequency distribution of all their values would make up the *sampling distribution* of $\overline{x}$. That is, this is the collection of all the possible sample mean values and their probabilities. Since n is relatively large ($n = 36$), the sampling distribution is approximately a normal distribution, by the central limit theorem. The mean of this distribution is the same as the population mean, $\mu = 5.2$. The standard error is $\sigma / \sqrt{n} = 3.0/\sqrt{36} = 3.0/6 = 0.50$. Unlike the population and data distributions, the sampling distribution is bell-shaped, and it is much less spread out. Sample means vary much less than individual observations in the sample or population.

Insight

In practice, the anthropologist does not know the population mean. How does she know how near her sample mean of 4.6 falls to the population mean? In the next chapter, we'll see how to use the sample data distribution to estimate the standard error. This tells us how spread out the sampling distribution is. That is, the standard error helps us predict how close the sample mean is likely to fall to the population mean, which is at the center of that distribution (even if we don't know its value).

♦

To practice this concept, try Exercises 6.76 and 6.77.

This example has shown the distinction among these three distributions for a quantitative variable. Let's now look at an example for a categorical variable.

EXAMPLE 22: POPULATION, DATA, AND SAMPLING DISTRIBUTIONS FOR AN EXIT POLL

Picture the Scenario

The 2000 U.S. senatorial election in New York pitted Hillary Clinton against Rick Lazio. The exit poll on which TV networks relied for their projections found after sampling 2232 voters that 55.7% of the sample (1244 voters) voted for Clinton and 44.3% (988 voters) voted for Lazio. When all 6.2 million votes were tallied, 56.0% voted for Clinton and 44.0% voted for Lazio.

Questions to Explore

Let X = vote outcome, with $x = 1$ for Hillary Clinton and $x = 0$ for Rick Lazio. What are the (a) population distribution, (b) data distribution, and (c) sampling distribution of the sample proportion for this scenario?

Think It Through

a. The *population distribution* is the 6.2 million values of the x vote variable, 44.0% of which are 0 and 56.0% of which are 1. That is, if we look at the votes for all subjects

Voter	Vote	x
1	Clinton	1
2	Lazio	0
3	Lazio	0
...		
6,200,000	Clinton	1

56.0% of the time we'd see $x = 1$. The mean of all the 0 and 1 observations would be $p = 0.560$, the population proportion supporting Clinton. This value $p = 0.560$ is the parameter value describing this distribution. The distribution is described by a graph consisting of two stems, with mean 0.560, as the first plot in Figure 6.21 shows. It's not at all bell shaped.

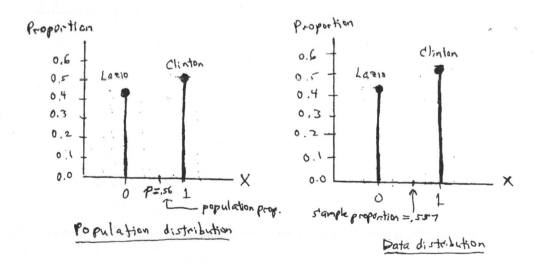

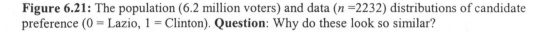

Figure 6.21: The population (6.2 million voters) and data (n =2232) distributions of candidate preference (0 = Lazio, 1 = Clinton). **Question**: Why do these look so similar?

b. The exit poll sampled *n* = 2232 voters. A graph of the 2232 votes in the sample provides a *data distribution*. We'd have a data file like the one shown above, but for 2232 voters rather than 6.2 million voters. Figure 6.21 also shows this distribution. Like the population distribution, it concentrates at *x* = 0 and *x* = 1. It is described by sample statistics. The sample proportion who voted for Clinton was 1244/2232 = 0.557. The sample data are summarized by the sample proportion of 0.557. So 55.7% of the data values are 1, and the height of the spike for Clinton in the data distribution figure in Figure 6.21 is 0.557. The larger the sample size, the more the data distribution tends to resemble the population distribution, since the data we observe are a subset of the population values. Here, they are very similar, since *n* is quite large.

c. The *sampling distribution* of the sample proportion, by the central limit theorem, is approximately normal. Its mean is *p* = 0.560 (the population proportion) with

$$\text{Standard error} = \sqrt{p(1-p)/n} = \sqrt{0.560(1-0.560)/2232} = 0.011.$$

With *n* = 2232, the sampling distribution has very little spread. Figure 6.22 portrays this sampling distribution, relative to the population distribution.

<table>
<tr><td>

RECALL

The *number* of successes has
mean = *np*
std. dev. = $\sqrt{np(1-p)}$

The *proportion* of successes has
mean = *p*
std.
error= $\sqrt{p(1-p)/n}$

</td></tr>
</table>

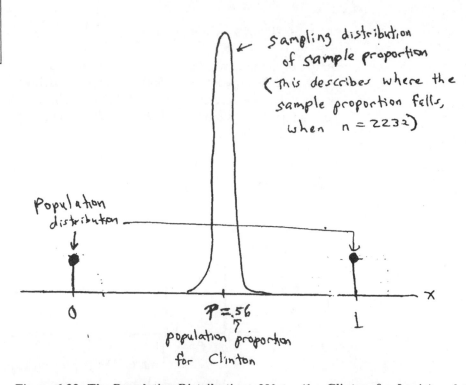

Figure 6.22: The Population Distribution of Votes (1 = Clinton, 0 = Lazio) and the Sampling Distribution of the Sample Proportion of Votes for Clinton. Both distributions have mean 0.560, namely the population proportion, but the sampling distribution is approximately normal. It looks quite different from the population distribution, and it gets narrower and more bell shaped as *n* increases. **Question**: Which of these two distributions resembles the data distribution?

416

Insight

The *population distribution* and the *data distribution* are not bell shaped. They are discrete, concentrated at only the two values 0 and 1. They describe the relative number of 0 and 1 values (votes for Lazio and votes for Clinton), in the population and in the sample.

By contrast, the *sampling distribution* describes the scatter of possible sample proportion values voting for Clinton when $n = 2232$. It looks very different from the population and data distributions. It is bell-shaped and much less spread out. You can think of the sampling distribution curve as describing what would happen if we could look at *every* possible sample of 2232 people from this population of voters, find the sample proportion for each such sample, and then plot the distribution of those sample proportions. In practice, of course, we'd never do this, but the curve helps us to visualize how much variation there can be among different exit polls in what they report.

♦

To practice this concept, try Exercise 6.78.

IN PRACTICE

Exit polls use a multistage type of random sampling, not simple random sampling. The standard error formulas in this chapter apply for simple random sampling. However, they provide good approximations for the standard errors with many multistage samples and are usually used for such sampling schemes also.

How Far Off Will a Sample Prediction Be?

In the Clinton-Lazio exit poll example, the sampling distribution of the sample proportion for $n = 2232$ is approximately a normal distribution. Its mean is 0.560 (the population proportion), and its standard error is 0.011. Based on the spread of the sampling distribution, what could we have concluded if we did *not* know the population proportion but did have the results of the exit poll and had to decide whether to predict a winner?

The sample proportion of the 2232 voting for Hillary Clinton was 0.557. Because the sampling distribution is approximately normal, we know that the sample proportion was very likely to fall within three standard errors of the population proportion, that is, within $3(0.011) = 0.033$ of p. Thus, using the sample proportion of 0.557, we would predict that the population proportion voting for Clinton was between about $0.557 - 0.033 = 0.524$ and $0.557 + 0.033 = 0.590$. Since all the values in the interval from 0.524 to 0.590 are above 0.50, this suggests that the population proportion was larger than 0.50. Based on results for these 2232 voters, we would predict that Hillary Clinton won the election.

This reasoning is similar to what we'll use in coming chapters to make inferences about population parameters. Fundamental to the logic is the central limit theorem result--that, with large random samples, sample proportions and sample means have approximately normal sampling distributions. *In fact, most statistics used to estimate population parameters have approximately normal sampling distributions, for large random samples. This is the primary reason for the key role of the normal distribution in statistics.*

We'll see in the next chapter that not all bell-shaped curves represent normal distributions. The normal distribution refers to a particular bell-shaped curve--The curve that comes from generating sampling distributions for many statistics with large random samples.

You may have noticed a slight flaw in the logic above about predicting the outcome of the Clinton--Lazio election. In practice, if we don't know the value of the parameter p, we also would not know the standard error, $\sqrt{p(1-p)/n}$. We'll have to approximate the standard error by substituting the sample proportion, for instance using $\sqrt{0.557(0.443)/2232} = 0.0105$. For most inference methods, we'll use standard error formulas that substitute sample values for unknown parameter values.

IN PRACTICE

Standard errors have exact values depending on parameter values, such as $\sqrt{p(1-p)/n}$ for a sample proportion and $\sigma/\sqrt{n}$ for a sample mean. In practice, these are unknown, so inference methods use standard errors that substitute sample values (such as sample proportions and sample standard deviations) in the exact formulas. These *estimated* standard errors are the numbers which we'll use in the next two chapters and that software reports for standard errors.

A final caveat: If the sampling is not random, the results we've used for sampling distributions no longer apply. For instance, in predicting the winner of an election, we should not merely use the first 3000 votes reported to a network, because they would not be a random sample of all voters. (They might all come from one county, for instance.)

Likewise, sometimes non-statistical factors adversely affect the accuracy of exit-poll predictions. A famous case when an exit-poll prediction was wrong occurred in the 2000 presidential election. About the time the polls closed in Florida, the major networks predicted that Al Gore would win that state based on exit polling. Later, a more complete count put George W. Bush slightly ahead, which resulted in Bush gaining enough electoral college votes to become president--even though he received half a million fewer votes than Gore nationwide. But, this was not necessarily the fault of the exit poll. In the 2000 presidential election, about 200,000 marked ballots in Florida were thrown out as invalid either because the tabulating machine could not read them or because the voter marked two names rather than one, often because of a confusing ballot design (see Example 6 in Chapter 3 and Exercise 2.86 in Chapter 2).

This source of error was so large that it negated the precision of a statistical prediction in a close election.

SECTION 6.6: PRACTICING THE BASICS

6.76 Household size: In 2000, according to the U.S. Census Bureau, the number of people in a household had a mean of 2.6 and a standard deviation of 1.5. This is based on census information for the population. Suppose the Census Bureau instead had estimated this mean using a random sample of 225 homes. Suppose the sample had a sample mean of 2.4 and standard deviation of 1.4.
a. Describe the center and spread of the population distribution.
b. Describe the center and spread of the data distribution.
c. Describe the center and spread (standard error) of the sampling distribution of the sample mean.

6.77 Sunshine City: Sunshine City was designed to attract retired people, and its current population of 50,000 residents has a mean age of 60 years and a standard deviation of 16 years. The distribution of ages is skewed to the left. A random sample of 100 residents of Sunshine City has $\bar{x}$ = 58.3 and s = 15.0.
a. Describe the center and spread of the population distribution.
b. Describe the center and spread of the data distribution. What shape does it probably have?
c. Describe the center and spread of the sampling distribution of the sample mean for n = 100. What shape does it have?
d. Explain why it would not be unusual to observe a person of age under 40 in Sunshine City, but it would be highly unusual to observe a *sample mean* under 40 for a random sample size of 100 people.

6.78 Gender distributions: At a university, 60% of the 7,400 students are female. The student newspaper reports results of a survey of a random sample of 50 students about various topics involving alcohol abuse, such as participation in binge drinking. They report that their sample contained 26 females.
a. Identify the population distribution of gender at this university.
b. Identify the data distribution of gender for this sample.
c. Identify the sampling distribution of the sample proportion of females in the sample. State its mean and standard error.

6.79 Predicting the Florida senate race: In the 2000 U.S. senatorial election in Florida, an exit poll indicated that 901 voted for Democrat Bill Nelson and 832 voted for Republican Bill McCollum. Would you have been willing to predict the winner? Use the reasoning shown in Example 21. That is, justify your answer by finding the sample proportion voting for Nelson and showing what you would expect for that proportion if the population proportion voting for him was p = 0.50, based on the standard error of the sample proportion for this size of sample when p = 0.50.

6.80 Did Schwarzenegger win?: In the 2003 California special election, of the candidates running to replace Gray Davis as governor, an exit poll indicated that 1081 voted for Arnold Schwarzenegger and 817 voted for Cruz Bustamonte. Would you predict a winner? Justify your answer by finding the sample proportion voting for Schwarzenegger and showing what you would expect for it if the population proportion voting for him was 0.50, based on the standard error of the sample proportion.

CHAPTER SUMMARY

- A **random variable** is a numerical measurement of the outcome of a random phenomenon. As with ordinary variables, random variables can be **discrete** (taking separate values) or **continuous** (taking an interval of values).

- A **probability distribution** specifies probabilities for the possible values of a random variable. Probability distributions have summary measures of the center and the spread, such as the mean μ and standard deviation σ. The mean (also called **expected value**) for a discrete random variable is

$$\mu = \Sigma \ x \, p(x),$$

where $p(x)$ is the probability of the outcome x and the sum is taken over all possible outcomes.

- The **binomial distribution** is the probability distribution of the discrete random variable that measures the number of successes X in n independent trials, with probability p of a success on a given trial. It has

$$p(x) = \frac{n!}{x!(n-x)!} p^{x} (1-p)^{n-x}, \quad x = 0, 1, 2, \ldots , n.$$

The mean is $\mu = np$ and standard deviation is $\sigma = \sqrt{np(1-p)}$.

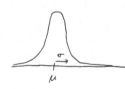

- The **normal distribution** is the probability distribution of a continuous random variable that has a symmetric bell-shaped graph specified by mean (μ) and standard deviation (σ) parameters. For any z, the probability falling within z standard deviations of μ is the same for every normal distribution.

- The **z-score** for an observation x equals

$$z = (x-\mu)/\sigma.$$

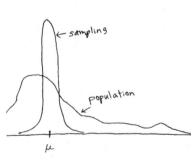

It measures the number of standard deviations that x falls from the mean μ. For a normal distribution, the z-scores have the **standard normal distribution**, which has mean = 0 and standard deviation = 1.

- A **sampling distribution** is a probability distribution of a sample statistic, such as the sample proportion or the sample mean. It specifies probabilities for the values of the statistic for all the possible random samples.

- The **sampling distribution of the sample mean** $\bar{x}$ is centered at the population mean μ. Its spread is described by its standard deviation, called the **standard error**. This is related to the standard deviation of the population by standard error $= \sigma / \sqrt{n}$. For binary data (0 or 1), with mean given by the population proportion p, the standard error simplifies to $\sqrt{p(1-p)/n}$.

- The **central limit theorem** states that for random samples of sufficiently large size (at least about 30 is usually enough), the sampling distribution of the sample mean is approximately normal. This holds *no matter what the shape of the population distribution*. It applies to sample proportions also, since the sample proportion is a sample mean of 0 and 1 values. In that case, the sampling distribution of the sample proportion is approximately normal whenever n is large enough that both np and $n(1-p)$ are at least 15. The bell-shaped appearance of the sampling distributions for most statistics is the main reason for the importance of the normal distribution.

SUMMARY OF NEW NOTATION IN CHAPTER 6

$p(x)$ Probability that a random variable takes value x

μ Mean of probability distribution or population distribution

σ Standard deviation of probability distribution or population distribution

p, $1-p$ Probabilities of the two possible outcomes of a binary variable

----

ANSWERS TO THE CHAPTER FIGURE QUESTIONS

Figure 6.1:   *The shape of the distribution is skewed to the left.*

Figure 6.2:   *The distribution with p = 0.1 is skewed to the right.*

Figure 6.3:   *The area under the curve from 0 to 15 should be shaded.*

Figure 6.4:   *The shape of the distribution is approximately symmetric and bell-shaped. The center of the distribution is about 65 inches. The spread as described by the standard deviation is about 3.5 inches.*

Figure 6.5:   *We evaluate 70 ± 3(4), giving as an interval for the men's heights from 58 inches to 82 inches.*

Figure 6.6:   *These probabilities are similar to the percentages stated for the Empirical Rule. The Empirical Rule states that for an approximately bell-shaped distribution, approximately 68% of the observations fall within one standard deviation of the mean, 95% within two standard deviations of the mean, and nearly all within three standard deviations of the mean.*

Figure 6.9:   For three standard deviations, the probability in one tail is 0.0015. By symmetry, this is also the probability in the other tail. The total tail area is 0.003, which subtracted from 1 gives an answer of 0.997.

Figure 6.10:   The 2nd percentile is the value located on the left side of the curve such that 2% of the distribution falls below it and 98% falls above.

Figure 6.11:   The SAT score with $z = -3$ is $500 - 3(100) = 200$. The SAT score with $z = 3$ is $500 + 3(100) = 800$.

Figure 6.12:   The values of -3 and 3.

Figure 6.13:   We can find the probabilities for each possible sample proportion value using the binomial distribution. The binomial formula is useful when p differs from 0.50. If p is a value other than 0.50, the 8 possible samples we found would not each have probability 1/8.

Figure 6.15:   The standard error is $\sqrt{p(1-p)/n} = \sqrt{(0.50)(0.50)/1000} = \sqrt{0.00025} = 0.0158$. Since the simulated sampling distribution is bell-shaped, the probability is close to 1.0 that the sample proportion falls within 3 standard errors of 0.50. Three standard errors equals $3(0.0158) = 0.047$, or approximately 0.05.

Figure 6.17:   The larger the sample size, the smaller the standard error and thus, the closer the sample mean tends to fall to the population mean.

Figure 6.18:   The probability distribution is highly skewed left. However, the sampling distribution of sample means from this probability distribution is bell-shaped. The mean of the sampling distribution is similar to the mean of the probability distribution, 4.5. The standard error of the sampling distribution is smaller than the standard deviation of the probability distribution. The standard error equals the standard deviation of the probability distribution divided by the square root of the sample size n, that is, $1.4/\sqrt{30} = 0.26$.

Figure 6.20:   The probability distribution and the sampling distribution do not have the same shape. The sampling distribution is bell shaped. The probability distribution is discrete, concentrated at two values, $0 with height of 0.9 and $5 with height of 0.1.

Figure 6.21:   The larger the sample size, the more we expect the data distribution to resemble the population distribution.

Figure 6.22:   The population distribution resembles the data distribution.

CHAPTER PROBLEMS: PRACTICING THE BASICS

6.81 Grandparents: Let $X =$ the number of living grandparents that a randomly selected adult American has. According to recent General Social Surveys, its probability distribution is approximately $p(0) = 0.71$, $p(1) = 0.15$, $p(2) = 0.09$, $p(3) = 0.03$, $p(4) = 0.02$.
a. Does this refer to a discrete, or a continuous, random variable? Why?
b. Show that the probabilities satisfy the two conditions for a probability distribution.

c. Find the mean of this probability distribution.

6.82 Playing the lottery in NJ: New Jersey has a "Pick 3" game in which you pick one of the 1000 3-digit numbers between 000 and 999, receiving $275 for a winning $1 bet and nothing otherwise.
a. Construct the probability distribution for the winnings.
b. Find the mean of the distribution.
c. Based on the mean in (b) and the $1 cost to play the game, on average, how much can you expect to lose each time you play this lottery?

6.83 More NJ Lottery: New Jersey has a "six-way combination" lottery game in which you pick three digits (each can be any of 0, 1, 2,..., 9) and you win if the digits the lottery picks are the same as yours in any of the six possible orders. For a dollar bet, a winner receives $45.50. Two possible strategies are (a) to pick three different digits, or (b) to pick the same digit three times. Let $X =$ winnings for a single bet.
a. Find the probability distribution of X and its mean for strategy (a).
b. Find the probability distribution of X and its mean for strategy (b).
c. Which is the better strategy? Why?

6.84 Are you risk averse? You need to choose between two alternative programs for dealing with the outbreak of a deadly disease. In program 1, 200 people are saved. In program 2, there is a 2/3 chance that no one is saved and a 1/3 chance that 600 people are saved.
a. Which program would you choose? Why? (There is no "correct" answer.)
b. Find the expected number of lives saved with each program.
c. Now you need to choose between program 3, in which 400 people will die, and program 4, in which there is a 1/3 chance that no one will die and a 2/3 chance that 600 people would die. Which would you choose? Find the expected number of deaths with each program.
d. Explain why programs 1 and 3 are similar, and 2 and 4 are similar. Were your choices consistent, or were you like most people and risk averse in (a), choosing program 1, and risk taking in (b), choosing program 4?

6.85 Flyers' insurance: An insurance company sells a policy to airline passengers for $1. If a flyer dies on a given flight (from a plane crash), the policy gives $100,000 to the chosen beneficiary. Otherwise, there is no return. Records show that a passenger has about a 1 in a million chance of dying on any given flight. You buy a policy for your next flight.
a. Specify the probability distribution of the amount of money the beneficiary makes from your policy.
b. Find the mean of the probability distribution in (a). Interpret.
c. Explain why the company is very likely to make money in the long run.

6.86 Bride's choice of surname: *Time* magazine reported (June 17, 2002) that 80% of brides take the surname of their new husband. Ann notes that of her 4 best friends who recently married, none kept her own name. If they had been a random sample of brides, how likely would this have been to happen?

6.87 Football wins: A football team has probability 0.50 of winning any particular game. In a season, it plays six conference games.

a. Using the binomial distribution, identify n and p, and find the probability that the team wins 3 of its conference games.
b. Find the probability that the team wins all six of its conference games.
c. What assumptions do the calculations in (a) and (b) require? Ordinarily, would you expect that a team would have exactly the same probability of winning in each game?

6.88 Weather: A weather forecaster states, "The chance of rain is 50% on Saturday and 50% again on Sunday. So, there's a 100% chance of rain sometime over the weekend." If whether or not it rains on Saturday is independent of whether or not it rains on Sunday, find the actual probability of rain *at least once* during the weekend.
a. Answer using methods from the previous chapter, such as by listing equally likely sample points or using the formula for the probability of a union of two events.
b. Answer using the binomial distribution.

6.89 Player choke?: A college basketball player has probability 0.50 of making any given shot, and her successive shots are independent. In overtime of an important game, she misses all three shots that she takes. Explain why this does not mean that she "choked," because it could happen by chance with reasonable probability.

6.90 Dating success: Based on past experience, Julio believes he has a 60% chance of success when he calls a woman and asks for a date.
a. State assumptions needed for the binomial distribution to apply to the number of times he is successful on his next five requests.
b. If he asks the same woman each of the five times, is it sensible to treat these requests as independent trials?
c. Under the binomial assumptions, state n and p, and find the probability he is successful exactly three times.

6.91 Revisiting Canada Lottery: In one Canadian lottery option, you bet on one of the million six-digit numbers between 000000 and 999999. For a $1 bet, you win $100,000 if you are correct. In playing n times, let X be the number of times you win.
a. Find the mean of the distribution of X, in terms of n.
b. How large an n is needed for you to expect to win once (that is, $np = 1$)?
c. If you play the number of times n that you determined in (b), show that your expected winnings is $100,000 but your expected profit is $-\$900,000$.

6.92 Female driving accidents: In a given year, the probability that an adult American female dies in a motor vehicle accident equals 0.0001 (*Statistical Abstract of the United States*, 2001}.
a. In a city having 1 million adult American females, state assumptions for a binomial distribution to apply to $X =$ the number of them who die in the next year from motor vehicle accidents. Identify n and p for that distribution.
b. If the binomial applies, find the mean and standard deviation of X.
c. Based on the Empirical Rule, find an interval of possible outcomes for the number of deaths that has probability about 0.95 of occurring.
d. Refer to the assumptions for the analysis in (a). Explain at least one way they may be violated.

6.93 Males drive more poorly?: Refer to the previous exercise. The probability of a motor vehicle death for adult American males is 0.0002. Repeat (b) and (c) for a city having 1 million of them, and compare results to those for females.

6.94 No royal family?: A recent poll of 1000 British adults asked, "If there were a referendum on the issue, would you favor Britain becoming a republic or remaining a monarchy?" (www.mori.com/polls). Suppose that the population proportion favoring the monarchy equals 0.70. (This was, in fact, the value for the sample proportion.) For a random sample of 1000 residents, let X denote the number in this category.
a. Find the mean of the probability distribution of X.
b. Find the standard deviation of the probability distribution of X.
c. What range of values falls within 3 standard deviations of the mean? Explain why it is unlikely that X will fall outside this interval.

6.95 Effect of sample size: Refer to the previous exercise.
a. Convert the limits in (c) to limits on the proportion scale. Interpret.
b. Answer (a)-(c) if $n = 250$ instead of 1000, and find the limits from (c) on the proportion scale.
c. Compare the proportion limits for $n - 250$ and 1000. How does n affect how near sample proportions fall to population proportions?

6.96 Normal probabilities: For a normal distribution, find the probability that an observation is:
a. Within 1.96 standard deviations of the mean.
b. More than 2.33 standard deviations from the mean.

6.97 z scores: Find the z score such that the interval within z standard deviations of the mean contains probability (a) 0.95, (b) 0.99 for a normal distribution. Sketch the two cases on a single graph.

6.98 z-score and tail probability:
a. Find the z-score for the number that is less than only 1% of the values of a normal distribution. Sketch a graph to show where this value is.
b. Find the z-scores corresponding to the (a) 90th, (b) 99th percentiles of a normal distribution.

6.99 Central probabilities and percentiles:
a. If z is the positive number such that the interval within z standard deviations of the mean contains 90% of a normal distribution, show that $\mu + z\sigma$ equals the 95th percentile. (Hint: Sketch a graph showing the regions to which the 90% and the 95% refer.)
b. If z is the positive number such that the interval within z standard deviations of the mean contains 80% of a normal distribution, show that $\mu - z\sigma$ equals the 10th percentile. Sketch a graph showing your solution.

6.100 Quartiles: If z is the positive number such that the interval within z standard deviations of the mean contains 50% of a normal distribution, then:
a. Explain why this value of z is about 0.67.

b. Explain why the first quartile equals $\mu - 0.67\sigma$ and the third quartile equals $\mu + 0.67\sigma$.

c. The interquartile range, IQR, relates to σ by $IQR = 2 \times 0.67\sigma$. Explain why.

6.101 Cholesterol: The American Heart Association reports that a total cholesterol score of 240 or higher represents high risk of heart disease. A study of post-menopausal women (*Clin. Drug. Invest.*, 2000, vol. 20, pp. 207-214) reported a mean of 220 and standard deviation of 40. If the total cholesterol scores have a normal distribution, what proportion of the women fall in the high-risk category?

6.102 Rental rates: Monthly rental rates for one-bedroom unfurnished apartments in Eugene, Oregon, have approximately a normal distribution with a mean of $500 and a standard deviation of $150.
a. What proportion of the rental rates is at least $700 per month?
b. What proportion of the rental rates is less than $400 per month?
c. What proportion of the rental rates falls between $400 and $700 per month?

6.103 Female height: For the normal height distribution of adult females in North America ($\mu = 65.0$ inches, $\sigma = 3.5$ inches), find the proportion who are
a. under five feet (60 inches).
b. over six feet (72 inches).
c. between 60 and 70 inches.

6.104 Male height: Repeat the previous exercise using the normal height distribution of adult males in North America ($\mu = 70.0$, $\sigma = 4.0$).

6.105 Gestation times: For 5459 pregnant women using Aarhus University Hospital in Denmark in a two-year period who reported information on length of gestation until birth, the mean was 281.9 days, with standard deviation 11.4 days. A baby is classified as premature if the gestation time is 258 days or less (*British Medical Journal*, July 24, 1993, p. 234).
a. If gestation times are normally distributed, what's the proportion of babies who are born prematurely?
b. The actual proportion born prematurely during this period was 0.036. Based on this information, explain why the distribution of gestation time is not normal. How would you expect it to differ from normal?

6.106 Energy use: An energy study in Gainesville, Florida, found that household use of electricity has a mean of 1200 kWh (kilowatt-hours) and a standard deviation of 500 kWh in a typical summer month. In June 2002, to encourage conservation, Gainesville Regional Utilities announced that the base cost would be $0.05 per kWh up to 1500 kWh of use and then $0.06 above that.
a. If the distribution were approximately normal, about what proportion of users would have their summer bill computed entirely based on the lower rate?
b. Suppose that actually for summer months, $\mu = 1000$ and $\sigma = 1000$. Do you think that this distribution of use is truly normal? Why or why not?

6.107 Winter energy use: Refer to the previous exercise. Electricity use is much less in the winter in Florida. To make electricity use values comparable from different seasons, in a given month each home's use is converted to a z-score. For each home with a z-score greater than 1.5 (electricity use more than 1.5 standard deviations above the mean), their bill contains a note suggesting a reduction in electricity use, to conserve resources. If the distribution of electricity use is normal, what proportion of households receives this note?

6.108 Gas use: Suppose that weekly use of gasoline for motor vehicle travel by adults in North America has approximately a normal distribution with a mean of 20 gallons and a standard deviation of 6 gallons. What proportion of adults uses more than 35 gallons per week?

6.109 Global warming: Refer to the previous exercise. Many people who worry about global warming believe that Americans should pay more attention to energy conservation. Assuming that the standard deviation and the normal shape are unchanged, to what level must the mean reduce so that 20 gallons per week is the third quartile rather than the mean?

6.110 Fast-food profits: Mac's fast-food restaurant finds that its daily profits have a normal distribution with mean $140 and standard deviation $80.
a. Find the probability that restaurant loses money on a given day (that is, daily profit less than 0).
b. Find the probability that the restaurant makes money for the next seven days in a row. What assumptions must you make for this calculation to be valid?

6.111 Metric height: A Dutch researcher reads that male height in the Netherlands has a normal distribution with $\mu = 72.0$ and $\sigma = 4.0$ inches. She prefers to convert this to the metric scale (1 inch = 2.54 centimeters). The mean and standard deviation then have the same conversion factor.
u. In centimeters, would you expect the distribution still to be normal? Explain.
b. Find the mean and standard deviation in centimeters. (Hint: What does 72.0 inches equal in centimeters?)
c. Find the probability that height exceeds 200 centimeters.

6.112 Manufacturing tennis balls: According to the rules of tennis, a tennis ball is supposed to weigh between 56.7 grams (2 ounces) and 58.5 grams (2 1/16 ounces). A machine for manufacturing tennis balls produces balls with a mean of 57.6 grams and a standard deviation of 0.3 grams, when it is operating correctly. Suppose that the distribution of the weights is normal.
a. If the machine is operating properly, find the probability that a ball manufactured with this machine satisfies the rules.
b. After the machine has been used for a year, the process still has a mean of 57.6, but because of wear on certain parts the standard deviation increases to 0.6 grams. Find the probability that a manufactured ball satisfies the rules.

6.113 Happy with euros?: *Eurobarometer* recently took a poll to gauge the approval of adults for the European currency. Of the 1000 people sampled in a particular country in the European Union, consider the sample *proportion* of people who indicate approval of the euro.
a. Find the mean and standard error of the sampling distribution for this sample proportion, if the population proportion equals 0.67.
b. What shape would you expect this sampling distribution to have? Explain.
c. Within what limits would the sample proportion almost certainly fall? Explain.

6.114 Fear of terrorism: According to the December 2001 Eurobarometer, 86% of Europeans personally fear terrorism. A survey is planned of Canadians to see if similar results occur there. Suppose that 86% of all Canadians also fear terrorism. If the survey uses a random sample of 400 Canadians, describe the sampling distribution of the sample proportion fearing terrorism in terms of the
a. mean.
b. standard error.
c. shape.

6.115 Range for proportions: Refer to the previous exercise. Give an interval such that the probability is about 0.95 that the sample proportion falls in that interval.

6.116 Weapons of mass destruction: Repeat the previous two exercises using results of the 400 Canadians using results of the 2001 Eurobarometer that indicated that 79% of Europeans fear the proliferation of weapons of mass destruction.

6.117 Terrorism: In June 2002, the Gallup organization asked a sample of 1000 Americans, "What do you think is the most important problem facing this country today?" A proportion of 0.46 answered terrorism. The Gallup report indicated that this had a margin of error of plus or minus 0.03 as an estimate of the population proportion. Usually the margin of error is defined as a distance of two standard errors on each side of the sample proportion, using the sample proportion in place of the unknown population proportion in the standard error formula. Show how Gallup got 0.03 as the margin of error (which they expressed as plus or minus 3%).

6.118 Governor exit poll: In a governor's race, the winner receives a population proportion of $p = 0.70$ of the votes. On the day of the election, an exit poll is taken of 1000 randomly selected voters. Consider the sampling distribution of the sample proportion of these 1000 voters who voted for the winner.
a. Find the mean and standard error of the sampling distribution.
b. What do you expect for the shape of the sampling distribution? Why?
c. If the population proportion is truly $p = 0.70$, would it be surprising if the exit poll had a sample proportion of 0.60? Justify your answer by using the central limit theorem to approximate the probability of a sample proportion of 0.60 or less.

6.119 Exit poll: In the New York exit poll of Example 22, it is possible that all 2232 voters sampled happened to be Hillary Clinton supporters. Investigate how surprising this would be, if actually 56% of the population voted for Clinton, by
a. Finding the probability that all 2232 people voted for Clinton. (*Hint*: Use the binomial.)
b. Finding the number of standard errors that the sample proportion of 1.0 falls from the population proportion of 0.56.

6.120 Physicians' assistants: The 2000 AAPA (www.aapa.org) survey of the population of 1434 physicians assistants who were working full-time after graduating in 1999 reported a mean annual income of $58,000 and standard deviation of $11,000.
a. Suppose the AAPA had randomly sampled 100 physicians' assistants instead of collecting data for all 1434 of them. Describe the mean, standard error, and shape of the sampling distribution of the sample mean.

b. Using this sampling distribution, find the z-score for a sample mean of $60,000.
c. Using (b), find the probability that the sample mean would fall within $2000 of the population mean.

6.121 More physicians' assistants: Repeat (a)-(c) of the previous exercise, using $n = 50$. Summarize the effect of the sample size on the probability that the sample mean falls near the population mean.

6.122 ATM withdrawals: An executive in an Australian savings bank decides to estimate the mean amount of money withdrawn in ATM transactions. From past experience, she believes that $50 (Australian) is a reasonable guess for the standard deviation of the distribution of withdrawals. She would like her sample mean to be within $10 of the population mean. Estimate the probability that this happens if:
a. She randomly samples 25 withdrawals. (Hint: How many standard errors is $10?)
b. She randomly samples 100 withdrawals.

6.123 PDI: The scores on the Psychomotor Development Index (PDI), a scale of infant development, are approximately normally distributed with mean 100 and standard deviation 15. An infant is selected at random.
a. Find the probability that the infant's PDI score exceeds 103.
b. Find the probability that PDI is between 97 and 103.
c. Find the z-score for a PDI value of 90.

6.124 Mean PDI: Refer to the previous exercise. A study uses a random sample of 225 infants. Using the sampling distribution of the sample mean PDI:
a. Find the probability that the sample mean exceeds 103. (Hint: You need to find the standard error of the sampling distribution of the sample mean.)
b. Find the probability that the sample mean falls between 97 and 103.
c. Find the z-score from the sampling distribution corresponding to a sample mean of 90. A PDI value of 90 is not surprising, but would you be surprised to observe a sample mean PDI score of 90? Why?
d. Compare the results of parts (a)-(c) with those in the previous exercise, and interpret the differences.

6.125 PDI and effect of n: Refer to the previous exercise.
a. Repeat parts (a)-(c) for a random sample size of $n = 25$ and compare the results to those for $n = 225$.
b. Sketch the population distribution for PDI. Superimpose a sketch of the sampling distribution of the sample mean for a random sample of 25 adults and a sketch of the sampling distribution of the sample mean for a random sample of 225 adults.

6.126 Ohio Lottery: In the "Pick 3" option in the Ohio lottery, you pick one of the 1000 3-digit numbers between 000 and 999, and the lottery selects a 3-digit number at random. With a bet of $1, you win $500 if your number is selected and nothing otherwise.
a. Construct the probability distribution of your winnings.
b. Find the mean of the probability distribution of your winnings.
c. If you play in two different games, find the probability that you win both times. Compare this to the probability of being hit by lightning in your lifetime (which an Associated Press story (4/16/02) estimated to be 1/9100) and the probability of

drowning in your bath in the next year (which the National Safety Council (NSC) estimates to be 1 in a million).

6.127 Mean winnings in Ohio: Refer to the previous exercise. When you play "Pick 3" twice:
a. Find the probability distribution of your total winnings.
b. Construct the sampling distribution of your sample mean winnings.
c. Find the mean of the sampling distribution. How does it compare to the mean of the probability distribution in Exercise 6.126(b) for a single play? Why does this happen?

6.128 Number of sex partners: According to recent General Social Surveys, in the United States, the distribution for adults of X = number of sex partners in the past 12 months has a population mean of about 1.0 and a standard deviation of about 1.0.
a) Does X have a normal distribution? Explain.
b) For a random sample of 100 adults, describe the sampling distribution of $\bar{x}$ and give its mean and standard error. What is the effect of X not having a normal distribution?
c) Refer to (b). Report an interval within which the sample mean would have probability of about 0.95 of falling.

6.129 Ranch size: To estimate the mean acreage of ranches in Alberta, Canada, a researcher plans to obtain the acreage for a random sample of 64 farms. Results from an earlier study suggest that 800 acres is a reasonable guess for the standard deviation of ranch size.
a. Find the probability that the sample mean acreage falls within 100 acres of the population mean acreage. (Hint: How many standard errors from the mean is the value that is 100 units away?)
b. If the researcher can increase n above 64, will the probability that the sample mean falls within 100 acres of the population mean increase or decrease? Why?

◆◆ **6.130 Using control charts to assess quality**: In many industrial production processes, measurements are made periodically on critical characteristics to ensure that the process is operating properly. Observations vary from item to item being produced, perhaps reflecting variability in material used in the process and/or variability in the way a person operates machinery used in the process. There is usually a target mean for the observations, which represents the long-run mean of the observations when the process is operating properly. There is also a target standard deviation for how observations should vary around that mean if the process is operating properly. A **control chart** is a method for plotting data collected over time to monitor whether the process is operating within the limits of expected variation. A control chart that plots *sample means* over time is called an $\bar{x}$ -chart. As shown in the margin, the horizontal axis is the time scale and the vertical axis shows possible sample mean values. The horizontal line in the middle of the chart shows the target for the true mean. The upper and lower lines are called the **upper control limit** and **lower control limit**, denoted by **UCL** and **LCL**. These are usually drawn 3 standard errors above and below the target value. The region between the LCL and UCL contains the values that theory predicts for the sample mean when the process is in control. When a sample mean falls above the UCL or below the LCL, it indicates that something may have gone wrong in the production process.

$\bar{x}$ **-chart**

430

a. Walter Shewhart invented this method in 1924 at Bell Labs. He suggested using 3 standard errors in setting the UCL and LCL to achieve a balance between having the chart fail to diagnose a problem and having it indicate a problem when none actually existed. If the process is working properly ("in statistical control") and if n is large enough that $\bar{x}$ has approximately a normal distribution, what is the probability that it indicates a problem when none exists? (That is, what's the probability a sample mean will be at least three standard errors from the target, when that target is the true mean?)

b. What would the probability of falsely indicating a problem be if we used 2 standard errors instead for the UCL and LCL?

6.131 Too little or too much Coke?: Refer to the previous exercise. When a machine for dispensing a cola drink into bottles is in statistical control, the amount dispensed has a mean of 500 ml (milliliters) and a standard deviation of 4 ml.

a. In constructing a control chart to monitor this process with periodic samples of size 4, how would you select the target line and the upper and lower control limits?

b. If the process actually deteriorates and operates with a mean of 491 ml and a standard deviation of 6 ml, what is the probability that the next value plotted on the control chart indicates a problem with the process, falling more than three standard errors from the target? What do you assume in making this calculation?

6.132 Lack of control: Exercise 6.130 introduced control charts. When about 9 sample means in a row fall on the same side of the target for the mean in a control chart, this is an indication of a potential problem, such as a shift up or a shift down in the true mean relative to the target value. If the process is actually in control and has a normal distribution around that mean, what is the probability that the next 9 sample means in a row would (a) all fall above the mean, (b) all fall above or all fall below the mean? (Hint: Use the binomial distribution, treating the successive observations as independent.)

CHAPTER PROBLEMS: CONCEPTS AND INVESTIGATIONS

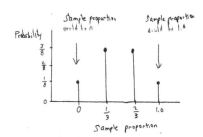

6.133 Simulating pizza preference: Using the *sampling distribution* applet (www.prenhall.com/TBA),

a. For Example 15, in which $n = 3$ people indicate their preferred pizza, simulate the sampling distribution of the count preferring pizza A when $p = 0.50$, by taking 1000 samples of size 3. Sketch a plot of the sampling distribution that you generate.

b. Simulate the sampling distribution of sample proportions for Example 15 with $n = 3$ subjects, by taking 1000 samples of size 3. Sketch a plot of the sampling distribution that you generate. Compare this generated sampling distribution to the sampling distribution of counts generated in part (a)? Comment on the shape, means, and standard deviations. Does your generated sampling distribution for sample proportions look roughly like Figure 6.13?

c. Repeat parts (a) and (b) using a sample size $n = 3$ and $p = 0.60$. Compare the sampling distribution that you generated to the sampling distributions generated in parts (a) and (b). Comment on similarities and differences.

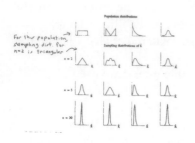

For this population, sampling dist. for n=2 is triangular

🖥 **6.134 CLT through simulation:** Use the *sampling distribution* applet (www.prenhall.com/?TBA) to see how

a. The first population distribution shown in Figure 6.19 (reproduced in the margin) has a more nearly normal sampling distribution for the mean as *n* increases and how the spread of that sampling distribution changes as *n* increases. You will need to select *uniform* for the parent population and select your range of values for this population distribution. Use the applet to create sampling distributions for the sample mean using different sample sizes *n*. Carry out each simulation with a different sample size taking $N = 10000$ repeated samples. Write a short report in which you describe the investigation you carried out, and the results you obtained.

b. Repeat part (a) using the fourth population distribution shown in Figure 6.19. This is a normal distribution. You will need to select your desired mean and standard deviation for the parent population. In your report, compare the effect of the sample size *n* on the sampling distribution shape based on whether the parent population is a normal distribution or a non normal distribution (such as the uniform distribution).

6.135 Family size in Gaza: The Palestinian Central Bureau of Statistics (www.pcbs.org) asked mothers of age 20-24 about the ideal number of children. For those living on the Gaza Strip, the probability distribution is approximately $p(1) = 0.01$, $p(2) = 0.10$, $p(3) = 0.09$, $p(4) = 0.31$, $p(5) = 0.19$, and $p(6 \text{ or more}) = 0.29$. Because the last category is open-ended, it is not possible to calculate the mean exactly. Explain why you can find the *median* of the distribution, and find it.

6.136 Longest streak made: In basketball, when the probability of making a free throw is 0.50 and successive shots are independent, the probability distribution of the longest streak of shots made has $\mu = 4$ for 25 shots, $\mu = 5$ for 50 shots, $\mu = 6$ for 100 shots, and $\mu = 7$ for 200 shots.

a. How does the mean change for each doubling of the number of shots taken? Interpret.

b. What would you expect for the longest number of consecutive shots made in a sequence of (i) 400 shots, (ii) 3200 shots?

c. For a long sequence of shots, the probability distribution of the longest streak is approximately bell shaped and σ equals approximately 1.9, no matter how long the sequence (Schilling, 1990). Explain why the longest number of consecutive shots made is quite likely to fall within about 4 of the mean, whether we consider 200 shots, 3200 shots, or a million shots.

6.137 Stock market randomness: Based on the previous exercise and what you have learned in this and the previous chapter, if you are a serious investor, explain why you should not get too excited if sometime in the next year the stock market goes up for seven days in a row.

6.138 Airline overbooking: For the Boston to Chicago route, an airline flies a Boeing 737-800 with 170 seats. Based on past experience, the airline finds that people who purchase a ticket for this flight have 0.80 probability of showing up for the flight. They routinely sell 190 tickets for the flight, claiming it is unlikely that more than 170 people show up to fly.

a. Provide statistical reasoning they could use for this decision.

b. Describe a situation in which the assumptions on which their reasoning is based may not be satisfied.

6.139 Babies in China: The sex distribution of new babies is close to 50% each, with the percentage of males usually being just slightly higher. In China in recent years, the percentage of female births seems to have dropped, a combination of policy limiting family size, the desire of most families to have at least one male child, the possibility of determining sex well before birth, and the availability of abortion. Suppose that historically 49% of births in China were female but birth records in a particular town for the past year show 800 females born and 1200 males. Conduct an investigation to determine if the current probability of a female birth in this town is less than 0.49. (*Hint*: Use the mean and standard deviation of the probability distribution of what you would observe with 2000 births if it were still 0.49.)

6.140 Cancer deaths: According to the National Center of Health Statistics (NCHS), 22% of all deaths are due to cancer. For various reasons (genetic, socioeconomic, environmental), some individuals may have a greater tendency for this to happen. Of your four closest relatives who have died, suppose all died from cancer. Does this suggest that the probability may be higher than 0.22 for members of your family? Provide supporting evidence.

6.141 Toxic waste: Residents of a particular city worry that a waste dump next to the city's water supply may elevate the chance of cancer for its residents. Of the past 80 deaths of residents, 50 were due to cancer. Investigate whether this number is unusually high or if it could it be explained by random variation, if actually the probability of a death being due to cancer was 0.22 in this town. Answer by finding the mean and standard deviation of the probability distribution of the number of deaths due to cancer and the z-score of the observed result if actually the probability equals 0.22. Explain your reasoning, stating all assumptions.

6.142 Come out ahead in lottery: Example 20 used the central limit theorem to approximate the probability of coming out ahead if you play a lottery 52 times. Each game cost $1, with winnings equal to $5 with probability 0.1 and $0 with probability 0.9.
a. Explain why you would have to win at least 11 times to come out ahead after playing 52 times.
b. Explain why the exact probability in (a) equals the binomial probability of at least 11 successes in 52 trials, with success probability 0.1. (This equals 0.013.)

6.143 Shapes of distributions:
a. With random sampling, does the shape of the data distribution tend to resemble more closely the sampling distribution or the population distribution? Explain.
b. Explain carefully the difference between a *data distribution* of sample observations and the *sampling distribution* of the sample proportion for a binary variable that can take values only of 0 and 1.

6.144 Smoking and illicit drug use: The 2000 National Household Survey on Drug Abuse (NHSDA) reported that for persons aged 18 to 25, the proportion who reported use of illicit drugs within the past month was 0.31 for those who smoke cigarettes daily and 0.07 for those who were nonsmokers.
a. If the sample sizes were equal for smokers and nonsmokers, which sample proportion would have a smaller standard error? (Hint: According to the values given, which has the smaller numerator for the standard error formula?)

b. For a given sample size, the standard error of a sample proportion is smaller when a proportion is near 0 or 1 than when it is near 0.50. Explain why this indicates that it is harder to estimate accurately a population proportion when it is near 0.50 than when it is near 0 or 1.

6.145 True or False: As the sample size increases, the standard error of the sampling distribution of $\bar{x}$ increases. Explain your answer.

Multiple choice: Select the correct response in the following multiple-choice questions (6.146 – 6.151).

6.146 Guess answers: A question has four possible answers, only one of which is correct. You randomly guess the correct response. With 20 such questions, the distribution of the number of correct answers
a. is binomial with $n = 20$ and $p = 0.25$.
b. is binomial with $n = 20$ and $p = 0.50$.
c. has mean equal to 10.
d. has probability $(.75)^{20}$ that all 20 guesses are incorrect.

6.147 Terrorist coincidence?: On 9/11/2002, the first anniversary of the terrorist destruction of the World Trade Center in New York city, the winning three-digit New York State Lottery number came up 9-1-1. The probability of this happening was
a. 1/1000.
b. $(1/1000)^2 = 0.000001$.
c. 1 in a billion.
d. 3/10.

6.148 Table probabilities: The numbers in the body of Table A provide cumulative probabilities:
a. For the normal distribution.
b. For any probability distribution.
c. For any symmetric probability distribution.
d. For any symmetric probability distributions of a continuous random variable.

6.149 Standard error: Which of the following is *not* correct? The standard error of a statistic describes:
a. The standard deviation of the sampling distribution of that statistic.
b. The standard deviation of the sample data measurements.
c. How close that statistic falls to the parameter that it estimates.
d. The variability in the values of the statistic for repeated random samples of size n.

6.150 Mean and standard error: A population distribution has mean 50 and standard deviation 20. For a random sample of size 100, the sampling distribution of the sample mean has
a. mean 50 and standard error 2.
b. mean 50 and standard error 0.20.
c. mean 50/100 and standard error 20/100.
d. mean and standard error that are unknown unless we know the exact shape of the population distribution.

6.151 CLT: The central limit theorem implies:
a. All variables have approximately bell-shaped data distributions if a random sample contains at least about 30 observations.
b. Population distributions are normal whenever the population size is large.
c. For sufficiently large random samples, the sampling distribution of $\bar{x}$ is approximately normal, regardless of the shape of the population distribution.
d. The sampling distribution of the sample mean looks more like the population distribution as the sample size increases.

♦♦ **6.152 SAT and ethnic groups:** Lake Wobegon Junior College admits students only if they score above 800 on the sum of their verbal and math SAT scores. Applicants from ethnic group A have a mean of 1000 and a standard deviation of 200 on this test, and applicants from ethnic group B have a mean of 900 and a standard deviation of 200. Both distributions are approximately normal, and both ethnic groups have the same size.
a. Find the proportion not admitted for each ethnic group.
b. Of the students who are not admitted, what proportion is from group B?
c. A state legislator proposes that the college lower the cutoff point for admission to 600, thinking that of the students who are not admitted, the proportion from ethnic group B would decrease. If this policy is implemented, determine the effect on the answer to (b), and comment.

♦♦ **6.153 Sample = population**: Let X = GPA for students in your school.
a. What would the sampling distribution of the sample mean look like if you sampled everyone in the population of students in the school, so the sample size equals the population size? (Hint: The sample mean then equals the population mean.)
b. How does the sampling distribution compare to the population distribution if we take a sample of size $n = 1$?

♦♦ **6.154 Binomial mean and standard deviation**: Let X denote a binomial random variable with $n = 1$ and $p = 0.50$, that is, $x = 1$ and $x = 0$ with probability 0.50 each.
a. Using the formula $\mu = \Sigma\ x\,p(x)$, show that $\mu = p = 0.5$.
b. If $x = 1$ with probability p and $x = 0$ with probability $1 - p$, show that $\mu = p$.

♦♦ **6.155 Standard deviation of a discrete probability distribution:** The **variance** of a probability distribution of a random variable is a weighted average of its squared distances from the mean μ. For discrete random variables, it equals

$$\sigma^2 = \sum (x-\mu)^2\, p(x).$$

Multiply each possible squared deviation $(x-\mu)^2$ by its probability $p(x)$, and then add. The **standard deviation** σ is the positive square root of the variance.
a. Suppose $x = 1$ or 0 with probability 0.50 each. Since $(x - \mu)^2 = (0 - 0.5)^2 = 0.25$ when $x = 0$ and $(1 - 0.5)^2 = 0.25$ when $x = 1$, show that $\sigma^2 = \Sigma (x - \mu)^2 p(x) = 0.25$ and $\sigma = 0.5$.

b. When $x = 1$ with probability p and $x = 0$ with probability $(1-p)$, so that $\mu = p$, show that $\sigma = \sqrt{p(1-p)}$, the special case of the binomial formula with $n = 1$.

c. With n trials, using the formula $\sigma/\sqrt{n}$ for a standard error, explain why the standard error of a sample proportion equals $\sqrt{p(1-p)/n}$.

◆◆ **6.156 Binomial probabilities**: Justify the $p^x(1\text{-}p)^{n-x}$ part of the binomial formula for the probability $p(x)$ of a particular sequence with x successes, using what you learned in Section 5.3 about probabilities for intersections of independent events.

◆◆ **6.157 Waiting time for doubles**: Most discrete random variables take on a finite number of values, such as 1, ..., 6 for the higher number on two dice. In *Monopoly*, rolling doubles (the same number on each die) guarantees another turn. Let $X =$ the number of rolls of two dice necessary until doubles first appears. The possible values for this discrete random variable (called the **geometric**) are 1, 2, 3, 4, 5, 6, 7, and so on, still separate values but now an infinite number of them.
a. Using intersections of independent events, explain why $p(1) = 1/6$ and $p(2) = (5/6)(1/6)$,
b. Find $p(3)$, $p(4)$, and explain how to find $p(x)$ for an arbitrary positive integer x.

◆◆ **6.158 Finite populations**: The formula $\sigma/\sqrt{n}$ for the standard error of $\bar{x}$ actually treats the population size as *infinitely* large relative to the sample size n. The formula for a *finite* population size N is

$$\text{Standard error} = \sqrt{\frac{N-n}{N-1}}\frac{\sigma}{\sqrt{n}}$$

The term $\sqrt{(N-n)/(N-1)}$ is called the **finite population correction**.
a. When $n = 300$ students are selected from a college student body of size $N = 30,000$, show that the standard error equals $0.995\,\sigma/\sqrt{n}$.
b. If $n = N$ (that is, we sample the entire population), show that the standard error equals 0. In other words, no sampling error occurs, since $\bar{x} = \mu$ in that case.

CHAPTER PROBLEMS: CLASS EXPLORATIONS

⌨ **6.159 Best of 7 games**: In professional baseball, basketball, and hockey in North America, the final two teams in the playoffs play a "best of seven" series of games. The first team to win 4 games is the champion. Use simulation with the *random number* applet at www.prenhall.com/TBA to approximate the probability distribution of the number of games needed to declare a champion, when (a) the teams are evenly matched, (b) the better team has probability 0.90 of winning any particular game. In each case, conduct 10 simulations. Then combine results with other students in your class and estimate the mean number of games needed in each case. In which case does the series tend to be shorter?

6.160 Simulate a sampling distribution: The table provides the ages of all 50 heads of households in a small Nova Scotian fishing village. The distribution of these ages is characterized by $\mu = 47.18$ and $\sigma = 14.74$.

NAME	AGE	NAME	AGE	NAME	AGE	NAME	AGE
Alexander	50	Griffith	66	McTell	49	Staines	33
Bell	45	Grosvenor	51	MacLeod	30	Stewart	36
Bok	23	Ian	57	Mayo	28	Stewart	25
Bok	28	Jansch	40	McNeil	31	Thames	29
Clancy	67	Kagan	36	Mitchell	45	Thomas	57
Cochran	62	Lavin	38	Morrison	43	Todd	39
Fairchild	41	Lunny	81	Muir	43	Trickett	50
Finney	68	MacColl	27	Muir	54	Trickett	64
Fisher	37	McCusker	37	Oban	62	Tyson	76
Fraser	60	McCusker	56	Reid	67	Watson	63
Fricker	41	McDonald	71	Renbourn	48	Young	29
Gaughan	70	McDonald	39	Rogers	32		
Graham	47	McTell	46	Rusby	42		

a. Each student should construct a graphical display (stem-and-leaf plot, dotplot, or histogram) of the population distribution of the ages.

b. Each student should select nine random numbers between 01 and 50, with replacement. Using these numbers, each student should sample nine heads of households and find their sample mean age. Collect all sample mean ages. Using technology, construct a graph (using the same graphical display as in part (a)) of the simulated sampling distribution of the $\overline{x}$-values for all the student samples. Compare it to the distribution in (a).

c. Find the mean of the $\overline{x}$-values generated in (b). How does this value compare to the value the class would expect for this mean in a long run of repeated samples of size 9?

d. Find the standard deviation of the $\overline{x}$-values generated in (b). How does this value compare to the value the class would expect for this standard deviation in a long run of repeated samples of size 9?

6.161 Coin-tossing distributions: For a single toss of a balanced coin, let $x = 1$ for a head and $x = 0$ for a tail.

a. Construct the probability distribution for x, and calculate its mean.

b. The coin is flipped ten times, yielding six heads and four tails. Construct the data distribution.

c. Each student in the class should flip a coin ten times and find the proportion of heads. Collect the sample proportion of heads from each student. Summarize the simulated sampling distribution by constructing a plot of all the proportions obtained by the students. Describe the shape and spread of the sampling distribution compared to the distributions in (a) and (b). Compute the mean and standard deviation of the sample proportion values.

d. If each student performed the experiment in (c) an indefinitely large number of times, what values would we expect to get for the (i) mean, (ii) standard error of the sample proportions?

6.162 Sample vs. sampling: Each student should bring ten coins to class. Make the following observations for each coin -- its age, and the difference between the current year and the year on the coin.

a. Using all the students' observations, the class should construct a histogram of the sample ages. What is its shape?

b. Now each student should find the mean for that student's ten coins, and the class should plot the means of all the students. What type of distribution is this, and how does it compare to the one in (a)?

c. What concepts does this exercise illustrate?

ACTIVITY 2: WHAT "HOT STREAKS" SHOULD WE EXPECT IN BASKETBALL?

In basketball games, TV commentators and media reporters often describe a player as being "hot" if he or she makes several shots in a row. Yet statisticians have shown that for players at the professional level, the frequency and length of streaks of good (or poor) shooting are similar to what we'd expect if the success of the shots were random, with the outcome of a particular shot being independent of previous shots.[11]

Shaquille O'Neal is one of the top players in the National Basketball Association, but he is a poor free-throw shooter. He makes about 50% of his free-throw attempts over the course of a season. Let's suppose that he has probability 0.50 of making any particular free throw that he takes in a game. Suppose also whether or not he makes any particular free throw is independent of his previous ones. He takes 20 free throws during the game. Let X denote the longest streak he makes in a row during the game. This is a discrete random variable, taking the possible values $x = 0, 1, 2, 3, \ldots, 20$. Here, $x = 0$ if he makes none of the 20, $x = 1$ if he never makes more than 1 in a row, $x = 2$ if his longest streak is 2 in a row, and so forth.

What is the probability distribution of X? This is difficult to find using probability rules, so let's approximate it by simulating 10 games with $n = 20$ free throws in each, using either coin flipping or the *random number* applet at www.prenhall.com/TBA. Representing each of O'Neal's shots by the flip of a coin, we treat a head as making a shot and a tail as missing it. We simulate O'Neal's 20 free throws by flipping a coin 20 times. We did this and got

TTHHHTHTTTHTHHHTHTTHT

This corresponds to missing shots 1 and 2, then making shots 3, 4, and 5, missing shot 6, and so forth. The simulated value of X = the longest streak of shots made is the longest sequence of heads in a row. In the sequence just shown, the longest streak made is $x = 3$, corresponding to the heads on flips 3, 4, and 5.

You would do the 20 coin flips 10 separate times to simulate what would happen in 10 games with 20 free throws in each. This would be a bit tedious. Instead, we suggest that you use your own judgment to write down quickly 10 sets of 20 H and T

---

[11] For instance, see articles by A. Tversky and T. Gilovich, *Chance*, vol. 2 (1989), pp. 16-21 and 31-34.

symbols (as shown above for one set) on a sheet of paper to reflect the sort of results you would expect for 10 games with 20 free throws in each. After doing this, find the 10 values of X = longest streak of H's, using each set of 20 symbols. Do you think your instructor would be able to look at your 200 H's and T's and figure out that you did not actually flip the coin?

A more valid way to do the simulation uses software, such as the *random number* applet at www.prenhall.com/TBA. For each digit, we could let 0-4 represent making a shot (H) and 5-9 as missing it (T). Use this to simulate 10 games with n = 20 free throws in each. When we did this, we got the following results for the first 5 games:

Coin flips	x
HTHHTHTTTHTHHHHTTHTT	4
HTTTHTHHTHHHTTHTHHTH	3
TTHTHHHTHTTHTTTTHTHH	3
THTTTHHTTHHTHTTTHTHH	2
HHTHTTTHTHHHHTTHTTTH	4

In practice, to get accurate results you have to simulate a *huge* number of games (at least a thousand), for each set of 20 free throws observing the longest streak of successes in a row. Your results for the probability distribution would then approximate those shown in Table 6.10. The most likely value for the longest streak is 3 in a row (This is the **mode** of the probability distribution).

Table 6.10: Probability Distribution of X = Longest Streak of Successful Free Throws. The distribution refers to 20 free throws with a 0.50 chance of success for each. All potential x values higher than 9 had a probability of 0.00 to two decimal places and a total probability of only 0.006.

x	$p(x)$	x	$p(x)$
0	0.00	5	0.13
1	0.02	6	0.06
2	0.20	7	0.03
3	0.31	8	0.01
4	0.23	9	0.01

The probability that O'Neal never makes more than 4 free throws in a row equals $p(0) + p(1) + p(2) + p(3) + p(4) = 0.76$. This would usually be the case. The mean of the probability distribution is μ = 3.7.

The longest streak of successful shots tends to be longer, however, with a larger number of total shots. For instance, with 200 shots, the distribution of the longest streak has a mean of μ = 7. Although it *would* be a bit unusual for O'Neal to make his *next* 7 free throws in a row, it would not be at all unusual if he makes 7 in a row *sometime* in his next 200 free throws. Making 7 in a row is then not really a hot streak, but merely what we expect by random variation. Sports announcers often get excited by streaks that they think represent a "hot hand" but merely represent random variation.

With 200 shots, the probability is only 0.03 that the longest streak equals 4 or less. Look at the 10 sets of 20 H and T symbols that you wrote on a sheet of paper to reflect the results you expected for 10 games of 20 free throws each. Find the longest streak of H's out of the string of 200 symbols. Was your longest streak 4 or less? If so, your instructor could predict that you faked the results rather than used random numbers, because there's only a 3% chance of this if they were truly generated randomly. Most students underestimate the likelihood of a relatively long streak.

This concept relates to the discussion in Section 5.4 about coincidences. By itself, an event may seem unusual. But when you think of all the possible coincidences and all the possible times they could happen, it is probably not so unusual.

BIBLIOGRAPHY

Schilling, M. F. (1990). The longest run of heads. *The College Mathematics Journal* vol. 21, pp. 196-207.

Chapter 7   Statistical Inference:
Confidence Intervals

EXAMPLE 1: ANALYZING DATA FROM THE GENERAL SOCIAL SURVEY

Picture the Scenario
For more than 30 years, the National Opinion Research Center at the University of Chicago (www.norc.uchicago.edu) has conducted each year or every other year an opinion survey called the General Social Survey (GSS). The survey randomly samples about 2000 adult Americans. In a 90-minute in-person interview, the interviewer asks a long list of questions about opinions and behavior for a wide variety of issues. Other nations have similar surveys. For instance, every five years Statistics Canada conducts its own General Social Survey. *Eurobarometer* regularly samples about 1000 people in each country in the European Union.

Analyzing data from such databases helps researchers to learn about how people think and behave at a given time and to track opinions over time. Activity 1 in Chapter 1 showed how to access the GSS data at www.icpsr.umich.edu/GSS. Some examples and exercises in previous chapters have used GSS data.

Questions to Explore
Based on data from a recent GSS, how can you make an inference about:

- The proportion of Americans who are willing to pay higher prices in order to protect the environment?

- The proportion of Americans who believe a wife should sacrifice her career for her husband's?

- The mean number of hours that Americans watch TV per day?

Thinking Ahead
We will analyze data from the GSS in examples and exercises throughout this chapter. For instance, in Example 3 we'll see how to estimate the proportion of Americans who are "environmentally friendly" in terms of being willing to pay higher prices in order to protect the environment. We haven't shown any data yet, but do you have a guess for how high this proportion is?

We'll answer the other two questions above in Examples 2 and 7, and in exercises we'll explore opinions on such issues as whether it should or should not be the government's responsibility to reduce income differences between the rich and the poor, whether a preschool child is likely to suffer if his or her mother works, and how conservative or liberal Americans are (on the average) in political ideology.

A sample of about 2000 people (as the GSS takes) may seem large, but this number is usually relatively small. For instance, in the United States, a survey of this size gathers data for less than 1 out of every 100,000 people. How can we possibly make reliable predictions about the entire population with so few people?

We now have the tools to see how this is done. We're ready to study a powerful use of statistics: **statistical inference** about population parameters using sample data. Inference methods help us to determine how close a sample statistic is likely to fall to the population parameter. We can make decisions and predictions about populations even if we have data for relatively few subjects from that population. A preview of statistical inference at the end of the previous chapter showed it's often possible to predict the winner of an election in which millions of people voted, knowing only how a couple of thousand people voted.

For statistical inference methods, you may wonder, what's the relevance of the material you've been learning about the role of randomization in gathering data (Chapter 4), concepts of probability (Chapter 5), and the normal distribution and its use as a sampling distribution (Chapter 6)? They're important because of two key aspects of statistical inference:

- *Statistical inference methods use probability calculations that assume that the data are gathered with a random sample or a randomized experiment.*

- *The probability calculations refer to a sampling distribution of a statistic, which is often approximately a normal distribution.*

In other words, statistical inference uses sampling distributions of statistics calculated from data gathered using randomization, and those sampling distributions are often approximately normal.

The next two chapters present the two types of statistical inference methods-- **estimation** of population parameters and **testing hypotheses** about the parameter values. This chapter discusses the first type, estimating population parameters. We'll learn how to estimate population proportions for categorical variables and population means for quantitative variables. For instance, a study dealing with how college students pay for their education might estimate the proportion of college students who work at least parttime and the mean annual income for those who do work. The most informative estimation method constructs an interval of numbers, called a **confidence interval**, within which the unknown parameter value is believed to fall.

7.1   WHAT ARE POINT AND INTERVAL ESTIMATES OF POPULATION PARAMETERS?

Population parameters have two types of estimates:

RECALL

A **statistic** describes a **sample**. Examples are the sample mean $\overline{x}$ and standard deviation s.

A **parameter** describes a **population**. Examples are the population mean μ and standard deviation σ.

Statistical inference uses sample statistics to make decisions and predictions about population parameters.

RECALL

Sections 6.4-6.6 introduced the **sampling distribution**, which specifies the possible values a statistic can take and their probabilities.

> **Point Estimate** and **Interval Estimate**
>
> A **point estimate** is a *single number* that is our "best guess" for the parameter.
>
> An **interval estimate** is an *interval of numbers* within which the parameter value is believed to fall.

For example, the 1998 General Social Survey asked "Do you believe in Hell?" From the sample data, the point estimate for the proportion of adult Americans who would respond "yes" equals 0.74--more than 7 out of 10. The adjective "point" in point estimate refers to using a single number or *point* as the parameter estimate.

An interval estimate, found with a method introduced in the next section, predicts that the proportion of *all* adult Americans who believe in Hell falls between 0.71 and 0.77. That is, it predicts that the sample point estimate of 0.74 falls within a *margin of error* of 0.03 of the population proportion. Figure 7.1 illustrates.

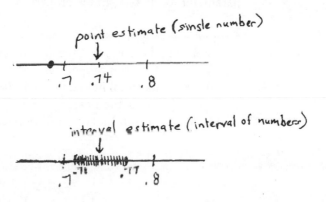

Figure 7.1: A **point estimate** predicts a parameter by a single number. An **interval estimate** is an interval of numbers that are believable values for the parameter. **Question**: Why is a point estimate alone usually not sufficiently informative?

Usually, a point estimate by itself is not sufficient, because it doesn't tell us *how close* the estimate is likely to be to the parameter. An interval estimate is more useful. It incorporates a margin of error, so it helps us to gauge the accuracy of the point estimate.

Point Estimation: How Do We Make a Best Guess for a Population Parameter?

Once we've collected the data, how do we find a point estimate, representing our "best guess" for a parameter value? The answer is straightforward--we can simply use an appropriate sample statistic. For instance, for a population mean μ, the sample mean $\bar{x}$ is a point estimate of μ. For the population proportion, the sample proportion is a point estimate.

Point estimates are the most common form of inference reported by the mass media. For example, every month the Gallup organization conducts a survey to estimate the U.S. president's popularity, and the mass media report results. At the end of December 2004, this survey reported that 49% of the American public approved of President Bush's performance in office. By contrast, surveys in Canada and in European countries at that time gave him lower approval ratings, for instance 32% in Canada, 30% in Great Britain, 19% in France, 19% in Spain, and 17% in Germany.[1] These were *point estimates* rather than parameters, since they used a sample of about 1000 people in each country rather than the entire populations. For simplicity, we'll usually use the term "estimate" in place of "point estimate" when there is no risk of confusing it with an interval estimate.

Properties of point estimators

For any particular parameter, there are several possible point estimates. For a normal distribution, for instance, the center is the mean μ and the median, since that distribution is symmetric. So, with sample data from a normal distribution, two possible estimates of that center value are the sample mean and the sample median. What makes a particular estimate better than others? *A good estimator* of a parameter has two desirable properties:

Property 1: A good estimator has a sampling distribution that is centered at the parameter, in the sense that the parameter is the mean of that sampling distribution. An estimator with this property is said to be **unbiased**. From Section 6.5, we know that for random sampling the mean of the sampling distribution of the sample mean $\bar{x}$ equals the population mean μ. So, the sample mean $\bar{x}$ is an unbiased estimator of μ. Figure 7.2 recalls this result from Section 6.5.

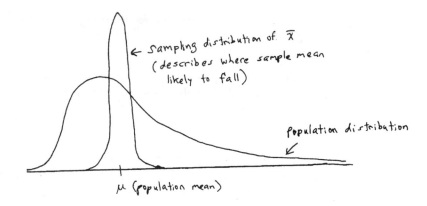

Figure 7.2: The Sample Mean $\bar{x}$ is an Unbiased Estimator. Its sampling distribution is centered at the parameter it estimates -- the population mean μ. **Question:** When is the sampling distribution bellshaped, as it is in this figure?

The sample proportion is a sample mean for binary data. It is an unbiased estimator of a population proportion. In polling about the president's popularity, we don't know whether any particular estimate of the population proportion supporting the

---

[1] Polling done for AP by Ipsos, an international polling company (reported Dec. 14, 2004).

president falls below or above the actual population proportion. We do know, however, that with random sampling the sample proportions tend to fall *around* the population proportion.

Property 2: A good estimator has a *small standard error* compared to other estimators. This means it tends to fall closer than other estimates to the parameter. For example, for estimating the center of a normal distribution, it turns out that the sample mean has a smaller standard error than the sample median. The sample mean is a better estimator of this parameter. In this text, we'll use estimators that are unbiased (or nearly so, in practical terms) and that have relatively small standard errors.

Interval Estimation: Constructing an Interval that Contains the Parameter (We Hope!)

A recent survey[2] of new college graduates estimated the mean salary for those who had taken a full-time job after graduation to equal $37,000. Does $37,000 seem plausible to you? Too high? Too low? Any one point estimate may or may not be close to the parameter it estimates. For the estimate to be useful, we need to know how close it is likely to fall to the actual parameter value. Is the estimate of $37,000 likely to be within $1000 of the actual population mean? Within $5,000? Within $10,000? Inference about a parameter should provide not only a point estimate but should also indicate its likely precision.

An **interval estimate** indicates precision by giving an interval of numbers around the point estimate. The interval is made up of numbers that are the most believable values for the unknown parameter. For instance, perhaps a survey of new college graduates predicts that the mean salary of all the graduates working fulltime falls somewhere between $35,000 and $39,000, that is, within a *margin of error* of $2000 of the point estimate of $37,000. An interval estimate is designed to contain the parameter with some chosen probability, such as 0.95. Because interval estimates contain the parameter with a certain degree of confidence, they are referred to as **confidence intervals**.

Confidence Interval

A **confidence interval** is an interval containing the most believable values for a parameter. The probability that this method produces an interval that contains the parameter is called the **confidence level**. This is a number chosen to be close to 1, most commonly 0.95.

The interval from $35,000 to $39,000 is an example of a confidence interval. It was constructed using a confidence level of 0.95. This is often expressed as a percentage, and we say that we have "95% confidence" that the interval contains the parameter. It is a **95% confidence interval**.

---

[2] By the National Association of Colleges and Employers, www.naceweb.org

RECALL

From Sections 6.4-6.6, a **standard error** of a statistic is the standard deviation of the sampling distribution of the statistic. It describes the variability in the possible values of the statistic, for the given sample size. It also tells us how much the statistic would vary from sample to sample of that size.

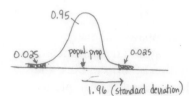

How can we construct a confidence interval? The key is the *sampling distribution* of the point estimate. This distribution tells us the probability that the point estimate falls within any certain distance of the parameter.

What's the Logic behind Constructing a Confidence Interval?

To construct a confidence interval, we'll put to work some results about sampling distributions that we learned in the previous chapter. Let's do this for estimating a proportion. We saw that the sampling distribution of a sample proportion:

- Gives the possible values for the sample proportion and their probabilities.

- Is approximately a normal distribution, for large random samples, because of the central limit theorem.

- Has mean equal to the population proportion.

- Has standard deviation called the **standard error**.

Let's use these results to construct a 95% confidence interval for a population proportion. From Chapter 6, approximately 95% of a normal distribution falls within two standard deviations of the mean. More precisely, on p. 376 we saw that plus and minus 1.96 standard deviations includes exactly 95% of a normal distribution. Since the sampling distribution of the sample proportion is approximately normal, with probability 0.95 the sample proportion falls within about 1.96 standard errors of the population proportion. The distance of 1.96 standard errors is the **margin of error**. We've been using this term since Chapter 4. Let's take a closer look at how it's calculated:

> **RECALL**
>
> Using normal cumulative probabilities (Table A or a calculator), $z = 1.96$ has cumulative probability 0.975, right-tail probability 0.025, two-tail probability 0.05, and central probability 0.950.

> **RECALL**
>
> Example 6 in Chapter 4 showed how the **margin of error** is reported in practice for a sample proportion. We approximated it there by $1/\sqrt{n}$. This is a rough approximation for 1.96(standard error), as shown following Example 10 in Section 7.4.

Margin of Error

The **margin of error** measures how accurate the point estimate is likely to be in estimating a parameter. It is a multiple of the standard error of the sampling distribution of the estimate, such as $1.96 \times$ (standard error) for a normal sampling distribution.

> **IN WORDS**
>
> "Sample proportion $\pm$ 1.96(standard error)" represents taking the sample proportion and adding and subtracting 1.96 standard errors.

Once the sample is selected, if the sample proportion *does* fall within 1.96 standard errors of the population proportion, then the interval from

[sample proportion − 1.96(standard error)] to [sample proportion + 1.96(standard error)]

contains the population proportion. In other words, with probability about 0.95, a sample proportion value occurs such that the interval

$$\text{sample proportion} \pm 1.96(\text{standard error})$$

contains the unknown population proportion. This interval of numbers is an approximate **95% confidence interval** for the population proportion.

EXAMPLE 2: SHOULD A WIFE SACRIFICE HER CAREER FOR HER HUSBAND'S?

Picture the Scenario
One question on the General Social Survey asks whether you agree or disagree with the following statement: "It is more important for a wife to help her husband's career than to have one herself." The last time this was asked (in 1998), 19% of 1823 respondents agreed. So the sample proportion agreeing was 0.19. From a formula in the next section, we'll see that this estimate has standard error = 0.01.

Questions to Explore
a. Find and interpret the margin of error for a 95% confidence interval for the population proportion who agreed with the statement about a woman's role.
b. Construct the 95% confidence interval, and interpret it in context.

Think It Through
a. The margin of error for a 95% confidence interval for a population proportion equals $1.96 \times$ (standard error), or $1.96(0.01) = 0.02$. This means that with probability 0.95, the error in using the point estimate to predict the population proportion (in 1998) is no greater than 0.02.

b. The approximate 95% confidence interval is

$$\text{Sample proportion} \pm 1.96(\text{standard error}),$$

$$\text{which is } 0.19 \pm 1.96(0.01), \text{ or } 0.19 \pm 0.02.$$

This gives the interval of proportions from 0.17 to 0.21, denoted by (0.17, 0.21). In summary, using this 95% confidence interval, we predict that the population proportion (in 1998) who believed it is more important for a wife to help her husband's career than to have one herself is somewhere between 0.17 and 0.21.

Insight
By contrast, in 1977 when this question was first asked on the GSS, the point estimate was 0.57 and the 95% confidence interval was (0.55, 0.59). The proportion of Americans who agree with this statement decreased considerably during the 1980s and 1990s.

♦

To practice this concept, try Exercise 7.7.

In summary, *a confidence interval is constructed by adding and subtracting a margin of error to a given point estimate. The margin of error is based on the standard error of the sampling distribution of that point estimate.* When the sampling distribution is approximately normal, a 95% confidence interval has margin of error equal to 1.96 standard errors.

The sampling distribution of most point estimates is approximately normal when the random sample size is relatively large. Thus, this logic of taking the margin of error for a 95% confidence interval to be approximately two standard errors applies with large random samples, such as those found in the General Social Survey. The next two sections show more precise details for estimating proportions and means.

ACTIVITY 1: DOWNLOAD DATA FROM THE GENERAL SOCIAL SURVEY

We saw in Chapter 1 that it's easy to download data from the GSS. Let's recall how.

- Go to the Web site, `www.icpsr.umich.edu/GSS/`. Click on the *Analyze* tab. In the *Select an action* menu, select the *Frequencies and crosstabulation* option and click on *Start*.

- The GSS name for the variable in Example 2 (whether a wife should sacrifice her career for her husband's) is FEHELP. Type FEHELP as the *Row variable* name. Click on *Run the table*. Now you'll see category counts and the percentages for all the years combined in which this question was asked.

- To download results only for the year 1998, go back to the previous menu and enter YEAR(1998) in the *Selection Filter* space. When you click again on *Run the table*, you'll see that in 1998 2.4% strongly agreed and 16.8% agreed, a total of about 19% (a proportion of 0.19) in the two agree categories.

Now, create other sample results. Click on *Subject* under Codebook Indexes. Look up a subject that interests you and find the GSS code name for a variable. For instance, to see the percentage sampled who believe in hell, enter HELL as the code name.

SECTION 7.1: PRACTICING THE BASICS

7.1   Health care: A study dealing with health care issues plans to take a sample survey of 1500 Americans to estimate the proportion who have health insurance and the mean dollar amount that Americans spent on health care this past year.
a.   Identify two population parameters that this study will estimate.
b.   Identify two statistics that can be used to estimate these parameters.

7.2   Types of estimates: Explain the difference between a point estimate and an interval estimate.

7.3   Believe in hell?: Using the General Social Survey website www.icpsr.umich.edu/GSS, find the point estimate of the population proportion of Americans who would have answered "yes, definitely" in 1998 when asked whether they believe in hell (variable HELL).

7.4   Help the poor?: One question (coded EQUALIZE) on the 1998 General Social Survey asked, "Do you think that it should be government's responsibility to reduce income differences between the rich and the poor?" Of the possible responses, 548 picked "definitely or probably should be," and 620 picked "probably or definitely should not be."

a.　Find the point estimate of the population proportion who would answer "definitely or probably should be."

b.　The margin of error of this estimate is 0.03. Explain what this represents.

7.5　　**Watching TV**: In response to the GSS question in 2002 about the number of hours daily spent watching TV, the responses by the seven subjects who identified themselves as Buddhists were 2, 2, 1, 3, 2, 3, 2.

a.　Find a point estimate of the population mean for Buddhists.

b.　The margin of error for this point estimate is 0.64. Explain what this represents.

7.6　　**Starting salaries**: The starting salaries of a random sample of three students who graduated from North Carolina State University last year with majors in the mathematical sciences are $45,000, $35,000, and $55,000.

a.　Find a point estimate of the population mean of the starting salaries of all math science graduates at that university last year.

b.　A method that we'll study in Section 7.3 provides a margin of error of $24,800 for a 95% confidence interval for the population mean starting income. Construct that interval.

c.　Use this example to explain why a point estimate alone is usually insufficient for statistical inference.

7.7　　**Believe in heaven?**: When the 1998 GSS asked 1158 subjects, "Do you believe in Heaven?" (coded HEAVEN), the proportion who answered yes was 0.86. From results in the next section, the standard error of this estimate is 0.01.

a.　Find and interpret the margin of error for a 95% confidence interval for the population proportion of Americans who believe in heaven.

b.　Construct the 95% confidence interval. Interpret it in context.

7.8　　**Feel lonely often?**: The 1996 GSS asked 1450 subjects, "On how many days in the past 7 days have you felt lonely?" At www.icpsr.umich.edu/GSS, enter LONELY as the variable, check *Statistics* in the menu of *Other Options*, and click on *Run the Table* to see the responses.

a.　Report the percentage making each response and the mean and standard deviation of the responses. Interpret.

b.　Note that the output also reports a standard error value of 0.06. What does this mean?

7.9　　**7.9 CI for loneliness**: Refer to the previous exercise. The margin of error for a 95% confidence interval for the population mean is 0.12. Construct that confidence interval, and interpret it.

7.10　　**7.10 Newspaper article:** Conduct a search to find a newspaper or magazine article that reported a margin of error. If you prefer, use the Internet for the search.

7.11　　Specify the population parameter, the value of the sample statistic, and the size of the margin of error.

7.12　　Explain how to interpret the margin of error.

HOW CAN WE CONSTRUCT A CONFIDENCE INTERVAL TO ESTIMATE A POPULATION PROPORTION?

Let's now see the details of how to construct a confidence interval for a population proportion. We'll apply the ideas just discussed at the end of Section 7.1. The data are binary, with each observation either falling or not falling in the category of interest. We'll use the generic terminology "success" and "failure" for these two possible outcomes (as in the discussion of the binomial distribution in Section 6.2). We summarize the data by the sample proportion of successes and construct a confidence interval for the population proportion.

Finding the 95% Confidence Interval for a Population Proportion

The 2000 General Social Survey asked respondents whether they would be willing to pay much higher prices in order to protect the environment. Of $n = 1154$ respondents, 518 said yes. The sample proportion of yes responses was $518/1154 = 0.45$, less than half. How can we construct a confidence interval for the population proportion that would respond yes?

We symbolize a population proportion by p. The point estimate of the population proportion is the *sample proportion*. We symbolize the sample proportion by $\hat{p}$, called "p-hat." In statistics, the circumflex ("hat") symbol over a parameter symbol represents a point estimate of that parameter. Here, the sample proportion $\hat{p}$ is a point estimate, such as $\hat{p} = 0.45$ for the proportion willing to pay much higher prices to protect the environment.

For large random samples, the central limit theorem tells us that the sampling distribution of the sample proportion $\hat{p}$ is approximately normal. As discussed in the previous section, the z-score for a 95% confidence interval with the normal sampling distribution is 1.96. So there is about a 95% chance that $\hat{p}$ falls within 1.96 standard errors of the population proportion p. A 95% confidence interval uses margin of error = 1.96(standard error). The formula

[point estimate $\pm$ margin of error] becomes $\hat{p} \pm 1.96$(standard error).

The exact standard error of a sample proportion equals $\sqrt{p(1-p)/n}$. This formula depends on the unknown population proportion, p. In practice, we don't know p, and we need to estimate the standard error.

IN WORDS

It is traditional in statistics to use Greek letters for most parameters. Some books use the Greek letter π ("pi") for a proportion. Since π already has its own meaning in math and statistics, we use the symbol p. This is also the parameter for the probability of success in the binomial distribution (Section 6.2). The sample proportion $\hat{p}$ is read as "p-hat."

RECALL

The standard error of a sample proportion was introduced in Section 6.4.

IN PRACTICE

The exact values of standard errors depend on parameter values. In practice, these are unknown, so we find standard errors by substituting estimates of parameters. The term "standard error" is then used as short-hand for what is actually an "estimated standard deviation of a sampling distribution." Beginning here, we'll use **standard error** to refer to this estimated value, since that's what we'll use in practice. When we need to refer to the exact value based on the parameter value, we'll call it the "exact standard error."

<table>
<tr><td>

IN WORDS

se = standard error

</td></tr>
</table>

In summary, in practice *a standard error is an estimated standard deviation of a sampling distribution.* We will use *se* as shorthand for *standard error.* For example, for finding a confidence interval for a population proportion *p*, the standard error is

$$se = \sqrt{\hat{p}(1-\hat{p})/n} \ .$$

<table>
<tr><td>

IN WORDS

To find the **95% confidence interval**, you take the sample proportion and add and subtract 1.96 standard errors.

</td></tr>
</table>

<table>
<tr><td>

A 95% confidence interval for a population proportion *p* is

$$\hat{p} \pm 1.96(se), \text{ with } se = \sqrt{\hat{p}(1-\hat{p})/n} \ ,$$

where $\hat{p}$ denotes the sample proportion based on *n* observations.

</td></tr>
</table>

Figure 7.3 shows the sampling distribution of $\hat{p}$ and how there's about a 95% chance that $\hat{p}$ falls within 1.96(*se*) of the population proportion *p*. This confidence interval is designed for large samples. We'll be more precise about what "large" means after the following example.

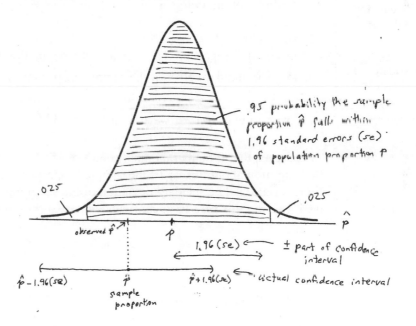

Figure 7.3: Sampling Distribution of Sample Proportion $\hat{p}$ **.** For large random samples, the sampling distribution is normal around the population proportion *p*, so $\hat{p}$ has probability 0.95 of falling within 1.96(*se*) of *p*. As a consequence, $\hat{p} \pm 1.96(se)$ is a 95% confidence interval for *p*. **Question**: Why is the confidence interval $\hat{p} \pm 1.96(se)$ instead of $p \pm 1.96(se)$?

EXAMPLE 3: WOULD YOU PAY HIGHER PRICES TO PROTECT THE ENVIRONMENT?

Picture the Scenario
Many people consider themselves "green," meaning they are supportive (in theory) when it comes to environmental issues. But how do they act in practice? How would you answer the question that the GSS asked about being willing to pay much higher prices in order to protect the environment? It might be more informative, and easier for people to answer, if the question were more specific. For instance, Americans' per capita use of energy is roughly double that of Western Europeans. If you live in North America, would you be willing to pay the same price for gas that Europeans do (often about $5 or more per gallon), if the government proposed a significant price hike as an incentive for conservation and for driving more fuel-efficient cars to reduce air pollution and global warming?

Questions to Explore
In 2000, the GSS asked the question about paying higher prices to protect the environment. Of $n = 1154$ respondents, 518 were willing to do so.
a. Find a 95% confidence interval for the population proportion of adult Americans willing to do so at the time of that survey.
b. Interpret that interval.

> **ACTIVITY**
>
> To obtain these data for yourself, go to the GSS Web site www.icpsr.umich.edu/GSS and download data on the variable GRNPRICE for the 2000 survey.

Think It Through
a. The sample proportion that estimates the population proportion p is $\hat{p} = 518/1154 = 0.45$. The standard error of the sample proportion $\hat{p}$ equals

$$se = \sqrt{\hat{p}(1-\hat{p})/n} = \sqrt{(0.45)(0.55)/1154} = \sqrt{0.0002144} = 0.015.$$

This means that if many random samples of 1154 people each were taken to gauge their opinion about this issue, the sample proportions from those studies would show little variation: The standard deviation of the sample proportions would be about 0.015. Using this se, a 95% confidence interval for the population proportion is

$$\hat{p} \pm 1.96(se), \text{ which is } 0.45 \pm 1.96(0.015)$$
$$= 0.45 \pm 0.03, \text{ or } (0.42, 0.48).$$

b. At the 95% confidence level, we estimate that the population proportion of adult Americans willing to pay much higher prices to protect the environment was at least 0.42 but no more than 0.48, that is, between 42% and 48%. The point estimate of 0.45 has a margin of error of 0.03. None of the numbers in the confidence interval (0.42, 0.48) fall above 0.50. So we infer that less than half the population was willing to pay much higher prices to protect the environment.

Insight
As usual, results depend on the question's wording. For instance, when asked whether the government should impose strict laws to make industry do less damage to the environment, a 95% confidence interval for the population proportion responding yes is (0.87, 0.90). (See Exercise 7.14.)

◆

To practice this concept, try Exercise 7.15.

IN PRACTICE

The **standard errors** reported in this book and by most software assume **simple random sampling**. The GSS uses a form of multistage random sampling, but this form of sampling is similar to simple random sampling for purposes of forming standard errors and conducting statistical inference.

RECALL

From Section 6.2, the **binomial** random variable X counts the *number* of successes, and the sample proportion equals x/n, for instance, $518/1154 = 0.45$.

Table 7.1 shows how the Minitab software reports the data summary and confidence interval. (We added the italicized lines as annotation.) Here, X represents the number that *support* paying much higher prices. The notation reflects that this is the outcome of a random variable, specifically a binomial random variable. The heading "Sample p" stands for the sample proportion $\hat{p}$, and "CI" stands for confidence interval. In reporting results from such output, you should use only the first two or three significant digits. Report the sample proportion as 0.45 or 0.449 rather than 0.448873. Software's extra precision helps to provide accurate calculations in finding the standard error and the confidence interval. However, the extra digits are distracting when reported to others and do not tell them anything extra in a practical sense about the population proportion.

Table 7.1: MINITAB Output for 95% Confidence Interval for a Proportion for Example 3.

```
     X          N        Sample p        95.0% CI
    518        1154      0.448873      (.420176, 0.477571)
     ↑      Sample Size     ↑         Endpoints of ConfidenceInterval
Category Count          Sample Proportion
```

Annotation of Output →

Sample Size Needed for Validity of Confidence Interval for a Proportion

RECALL

From Section 6.2, the binomial distribution is bell-shaped when the *expected* counts np and $n(1-p)$ of successes and failures are both at least 15. Here, we don't know p, and we use the guideline with the *observed* counts of successes and failures.

The confidence interval formula $\hat{p} \pm 1.96(se)$ applies with *"large" random samples*. This is because the sampling distribution of the sample proportion $\hat{p}$ is then approximately normal and the *se* estimate also tends to be good. This is what allows us to use the *z*-score of 1.96 from the normal distribution.

In practice, "large" means that *you should have at least 15 "successes" and at least 15 "failures"* for the binary outcome.[3] Later in this section, we'll see how the large-sample method can fail when this guideline is not satisfied.

---

[3] Many statistics texts use 5 or 10 as the minimum instead of 15, but recent research suggests that's too small (e.g., L. Brown et al., *Statistical Science*, vol. 16, pp. 101 - 133, 2001).

Sample Size Needed for Large-Sample Confidence Interval for a Proportion

For the 95% confidence interval $\hat{p} \pm 1.96(se)$ for a proportion p to be valid, you should have at least 15 successes and 15 failures. This can also be expressed as

$$n\hat{p} \geq 15 \text{ and } n(1-\hat{p}) \geq 15.$$

This guideline was easily satisfied in Example 3. The binary outcomes had counts 518 willing to pay much higher prices and 636 (= 1154 − 518) unwilling, both much larger than 15.

EXAMPLE 4: WHAT PROPORTION *WON'T* PAY HIGHER PRICES TO PROTECT THE ENVIRONMENT?

Picture the Scenario
Example 3 found a confidence interval for the population proportion p that will pay higher prices to support the environment, based on a sample proportion from the GSS of 518/1154 = 0.45. The population proportion that *won't* pay higher prices to support the environment is then $1-p$.

Question to Explore
How can we construct a 95% confidence interval for the population proportion $1-p$?

Think It Through
We can use the formula for a 95% confidence interval, because the GSS uses random sampling and because there's at least 15 successes and 15 failures (the counts are 518 and 636, with $n = 1154$). The estimate of $1-p$ is 636/1154 = 0.55, which is $1-\hat{p} = 1 - 0.45$. When $n = 1154$ and $\hat{p} = 0.45$, we saw that the standard error of $\hat{p}$ is $se = \sqrt{\hat{p}(1-\hat{p})/n} = 0.015$. Similarly, the standard error for $1-\hat{p} = 0.55$ is

$$\sqrt{\text{proportion}(1-\text{proportion})/n} = \sqrt{(0.55)(0.45)/1154} = 0.015.$$

This is necessarily the same as the standard error of $\hat{p}$. A 95% confidence interval for the population proportion that won't pay higher prices is

sample proportion $\pm$ 1.96(*se*), or 0.55 $\pm$ 1.96(0.015),
which is 0.55 $\pm$ 0.03, or (0.52, 0.58).

In summary, we can be 95% confident that the population proportion that won't pay higher prices to support the environment falls between 0.52 and 0.58.

Insight
Now 0.52 = 1− 0.48 and 0.58 = 1− 0.42, where (0.42, 0.48) is the 95% confidence interval for the population proportion who *will* pay higher prices. For a binary variable, inferences about the second category (*won't* pay higher prices) follow directly from those for the first category (*will* pay higher prices) by subtracting each

endpoint of the confidence interval from 1.0. We do not need to construct the confidence interval separately for both categories. The confidence interval for one determines the confidence interval for the other.

◆

To practice this concept, try Exercises 7.17 and 7.18.

How Can We Use Confidence Levels Other than 95%?

So far we've used a confidence level of 0.95, that is, "95% confidence." This means that with probability 0.95, a sample proportion value $\hat{p}$ occurs such that the confidence interval $\hat{p} \pm 1.96(se)$ contains the population proportion p. With probability 0.05, however, the method produces a confidence interval that misses p. The population proportion then does *not* fall in the interval, and the inference is incorrect.

In practice, the confidence level 0.95 is the most common choice. But some applications require greater confidence. This is often true in medical research, for example. For estimating the probability p that a new treatment for a deadly disease works better than the treatment currently used, we would want to be extremely confident about any inference we make. To increase the chance of a correct inference, we use a larger confidence level, such as 0.99.

Now, 99% of the normal sampling distribution for the sample proportion $\hat{p}$ occurs within 2.58 standard errors of the population proportion p. So, with probability 0.99, $\hat{p}$ falls within 2.58(se) of p. A 99% confidence interval for p is $\hat{p} \pm 2.58(se)$.

RECALL

99% confidence interval:
From Section 6.3, for central probability 0.99, you look up the cumulative probability of 0.005 or $1 - 0.005 = 0.995$ in Table A, or use software or a calculator, to find $z = 2.58$.

EXAMPLE 5: SHOULD A HUSBAND BE FORCED TO HAVE CHILDREN?

Picture the Scenario
A recent GSS asked "If the wife in a family wants children, but the husband decides that he does not want any children, is it all right for the husband to refuse to have children?" Of 598 respondents, 366 said yes and 232 said no.

Questions to Explore
a. Find a 99% confidence interval for the population proportion who would say yes.
b. How does it compare to the 95% confidence interval?

Think It Through
a. The assumptions for the method are satisfied in that the GSS sample was randomly selected, and there were at least 15 successes and 15 failures (366 yes and 232 no). The sample proportion who said yes is $\hat{p} = 366/598 = 0.61$. Its standard error is $se = \sqrt{\hat{p}(1-\hat{p})/n} = \sqrt{(0.61)(0.39)/598} = 0.020$. The 99% confidence interval is

$$\hat{p} \pm 2.58(se) = 0.61 \pm 2.58(0.020),$$
$$\text{which is } 0.61 \pm 0.05, \text{ or } (0.56, 0.66).$$

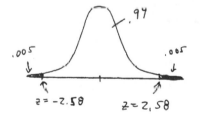

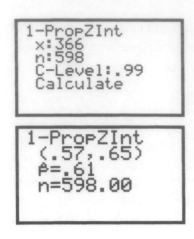

In summary, we can be 99% confident that the population proportion who agree it's all right for the husband to refuse to have children falls between 0.56 and 0.66. Statistical software or a statistical calculator can calculate confidence intervals. Screen shots from the TI-83+/84 are shown in the margin.

b. The 95% confidence interval is $0.61 \pm 1.96(0.020)$, which is 0.61 ± 0.04, or (0.57, 0.65). This is a bit narrower than the 99% confidence interval of (0.56, 0.66). Figure 7.4 illustrates.

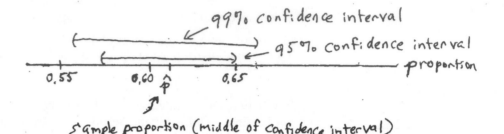

Figure 7.4: A 99% Confidence Interval Is Wider than a 95% Confidence Interval.
Question: If you want greater confidence, why would you expect a wider interval?

Insight

The inference with a 99% confidence interval is less precise, with margin of error equal to 0.05 instead of 0.04. Having a greater margin of error is the sacrifice for gaining greater assurance of a correct inference. A medical study that uses a higher confidence level will be less likely to make an incorrect inference, but it will not be able to "narrow in" as well on where the true parameter value falls.

To practice this concept, try Exercise 7.23.

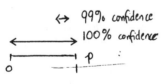

Why settle for anything less than 100% confidence? To be absolutely 100% certain of a correct inference, the confidence interval must contain *all* possible values for the parameter. For example, a 100% confidence interval for the population proportion believing it is all right for a husband to refuse to have children goes from 0.0 to 1.0. This inference would tell us only that some number between 0.0% and 100.0% of Americans feel this way. This is obviously not helpful. In practice, we settle for a little less than perfect confidence so we can estimate the parameter value more precisely. It is far more informative to have 99% confidence that the population proportion is between 0.56 and 0.66 than to have 100% confidence that it is between 0.0 and 1.0.

In using confidence intervals, *we must compromise between the desired margin of error and the desired confidence of a correct inference.* As one gets better, the other gets worse. This is why you would probably not use a 99.9999% confidence interval. It would usually have too large a margin of error to tell you much about where the parameter falls (its *z*-score is 4.9). In practice, 95% confidence intervals are the most common.

IN PRACTICE

When the news media report a **margin of error**, it is the margin of error for a 95% confidence interval.

What Is the Error Probability for the Confidence Interval Method?

The general formula for the confidence interval for a population proportion is

sample proportion $\pm$ (*z*-score from normal table)(standard error),
which in symbols is $\hat{p} \pm z(se)$.

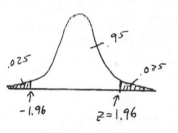

The *z*-score depends on the confidence level. Table 7.2 shows the *z*-scores for the confidence levels usually used in practice. There is no need to memorize them. You can find them yourself, using Table A or a calculator or software. Let's review how by finding the *z*-score for a 95% confidence interval. When 0.95 probability falls within *z* standard errors of the mean, then 0.05 probability falls in the two tails and 0.05/2 = 0.025 falls in each tail. Looking up 0.025 in the body of Table A, we find *z* = -1.96, or we find *z* = 1.96 if we look up the cumulative probability 1 – 0.025 = 0.975 corresponding to tail probability 0.025 in the right tail. The 95% confidence interval equals $\hat{p} \pm 1.96(se)$. Try this again for 99% confidence, using either the table or a calculator or software (you should get *z* = 2.58).

Table 7.2: *z*-Scores for the Most Common Confidence Levels. The large-sample confidence interval for the population proportion is $\hat{p} \pm z(se)$.

Confidence Level	Error Probability	*z*-Score	Confidence Interval
0.90	0.10	1.645	$\hat{p} \pm 1.645(se)$
0.95	0.05	1.96	$\hat{p} \pm 1.96(se)$
0.99	0.01	2.58	$\hat{p} \pm 2.58(se)$

Table 7.2 contains a column labeled **error probability.** This is the probability that the method results in an incorrect inference, with the confidence interval *not* containing the population proportion. The error probability equals 1 minus the confidence level. For example, when the confidence level equals 0.95, the error probability equals 0.05. The error probability is the two-tail probability under the normal curve for the given *z*-score. Half the error probability falls in each tail. For 95% confidence with its error probability of 0.05, the *z*-score of 1.96 is the one with probability 0.05/2 = 0.025 in each tail.

Summary: Confidence Interval for a Population Proportion *p*

A confidence interval for a population proportion *p*, using the sample proportion $\hat{p}$ and the standard error $se = \sqrt{\hat{p}(1-\hat{p})/n}$ for sample size *n*, is

$$\hat{p} \pm z(se), \text{ which is } \hat{p} \pm z\sqrt{\hat{p}(1-\hat{p})/n}.$$

For 95% and 99% confidence intervals, *z* equals 1.96 and 2.58. To use this method, you need

- Data obtained by randomization (such as a random sample)
- A large enough sample size *n* so that the number of successes and the number of failures, that is, $n\hat{p}$ and $n(1-\hat{p})$, are both at least 15.

What's the Effect of the Sample Size?

We'd expect that estimation should be more precise with larger sample sizes. With more data, we know more about the population. The margin of error is $z(se) = z\sqrt{\hat{p}(1-\hat{p})/n}$. This margin decreases as the sample size *n* increases, for a given value of $\hat{p}$. The larger the value of *n*, the narrower is the interval.

To illustrate, in Example 5, suppose the survey result of $\hat{p} = 0.61$ for the proportion believing a husband can refuse to have children had resulted from a sample of size *n* = 150, only one-fourth the actual sample size of *n* = 598. Then the standard error would be

$$se = \sqrt{\hat{p}(1-\hat{p})/n} = \sqrt{0.61(0.39)/150} = 0.040,$$

twice as large as the *se* = 0.020 we got for *n* = 598. The 99% confidence interval would be

$$\hat{p} \pm 2.58(se) = 0.61 \pm 2.58(0.040), \text{ which is } 0.61 \pm 0.10, \text{ or } (0.51, 0.71).$$

The margin of error of 0.10 is twice as large as the margin of error of 0.05 in the 99% confidence interval with the sample size *n* = 598 in Example 5. The confidence interval with larger *n* is narrower.

Because the standard error has the square root of *n* in the denominator, and because $\sqrt{4n} = 2\sqrt{n}$, *quadrupling* the sample size *halves* the standard error. That is, we must quadruple *n*, rather than double it, to halve the margin of error.

In summary, we've observed the following properties of a confidence interval:

> **Summary: Effects of Confidence Level and Sample Size on Margin of Error**
>
> The **margin of error** for a confidence interval:
> * Increases as the confidence level increases
> * Decreases as the sample size increases

For instance, a 99% confidence interval is wider than a 95% confidence interval, and a confidence interval with 200 observations is narrower than one with 100 observations. These properties apply to *all* confidence intervals, not just the one for the population proportion.

Interpretation of the Confidence Level

In Example 3 the 95% confidence interval for the population proportion p willing to pay higher prices to protect the environment was (0.42, 0.48). The value of p is unknown to us, so we don't know that it actually falls in that interval. It may actually equal 0.41 or 0.49, for example.

So what does it mean to say that we have "95% confidence"? The meaning refers to a *long-run* interpretation--how the method performs when used over and over with many different random samples. If we used the 95% confidence interval method over time to estimate many population proportions, then *in the long run about 95% of those intervals would give correct results, containing the population proportion.* This happens because 95% of the sample proportions would fall within $1.96(se)$ of the population proportion. A graphical example is the $\hat{p}$ in line 1 of Figure 7.5.

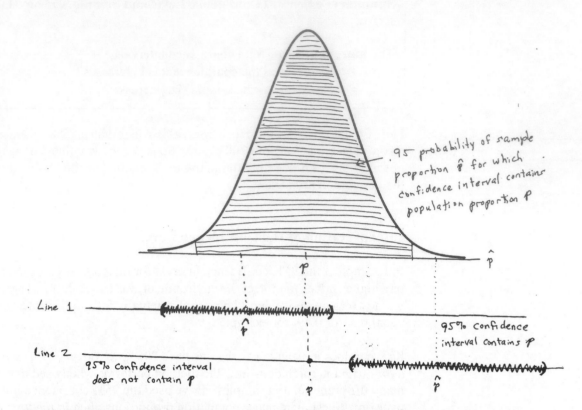

.95 probability of sample proportion $\hat{p}$ for which confidence interval contains population proportion p

95% confidence interval contains p

95% confidence interval does not contain p

Line 1

Line 2

Figure 7.5: The Sampling Distribution of the Sample Proportion $\hat{p}$. The lines below it show two possible $\hat{p}$ values and the corresponding 95% confidence intervals for the population proportion p. The interval on line 1 provides a correct inference, but the one on line 2 provides an incorrect inference. **Question:** Can you identify on the figure all the $\hat{p}$ values for which the 95% confidence interval would not contain p?

Saying that a particular interval contains p with "95% confidence" signifies that in the long run, 95% of such intervals would provide a correct inference. On the other hand, in the long run, 5% of the time the sample proportion $\hat{p}$ *does not* fall within 1.96(*se*) of p. If that happens, then the confidence interval *does not* contain p as is seen for $\hat{p}$ in line 2 of Figure 7.5.

By our choice of the confidence level, we can control the chance that we make a correct inference. If an error probability of 0.05 makes us too nervous, we can instead form a 99% confidence interval. This is in error only 1% of the time in the long run. But then we must settle for a wider confidence interval and less precision.

Long-Run versus Subjective Probability Interpretation of Confidence

You might be tempted to interpret a statement such as, "We can be 95% confident that the population proportion p willing to pay higher prices to protect the environment falls between 0.42 and 0.48," as meaning that the *probability* is 0.95 that p falls between 0.42 and 0.48. However, probabilities apply to statistics (such as in sampling distributions of the sample mean or sample proportion), not to parameters.

The estimate $\hat{p}$, not the parameter p, is the random variable having a sampling distribution and probabilities. The 95% confidence refers not to a probability for the population proportion p but rather to a probability that applies to the confidence interval *method* in its relative frequency sense: If we use it over and over for various samples, in the long run we make correct inferences 95% of the time.

Section 5.1 mentioned a *subjective* definition of probability that's an alternative to the relative frequency definition. This approach treats the *parameter* as a random variable. Statistical inferences based on the subjective definition of probability *do* make probability statements about parameters. For instance, with it you *can* say that the probability is 0.95 that p falls between 0.42 and 0.48. Statistical inference based on the subjective definition of probability is called **Bayesian statistics.**[4] It has gained in popularity in recent years, but it is beyond the scope of this text.

Confidence Interval for a Proportion with Small Samples

How do we form confidence intervals when there are fewer than 15 successes or failures? The confidence interval formula we've used is not valid. Why? One problem is that the central limit theorem no longer applies. From Sections 6.2 and 6.4, the expected numbers of successes and failures must both be at least 15 for the normal distribution to approximate well the binomial distribution or the sampling distribution of a sample proportion.

An even more serious difficulty is that the exact standard error depends on the parameter we're trying to estimate. If the estimate $\hat{p}$ of p is poor, as often happens for small n, then so is the estimate $se = \sqrt{\hat{p}(1-\hat{p})/n}$ of the exact standard error. As a result, the confidence interval formula works poorly, as we'll see in the next example.

EXAMPLE 6: WHAT PROPORTION OF STUDENTS ARE VEGETARIANS?

Picture the Scenario
In recent years, vegetarianism has gone from being viewed as a slightly eccentric fringe practice to receiving mainstream attention. Seemingly more and more people cite health, animal rights, and religious reasons for becoming vegetarians. Yet, a cover story of the subject in *Time* magazine (July 15, 2002) quoted a poll of 10,000 Americans in which only 4% said they were vegetarians.

For a recent class project, a student randomly sampled 20 fellow students at the University of Florida (UF) to estimate the proportion at that university who were vegetarians. Of the 20 students she sampled, none were vegetarians.

Question to Explore
What does a large-sample 95% confidence interval tell her about the proportion of vegetarians among all 49,000 students at UF?

---

[4] The name refers to **Bayes's theorem**, which can generate probabilities of parameter values, given the data, from probabilities of the data, given the parameter values.

Think It Through

Let p denote the population proportion of vegetarians at the university. The sample proportion was $\hat{p} = 0/20 = 0.0$. When $\hat{p} = 0.0$, then $se = \sqrt{\hat{p}(1-\hat{p})/n} = \sqrt{0.0(1.0)/20} = 0.0$. The large-sample 95% confidence interval for the population proportion of vegetarians is

$$\hat{p} \pm z(se) = 0.0 \pm 1.96(0.0), \text{ which is } 0.0 \pm 0.0, \text{ or } (0.0, 0.0).$$

This investigation told the student that she can be 95% confident that p falls between 0 and 0, that is, that $p = 0$. She was surprised by this result. It seemed unrealistic to conclude that *no* student at the university is a vegetarian.

Insight

Do you trust this inference? Just because no one in a small sample is a vegetarian, would you conclude that no one in the much larger *population* is a vegetarian? We doubt it. Perhaps the population proportion p is close to 0, but it is almost surely not exactly equal to 0.

The confidence interval formula $\hat{p} \pm z\sqrt{\hat{p}(1-\hat{p})/n}$ is valid only if the sample has at least 15 vegetarians and at least 15 nonvegetarians. This is not true here, so we have to use a different method.

$\blacklozenge$

Constructing a Small-Sample Confidence Interval for a Proportion p

Suppose a random sample does *not* have at least 15 successes and 15 failures. The confidence interval formula $\hat{p} \pm z\sqrt{\hat{p}(1-\hat{p})/n}$ still is valid if we use it after adding 2 to the original number of successes and 2 to the original number of failures. This results in adding 4 to the sample size n.

The sample in Example 6 of size $n = 20$ had 0 vegetarians and 20 nonvegetarians. We can apply the confidence interval formula with $0 + 2 = 2$ vegetarians and $20 + 2 = 22$ nonvegetarians. The value of the sample size for the formula is then $n = 2 + 22 = 24$. *Now we can use the formula, even though we don't have at least 15 successes and 15 failures*. We then get

$$\hat{p} = 2/24 = 0.083, \quad se = \sqrt{\hat{p}(1-\hat{p})/n} = \sqrt{(0.083)(0.917)/24} = 0.056.$$

The resulting 95% confidence interval is

$$\hat{p} \pm 1.96(se), \text{ which is } 0.083 \pm 1.96(0.056), \text{ or } (-0.03, 0.19).$$

A proportion cannot be negative, so we report the interval as (0.0, 0.19). We can be 95% confident that the proportion of vegetarians at the University of Florida is no greater than 0.19.

This approach enables us to use a large-sample method even when the sample size is small. With it, the point estimate moves the sample proportion a bit toward ½ (e.g., from 0.0 to 0.083). This is particularly helpful when the ordinary sample proportion is 0 or 1, which we would not usually expect to be a believable estimate of a population proportion. Why do we add 2 to the counts of the two types? The reason is that the confidence interval then approximates one based on a more complex method (described in Exercise 7.136) that does not require estimating the exact standard error.[5]

To practice this concept, try Exercise 7.25.

ACTIVITY 2: LET'S SIMULATE THE PERFORMANCE OF CONFIDENCE INTERVALS

Let's get a feel for how confidence intervals sometimes provide an incorrect inference. To do this, for a given population proportion value we will simulate taking many samples and forming a confidence interval for each sample. We can then check how often the intervals provide an incorrect inference. We can conduct the simulation using statistical software (such as Minitab) or using an applet in which we can control the parameter value, the sample size, and the confidence level.

Try this by going to the *confidence interval for a proportion* applet at www.prenhall.com/TBA. We'll set the population proportion $p = 0.50$, and see what happens when we take samples of size 50 and form 95% confidence intervals. At the menu, set the proportion value to 0.50, set the confidence level to 0.95, and set the sample size to 50. You will see that a certain number of outcomes occurred of each type, much as if you flipped a coin 50 times and counted the numbers of heads and tails. When we did this, we got 23 outcomes of one type and 27 of the other. You will probably get something different, because the process is random. After generating the sample, the applet calculates the sample proportion (such as $\hat{p} = 23/50 = 0.46$) and the 95% confidence interval. For our sample, we got the 95% confidence interval (0.32, 0.60). This gives us a correct inference, since it contains the parameter value of $p = 0.50$. Does your confidence interval contain 0.50? For about 5% of you, it won't.

Now collect a separate sample of size 50 and construct another 95% confidence interval. This time we got 28 outcomes of one type and 22 of the other, a sample proportion of 28/50 = 0.56, and a 95% confidence interval of (0.42, 0.70). This again gives a correct inference. Is your inference correct?

To get a feel for what happens "in the long run," do this simulation 100 times. (Do this with the applet by setting the number of simulations = 100.) You will then form 100 separate 95% confidence intervals. Perhaps not exactly 95 of the 100 were correct (containing the true proportion value of 0.50), since this is a random process, but probably somewhere between 90% and 99% were correct. Now do this a much larger number of times (say, 10,000 times). You will see that close to 95% of the confidence intervals contained the true parameter.

---

[5] See article by A. Agresti and B. Coull (who proposed this small-sample confidence interval), *The American Statistician*, vol. 52, pp. 119-126, 1998.

To practice this activity, try Exercises 7.26 and 7.27.

SECTION 7.2: PRACTICING THE BASICS

7.11   Drug abuse: In 2004 the National Center on Addiction and Substance Abuse at Columbia University (www.casacolumbia.org) took a random telephone survey of 1000 students age 12-17. Among other things, students were asked, "What is the most important problem facing people your age?" with options including drugs, crime, peer pressure, sexuality, and the environment. The most common response was drugs, cited by 29%. A story about the survey stated, "It has a margin of error of plus or minus 3 percentage points." Explain how they got this.

7.12   Crime victims: A recent General Social Survey asked, "During the last year, did anyone take something from you by using force -- such as a stickup, mugging, or threat?" Of 987 subjects, 17 answered yes and 970 answered no.
a.   Find the point estimate of the proportion of the population who were victims.
b.   Find the standard error of this estimate.
c.   Find the margin of error for a 95% confidence interval.
d.   Construct the 95% confidence interval for the population proportion. Can you conclude that fewer than 10% of all adults in the U.S. were victims?

7.13   How green are you?: When the 2000 GSS asked subjects (variable GRNSOL) whether they would be willing to accept cuts in their standard of living to protect the environment, 344 of 1170 subjects said yes.
a.   Estimate the population proportion who would answer yes.
b.   Find the margin of error for this estimate.
c.   Find a 95% confidence interval for that proportion. What do the numbers in this interval represent?
d.   State and check the assumptions needed for the interval in (c) to be valid.

7.14   Make industry help environment?: When the GSS recently asked subjects whether it should or should not be the government's responsibility to impose strict laws to make industry do less damage to the environment (variable GRNLAWS), 1093 of 1232 subjects said yes.
a.   What assumptions are made to construct a 95% confidence interval for the population proportion who would say yes? Do they seem satisfied here?
b.   Construct the 95% confidence interval. Interpret in context. Can you conclude whether a majority or minority of the population would answer yes?

7.15   Favor death penalty: In the 2000 General Social Survey, respondents were asked whether they favored or opposed the death penalty for people convicted of murder. Software shows results

```
     X        N        Sample p        95.0% CI
   1764     2565        0.6877       (0.670, 0.706)
```

where X refers to the number of the respondents who were in favor.
a.   Show how to obtain the value reported under "Sample p".
b.   Interpret the confidence interval reported, in context.

c.   Explain what the "95% confidence" refers to, by describing the "long-run" interpretation.

d.   Can you conclude that more than half of all American adults were in favor? Why?

7.16   Oppose death penalty: Refer to the previous exercise. Show how you can get a 95% confidence interval for the proportion of American adults who were *opposed* to the death penalty from the confidence interval stated in the previous exercise for the proportion in favor. (*Hint*: The proportion opposed is 1 minus the proportion in favor.)

7.17   Reincarnation: A Harris poll (www.harrisinteractive.com) of a random sample of 2201 adults in the U.S. in 2003 reported that 90% believe in God, 82% believe in heaven, 69% believe in hell (although only 2% of those who did so expected to go there), and 51% believe in ghosts. The results on belief in reincarnation, using MINITAB software, are:

```
    X     N    Sample p        95% CI
   594  2201  0.269877  (0.251333, 0.288433)
```

where X denotes the number who believed in reincarnation. Explain how to interpret "Sample p" and "95% CI" on this printout.

7.18   *z*-score and confidence level: Which *z*-score is used in a (a) 90%, (b) 98%, (c) 99.9% confidence interval for a population proportion?

7.19   Fear of breast cancer: A recent survey of 1000 American women between the ages of 45 and 64 asked them what medical condition they most feared. Of those sampled, 61% said breast cancer, 8% said heart disease, and the rest picked other conditions. By contrast, currently about 3% of female deaths are due to breast cancer whereas 32% are due to heart disease.[6]

a.   Construct a 90% confidence interval for the population proportion of women who most feared breast cancer. Interpret.

b.   Indicate the assumptions you must make for the inference in (a) to be valid.

7.20   Wife doesn't want kids: The 1996 GSS asked, "If the husband in a family wants children, but the wife decides that she does not want any children, is it all right for the wife to refuse to have children?" Of 708 respondents, 576 said yes.

a.   Find a 99% confidence interval for the population proportion who would say yes. Can you conclude that the population proportion exceeds 75%? Why?

b.   Without doing any calculation, explain whether the interval in (a) would be wider or narrower than a 95% confidence interval for the population proportion who would say yes.

7.21   Exit poll predictions: A national television network takes an exit poll of 1400 voters after each has cast a vote in a state gubernatorial election. Of them, 660 say they voted for the Democratic candidate and 740 say they voted for the Republican candidate.

---

[6]   See B. Lomborg, *The Skeptical Environmentalist*, Cambridge Univ. Press, 2001, p. 222.

a.   Treating the sample as a random sample from the population of all voters, would you predict the winner? Base your decision on a 95% confidence interval.
b.   Base your decision on a 99% confidence interval. Explain why you need stronger evidence to make a prediction when you want greater confidence.

7.22   Exit poll with smaller sample: In the previous exercise, "Exit poll predictions", suppose the same proportions resulted from a much smaller sample, with counts 66 and 74 instead of 660 and 740.
a.   Now does a 95% confidence interval allow you to predict the winner? Explain.
b.   Explain why the same proportions but with smaller samples provide less information. (*Hint*: What effect does *n* have on the standard error?)

7.23   Do you like tofu? You randomly sample five students at your school to estimate the proportion of students who like tofu. All five students say they like it.
a.   Find the sample proportion who like it.
b.   Find the standard error. Does its usual interpretation make sense?
c.   Find a 95% confidence interval, using the large-sample formula. Is it sensible to conclude that *all* students at your school like tofu?
d.   Why is it not appropriate to use the large-sample confidence interval in (c)? Use a more appropriate approach, and interpret the result.

7.24   Alleviate PMS?: A pharmaceutical company proposes a new drug treatment for alleviating symptoms of PMS (pre-menstrual syndrome). In the first stages of a clinical trial, it was successful for 7 out of 10 women.
a.   Construct a 95% confidence interval for the population proportion.
b.   Is it plausible that it's successful for only half the population? Explain.

7.25   Accept a credit card?: A bank wants to estimate the proportion of people who would agree to take a credit card they offer if they send a particular mailing advertising it. For a trial mailing to a random sample of 100 potential customers, 0 people accept the offer. Can they conclude that fewer than 10% of their population of potential customers would take the credit card? Answer by finding a 95% confidence interval.

7.26   Simulating confidence intervals: Repeat the simulation activity at the end of the section, but this time forming 1000 90% confidence intervals. What percentage of the 1000 confidence intervals contained *p* = 0.50? What percentage did you expect?

7.27   Simulating poor confidence intervals: Using the *confidence interval on a proportion* applet at www.prenhall.com/TBA, let's check that the large-sample confidence interval for a proportion may work poorly with small samples. Set *n* = 10 and *p* = 0.10. Generate 100 random samples, each of size 10, and for each one, form a 95% confidence interval for *p*.
a.   How many of the intervals fail to contain the true value, *p* = 0.10?
b.   How many would you expect not to contain the true value? What does this suggest?
c.   To see that this is not a fluke, now take 1000 samples and see what percentage of 95% confidence intervals contain 0.10. (Note: For every interval formed, the number of successes is smaller than 15, so the large-sample formula is not adequate.)

d.   Using the *sampling distribution* applet, generate 1000 random samples of size 10 when $p = 0.10$. Plot the empirical sampling distribution of the sample proportion values. Is it bellshaped and symmetric? Use this to help explain why the large-sample confidence interval performs poorly in this case.

7.3 HOW CAN WE CONSTRUCT A CONFIDENCE INTERVAL TO ESTIMATE A POPULATION MEAN?

We've learned how to construct a confidence interval for a population proportion -- a summary parameter for a categorical variable. Next we'll learn how to construct a confidence interval for a population mean--a summary parameter for a quantitative variable. We'll analyze GSS data to estimate the mean number of hours per day that Americans watch television. The method resembles that for a proportion. The margin of error again equals a multiple of a standard error. The confidence interval again has the form

$$\text{point estimate} \pm \text{margin of error}.$$

To apply this formula, what do you think plays the role of the point estimate and the role of the standard error (*se*)?

How to Construct a Confidence Interval for a Population Mean

RECALL

Section 6.5 introduced the standard error of the sampling distribution of $\bar{x}$, which describes how the sample mean varies from sample to sample of a given size n.

The sample mean $\bar{x}$ is the point estimate of the population mean μ. In Section 6.5, we learned that the exact standard error of the sample mean equals $\sigma / \sqrt{n}$, where σ is the population standard deviation. Like the exact standard error of the sample proportion, the exact standard error of the sample mean depends on a parameter whose value is unknown, in this case σ. In practice, we estimate σ by the sample standard deviation s. Then the standard error used in confidence intervals is

$$se = s/\sqrt{n} \,.$$

We again get a 95% confidence interval by taking the point estimate (the sample mean) and adding and subtracting the margin of error. We'll see the details following the next example.

EXAMPLE 7: ON THE AVERAGE, HOW MUCH TELEVISION DO YOU WATCH?

ACTIVITY

Try finding n, the mean, standard deviation, and *se* at the GSS website by entering the variable TVHOURS for the 2002 survey.

Picture the Scenario
Are you a couch potato? How much of the typical American's day is spent in front of the TV set? In recent years excessive TV watching has been claimed to be one factor, other than diet, for the increasing proportion of Americans who suffer from obesity. A recent General Social Survey asked respondents, "On the average day, about how many hours do you personally watch television?" A computer printout (from MINITAB) summarizes the results for the variable TV:

```
Variable    N      Mean    StDev    SE Mean    95.0% CI
TV         905    2.983   2.361    .0785     (2.83, 3.14)
```

We see that for the variable denoted by TV, the sample size was 905, the sample mean was 2.98, the sample standard deviation was 2.36, the standard error of the sample mean was 0.0785, and the 95% confidence interval for the population mean time μ spent watching TV goes from 2.83 to 3.14 hours per day.

Questions to Explore
a. What do the sample mean and standard deviation suggest about the likely shape of the population distribution?
b. How did the software get the standard error? What does it mean?
c. Interpret the 95% confidence interval reported by software.

Think It Through
a. For the sample mean of $\bar{x}$ = 2.98 and standard deviation of s = 2.36, the lowest possible value of 0 falls only a bit more than one standard deviation below the mean. This information suggests that the population distribution of TV watching may be skewed to the right. Figure 7.6 shows a histogram of the data, which suggests the same thing. The median was 2, the lower and upper quartiles were 1 and 4, the 95th percentile was 7, yet some subjects reported much higher values. One subject reported watching 24 hours a day!

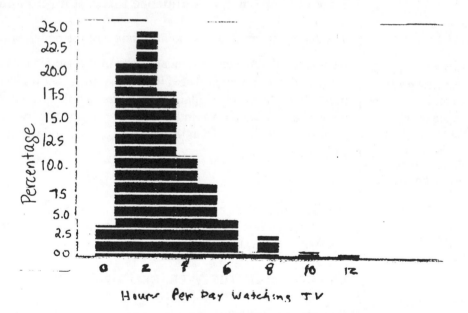

Figure 7.6: **Histogram of Number of Hours a Day Watching Television. Question:** Does this skew affect the validity of a confidence interval for the population mean?

b. With sample standard deviation s = 2.36 and sample size n = 905, the standard error of the sample mean is

$$se = s/\sqrt{n} = 2.36/\sqrt{905} = 0.0785 \text{ hours.}$$

If many studies were conducted about TV watching, with $n = 905$ for each, the sample mean would vary little among those studies.

c. A 95% confidence interval for the population mean report μ of TV watching in the U.S. is (2.83, 3.14) hours. The point estimate is 2.98 hours, and the confidence interval predicts that μ is no smaller than 2.83 hours and no greater than 3.14 hours. On the average, we infer that Americans watch about 3 hours of TV a day.

Insight
Because this GSS question produced a relatively large sample, the estimation is precise and the confidence interval is quite narrow. The larger the sample size, the smaller the standard error and the subsequent margin of error.

◆

To practice this concept, try Exercise 7.28.

We have not yet seen how software found the confidence interval in Example 7. Is this inference valid, given the evidence that the population distribution is skewed? We'll see that skew is usually not a problem when n is large, as it is here, since by the central limit theorem the sampling distribution of the sample mean is bell shaped. However, some studies use small sample sizes. For example, each observation may result from an expensive experimental procedure. A consumer group that estimates the mean repair cost after a new-model automobile crashes into a concrete wall at 30 miles per hour would probably not want to have to crash a large sample of cars to invoke the central limit theorem!

We'll now learn how to construct a confidence interval for a population mean that holds regardless of the sample size n. As with the proportion, the margin of error for a 95% confidence interval will be roughly two standard errors. However, we will need to introduce a new distribution similar to the normal distribution to give us a more precise margin of error. We'll find the margin of error by multiplying se by a score that is a bit different from the z-score when n is small but very similar to it when n is large.

The *t*-Distribution and Its Properties

To form a confidence interval that applies with any n, we must make the additional assumption that the *population distribution of the variable is normal*. In that case, the sampling distribution of $\overline{x}$ is normal even for small sample sizes. (The right panel of Figure 6.19, which showed sampling distributions for various shapes of population distributions, illustrated this, as does the figure here in the margin.) When the population distribution is normal, the sampling distribution of $\overline{x}$ is normal for all n, not just large n.

Suppose we knew the exact standard error, $\sigma/\sqrt{n}$, of the sample mean. Then, with the additional assumption that the population is normal, with small n we could use the formula $\overline{x} \pm z(se)$ with that exact standard error, for instance with $z = 1.96$ for 95% confidence. In practice, however, we don't know the population standard deviation σ. Substituting the sample standard deviation s for σ to get $se = s/\sqrt{n}$ then introduces extra error. This error can be sizeable when n is small. To account for this increased error, we must replace the z-score by a slightly larger score, called a **t-score**. The confidence interval is then a bit wider. *The t-score is like a z-score but it comes from a bell shaped distribution that has slightly thicker tails than a normal distribution.* This distribution is called the **t-distribution**.

IN PRACTICE

In practice, we estimate the standard error of the sample mean by $se = s/\sqrt{n}$. Then, we multiply *se* by a *t*-score from the **t distribution** to get the margin of error for a confidence interval for the population mean.

RECALL

From Section 6.3, the **standard normal** distribution has mean 0 and standard deviation 1.

The *t*-distribution resembles the *standard normal* distribution, being bell shaped around a mean of 0. Its standard deviation is a bit larger than 1, the precise value depending on what is called the **degrees of freedom**, denoted by *df*. For inference about a population mean, the degrees of freedom equal $df = n - 1$, one less than the sample size. Before presenting this confidence interval for a mean, we list the major properties of the *t*-distribution.

Properties of the *t*-distribution

- The *t*-distribution is bell shaped and symmetric about 0.

- The probabilities depend on the degrees of freedom, *df*. The *t* distribution has a slightly different shape for each distinct value of *df*, and different *t*-scores apply for each *df* value.

- The *t*-distribution has thicker tails and is more spread out than the standard normal distribution. The larger the *df* value, however, the closer it gets to the standard normal. Figure 7.7 illustrates this point. When *df* is about 30 or more, the two distributions are nearly identical.

- A *t*-score multiplied by the standard error gives the margin of error for a confidence interval for the mean.

Table B at the end of the text lists *t*-scores from the *t* distribution for the right-tail probabilities of 0.100, 0.050, 0.025, 0.010, 0.005, and 0.001. The table labels these by $t_{.100}$, $t_{.050}$, $t_{.025}$, $t_{.010}$, $t_{.005}$, and $t_{.001}$. For instance, $t_{.025}$ has probability 0.025 in the right tail, a two-tail probability of 0.05, and is used in 95% confidence intervals. Statistical software reports *t*-scores for any tail probability.

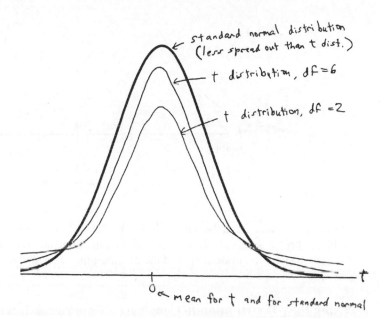

standard normal distribution
(less spread out than t dist.)

t distribution, df = 6

t distribution, df = 2

0 ← mean for t and for standard normal

Figure 7.7: The *t*-Distribution Relative to the Standard Normal Distribution. The *t* distribution gets closer to the standard normal as the degrees of freedom (*df*) increase. The two are practically identical when $df \geq 30$. **Question:** Can you find any *z*-scores (such as 1.96) for a normal distribution on the *t* table (Table B)?

Table 7.3 is an excerpt from the *t*-table (Table B). To illustrate its use, suppose the sample size is 7. Then, the degrees of freedom $df = n - 1 = 6$. Row 6 of the *t* table shows the *t*-scores for $df = 6$. The column labeled $t_{.025}$ contains *t*-scores with right-tail probability equal to 0.025. With $df = 6$, this *t*-score is $t_{.025} = 2.447$. This means that 2.5% of the *t* distribution falls in the right tail above 2.447. By symmetry, 2.5% also falls in the left tail below $-t_{.025} = -2.447$. Figure 7.8 illustrates. When $df = 6$, the probability equals 0.95 between -2.447 and 2.447. This is the *t*-score for a 95% confidence interval when $n = 7$. The interval is $\overline{x} \pm 2.447(se)$.

Table 7.3: Part of Table B Displaying *t*-Scores. The scores have right-tail probabilities of 0.100, 0.050, 0.025, 0.010, 0.005, and 0.001. When $n = 7$, $df = 6$, and $t_{.025} = 2.447$ is the *t*-score with right-tail probability = 0.025 and two-tail probability = 0.05. It is used in a 95% confidence interval, $\overline{x} \pm 2.447(se)$.

	Confidence Level					
	80%	90%	95%	98%	99%	99.8%
df	$t_{.100}$	$t_{.050}$	$t_{.025}$	$t_{.010}$	$t_{.005}$	$t_{.001}$
1	3.078	6.314	12.706	31.821	63.657	318.3
...						
6	1.440	1.943	2.447	3.143	3.707	5.208
7	1.415	1.895	2.365	2.998	3.499	4.785

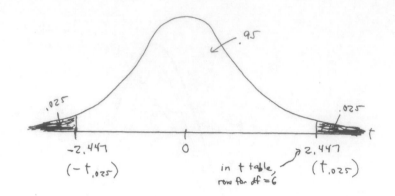

Figure 7.8: The *t*-Distribution with *df* = 6. 95% of the distribution falls between -2.447 and 2.447. These *t*-scores are used with a 95% confidence interval when *n* = 7. **Question:** Which *t*-scores contain the middle 99% of the distribution, when *df* = 6?

Using the *t*-Distribution to Construct a Confidence Interval for a Mean

The confidence interval for a mean has margin of error that equals a *t*-score times the standard error.

Summary: 95% Confidence Interval for a Population Mean

A 95% confidence interval for the population mean μ is

$$\bar{x} \pm t_{.025}\,(se), \text{ where } se = s/\sqrt{n}\,.$$

Here, $df = n - 1$ for the *t*-score $t_{.025}$ that has right-tail probability 0.025 (total probability 0.05 in the two tails and 0.95 between $-t_{.025}$ and $t_{.025}$). To use this method, you need

- Data obtained by randomization (such as a random sample)
- An approximately normal population distribution.

EXAMPLE 8: EBAY AUCTIONS OF PALM HANDHELD COMPUTERS

Picture the Scenario
eBay is a popular Internet company for personal auctioning of just about anything. When you list an item to sell on eBay, there is an online auction format in which the product sells for the highest price bid over a set period of time (1, 3, 5, 7, or 10 days). In addition, you can offer potential buyers a "buy-it-now" option, whereby they can buy the product immediately at a fixed price that you set.

Do you tend to get a higher, or a lower, price if you give bidders the "buy-it-now" option? Let's consider some data from sales of the Palm M515 PDA (personal digital assistant), a popular handheld computer, during the first week of May 2003. During that week, 25 of these handheld computers were auctioned off, 7 of which had the "buy-it-now" option. Here are the final prices (in dollars) at which the item sold:

Buy-it-now option: 235, 225, 225, 240, 250, 250, 210
Bidding only: 250, 249, 255, 200, 199, 240, 22 8, 255, 232,
                 246, 210, 178, 246, 240, 245, 225, 246, 225

Questions to Explore
a. Use numerical and graphical descriptive statistics to summarize selling prices for the two types of auctions.
b. Consider the probability distribution of selling prices with the buy-it-now option. State and check the assumptions for using these data to find a 95% confidence interval for the mean of that distribution.
c. Find the 95% confidence interval for the buy-it-now option, and interpret it. How does it compare to the 95% confidence interval for the mean sales price for the bidding only option, which in Exercise 7.32 you'll see is (220.70, 242.52)?

Think It Through
a. When we use MINITAB and request descriptive statistics, we get the mean and standard deviation and the five-number summary using quartiles:

buy_now	N	Mean	StDev	Minimum	Q1	Median	Q3	Maximum
no	18	231.61	21.94	178.0	221.25	240.0	246.75	255.0
yes	7	233.57	14.64	210.0	225.00	235.0	250.00	250.0

The sample mean selling price is a bit higher for items with the buy-it-now option, $233.57 compared to $231.61, but the medians have the reverse ordering. There's a bit less variability in the selling prices with the buy-it-now option.

Figure 7.9 shows a MINITAB dot plot for each set of selling prices. The plot does show the greater variability for the 18 observations not having the buy-it-now option ass well as evidence of skew to the left, but this is partly caused by a single quite low observation at $178.

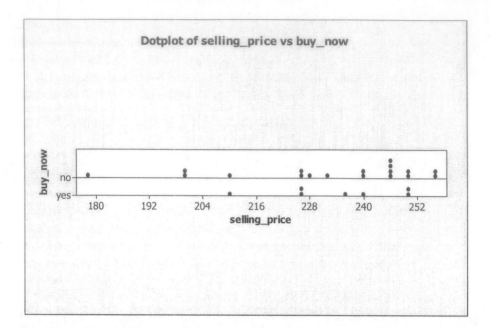

Figure 7.9: MINITAB Dot Plot for eBay Selling Prices for Palm Handheld Computers. Data are shown for 7 observations in which buyers had the buy-it-now option and 18 observations in which they did not have this option. **Question**: What does this plot tell you about how the selling prices compare for the two cases?

b. To construct a confidence interval using the *t*-distribution, we must assume a random sample from an approximately normal population distribution of selling prices. Unlike a survey such as the GSS, with the eBay selling prices, the sample and population distinction is not so clear. We'll regard the observed selling prices as independent observations from a probability distribution for selling price at that point in time (May 2003). That probability distribution reflects the variability in selling price from auction to auction. Inference treats that probability distribution like a population distribution and treats the observed selling prices as a random sample from it.

We can get some information about the shape of the probability distribution of selling price by looking at Figure 7.9. With only 7 observations for the buy-it-now option, it's hard to tell much. A bell-shaped, discrete population distribution could well generate sample data such as shown in the second line of Figure 7.9. Later in this section, we'll discuss this assumption further.

c. Let μ denote the population mean for the buy-it-now option. The estimate of μ is the sample mean of 235, 225, 225, 240, 250, 250, 210, which is $\bar{x}$ = \$233.57. This and the sample standard deviation of s = \$14.64 are reported in the above table. The standard error of the sample mean is $se = s/\sqrt{n}$ = 14.64/$\sqrt{7}$ = 5.53. Since $n = 7$, the degrees of freedom are $df = n\text{-}1 = 6$. For a 95% confidence interval, from Table 7.3 we use $t_{.025}$ = 2.447. The 95% confidence interval is

$$\bar{x} \pm t_{.025}(se) = 233.57 \pm 2.447(5.53),$$
which is 233.57 $\pm$ 13.54, or (220.03, 247.11).

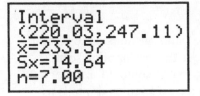

TI-83+/84 output

With 95% confidence, the range of believable values for the mean selling price for this probability distribution is $220.03 to $247.11. In the margin, a screen shot shows how the TI-83+/84 reports this confidence interval. Table 7.4 shows the way MINITAB reports it, with "SE Mean" being the standard error of the sample mean.

Table 7.4: MINITAB Output for 95% Confidence Interval for Mean

Variable	N	Mean	StDev	SE Mean	95.0% CI
selling_price	7	233.57	14.64	5.53	(220.03, 247.11)

This confidence interval is not much different from the confidence interval (220.70, 242.52) for the mean without the buy-it-now option. The intervals overlap a lot. There's not enough information for us here to conclude that one probability distribution clearly has a higher mean than the other.

Insight

With small samples, we usually must sacrifice precision. In Exercise 7.36 we'll make a similar comparison using a larger data set, and we will be able to infer that one option is better. In Chapter 9 we'll learn about inferential methods directed specifically toward comparing population means or proportions.

♦

To practice this concept, try Exercise 7.31.

How Do We Find a *t* Confidence Interval for Other Confidence Levels?

The 95% confidence interval uses $t_{.025}$, the *t*-score for a right-tail probability of 0.025, since 95% of the probability falls between $-t_{.025}$ and $t_{.025}$. For 99% confidence, the error probability is 0.01, the probability is $0.01/2 = 0.005$ in each tail, and the appropriate *t*-score is $t_{.005}$. The top margin of Table B and Table 7.3 show both the *t*-subscript notation and the confidence level.

For instance, the buy-it-now sample in the above example has $n = 7$, so $df = 6$ and from Table 7.3, $t_{.005} = 3.707$. A 99% confidence interval for the mean selling price is (again using $\bar{x} = 233.57$ and $se = 5.53$, as we found in Example 8)

$$\bar{x} \pm t_{.005}(se) = 233.57 \pm 3.707(5.53),$$
which is 233.57 ± 20.50, or (213.07, 254.07).

With a larger confidence level, the confidence interval is wider than the 95% confidence interval of (220.03, 247.11).

♦

To practice this concept, try Exercise 7.33.

If the Population Is Not Normal, is the Method "Robust"?

A basic assumption of the confidence interval using the t distribution is that the population distribution is normal. This is worrisome, because many variables have distributions that are far from a bell shape. How problematic is it if we use the t confidence interval even if the population distribution is not normal? For large random samples, it's not problematic, because of the central limit theorem. The sampling distribution is bell shaped even when the population is not. But what about for small n?

For the confidence interval in Example 8 with $n = 7$ to be valid, we must assume that the probability distribution of selling price is normal. Does this assumption seem plausible? A dot plot, histogram, or stem-and-leaf plot gives us some information about the population distribution, but it is not precise when n is small and it tells us little when $n = 7$. Fortunately, the confidence interval using the t-distribution is a **robust** method in terms of the normality assumption.

Robust Statistical Method

A statistical method is said to be **robust** with respect to a particular assumption if it performs adequately even when that assumption is violated.

Even if the population distribution is not normal, confidence intervals using t-scores usually work quite well. The actual probability that the 95% confidence interval method provides a correct inference is close to 0.95 and gets closer as n increases.

RECALL

Section 2.5 identified an observation as a potential **outlier** if it falls more than $1.5 \times IQR$ below the first quartile or above the third quartile, or if it falls more than 3 standard deviations from the mean.

The most important case when the t confidence interval method does *not* work well is with binary data, in which case the mean is a proportion. Section 7.2 presented a separate method for binary data. Another case that calls for caution is when the data contain extreme outliers. Partly this is because of the effect on the method, but also because the mean itself may not then be a relevant summary. In Example 8 with the 7 observations in the "buy-it-now" sample, you can check that no potential outliers are identified.

The t confidence interval method is not robust to violations of the random sampling assumption. The t method, like all inferential statistical methods, has questionable validity if the method for producing the data did not use randomization.

IN PRACTICE

Knowing that a statistical method is **robust** (that is, it still performs adequately) even when a particular assumption is violated is important, because in practice assumptions are rarely perfectly satisfied. Confidence intervals for a mean using the *t* **distribution** are robust against most violations of the normal population assumption. However, you should check the data graphically to identify **outliers** that could affect the validity of the mean or its confidence interval. Also, unless the data production used **randomization**, statistical inference may be inappropriate.

The Standard Normal Distribution is the *t*-Distribution with *df* = ∞

Look at the table of *t* scores (Table B in the Appendix), part of which is shown in Table 7.5 here. As *df* increases, you move down the table. The *t*-score decreases toward the *z*-score for a standard normal distribution. For instance, when *df* increases from 1 to 100 in Table B, the *t*-score $t_{.025}$ that has right-tail probability equal to 0.025 decreases from 12.706 to 1.984. This reflects the *t* distribution becoming less spread out and more similar in appearance to the standard normal distribution as *df* increases. The *z*-score with right-tail probability of 0.025 for the standard normal distribution is $z = 1.96$. When *df* is above about 30, the *t*-score is similar to this *z*-score. For instance, they both round to 2.0. The *t*-score gets closer and closer to the *z*-score as *df* keeps increasing. *You can think of the standard normal distribution as a t distribution with df = infinity.*

Table 7.5: Part of Table B Displaying *t*-scores for large *df* Values. The *z*-score of 1.96 is the *t*-score $t_{.025}$ with right-tail probability of 0.025 and *df* = ∞.

	Confidence level					
	80%	90%	95%	98%	99%	99.8%
df	$t_{.100}$	$t_{.050}$	$t_{.025}$	$t_{.010}$	$t_{.005}$	$t_{.001}$
1	3.078	6.314	12.706	31.821	63.657	318.3
...						
28	1.313	1.701	2.048	2.467	2.763	3.408
29	1.311	1.699	2.045	2.462	2.756	3.396
30	1.310	1.697	2.042	2.457	2.750	3.385
50	1.299	1.676	2.009	2.403	2.678	3.261
100	1.290	1.660	1.984	2.364	2.626	3.174
∞	1.282	1.645	1.960	2.326	2.576	3.090

DID YOU KNOW?

z-score = *t*-score with df = ∞ (infinity)

Table 7.5 shows the first row and the last several rows of Table B. The last row lists the *z*-scores for various confidence levels, opposite *df* = ∞ (infinity). The *t*-scores are not printed for *df* > 100, but they are close to the *z*-scores. For instance, to get the confidence interval about TV watching in Example 7, for which the GSS sample had $n = 2337$, software uses the $t_{.025}$ score for $df = 2337 - 1 = 2336$, which is 1.961. This is nearly identical to the *z*-score of 1.960 from the standard normal distribution.

Recall that the reason we use a t-score instead of a z-score in the confidence interval for a mean is that it accounts for the extra error due to estimating σ by s. You can get t-scores for *any df* value using software and many calculators, so you are not restricted to Table B. (For instance, MINITAB provides percentile scores for various shapes of distributions under the *CALC* menu.) If you don't have access to software, you won't be far off if you use a z-score instead of a t-score for *df* values larger than shown in Table B (above 100). For a 95% confidence interval you will then use

$$\bar{x} \pm 1.96(se) \text{ instead of } \bar{x} \pm t_{.025}(se).$$

You will not get exactly the same result that software would give, but it will be very, very close.

IN PRACTICE

Statistical software and calculators use the t distribution for *all* cases when the sample standard deviation s is used to estimate the population standard deviation σ. The normal population assumption is mainly relevant for small n, but even then the t confidence interval is a robust method, working well unless there are extreme outliers or the data are binary.

On the Shoulders of.....William S. Gosset

How do you find the best way to brew beer if you have only small samples?

The statistician and chemist William S. Gosset was employed as a Brewer in charge of the experimental unit of Guinness Breweries in Dublin, Ireland. The search for a better stout in 1908 led him to the discovery of the t distribution. Only small samples were available from several of his experiments pertaining to the selection, cultivation, and treatment of barley and hops for the brewing process. The established statistical methods at that time relied on large samples and the normal distribution. Because of company policy forbidding the publication of company work in one's own name, Gosset used the pseudonym "Student" in articles he wrote about his discoveries. The t-distribution became known as *Student's* t-distribution, a name sometimes still used today.

PHOTO of Gosset and photo of bottle of Guinness

((take picture from biography by Pearson and Plackett))

SECTION 7.3: PRACTICING THE BASICS

7.28 Females' ideal number of children: The 2002 General Social Survey asked, "What do you think is the ideal number of children for a family to have?" The 497 females who responded had a median of 2, mean of 3.02, and standard deviation of 1.81.
a. What is the point estimate of the population mean?
b. Find the standard error of the sample mean.
c. The 95% confidence interval is (2.89, 3.21). Interpret.
d. Is it plausible that the population mean $\mu = 2$? Explain.

7.29 Males' ideal number of children: Refer to the previous exercise. For the 397 males in the sample, the mean was 2.89 and the standard deviation was 1.77.
a. Find the point estimate of the population mean, and show that its standard error is 0.089.
b. The 95% confidence interval is (2.72, 3.06). Explain what "95% confidence" means for this interval.

7.30 Using *t* table: Using Table B or software or a calculator, report the *t*-score which you multiply by the standard error to form the margin of error for a
a. 95% confidence interval for a mean with 5 observations.
b. 95% confidence interval for a mean with 15 observations.
c. 99% confidence interval for a mean with 15 observations.

▣7.31 Anorexia in teenage girls: A study[7] compared various therapies for teenage girls suffering from anorexia, an eating disorder. For each girl, weight was measured before and after a fixed period of treatment. The variable measured was the change in weight, X = weight at the end of the study minus weight at the beginning of the study. The therapies were designed to aid weight gain, corresponding to positive values of X. For the sample of 17 girls receiving the family therapy, the changes in weight during the study were:

$$11, 11, 6, 9, 14, -3, 0, 7, 22, -5, -4, 13, 13, 9, 4, 6, 11.$$

a. Plot these with a dot plot or box plot, and summarize.
b. Verify that the weight changes have $\bar{x} = 7.29$ and $s = 7.18$ pounds.
c. Verify that the standard error of the sample mean was $se = 1.74$.
d. To use the *t*-distribution, explain why $df = 16$ and a 95% confidence interval uses the *t*-score equal to 2.120.
e. Let μ denote the population mean change in weight for this therapy. Verify that the 95% confidence interval for μ is (3.6, 11.0). Explain why this suggests that the true mean change in weight is positive, but possibly quite small.

7.32 eBay without buy-it-now option: From Example 8, selling prices of the Palm M515 PDA for a one-week selling period in May 2003 when the buy-it-now option was not used were:

250, 249, 255, 200, 199, 240, 228, 255, 232, 246, 210, 178, 246, 240, 245, 225, 246, 225.

[7] Data courtesy of Prof. Brian Everitt, Institute of Psychiatry, London.

These data have $\bar{x} = 231.61$, $s = 21.94$, Q1 = 221.25, Median = 240.0, Q3 = 246.75.

a. What assumptions are needed to construct a 95% confidence interval for μ? Point out any that seem questionable, based on a histogram or the dot plot in Figure 7.9.

b. Show that the 95% confidence interval is (220.70, 242.52). Interpret it in context.

c. Check whether this data set has any potential outliers according to the criterion of (i) $1.5 \times$ IQR below Q1 or above Q3, (ii) 3 standard deviations from the mean.

d. Figure 7.9 shows that the $178 observation is quite a bit lower than the others. Deleting this observation, find the mean and standard deviation, and construct the 95% confidence interval for μ. How does it compare to the 95% confidence interval (220.70, 242.52) using all the data?

7.33 New graduates' income: Exercise 7.6 reported annual incomes of $45,000, $35,000, and $55,000 for three new college graduates with math majors.

a. Using software or a calculator, verify that the 95% confidence interval for the population mean goes from $20,159 to $69,841.

b. Name two things you could do to get a narrower interval than the one in (a).

c. Construct a 99% confidence interval. Why is it wider than the 95% interval?

d. On what assumptions are the interval in (a) based? Explain how important each assumption is.

7.34 TV watching for Muslims: Having estimated the mean amount of time spent watching TV in Example 7, we might want to estimate the mean for various groups, such as different religious groups. Let's consider Muslims. A recent GSS had responses on TV watching from 7 subjects who identified their religion as Muslim. Their responses on the number of hours of TV watching were 0, 0, 2, 2, 2, 4, 6, shown also in the accompanying dot plot.

TV Watching for Muslims in GSS sample

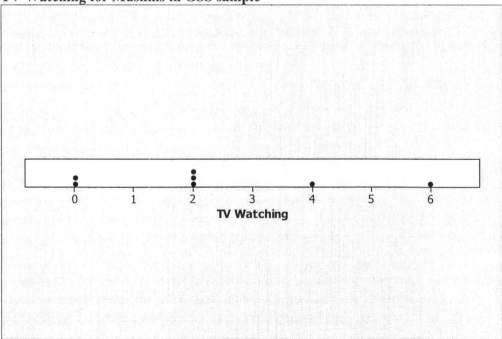

Minitab output for Muslim TV watching

```
Variable     N    Mean   StDev   SE Mean   95.0% CI
television   7    2.286  2.138   0.808     (0.308,   4.264)
```
---

a. What must we assume to use these data to find a 95% confidence interval for the mean amount of TV watched by the population of American Muslims?
b. The table shows the way Minitab reports results. Explain how to interpret the 95% confidence interval in context.
c. What is the main factor causing the confidence interval to be so wide?
d. What t-score would you multiply the standard error by to get the margin of error for a 99% confidence interval? That interval is (-0.7, 5.3). Explain why, in practice, you would report this as (0.0, 5.3).

7.35 Wage discrimination?: According to a union agreement, the mean income for all senior-level assembly-line workers in a large company equals $500 per week. A representative of a women's group decides to analyze whether the mean income for female employees matches this norm. For a random sample of nine female employees, using software she obtains a 95.0% confidence interval of (371, 509). Explain what is wrong with each of the following interpretations of this interval.
a. We infer that 95% of the women in the population have income between $371 and $509 per week.
b. If random samples of 9 women were repeatedly selected, then 95% of the time the sample mean income would be between $371 and $509.
c. We can be 95% confident that $\bar{x}$ is between $371 and $509.
d. If we repeatedly sampled the entire population, then 95% of the time the population mean would be between $371 and $509.

7.36 More eBay selling prices: Example 8 analyzed selling prices of the Palm M515 PDA for a one-week selling period in May 2003. When prices are analyzed for a three-month period in 2003, we get the results shown in the table.

```
Variable   Buy now   N     Mean     SE Mean   StDev    Median
Price         No     136   223.36   2.04      23.74    225.01
              Yes    132   235.45   1.79      20.57    239.99
```

The 95% confidence intervals for the mean are (231.91, 238.99) with the buy-it-now option and (219.32, 227.40) without it. Interpret each of these intervals, and explain what you learn by comparing them.

7.37 Heart rate of infants: Recent findings have suggested that infant sex differences exist in behavioral and physiological reactions to stress. One study (M. Davis and E. Emory, *Child Development*, Vol. 66, 1995, pp. 14-27) evaluated changes in the heart rate for a sample of infants placed in a stressful situation. For the 15 female infants, a printout for the data on the change in heart rate shows:

```
Variable   N   df   Mean   StDev   SE Mean    95.0% CI
change     15  14   10.70  17.70   4.570      (0.90,  20.50)
```

a.   Show how the software obtained the value for "SE Mean." Explain what this represents.

b.   Explain how software obtained the value of *df*, and indicate which *t*-score was used in constructing the 95% confidence interval.

c.   From the confidence interval shown, can you conclude that the true mean change in heart rate is positive? Explain.

d.   Explain the implications of the term "robust" regarding the normality assumption made to conduct this analysis.

7.38 How often read a newspaper?: For the "Florida student survey" data file on the text CD, software reports the results for responses on the number of times a week the subject reads a newspaper:

```
Variable   N    Mean   Std Dev   SE Mean    95.0% CI
News       60   4.1    3.0       0.387      (3.325,  4.875)
```

a.   Is it plausible that $\mu = 7$, where μ is the population mean for all Florida students? Explain.

b.   Suppose that the sample size had been 240, with $\bar{x} = 4.1$ and $s = 3.0$. Find a 95% confidence interval, and compare it to the one reported. Describe the effect of sample size on the margin of error.

c.   Does it seem plausible that the population distribution of this variable is normal? Explain your reasoning, and indicate what effect this has on the validity of the analysis reported above.

7.39 Political views: The General Social Survey asks respondents to rate their political views on a seven-point scale, where 1 = extremely liberal, 4 = moderate, and 7 = extremely conservative. A researcher analyzing data from the 2002 GSS gets MINITAB output:

```
Variable   N      Mean   StDev   SE Mean    95% CI
Polviews   1331   4.12   1.39    0.0381     (4.05,  4.19)
```

a.   Show how to construct the confidence interval from the other information provided.

b.   Can you conclude that the population mean is higher than the moderate score of 4.0? Explain.

c.   Would the confidence interval be wider, or narrower, (i) if you constructed a 99% confidence interval, (ii) if $n = 500$ instead of 1331?

7.40 Length of hospital stays: A hospital administrator wants to estimate the mean length of stay for all inpatients using that hospital. Using a random sample of 100 records of patients for the previous year, she reports that "The sample mean was 5.3. In repeated random samples of this size, the sample mean could be expected to fall within 1.0 of the true mean about 95% of the time." Explain the meaning of this sentence from the report, showing what it suggests about the 95% confidence interval.

7.41 Effect of *n*: Find the margin of error when the sample standard deviation equals 100, with a sample size of (a) 400, (b) 1600. What is the effect of the sample size?

7.42 Effect of confidence level: Find the margin of error when the sample standard deviation equals 100 for a sample size of 400, using confidence level (a) 95%, (b) 99%. What is the effect of the choice of confidence level?

7.43 Catalog mail-order sales: A company that sells its products through mail-order catalogs wants information about the success of its most recent catalog. The company decides to estimate the mean dollar amount of items ordered from those who received the catalog. For a random sample of 100 customers from their files, only 5 made an order, so 95 of the response values were $0. The overall mean of all 100 orders was $10, with a standard deviation of $10.
a.   Is it plausible that the population distribution is normal? Explain, and discuss how much this affects the validity of a confidence interval for the mean.
b.   Find a 95% confidence interval for the mean dollar order for the population of all customers who received this catalog. Normally, the mean of their sales per catalog is about $15. Can we conclude that it declined with this catalog? Explain.

7.44 Number of children: For the question "How many children have you ever had?" use the GSS website www.icpsr.umich.edu/GSS with the variable NUMKIDS to find the sample mean and standard deviation for the 2000 survey.
a.   Show how to get the standard error of 0.03 that the GSS reports for a random sample of 2801 adults.
b.   Construct a 95% confidence interval for the population mean. Can you conclude that the population mean is less than 2.0? Explain.
c.   Discuss the assumptions for the analysis in (b) and whether that inference seems to be justified.

7.45 Simulating the confidence interval: Go to the *confidence interval for a mean* applet at www.prenhall.com/TBA. Choose the sample size of 50 and the skewed population distribution with $\mu = 100$. Generate 100 random samples, each of size 50, and for each one form a 95% confidence interval for the mean.
a.   How many of the intervals fail to contain the true value?
b.   How many would you expect not to contain the true value?
c.   Now repeat the simulation using 1000 random samples of size 50. Why do close to 95% of the intervals contain μ, even though the population distribution is quite skewed?

7.4 HOW DO WE CHOOSE THE SAMPLE SIZE FOR A STUDY?

Have you ever wondered how the sample sizes are determined for polls? How does a polling organization know whether it needs 10,000 people, or only 100 people, or some odd number such as 745? The simple answer is that this depends on how much precision is needed, as measured by the margin of error. The smaller the margin of error, the larger the sample size must be. We'll next learn how to determine which sample size has the desired margin of error. For instance, we'll find out how large an exit poll must be so that a 95% confidence interval for the population proportion voting for a candidate has a certain margin of error, such as 0.04.

The key results for finding the sample size for a random sample are:

- The *margin of error* depends on the *standard error* of the sampling distribution of the point estimate.

- The *standard error* itself depends on the *sample size*.

So one of the main components of the margin of error is the sample size *n*. Once we specify a margin of error with a particular confidence level, we can determine the value of *n* that has a standard error giving that margin of error.

Choosing the Sample Size for Estimating a Population Proportion

How large should *n* be to estimate a population proportion? First we must decide on the desired *margin of error*--how close the sample proportion should be to the population proportion. Second, we must choose the *confidence level* for achieving that margin of error. In practice, 95% confidence intervals are most common. If we specify a margin of error of 0.04, this means that a 95% confidence interval should equal the sample proportion plus and minus 0.04.

EXAMPLE 9: WHAT SAMPLE SIZE DO YOU NEED FOR AN EXIT POLL?

Picture the Scenario
A television network plans to predict the outcome of an election between Jacalyn Levin and Roberto Sanchez. They will do this with an exit poll that randomly samples voters on election day. They want a reasonably accurate estimate of the population proportion that voted for Levin. The final poll a week before election day estimated her to be well ahead, 58% to 42%, so they do not expect the outcome to be close. Since their finances for this project are limited, they don't want to collect a large sample if they don't need it. They decide to use a sample size for which the margin of error is 0.04, rather than their usual margin of error of 0.03.

Question to Explore
What is the sample size for which a 95% confidence interval for the population proportion has margin of error equal to 0.04?

Think It Through
The 95% confidence interval for a population proportion p is $\hat{p} \pm 1.96(se)$. So, if the sample size is such that $1.96(se) = 0.04$, then the margin of error will be 0.04. See Figure 7.10.

RECALL

From Section 7.2, a 95% confidence interval for a population proportion p is

$$\hat{p} \pm 1.96(se),$$

where $\hat{p}$ denotes the sample proportion and the standard error (*se*) is

$$se = \sqrt{\hat{p}(1 - \hat{p})/n} \,.$$

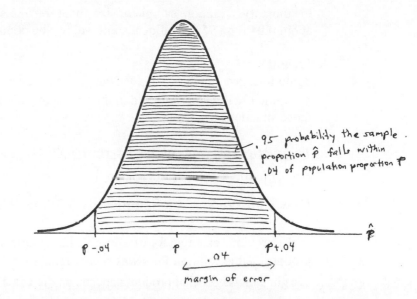

.95 probability the sample proportion $\hat{p}$ falls within .04 of population proportion p

margin of error

Figure 7.10: Sampling Distribution of Sample Proportion $\hat{p}$ **such that a 95% Confidence Interval Has Margin of Error 0.04.** We need to find the value of n that has this margin of error. **Question:** What must we assume for this distribution to be approximately normal?

Let's find the value of the sample size n for which $0.04 = 1.96(se)$. For a confidence interval for a proportion, the standard error is $\sqrt{\hat{p}(1-\hat{p})/n}$. So the equation $0.04 = 1.96(se)$ becomes

$$0.04 = 1.96\sqrt{\hat{p}(1-\hat{p})/n}.$$

To find the answer, we solve algebraically for n. Using a bit of algebra, you can check that

$$n = (1.96)^2\, \hat{p}(1 - \hat{p})/(0.04)^2.$$

(If your algebra is rusty, don't worry--we'll show a general formula below.)

Now, we face a problem. We're doing this calculation *before* gathering the data, so we don't yet have a sample proportion $\hat{p}$. The formula for n depends on $\hat{p}$ because the standard error depends on it. Since $\hat{p}$ is unknown, we must substitute an educated guess for what we'll get once we gather the sample and analyze the data.

Since the latest poll *before* election day predicted that 58% of the voters preferred Levin, it is sensible to substitute 0.58 for $\hat{p}$ in this equation. Then we find

$$n = (1.96)^2\, \hat{p}(1 - \hat{p})/(0.04)^2 = (1.96)^2(0.58)(0.42)/(0.04)^2 = 584.9.$$

In summary, a random sample of size about $n = 585$ should give a margin of error of about 0.04 for a 95% confidence interval for the population proportion.

Insight

Sometimes, we may have no idea what to expect for $\hat{p}$. We may prefer not to guess the value it will take, as we did in this example. We'll next learn what we can do in those situations.

♦

To practice this concept, try Exercise 7.47.

How Can We Select a Sample Size Without Guessing a Value for $\hat{p}$?

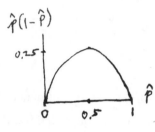

In the previous example for determining the sample size n for estimating a proportion, the solution for n was proportional to $\hat{p}(1 - \hat{p})$. The figure in the margin shows how that product depends on the value of $\hat{p}$. The largest possible value for $\hat{p}(1 - \hat{p})$ is 0.25, which occurs when $\hat{p} = 0.50$. You can check by plugging values of $\hat{p}$ into $\hat{p}(1 - \hat{p})$ that this product is near 0.25 unless $\hat{p}$ is quite far from 0.50. For example, $\hat{p}(1 - \hat{p}) = 0.24$ when $\hat{p} = 0.40$ or $\hat{p} = 0.60$.

In the formula for determining n, setting $\hat{p} = 0.50$ gives the largest value for n out of all the possible values to substitute for $\hat{p}$. So doing this is the "safe" approach that guarantees we'll have enough data. In the election exit-poll example above, for a margin of error of 0.04 we then get

$$n = (1.96)^2 \, \hat{p}(1 - \hat{p})/(0.04)^2$$
$$= (1.96)^2 (0.50)(0.50)/(0.04)^2 = 600.$$

This compares to $n = 585$ from guessing that $\hat{p} = 0.58$. Using the slightly larger value of $n = 600$ ensures that the margin of error for a 95% confidence interval will not exceed 0.04, no matter what value $\hat{p}$ takes once we collect the data.

This safe approach is not always sensible, however. Substituting $\hat{p} = 0.50$ gives us an overly large solution for n if $\hat{p}$ actually falls far from 0.50. Suppose that based on other studies we expect $\hat{p}$ to be about 0.10. Then an adequate sample size to achieve a margin of error of 0.04 is

$$n = (1.96)^2 \, \hat{p}(1 - \hat{p})/(0.04)^2$$
$$= (1.96)^2 (0.10)(0.90)/(0.04)^2 = 216.$$

A sample size of 600 would be much larger and more costly than needed.

General sample size formula for estimating a population proportion

A general formula exists for determining the sample size, based on solving algebraically for n by setting the margin of error formula equal to the desired value. Let m denote the desired margin of error. This is $m = 0.04$ in the above example. The general formula also uses the z-score for the confidence level.

Summary: Sample Size for Estimating a Population Proportion

The random sample size n for which a confidence interval for a population proportion p has margin of error m (such as $m = 0.04$) is

$$n = \frac{\hat{p}(1 - \hat{p})z^2}{m^2}.$$

The z-score is based on the confidence level, such as $z = 1.96$ for 95% confidence. You either guess the value you'd get for the sample proportion $\hat{p}$ based on other information or take the safe approach of setting $\hat{p} = 0.5$.

EXAMPLE 10: HOW LARGE SHOULD AN EXIT POLL BE IF A RACE IS CLOSE?

Picture the Scenario
An election is expected to be close. The pollsters who are planning an exit poll decide that a margin of error of 0.04 is too large.

Question to Explore
How large should the sample size be for the margin of error of a 95% confidence interval to equal 0.02?

Think It Through
Since the election is expected to be close, we expect $\hat{p}$ to be near 0.50. In the formula, we set $\hat{p} = 0.50$ to be safe. We also set the margin of error $m = 0.02$ and use $z = 1.96$ for a 95% confidence interval. The required sample size is

$$n = \frac{\hat{p}(1 - \hat{p})z^2}{m^2} = \frac{(0.50)(0.50)(1.96)^2}{(0.02)^2} = 2401.$$

Insight
The sample size of about 2400 is four times the sample size of 600 necessary to guarantee a margin of error of $m = 0.04$. *Reducing the margin of error by a factor of one-half requires quadrupling n.*

◆

To practice this concept, try Exercise 7.49.

Example 22 on p. 415 of the previous chapter described an exit poll for the U.S. senatorial race in 2002 between Hillary Clinton and Rick Lazio. That exit poll used a similar sample size ($n = 2232$) to the value of 2401 just determined. Did they achieve a margin of error of about 0.02? From that example, the sample proportion who voted for Clinton was 0.557. A 95% confidence interval for the population proportion p who voted for Clinton is $\hat{p} \pm 1.96(se)$, with $se = \sqrt{\hat{p}(1-\hat{p})/n} =$
$\sqrt{(0.557)(0.443)/2232} = 0.0105$. This gives the interval

$$0.557 \pm 1.96(0.0105), \text{ which is } 0.557 \pm 0.021, \text{ or } (0.536, 0.578).$$

In fact, the margin of error *was* 0.02 (rounded off). This was small enough to predict (correctly) that Clinton would win, since the confidence interval values (0.536, 0.578) all fell above 0.50.

Samples taken by polling organizations, such as the Gallup poll, typically contain 1000-2000 subjects. This is large enough to estimate a population proportion with a margin of error of about 0.02 or 0.03. At first glance, it seems astonishing that a sample of this size from a population of perhaps many millions or even hundreds of millions is adequate for predicting outcomes of elections, summarizing opinions on controversial issues, showing relative sizes of television audiences, and so forth. The basis for this inferential power lies in the standard error formulas for the point estimates, with random sampling. Good estimates result no matter how large the population size.[8]

Revisiting the approximation $1/\sqrt{n}$ for the margin of error

Chapter 4 introduced a margin of error approximation of $1/\sqrt{n}$ for estimating a population proportion. What's the connection between this approximation and the more exact margin of error we've used in this chapter? If we take the margin of error $1.96\sqrt{\hat{p}(1-\hat{p})/n}$ for a 95% confidence interval, round the z-score to 2 and replace the sample proportion by the value of 0.50 that gives the maximum possible standard error, we get margin of error

$$2\sqrt{0.50(0.50)/n} = 2(0.50)\sqrt{1/n} = 1/\sqrt{n}.$$

So for a 95% confidence interval, the margin of error is approximately $1/\sqrt{n}$ when $\hat{p}$ is not too far from 0.50.

---

[8] In fact, the mathematical derivations of these methods treat the population size as infinite. See Exercise 6.159.

Choosing the Sample Size for Estimating a Population Mean

To derive the sample size for estimating a population mean, you set the margin of error equal to its desired value and solve for n. Recall that a 95% confidence interval for the population mean is

$$\bar{x} \pm t_{.025}(se), \text{ where } se = s/\sqrt{n}$$

and s is the sample standard deviation. If you don't know n, you also don't know the degrees of freedom and the t-score. However, we saw in Table B than when $df > 30$ the t-score for 95% confidence intervals is about 2.0, eventually growing closer to the z-score of 1.96 as it keeps growing. Setting $2.0(s/\sqrt{n})$ equal to a desired margin of error m and solving for n yields the following result:

Sample Size for Estimating a Population Mean

The random sample size n for which a 95% confidence interval for a population mean has margin of error approximately equal to m is

$$n = \frac{4s^2}{m^2}.$$

To use this formula, you guess the value you'll get for the sample standard deviation s.

In practice, since you don't yet have the data, you don't know the value of the sample standard deviation s. You must substitute an educated guess for s. Sometimes you can use the sample standard deviation from a similar study already conducted. The next example shows another sort of reasoning to form an educated guess.

EXAMPLE 11: FINDING n TO ESTIMATE MEAN EDUCATION IN SOUTH AFRICA

Picture the Scenario

A social scientist plans a study of adult South Africans living in townships on the outskirts of Cape Town, to investigate educational attainment in the black community. Educational attainment is the number of years of education completed. Many of the study's potential subjects were forced to leave Cape Town in 1966 when the government passed a law forbidding blacks to live in the inner cities. Under the apartheid system, black South African children were not required to attend school, so some residents had very little education.

Question to Explore

How large a sample size is needed so that a 95% confidence interval for the mean number of years of attained education has margin of error equal to 1 year?

RECALL

Section 2.4 noted that for an approximately symmetric, bell shaped distribution, we can approximate the standard deviation by roughly a sixth of the range.

Think It Through

No prior information is stated about the standard deviation of educational attainment for the township residents. As a crude approximation, we might guess that the sample education values will fall within a range of about 18 years, such as between 0 and 18 years. If the data distribution is bell-shaped, the range from $\bar{x} - 3s$ to $\bar{x} + 3s$ will contain all or nearly all the distribution. Since the distance from $\bar{x} - 3s$ to $\bar{x} + 3s$ equals $6s$, the range of 18 years would equal about $6s$. Then, solving $18 = 6s$ for s, $18/6 = 3$ is a crude guess for s. So we'd expect a sample standard deviation value of about $s = 3$.

The desired margin of error is $m = 1$ year. The required sample size is

$$n = \frac{4s^2}{m^2} = \frac{4(3^2)}{1^2} = 36.$$

We need to randomly sample 36 subjects for a 95% confidence interval for mean educational attainment to have a margin of error of about 1 year.

Insight

A more cautious approach would select the *largest* value for the standard deviation that is plausible. This will give the largest sensible guess for how large n needs to be. For example, we could reasonably predict that s will be no greater than 4, since a range of six standard deviations then extends over 24 years. Then we get $n = 4(4^2)/(1^2) = 64$. If we collect the data and the sample standard deviation is actually less than 4, we will have more data than we need. The margin of error will be even less than 1.0.

◆

To practice this concept, try Exercise 7.52.

If we plan to use a confidence level other than 95%, we'd replace 4 in the formula for n by the square of the approximate *t*-score for that confidence. For large n, recall that the *t*-score is approximately equal to the *z*-score in the last row of Table B.

What Other Factors Affect the Choice of the Sample Size?

We've looked at two factors that play a role in determining a study's sample size.

- The first is the desired *precision*, as measured by the *margin of error m*.

- The second is the *confidence level,* which determines the *z*-score or *t*-score in the sample size formulas.

Other factors also play a role.

- A third factor is the *variability* in the data.

Let's look at the formula $n = 4s^2/m^2$ for the sample size for estimating a mean. The greater the value expected for the standard deviation s, the larger the sample size needed. If subjects have little variation (that is, s is small), we need fewer data than if they have substantial variation. Suppose a study plans to estimate the mean level of education in several countries. Western European countries have relatively little variation, as students are required to attend school until the middle teen years. To estimate the mean to within a margin of error of $m = 1$, we need fewer observations than in South Africa.

- A fourth factor is the *complexity of design and analyses* planned.

The sample size formulas we've used apply to *simple* random sampling. Other random sampling methods may need different sample sizes to achieve the same precision. Determining the sample size is then more difficult and requires guidance from a statistician. Also, the more variables one analyzes, the larger the sample needed for an adequate analysis.

- A fifth factor is *financial*.

> **RECALL**
>
> Section 4.4 discussed other random sampling methods, such as **cluster sampling, stratified sampling**, and **multistage samples**.

Larger samples are more time consuming to collect. They may be more expensive than a study can afford. Cost is often a major constraint. You may need to ask, "Should we go ahead with the smaller sample that we can afford, even though the margin of error will be greater than we would like?"

Finally, a word of caution: "Margin of error" refers to the size of error resulting from having data from a random sample rather than the population--what's called *sampling error*. This is the error that the sampling distribution describes in showing how close the estimate is likely to fall to the parameter. But that's not the only source of potential error. Data may be missing for a substantial percentage of the target sample, some observations may be recorded incorrectly by the data collector or data analyst, and some subjects in the study may not tell the truth. When errors like these occur, the actual confidence level may be much lower than advertised. Be skeptical about a claimed margin of error and confidence level unless you know that the study was well conducted and these other sources of error are negligible.

SECTION 7.4: PRACTICING THE BASICS

7.46 South Africa study: The researcher planning the study in South Africa also will estimate the population proportion having at least a high school education. No information is available about its value. How large a sample size is needed to estimate it to within 0.07 with 95% confidence?

7.47 Binge drinkers: A study in 2001 at the Harvard School of Public Health found that 44% of 10,000 sampled college students were binge drinkers. A student at the University of Minnesota plans to estimate the proportion of college students at that school who are binge drinkers. How large a random sample would she need to estimate it to within 0.05 with 95% confidence, if before conducting the study she uses the Harvard study results as a guideline?

7.48 Abstainers: The Harvard study mentioned in the previous exercise estimated that 19% of college students abstain from drinking alcohol. To estimate this proportion in your school, how large a random sample would you need to estimate it to within 0.05 with probability 0.95, if before conducting the study:

a. You are unwilling to predict the proportion value at your school.

b. You use the study as a guideline.

c. Use results to explain why strategy (a) is inappropriate if you are quite sure you'll get a sample proportion that is far from 0.50.

7.49 How many businesses fail?: A study is planned to estimate the proportion of businesses that started in the year 2000 that had failed within four years of their startup. How large a sample size is needed to guarantee estimating this proportion correct to within

a. 0.10 with probability 0.95?

b. 0.05 with probability 0.95?

c. 0.05 with probability 0.99?

d. Compare sample sizes for parts (a) and (b), and (b) and (c), and summarize the effects of decreasing the margin of error and increasing the confidence level.

7.50 Canada and the death penalty: A poll in Canada in 1998 indicated that 48% of Canadians favor imposing the death penalty (Canada does not have it). A report by Amnesty International on this and related polls (www.amnesty.ca) did not report the sample size but stated "Polls of this size are considered to be accurate within 2.5 percentage points 95% of the time." About how large was the sample size?

7.51 Farm size: An estimate is needed of the mean acreage of farms in Ontario, Canada. A 95% confidence interval should have a margin of error of 25 acres. A study ten years ago in this province had a sample standard deviation of 200 acres for farm size.

a. About how large a sample of farms is needed?

b. A sample is selected of the size found in (a). However, the sample has a standard deviation of 300 acres, rather than 200. What is the margin of error for a 95% confidence interval for the mean acreage of farms?

7.52 Income of Native Americans: How large a sample size do we need to estimate the mean annual income of Native Americans in Onondaga County, New York, correct to within $1000 with probability 0.99? No information is available to us about the standard deviation of their annual income. We guess that nearly all of the incomes fall between $0 and $120,000 and that this distribution is approximately bell shaped.

7.53 Population variability: Explain the reasoning behind the following statement: In studies about a very diverse population, large samples are often necessary, whereas for more homogeneous populations smaller samples are often adequate. Illustrate for the problem of estimating mean income for the entire adult population of the U.S. compared to the subpopulation of postal carriers with less than 5 years experience.

7.54 Web survey to get large *n*: A newspaper wants to gauge public opinion about legalization of marijuana. The sample size formula indicates that they need a random sample of 875 people to get the desired margin of error. But surveys cost money, and they can only afford to randomly sample 100 people. Here's a tempting alternative: If they place a question about that issue on their Web site, they will get more than

1000 responses within a day at little cost. Are they better off with the random sample of 100 responses or the web-site volunteer sample of more than 1000 responses? *Hint*: Think about the issues discussed in Section 4.2 about proper sampling of populations.

7.5   HOW DO COMPUTERS MAKE NEW ESTIMATION METHODS POSSIBLE?

We've seen how to construct point and interval estimates of a population proportion and a population mean. Confidence intervals are relatively simple to construct for these parameters. For some parameters, it's not so easy because it's difficult to derive the sampling distribution or the standard error of a point estimate. We'll now introduce a relatively new simulation method for constructing a confidence interval that statisticians often use for such cases.

The Bootstrap: Using Simulation to Construct a Confidence Interval

When it is difficult to derive a standard error or a confidence interval formula that works well, you can "pull yourself up by your bootstraps" to attack the problem without using mathematical formulas. A recent computational invention, actually called the **bootstrap,** does just that.

The bootstrap is a simulation method that re-samples from the observed data. It treats the data distribution as if it were the population distribution. You resample, *with replacement*, n observations from the data distribution. Each of the original n data points has probability 1/n of selection for each "new" observation. For this new sample of size n, you construct the point estimate of the parameter. You then re-sample another set of n observations from the original data distribution and construct another value of the point estimate. In the same way, you repeat this resampling process (using a computer) from the original data distribution a very large number of times, for instance, selecting 10,000 separate samples of size n and calculating 10,000 corresponding values of the point estimate.

This spread of resampled point estimates provides information about the accuracy of the original point estimate. For instance, a 95% confidence interval for the parameter is the 95% central set of the resampled point estimate values. These are the ones that fall between the 2.5th percentile and 97.5th percentile of those values. This takes a lot of computation, but it is simple with modern computing power.

EXAMPLE 12: HOW VARIABLE ARE YOUR WEIGHT READINGS ON A SCALE?

Picture the Scenario
Instruments used to measure physical characteristics such as weight and blood pressure do not give the same value every time they're used in a given situation. The measurements vary. One of the authors recently bought a scale (called "Thinner") that is supposed to give precise weight readings. To investigate how much the weight readings tend to vary, he weighed himself ten times, taking a 30 second break after each trial to allow the scale to reset. He got the values (in pounds):

160.2, 160.8, 161.4, 162.0, 160.8, 162.0, 162.0, 161.8, 161.6, 161.8.

These ten trials have a mean of 161.44 and standard deviation of 0.63.

Since weight varies from trial to trial, it has a probability distribution. You can regard that distribution as describing "long-run population" values that you would get if you could conduct a huge number of weight trials. The sample mean and standard deviation estimate the center and spread of that distribution. Ideally, you would like the scale to be precise and give the same value every time. Then the population standard deviation would be 0.0, but it's not that precise in practice.

How could we get a confidence interval for the population standard deviation? The sample standard deviation s has an approximate normal distribution for very large n. However, its standard error is highly sensitive to any assumption we make about the shape of the population distribution. It is safer to use the bootstrap method to construct a confidence interval.

Question to Explore
Using the bootstrap method, find a 95% confidence interval for the population standard deviation.

Think It Through
We sample from a distribution that has probability 1/10 at each of the values in the sample. For each new observation this corresponds to selecting a random digit and making the observation 160.2 if we get 0, 160.8 if we get 1, ..., 161.8 if we get 9. Of course, this can be done with software. The bootstrap with 100,000 resamples of the data uses the following steps:

- Randomly sample 10 observations from this sample data distribution. We did this and got the 10 new observations
    160.8, 161.4, 160.2, 161.8, 162.0, 161.8, 161.6, 161.8, 161.6, 160.2.

    For the 10 new observations, the sample standard deviation is $s = 0.67$.

- Proceed as in the preceding step, taking 100,000 separate re-samples of size 10. This gives us 100,000 values of the sample standard deviation. Figure 7.11 shows a histogram of their values.

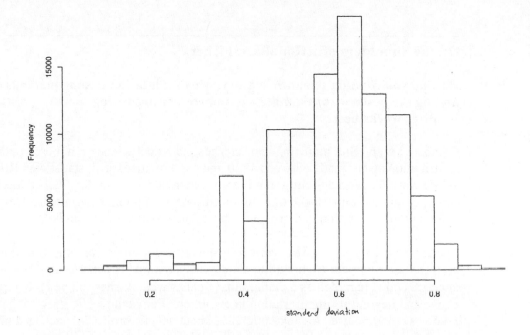

Figure 7.11: A Bootstrap Frequency Distribution of Standard Deviation Values. These were obtained by taking 100,000 samples of size 10 each from the sample data distribution. **Question:** What is the practical reason for using the bootstrap method?

• Now identify the middle 95% of these 100,000 sample standard deviation values. For the 100,000 samples we took, the 2.5th percentile was 0.26 and the 97.5th percentile was 0.80. In other words, 95% of the re-samples had sample standard deviation values between 0.26 and 0.80. This is our 95% bootstrap confidence interval for the population standard deviation.

In summary, the 95% confidence interval for σ is (0.26, 0.80). A typical deviation of a weight reading from the mean might be rather large, nearly a pound.

Insight
Figure 7.11 is skewed and is quite irregular. This appearance is because of the small sample size ($n = 10$). Such simulated distributions take a more regular shape when n is large, usually becoming symmetric and bell shaped when n is sufficiently large.

◆

To practice this concept, try Exercise 7.56.

The bootstrap method was invented by Brad Efron, a statistician at Stanford University, in 1979. It is now widely used, because of the power of modern computers. New statistical methods continue to be developed today. (The small-sample confidence interval for a proportion in Section 7.2 was developed in 1998 by one of the authors with his Ph.D. student at the time, Brent Coull.) The ever-increasing power of computers is making new methods feasible that would have been inconceivable when confidence intervals were first developed in the 1920s and 1930s.

On the Shoulders of...Ronald A. Fisher

How do you conduct scientific inquiry, whether it be developing methods of experimental design, finding the best way to estimate a parameter, or answering specific questions such as which fertilizer works best?

Compared with other mathematical sciences, statistical science is a mere youth. The most important contributions were made between 1920 and 1940 by the British statistician Ronald A. Fisher (1890-1962). While working at an agricultural research station north of London, Fisher was the first statistician to show convincingly the importance of randomization in designing experiments. He also developed the theory behind point estimation and proposed several new statistical methods.

Fisher was involved in a wide variety of scientific questions, ranging from finding the best ways to plant crops to controversies (in his time) about whether smoking is harmful. He also did fundamental work in genetics and is regarded as a giant in that field as well. Fisher showed the importance of natural selection, above and beyond its manifestation in evolution. For example, he showed that correlations between relatives could be used to make inference about genetic properties such as dominance.

Fisher had strong disagreements with others about the way statistical inference should be conducted. One of his main adversaries was Karl Pearson. Fisher corrected a major error Pearson made in proposing methods for contingency tables, and he also criticized Pearson's son's work on developing the theory of confidence intervals in the 1930s. Although Fisher often disparaged the ideas of other statisticians, he reacted strongly if anyone criticized him in return. Writing about Pearson, Fisher once said, "If peevish intolerance of free opinion in others is a sign of senility, it is one which he had developed at an early age."

(Use photo of Fisher from bio by Joan Fisher Box)

R. A. Fisher. He was the statistician most responsible for the statistical methods used to analyze data today.

SECTION 7.5: PRACTICING THE BASICS

7.55 Why bootstrap?: Explain the purpose of using the bootstrap method.

7.56 Estimating variability: Refer to Example 12 about weight readings of a scale. For 10 successive trials on the next day, the weight values were:

159.8, 159.8, 159.6, 159.0, 158.4, 159.2, 158.8, 158.4, 158.8, 159.0.

Explain how you could use the bootstrap to get a 95% confidence interval for a "long-run" standard deviation of such values.

🖳7.57 Bootstrap the proportion: We want a 95% confidence interval for the population proportion of students in a high school in Dallas, Texas who can correctly find Iraq on an unlabelled globe. For a random sample of size 50, 10 get the correct answer.

a.   Using software or the *sampling distribution* applet (www.prenhall.com/TBA), treat the sample proportion as the population proportion by setting the proportion parameter to 0.20 = 10/50. Take a random sample of size 50, and find the sample proportion of correct answers.

b.   Take 100 resamples like the one in (a), each time calculating the sample proportion. Now, construct a 90% confidence interval by identifying the 5th and 95th percentiles of the sample proportions. This is the 90% bootstrap confidence interval.

c.   To improve the bootstrap procedure, repeat (b) by taking 1000 resamples. Find the 90% bootstrap confidence interval.

d.   Explain why the sample proportion does not fall exactly in the middle of the bootstrap confidence interval. (*Hint*: Is the sampling distribution symmetric or skewed?)

Chapter Summary

We've now learned how to **estimate** the population proportion p for categorical variables and the population mean μ for quantitative variables.

- A **point estimate** (or "**estimate**," for short) is our best guess for the unknown parameter value. An estimate of the population mean μ is the sample mean $\bar{x}$. An estimate of the population proportion p is the sample proportion $\hat{p}$.

- A **confidence interval** contains the most plausible values for a parameter. Confidence intervals for most parameters have the form

  Estimate $\pm$ margin of error, which is estimate $\pm$ (z or t score)$\times$(se),

  where se is the standard error of the estimate. For the proportion, the score is a z-score from the normal distribution. For the mean, the score is a t-score from the t **distribution** with degrees of freedom $df = n - 1$. The t-score is similar to a z-score when $df \geq 30$. Table 7.6 summarizes the point and interval estimation methods.

Table 7.6: Estimation Methods for Means and Proportions

Parameter	Point Estimate	Standard Error	Confidence Interval	Sample Size for Margin of Error m
Proportion p	$\hat{p}$	$se = \sqrt{\hat{p}(1-\hat{p})/n}$	$\hat{p} \pm z(se)$	$n = [\hat{p}(1-\hat{p})z^2]/m^2$
Mean μ	$\bar{x}$	$se = s/\sqrt{n}$	$\bar{x} \pm t(se)$	$n = (4s^2)/m^2$

Note: The z or t-score depends on the confidence level. The t-score has $df = n - 1$. The sample size formula for a mean applies for a 95% confidence interval.

- The *z* or *t*-score depends on the **confidence level**, the probability that the method produces a confidence interval that contains the population parameter value. For a proportion, for instance, since a probability of 0.95 falls within 1.96 standard errors of the center of the normal sampling distribution, we use $z = 1.96$ for 95% confidence. *To achieve greater confidence, we make the sacrifice of a larger margin of error and wider confidence interval.*

- For **large samples,** the formulas rely on the central limit theorem. This guarantees that the estimate has a normal sampling distribution, for large random samples. *The formulas then make no assumption about the population distribution*, since the sampling distribution is approximately normal even if the population distribution is not.

- For estimating a **proportion** with **small samples** (fewer than 15 successes or fewer than 15 failures), the confidence interval formula $\hat{p} \pm z(se)$ still applies if we use it after adding 2 successes and 2 failures (and add 4 to *n*).

- For estimating a **mean**, the **t-distribution** accounts for the extra variability due to using the sample standard deviation *s* to estimate the population standard deviation in finding a standard error. When *n* is small, since the central limit theorem does not then apply, the *t* method assumes that the population distribution is normal. This ensures that the sampling distribution of $\bar{x}$ is bell-shaped even for small *n*.

- Before conducting a study, we can **determine the sample size *n*** having a certain margin of error. Table 7.6 shows the sample size formulas. To use them, we must (1) select the margin of error *m,* (2) select the confidence level, which determines the *z*-score or *t*-score, and (3) guess the value the data will have for the sample standard deviation *s* (to estimate a population mean) or the sample proportion $\hat{p}$ (to estimate a population proportion). In the latter case, substituting $\hat{p} = 0.50$ guarantees that the sample size is large enough regardless of the value the sample has for $\hat{p}$.

SUMMARY OF NEW NOTATION IN CHAPTER 7

se = standard error

$\hat{p}$ = sample proportion

m = margin of error

$t_{.025}$ = *t*-score with right-tail probability 0.025

df = degrees of freedom (= *n* - 1 for inference about a mean)

ANSWERS TO THE CHAPTER FIGURE QUESTIONS

Figure 7.1:   *A point estimate alone will not tell us how close the estimate is likely to be to the parameter.*

Figure 7.2:   *When the random sample size is relatively large, by the central limit theorem.*

Figure 7.3:   *We don't know the value of p, the population proportion, to form the interval p ± 1.96(se). p is what we're trying to estimate.*

Figure 7.4:   *Having greater confidence means that we want to have greater assurance of a correct inference. Thus, it is natural that we would expect a wider interval of believable values for the population parameter.*

Figure 7.5:   *The $\hat{p}$ values falling in the non-shaded left and right tails of the bell-shaped curve.*

Figure 7.6:   *This skew should not affect the validity of the confidence interval for the mean because of the large sample size (n = 905).*

Figure 7.7:   *Using Table B or Table 7.5, t = 1.96 when df = ∞ with right-tail probability = 0.025.*

Figure 7.8:   *t = -3.707 and t = 3.707*

Figure 7.9:   *The buy-it-now option has greater variability and more skew than the do not buy-it-now option. This is largely influenced by the value $178. The centers of the two distributions seem to be similar.*

Figure 7.10:   *The sample size n is sufficiently large such that np ≥ 15 and n(1-p) ≥ 15.*

Figure 7.11:   *The bootstrap method is used when it is difficult to derive a standard error or confidence interval formula by using mathematical techniques.*

CHAPTER PROBLEMS: PRACTICING THE BASICS

7.58 Divorce and age of marriage: A July 2002 report published by the Centers for Disease Control estimated that for non-Hispanic white women, the probability that their first marriage breaks up within 10 years is 0.50 for those whose marriage occurred before their 18th birthday, but only 0.22 for those whose marriage occurred at age 25 and over.
a.   Are these point estimates or interval estimates?
b.   Is the information given here sufficient to allow you to construct confidence intervals? Why or why not?

7.59 *z*-scores:
a.   Find the *z*-score for a confidence interval for a proportion that has confidence level (i) 0.50, (ii) 0.68, (iii) 0.9973.
b.   Why are confidence levels 0.50 and 0.68 not commonly used?

7.60 Snoring and insomnia: In 2002, the National Sleep Foundation (www.sleepfoundation.org) conducted 1010 phone interviews using a random sampling of phone numbers. Of the respondents, 37% reported problems with snoring and 58% reported problems with insomnia. One way to report the latter result says, "It is estimated that 58% of people have problems with insomnia. This estimate has a margin of error of plus or minus 3%." How could you explain what this means to someone who has not taken a statistics course?

7.61 British monarchy: In February 2002, the Associated Press quoted a survey of 3000 British residents conducted by YouGov.com. It stated, "Only 21 percent wanted to see the monarchy abolished, but 53% felt it should become more democratic and approachable. No margin of error was given." If the sample was random, find the 95% margin of error for each of these estimated proportions.

7.62 Born again: A poll of a random sample of $n = 2000$ Americans by the Pew Research Center (www.people-press.org) in March 2002 indicated that 36% considered themselves "born-again" or evangelical Christians. How would you explain to someone who has not studied statistics:
a.   What it means to call this a point estimate.
b.   Why this does not mean that *exactly* 36% of *all* Americans are born-again or evangelical Christians.

7.63 Life after death: The variable POSTLIFE in the 2002 General Social Survey asked "Do you believe in life after death?" Of 1211 respondents, 975 answered yes. A report based on these data stated that "81% of Americans believe in life after death. The margin of error for this result is plus or minus 2%." Explain how you could form a 95% confidence interval using this information, and interpret that confidence interval in context.

7.64 Female belief in life after death: Refer to the previous exercise. The following printout shows results for the females in the sample, where $X =$ the number answering yes. Explain how to interpret each item, in context.

```
Sample        X      N     Sample p      95.0% CI
Females      550    648    0.84844     0.821, 0.876)
```

7.65 Marijuana for medical purposes: A poll in 2000 of 500 Canadians by the *National Post* asked whether marijuana should be legalized for medical purposes. 72% said definitely yes, 20% said probably, 2% said probably not, 5% said definitely not, and 2% had no opinion.
a.   Assuming that this was a random sample, construct a 95% confidence interval for the population proportion who would answer definitely yes or probably. Can you conclude that a majority of all Canadians would answer this way? Explain.
b.   Check that the sample size was large enough to construct the interval in (a).

7.66 Vegetarianism: *Time* magazine (July 15, 2002) quoted a poll of 10,000 Americans in which only 4% said they were vegetarians
a.   What has to be assumed about this sample in order to construct a confidence interval for the population proportion of vegetarians?
b.   Check that the sample size is large enough to construct the large-sample confidence interval.

c.   Construct a 99% confidence interval for the population proportion. Explain why the interval is so narrow, even though the confidence level is high.

d.   In interpreting this confidence interval, can you conclude that fewer than 10% of Americans are vegetarians? Explain your reasoning.

7.67 Trust in other people: A Gallup poll of 1013 people conducted in July 2002 for *USA Today* and *CNN* indicated that 41% of Americans said that they could trust most people. Can we conclude that less than half of all Americans feel this way? Explain your reasoning.

7.68 Alternative therapies: The Department of Public Health at the University of Western Australia conducted a survey in which they randomly sampled general practitioners in Australia.[9] One question asked whether the GP had ever studied alternative therapy, such as acupuncture, hypnosis, homeopathy, and yoga. Of 282 respondents, 132 said yes. Is the interpretation, "We are 95% confident that the percentage of all GPs in Australia who have ever studied alternative therapy equals 46.8%" correct or incorrect? Explain.

7.69 Population data: You would like to find the proportion of bills passed by Congress that were vetoed by the president in the last congressional session. After checking congressional records, you see that for the population of all 40 bills passed, 5 were vetoed. Does it make sense to construct a confidence interval using these data? Explain. (*Hint*: Identify the sample and population.)

7.70 Seat belt use: Of all 577,006 people involved in motor vehicle accidents in Florida in a recent year, 412,878 were wearing seat belts (Florida Department of Highway Safety and Motor Vehicles). Does it make sense to use these data to calculate a 95% confidence interval for the proportion of people wearing seat belts for all people involved in motor vehicle accidents that year in Florida? Explain.

⌨**7.71 Wife supporting husband**: Consider the statement that it is more important for a wife to help her husband's career than to have one herself, from the GSS (variable denoted FEHELP).

a.   Go to the Web site, www.icpsr.umich.edu/GSS. In 1998, find the number who agreed with that statement and the sample size.

b.   Show that the sample proportion was 0.189, with standard error 0.0092.

c.   Show that a 99% confidence interval for the population proportion who would agree is 0.189 ± 0.024, or (0.165, 0.213), and interpret it

d.   Without doing any calculation, explain whether the 95% confidence interval would be narrower, or wider.

⌨**7.72 Legalize marijuana?** The General Social Survey has asked respondents, "Do you think the use of marijuana should be made legal or not?" Go to the GSS Web site, www.icpsr.umich.edu/GSS. For the 2002 survey with variable GRASS:

a.   Of the $n = 851$ respondents, how many said "legal" and how many said "not legal"? Report the sample proportions.

b.   Is there enough evidence to conclude whether a majority or a minority of the population support legalization? Explain your reasoning.

---

[9] This was reported at www.internethealthlibrary.com/Surveys/

c.   Now look at the data on this variable for all years by entering YEAR as the column variable. Describe any trend you see over time in the proportion favoring legalization.

7.73 Smoking: A report in 2001 by the U.S. National Center for Health Statistics provided an estimate of 23% for the percentage of Americans over the age of 18 who were currently smokers. The sample size was 38,633. Assuming that this sample has the characteristics of a random sample, a 99.999999% confidence interval for the proportion of the population who were smokers is (0.23, 0.24). When the sample size is extremely large, explain why even confidence intervals with large confidence levels are narrow.

7.74 Nondrinkers: Refer to the previous exercise. The same study provided the following results for estimating the proportion of adult Americans who have had fewer than 12 drinks in their lifetime.

Sample	X	N	Sample p	95.0% CI
Nodrink	8654	38633	0.2240	(0.220, 0.228)

Explain how to interpret all results on this printout, in context.

7.75 No education: The variable EDUC in the 2002 General Social Survey asked "What is the highest grade that you finished and got credit for?" Of 2753 respondents, only 5 people said 0 years. You would like a 95% confidence interval for the population proportion with no formal education.
a.   Why is the ordinary large-sample confidence interval formula not valid?
b.   Construct a valid interval. Interpret in context.

7.76 Kicking accuracy: A football coach decides to estimate the kicking accuracy of a player who wants to join the team. Of 10 extra point attempts, the player makes all ten.
a.   Find a 95% confidence interval for the probability that the player makes any given extra point attempt.
b.   What's the lowest value that you think is plausible for that probability?
c.   How would you interpret the random sample assumption in this context? Describe a scenario such that it would not be sensible to treat these 10 kicks as a random sample.

7.77 Males go daily to a bar: When the 2002 General Social Survey asked a random sample of 309 males how often they go to a bar or tavern (variable SOCBAR), 8 answered "almost every day." Estimate and find a 95% confidence interval for the population proportion of males who go to a bar or tavern almost every day. Check the assumptions on which the confidence interval is based.

7.78 Females go daily to a bar: Repeat the previous exercise for the 506 females sampled, 3 of whom answered "almost every day."

7.79 Travel to work: As part of the 2000 census, the Census Bureau surveyed 700,000 households to study transportation to work. They reported that 76.3% drove alone to work, 11.2% carpooled, 5.1% took mass transit, 3.2% worked at home, 0.4% bicycled, and 3.8% took other means.

a.   With such a large survey, explain why the margin of error for any of these values is extremely small.

b.   The survey also reported that the mean travel time to work was 24.3 minutes, compared to 22.4 minutes in 1990. Explain why this is not sufficient information to construct a confidence interval for the population mean. What else would you need?

7.80 *t*-scores:

a.   Show how the *t*-score for a 95% confidence interval changes as the sample size increases from 10 to 20 to 30 to infinity.

b.   What does the answer in (a) suggest about how the *t*-distribution compares to the standard normal distribution?

7.81 Buddhists Watching TV: In a recent GSS, the responses about the number of hours daily spent watching TV for the five subjects who identified themselves as Buddhists were 0, 5, 0, 1, 2.

a.   Find the mean, standard deviation, and standard error.

b.   Construct a 95% confidence interval for the population mean.

c.   What is the main factor that causes the interval in (b) to be so wide?

d.   Specify the assumptions for the method. What can you say about their validity for these data?

7.82 Male Buddhists Watching TV: Refer to the previous exercise. The two male Buddhists had responses 0 and 5. The 95% confidence interval for the population mean is then (-29.3, 34.3).

a.   Does it make sense to report the negative values in this interval? Why or why not?

b.   Suppose there was only one male Buddhist. Would there be enough information to find a confidence interval for the mean? (*Hint*. Review the formula for *s*, and consider whether you would be able to calculate *s* and the standard error.)

7.83 Psychologists' income: In 1999, the American Psychological Association conducted a survey of a random sample of psychologists to estimate mean incomes for psychologists of various types (research.apa.org). Of the 10 clinical psychologists with 5-9 years experience who were in a medical psychological group practice, the mean income was $78,500 with a standard deviation of $33,287.

a.   Construct a 95% confidence interval for the population mean. Interpret.

b.   What assumption about the population distribution of psychologists' incomes does the confidence interval method make?

c.   If the assumption about the shape of the population distribution is not valid, since *n* is small, does this invalidate the results? Explain.

7.84 More psychologists: Refer to the previous exercise. Interpret each item on the following printout that software reports for the clinical psychologists with 5-9 years experience who worked in a rehabilitation facility:

Variable	N	Mean	StDev	SE Mean	95.0% CI
income	8	57625	12614	4459.7	(47079.4, 68170.6)

7.85 Unrelated housemates: The 2002 GSS asked (variable UNRELAT), "How many persons in your household are not related to you in any way?" Of 1883 responses, the mean was 0.19 and the standard deviation was 0.64.
a. Is it plausible that this variable has a normal distribution? Explain.
b. Find the margin of error for estimating the population mean. Interpret.
c. Given your answer to (a), is it appropriate to use the margin of error in (b) to construct a 95% confidence interval? If not, explain why not. If so, do so and interpret.

7.86 How long lived in town?: The General Social Survey has asked subjects, "How long have you lived in the city, town or community where you live now?" The responses of 1415 subjects in one survey had a mode of less than one year, a median of 16 years, a mean of 20.3 and a standard deviation of 18.2.
a. Do you think that the population distribution is normal? Why or why not?
b. Based on your answer in (a), can you construct a 95% confidence interval for the population mean? If not, explain why not. If so, do so and interpret.

7.87 How often do women feel sad?: A recent GSS asked, "How many days in the past 7 days have you felt sad?" The 816 women who responded had a median of 1, mean of 1.81, and standard deviation of 1.98.
a. Find a 95% confidence interval for the population mean. Interpret.
b. Do you think that this variable has a normal distribution? Does this cause a problem with the confidence interval method in (a)? Explain.

7.88 How often do men feel sad?: Repeat the previous exercise for the 633 men who responded, who had a median of 1, mean of 1.42, and standard deviation of 1.83.

7.89 Happy often?: The 1996 GSS asked, "How many days in the past 7 days have you felt happy?"
a. Using the GSS variable HAPFEEL, verify that the sample had a mean of 5.27 and a standard deviation of 2.05. What was the sample size?
b. Find the standard error for the sample mean.
c. Stating assumptions, construct and interpret a 95% confidence interval for the population mean. Can you conclude that the population mean is at least 5.0?

7.90 Revisiting the Cereal Data: Example 4 in Chapter 2 presented data on cereal sodium values. The data are in the "Cereal" data file on the text CD and shown in the margin.
a. Form a 95% confidence interval for the mean sodium level in all breakfast cereals.
b. What assumptions are made in forming the interval in part (a)? State at least one important assumption that does not seem to be satisfied, and indicate its impact on this inference.

Cereal	Sodium
Frosted Mini Wheats	0
Raisin Bran	210
All Bran	260
Apple Jacks	125
Capt Crunch	220
Cheerios	290
Cinnamon Toast	210
Crackling Oat Bran	140
Crispix	220
Frosted Flakes	200
Fruit Loops	125
Grape Nuts	170
Honey Nut Cheerios	250
Life	150
Oatmeal Raisin Crisp	170
Sugar Smacks	70
Special K	230
Wheaties	200
Corn Flakes	290
Honeycomb	180

💻7.91 Revisiting mountain bikes: Use the "Mountain Bike" data file from the text CD, shown also in the table. The data were presented in Exercise 3.24.

Mountain bikes

Brand and Model	Price($)
Trek VRX 200	1000
Cannondale SuperV400	1100
GT XCR-4000	940
Specialized FSR	1100
Trek 6500	700
Specialized Rockhop	600
Haro Escape A7.1	440
Giant Yukon SE	450
Mongoose SX 6.5	550
Diamondback Sorrento	340
Motiv Rockridge	180
Huffy Anorak 36789	140

----

a.   Form a 95% confidence interval for the population mean price of all mountain bikes. Interpret.
b.   What assumptions are made in forming the interval in part (a)? State at least one important assumption that does not seem to be satisfied, and indicate its impact on this inference.

💻7.92 eBay selling prices: For eBay auctions of the Palm M515 PDA on the final day for which data were available for auctions of this product, the selling prices were:

$$\$212, \$249, \$250, \$240, \$210, \$230, \$195, \$193.$$

a.   Explain what a parameter μ might represent that you could estimate with these data.
b.   Find the point estimate of μ.
c.   Find the standard deviation of the data and the standard error of the sample mean. Interpret.
d.   Find the 95% confidence interval for μ. Interpret the interval in context.

💻7.93 Income for families in public housing: A survey is taken to estimate the mean annual family income for families living in public housing in Chicago. For a random sample of 29 families, the annual incomes (in hundreds of dollars) are as follows:

90 77 100 83 64 78 92 73 122 96 60 85 86 108
70 139 56 94   84 111 93 120 70 92 100 124 59 112 79

a.   Construct a box plot of the incomes. What do you predict about the shape of the population distribution? Does this affect the possible inferences?
b.   Using software, find point estimates of the mean and standard deviation of the family incomes of all families living in public housing in Chicago.

c.   Obtain and interpret a 95% confidence interval for the population mean.

7.94 Females watching TV: The GSS asked in 2002 "On the average day about how many hours do you personally watch television?" Software reports the results for females,

```
Variable    N     Mean   StDev    SE Mean     95.0% CI
TV         506    3.061  2.122    0.0944     (2.88, 3.25)
```

The following figure shows histograms of TV watching for females and for males.

Histograms of Television Watching, for Females and Males.

a.   Would you say that TV watching has a normal distribution? Why or why not?

b.   On what assumptions is the confidence interval shown based? Are any of them violated here? If so, is the reported confidence interval invalid? Explain.

c.   What's wrong with the interpretation, "In the long run, 95% of the time females watched between 2.88 and 3.25 hours of TV a day."

7.95 Males watching TV: Refer to the previous exercise. The 399 males had a mean of 2.885 and a standard deviation of 2.633.

a.   Based on these and the histogram, is the confidence interval method valid for these data? Explain.

b.   The 95% confidence interval for the population mean is (2.63, 3.14). Interpret in context.

7.96 Working mother: In response to the statement on a recent General Social Survey, "A preschool child is likely to suffer if his or her mother works," the response categories (Strongly agree, Agree, Disagree, Strongly disagree) had counts (91, 385, 421, 99). Scores (2, 1, -1, -2) were assigned to the four categories, to treat the variable as quantitative. Software reported:

```
Variable   N     Mean    StDev   SE Mean    95.0% CI
Response   996   -0.052  1.253   0.0397     (-0.130, 0.026)
```

a.   Explain what this choice of scoring assumes about relative distances between categories of the scale.
b.   Based on this scoring, how would you interpret the sample mean of -0.052?
c.   Explain how you could also make inference about proportions for these data.

7.97 Highest grade completed: The 2002 GSS asked "What is the highest grade that you finished and got credit for?" (variable EDUC). Of 2753 respondents, the mean was 13.4, the standard deviation was 3.0, and the proportion who gave responses below 12 (i.e., less than a high school education) was 0.16. Explain how you could analyze these data by making inferences about a population mean or about a population proportion, or both. Show how to implement one of these types of inference with these data.

7.98 Interpreting an interval for μ: Refer to the previous exercise. For the 429 African-Americans in the 2000 GSS, a 99% confidence interval for the mean of EDUC is (12.1, 12.6). Explain why the following interpretation is incorrect: 99% of all African-Americans have completed grades between 12.1 and 12.6.

7.99 Sex partners in previous year: The 2002 General Social Surveys asked respondents how many sex partners they had in the previous 12 months (variable PARTNERS). Software summarizes the results of the responses by:

```
Variable   N      Mean    StDev   SE Mean    95.0% CI
partners   2242   1.08    1.08    0.023      1.03, 1.13)
```

a.   Based on these results, explain why the distribution was probably highly skewed to the right.
b.   Explain why the skew need not cause a problem with constructing a confidence interval for the population mean, unless there are extreme outliers such as a reported value of 1000.

7.100 Sex partners for females: For 2002 data from the GSS on subjects' reports of the number of sex partners they had in the last 12 months, a computer printout reports the following information for female respondents:

```
Variable   N      Mean   StDev   SE Mean    95.0% CI
partners   1237   0.90   0.87    0.025      (0.85, 0.95)
```

a.   Based on the reported sample size and standard deviation, verify the reported value for the standard error.
b.   Would a 99% confidence interval for μ be narrower, or wider, than the one reported here? Explain why, without doing any calculations.
c.   For the 1005 males, the mean was 1.30 with standard deviation 1.26. What two statistical factors cause a 95% confidence interval for females to be narrower than a 95% confidence interval for males?

7.101 Men don't go to the doctor: A survey of 1084 men age 18 and older in 1998 for the Commonwealth Fund (www.cmwf.org) indicated that more than half did not have a physical exam or a blood cholesterol test in the past year. A medical researcher plans to sample men in her community randomly to see if similar results occur. How large a random sample would she need to estimate this proportion to within 0.05 with probability 0.95?

7.102 Driving after drinking: In December 2004, a report based on the National Survey on Drug Use and Health estimated that 20% of all Americans of ages 16 to 20 drove under the influence of drugs or alcohol in the previous year (AP, Dec. 30, 2004). A public health unit in Wellington, New Zealand, plans a similar survey for young people of that age in New Zealand. They want a 95% confidence interval to have a margin of error of 0.04.
a.   Find the necessary sample size if they expect results similar to those in the U.S.
b.   Suppose that in determining the sample size, they use the safe approach that sets $\hat{p} = 0.50$ in the formula for n. Then, how many records need to be sampled?
Compare to the answer in (a). Explain why it is better to make an educated guess about what to expect for $\hat{p}$, when possible.

7.103 Changing views of U.S.: The June 2003 report on *Views of a Changing World*, conducted by the Pew Global Attitudes Project (www.people-press.org), discussed changes in views of the U.S. by other countries. In the largest Muslim nation, Indonesia, a poll conducted in May 2003 after the Iraq war began reported that 83% had an unfavorable view of America, compared to 36% a year earlier. The 2003 result was claimed to have a margin of error of 3 percentage points. How can you approximate the sample size the study was based on?

7.104 Mean property tax: A tax assessor wants to estimate the mean property tax bill for all homeowners in Madison, Wisconsin. A survey ten years ago got a sample mean and standard deviation of $1400 and $1000.
a.   How many tax records should the tax assessor randomly sample in order for a 95% confidence interval for the mean to have margin of error $100? What assumption does your solution make?
b.   In reality, suppose that they'd now get a standard deviation equal to $1500. Using the sample size you derived in (a), without doing any calculation, explain whether the margin of error for a 95% confidence interval would be less than $100, equal to $100, or more than $100.
c.   Refer to (b). Would the probability that the sample mean falls within $100 of the population mean be less than 0.95, equal to 0.95, or greater than 0.95? Explain.

7.105 Distance from work: A study is conducted of the distance that employees at a large company live from the company to find out whether people tend to have longer commutes than in the past. For a random sample of 100 employees, the mean distance is 5.3 miles and the standard deviation is 4.0.
a.   Find the margin of error for a 95% confidence interval for the mean distance from the factory of all employees.
b.   About how large a sample would have been adequate if we merely needed a margin of error of 1.0?

CHAPTER PROBLEMS: CONCEPTS AND INVESTIGATIONS

⌨**7.106 Student survey**: Refer to the "Florida student survey" data set on the text CD. For the variable "belief in life after death," find a 95% confidence interval for a relevant parameter, stating the assumptions. Summarize and interpret your findings.

⌨**7.107 Women working and having children**: Go to www.icpsr.umich.edu/GSS and look at responses to the question WRKBABY, "Do you think a woman should work outside the home fulltime, parttime, or not at all when she has a child under school age?" Using data from the most recent survey where this was asked, find a 95% confidence interval for the population proportion for the "not at all" category. Write a short report (one page) summarizing descriptive statistics, assumptions for inference, results of the inference, and the interpretation.

7.108 Religious beliefs: A column by *New York Times* columnist Nicholas Kristof (August 15, 2003) discussed results of polls indicating that religious beliefs in the U.S. tend to be quite different from those in other Western nations. He quoted recent Gallup and Harris polls of random samples of about 1000 Americans estimating that 83% believe in the Virgin Birth of Jesus but only 28% believe in evolution. A friend of yours is skeptical, claiming that it's impossible to predict beliefs of over 200 million adult Americans based on interviewing only 1000 of them. Write a one-page report using this context to show how you could explain, with nontechnical language, about random sampling, the margin of error, how a margin of error depends on the sample size, and how you can control the chance of a correct inference by increasing that margin of error.

7.109 TV watching and race: For the number of hours of TV watching, the 2000 GSS reported a mean of 2.75 for the 1435 white subjects, with a standard deviation of 2.25. The mean was 4.14 for the 305 black subjects, with a standard deviation of 3.56. Analyze these data, preparing a short report in which you mention the methods used and the assumptions on which they are based, and summarize and interpret your findings.

7.110 Housework and gender: Using data from the National Survey of Families and Households, a study (by S. South and G. Spitze, *American Sociological Review*, Vol. 59, 1994, pp. 327-347) reported the descriptive statistics in the following table for the hours spent in housework and in employment per week. Analyze these data. Summarize results in a short report, including assumptions you made to perform the inferential analyses.

GENDER	SAMPLE SIZE	Housework Hours MEAN	STD DEV	Employment Hours MEAN	STD DEV
Men	4252	18.1	12.9	31.8	22.6
Women	6764	32.6	18.2	18.4	20.0

7.111 Cohabitation: A General Social Survey asked respondents, "Did you live with your husband/wife before you got married?" The responses were 176 yes, 566 no. Analyze and interpret these data in a one-half page report.

7.112 Women's role opinions: When subjects in a recent GSS were asked whether they agreed with the following statements, the (yes, no) counts under various conditions were as follows:

- Women should take care of running their homes and leave running the country up to men: (275, 1556)
- It is better for everyone involved if the man is the achiever outside the home and the woman takes care of the home and the family: (627, 1208)
- A preschool child is likely to suffer if her mother works: (776, 1054)

Analyze these data. Prepare a one-page report stating assumptions, showing results of description and inference, and summarizing conclusions.

7.113 Types of estimates: An interval estimate for a mean is more informative than a point estimate, because with an interval estimate you can figure out the point estimate, but with the point estimate alone you have no idea how wide the interval estimate is. Explain why this statement is correct, illustrating using the reported 95% confidence interval of (4.0, 5.6) for the mean number of dates in the previous month based on a sample of women at a particular college.

7.114 Width of a confidence interval: Why are confidence intervals wider when we use larger confidence levels but narrower when we use larger sample sizes, other things being equal?

7.115 99.9999% confidence: Explain why confidence levels are usually large, such as 0.95 or 0.99, but not extremely large, such as 0.999999. (*Hint*: What impact does the extremely high confidence level have on the margin of error?)

7.116 Estimating p near ½ is tough:
a.   Find the standard error of $\hat{p}$ for $n = 1000$ when $\hat{p} = 0.10, 0.30, 0.50, 0.70, 0.90$.
b.   Using these, explain why it is more difficult to estimate a proportion when it is near 0.50 than when it is near 0 or 1.

7.117 Need 15 successes and failures: To use the large-sample confidence interval for p, you need at least 15 successes and 15 failures. Show that the smallest value of n for which the method can be used is (a) 30 when $\hat{p} = 0.50$, (b) 50 when $\hat{p} = 0.30$, (c) 150 when $\hat{p} = 0.10$. That is, the overall n must increase as $\hat{p}$ moves toward 0 or 1. (When the true proportion is near 0 or 1, the sampling distribution can be highly skewed unless n is quite large.)

7.118 Study design: Give an example of a study in which it would be important to have:
a. A high degree of confidence
b. A high degree of precision (that is, a narrow confidence interval)

7.119 Margin of error for binary data: For a binary variable, explain why the margin of error is the same if we are estimating the population proportion of successes or the population proportion of failures.

7.120 Larger spread makes estimation tougher: It's easier to estimate a population mean precisely when the data have less spread. Explain why, using the standard error formula. Explain the implications for estimating population mean income for medical doctors and for entry-level employees in hamburger fast-food restaurants, based on random samples of $n = 100$ each.

📖**7.121 Outliers and CI**: For the observations 0, 0, 2, 2, 2, 4, 6 on TV watching for the seven Muslims in the sample in 1998, a 95% confidence interval for the population mean for that group is (0.3, 4.3). Suppose the observation of 6 for the seventh subject was incorrectly recorded as 60. What would have been obtained for the 95% confidence interval? Compare to the interval (0.3, 4.3). How does this warn you about potential effects of outliers when you construct a confidence interval for a mean?

7.122 What affects n?: Using the sample size formula $n = [\hat{p}(1-\hat{p})z^2]/m^2$ for a proportion, explain the effect on n of (a) increasing the confidence level, (b) decreasing the margin of error.

Multiple choice: Select the best response in Exercises 7.123 – 7.126.

7.123   CI property: Increasing the confidence level causes the margin of error of a confidence interval to (a) increase, (b) decrease, (c) stay the same.

7.124 CI property 2: Other things being equal, increasing the sample size causes the margin of error of a confidence interval to (a) increase, (b) decrease, (c) stay the same.

7.125 Number of close friends: Based on responses of 1467 subjects in a General Social Survey, a 95% confidence interval for the mean number of close friends equals (6.8, 8.0). Which of the following two interpretations are correct?
a.   We can be 95% confident that $\bar{x}$ is between 6.8 and 8.0.
b.   We can be 95% confident that μ is between 6.8 and 8.0.
c.   Ninety-five percent of the values of X = number of close friends (for this sample) are between 6.8 and 8.0.
d.   If random samples of size 1467 were repeatedly selected, then 95% of the time $\bar{x}$ would be between 6.8 and 8.0.
e.   If random samples of size 1467 were repeatedly selected, then in the long run 95% of the confidence intervals formed would contain the true value of μ.

7.126 Why z?: The reason we use a z-score from a normal distribution in constructing a large-sample confidence interval for a proportion is that:
a.   For large random samples the sampling distribution of the sample proportion is approximately normal.
b.   The population distribution is normal.
c.   For large random samples the data distribution is approximately normal.
d.   For any n we use the t distribution to get a confidence interval, and for large n the t distribution looks like the standard normal distribution.

7.127 Mean age at marriage: A random sample of 50 records yields a 95% confidence interval for the mean age at first marriage of women in a certain county of

21.5 to 23.0 years. Explain what is wrong with each of the following interpretations of this interval.

a.   If random samples of 50 records were repeatedly selected, then 95% of the time the sample mean age at first marriage for women would be between 21.5 and 23.0 years.

b.   Ninety-five percent of the ages at first marriage for women in the county are between 21.5 and 23.0 years.

c.   We can be 95% confident that $\bar{x}$ is between 21.5 and 23.0 years.

d.   If we repeatedly sampled the entire population, then 95% of the time the population mean would be between 21.5 and 23.5 years.

7.128 Interpret CI: For the previous exercise, provide the proper interpretation.

7.129 True or false 1: Suppose a 95% confidence interval for a population proportion of students at your school who regularly drink alcohol is (0.61, 0.67). The inference is that you can be 95% confident that the sample proportion falls between 0.61 and 0.67.

7.130 True or false 2: The confidence interval for a mean with a random sample of size $n = 2000$ is invalid if the population distribution is bimodal.

7.131 True or false 3: If you have a volunteer sample instead of a random sample, then a confidence interval for a parameter is still completely reliable as long as the sample size is larger than about 30.

7.132 Women's satisfaction with appearance: A special issue of *Newsweek* in March 1999 on women and their health reported results of a poll of 757 American women aged 18 or older. When asked, "How satisfied are you with your overall physical appearance?" 30% said very satisfied, 54% said somewhat satisfied, 13% said not too satisfied, and 3% said not at all satisfied. True or false: Since all these percentages are based on the same sample size, they all have the same margin of error.

♦♦**7.133 Opinions over time about the death penalty**: For many years, the General Social Survey has asked respondents whether they favor the death penalty for persons convicted of murder. Support has been quite high in the United States., one of few Western nations that currently has the death penalty. The following figure uses the 20 General Social Surveys taken between 1975 and 2000 and plots the confidence intervals for the population proportion in the U.S. who supported the death penalty in each of the 20 years of these surveys.

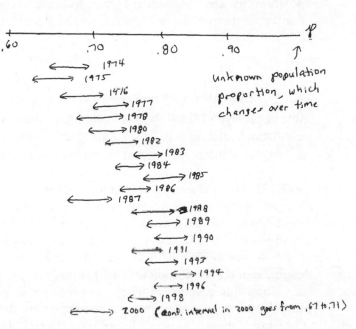

Twenty 95% Confidence Intervals for the Population Proportions Supporting the Death Penalty

a. When we say we have "95% confidence" in the interval for a particular year, what does this mean?

b. For 95% confidence intervals constructed using data for 20 years, let X = the number of the intervals that contain the true parameter values. Find the probability that $x = 20$, that is, all 20 inferences are correct. (*Hint*: You can use the binomial distribution to answer this.)

c. Find the mean of the probability distribution of X.

d. What could you do differently so it is more likely that all 20 inferences are correct?

♦♦7.134 Halving the error: Why does reducing the margin of error by a factor of one-half require quadrupling n?

♦♦7.135 Why called "degrees of freedom"?: You know the sample mean $\bar{x}$ of n observations. Once you know $(n - 1)$ of the observations, show that you can find the remaining one. In other words, for a given value of $\bar{x}$, the values of $(n - 1)$ observations determine the remaining one. In summarizing scores on a quantitative variable, having $(n - 1)$ *degrees of freedom* means that only that many observations are independent. (If you have trouble with this, try to show it for $n = 2$, for instance showing that if you know that $\bar{x} = 80$ and you know that one observation is 90, then you can figure out the other observation. The *df* value also refers to the divisor in $s^2 = \Sigma(x - \bar{x})^2/(n-1)$.)

♦♦7.136 Estimating p without estimating se: The large-sample confidence interval for a proportion substitutes $\hat{p}$ for the unknown value of p in the exact standard error of $\hat{p}$. A less approximate 95% confidence interval has endpoints determined by the *p* values that are 1.96 standard errors from the sample proportion, without estimating the standard error. To do this, you solve for *p* in the equation

$$|\hat{p} - p| = 1.96\sqrt{p(1-p)/n}\,.$$

a. For Example 6 with no vegetarians in a sample of size 20, substitute $\hat{p}$ and *n* in this equation and show that the equation is satisfied at $p = 0$ and at $p = 0.161$. So, the confidence interval is (0, 0.161), compared to (0, 0) with $\hat{p} \pm 1.96(se)$.
b. Which confidence interval seems more believable? Why?

♦♦**7.137 What if everyone is in one category?**: Does the sample proportion seem a sensible estimate if all observations fall in one category? For instance, if you flip a coin 4 times and get 4 heads, are you comfortable estimating that the probability of a head is 1.0? An alternative type of estimate, called a **Bayes estimate**, moves the estimate a bit toward 0.50. When *x* of *n* observations fall in a category, one Bayes estimate of the proportion is (*x* + 2)/(*n* + 4).
a. Show this equals 0.75 when *x* = 4 with *n* = 4.
b. Explain why this estimate is the midpoint of the small-sample confidence interval for a population proportion described at the end of Section 7.2.

♦♦**7.138 *m* and *n***: Consider the sample size formula $n = [\hat{p}(1-\hat{p})z^2]/m^2$ for estimating a proportion. When $\hat{p}$ is close to 0.50, for 95% confidence explain why this formula gives roughly $n = 1/m^2$.

♦♦**7.139 Median as point estimate**: When the population distribution is normal, the population mean equals the population median. How good is the sample median as a point estimate of this common value? For a random sample, the estimated standard error of the sample median equals $1.25(s/\sqrt{n}\,)$. If the population is normal, explain why the sample mean tends to be a better estimate than the sample median.

7.140 Trimmed mean: Usually we estimate a parameter by its sample value. This is not the only possible point estimate, however. For instance, a **trimmed mean** is an estimate of a population mean that ignores certain outliers and calculates the sample mean from the remaining observations, in order to keep the outliers from having too strong an influence. If the population distribution is assumed to be symmetric, what is another possible point estimate of the population mean?

CHAPTER PROBLEMS: CLASS EXPLORATIONS

♦♦**7.141 Randomized response**: To encourage subjects to make honest responses on sensitive questions, the method of *randomized response* is often used. Let's use your class to estimate the proportion who have had alcohol at a party. Before carrying out this method, the class should discuss what they would guess for the value of the proportion of students in the class who have had alcohol at a party. Come to a class consensus. Now each student should flip a coin, in secret. If it is a head, toss the coin once more and report the outcome, head or tails. If the first flip is a tail, report instead the response to whether you have had alcohol, reporting the response *head* if the true response is *yes* and reporting the response *tail* if the true response is *no*. Let *p* denote the true probability of the *yes* response on the sensitive question.

a. Explain why the numbers in the following table are the probabilities of the four possible outcomes.

b. Let $\hat{q}$ denote the sample proportion of subjects who report *head* for the second response. Explain why we can set $\hat{q} = 0.25 + p/2$ and hence use $\hat{p} = 2\hat{q} - 0.5$ to estimate *p*.

c. Using this approach with your class, estimate the probability of having had alcohol at a party. Is it close to the class guess?

Table for Randomized Response

	SECOND RESPONSE	
FIRST COIN	Head	Tail
Head	0.25	0.25
Tail	$p/2$	$(1-p)/2$

7.142 GSS project: The instructor will assign the class a theme to study. Download recent results for variables relating to that theme from www.icpsr.umich.edu/GSS. Find and interpret confidence intervals for relevant parameters. Prepare a two-page report summarizing results.

BIBLIOGRAPHY

Box, J. F. (1978). *R. A. Fisher, The Life of a Scientist*. New York: Wiley.

Pearson, E. S. (1990). *"Student: A Statistical Biography of William Sealy Gosset.* Oxford: Clarendon.

Chapter 8
Statistical Inference: Significance Tests about Hypotheses

EXAMPLE 1: ARE ASTROLOGY PREDICTIONS BETTER THAN GUESSING?

Picture the Scenario
Astrologers believe that the positions of the planets and the moon at the moment of your birth determine your personality traits. But have you ever seen any scientific evidence that astrology works? One scientific "test of astrology" used the following experiment[1]: Volunteers were asked to give their dates and times of birth. From this information, an astrologer prepared each subject's horoscope based on the positions of the planets and the moon at the moment of birth. Each volunteer also filled out a California Personality Index survey. Then the birth data and horoscope for one subject, together with the results of the personality survey for that individual and for two other participants randomly selected from the experimental group, were given to an astrologer. The astrologer was asked to predict which of the three personality charts matched the birth data and horoscope for the subject.

Let p denote the probability of a correct prediction by an astrologer. Suppose an astrologer actually has no special predictive powers, as would be expected by those who view astrology as "quack science." The predictions then merely correspond to random guessing, that is, picking one of the three personality charts at random, so $p = 1/3$. However, the participating astrologers claimed that $p > 1/3$. They felt they could predict better than with random guessing.

Questions to Explore
- How can we use data from such an experiment to summarize the evidence about the claim by the astrologers?
- How can we decide, based on the data, whether or not the claims are believable?

Thinking Ahead
In this chapter we'll learn how to use inferential statistics to answer these questions. An inferential method called a **significance test** analyzes the evidence that the data provide in favor of the astrologers' claim. We'll analyze data from the astrology experiment in Examples 3 and 13.

Therapeutic touch is another practice that many believe is "quack science." Its practitioners claim to be able to heal many medical conditions by using their hands to manipulate a "human energy field" they perceive above the patient's skin. We'll analyze data from an experiment to investigate this claim in Examples 2, 6, and 14.

The significance test is the second major method for making statistical inference about a population. Like a confidence interval for estimating a parameter (the first

---

[1] S. Carlson, *Nature,* vol. 318, pp. 419-425, 1985

major method), this method uses probability to determine the likely parameter values while controlling the chance of an incorrect inference. With this method, we'll be able to use data to answer questions about astrology, therapeutic touch, and many other topics, such as:

- Does a proposed diet truly result in weight loss, on the average?

- Is there evidence of discrimination against women in promotion decisions?

- Does one advertising method result in better sales, on the average, than another advertising method?

8.1 WHAT ARE THE STEPS FOR PERFORMING A SIGNIFICANCE TEST?

The main goal in many research studies is to check whether the data support certain statements or predictions. These statements are **hypotheses** about a population. They are usually expressed in terms of population parameters for variables measured in the study.

Hypothesis

In statistics, a **hypothesis** is a statement about a population, usually of the form that a certain parameter takes a particular numerical value or falls in a certain range of values.

For instance, the parameter might be a population proportion or a probability. Here's an example of a hypothesis, for the astrology experiment in Example 1:

- Hypothesis: Using a person's horoscope, the probability p that an astrologer can correctly predict which of three personality charts applies to that person equals 1/3. That is, astrologers' predictions correspond to random guessing.

A **significance test** is a method of using data to summarize the evidence about a hypothesis. For instance, if in the astrology experiment a large proportion of the astrologers' predictions are correct, the data might then provide strong evidence against the hypothesis that $p = 1/3$ in favor of an alternative hypothesis representing the astrologers' claim that $p > 1/3$.

A significance test (or "test" for short) about a hypothesis has five steps. We'll now introduce these steps. Sections 8.2 and 8.3 provide the details for a test about proportions and for a test about means.

Step 1: Assumptions

Each significance test makes certain assumptions or has certain conditions under which it applies. Foremost, a test assumes that the data production used randomization. Other assumptions may include ones about the sample size or about the shape of the population distribution.

Step 2: Hypotheses

Each significance test has two hypotheses about a population parameter.

<table>
<tr><td>

IN WORDS

H_0: null hypothesis
(read as "H zero" or
      "H naught")

H_a : alternative hypothesis
      (read as "H a")

In everyday English, *null* is an adjective meaning "of no consequence or effect, amounting to nothing."

</td><td>

Null Hypothesis, Alternative Hypothesis

The **null hypothesis** is a statement that the parameter takes a particular value.

The **alternative hypothesis** states that the parameter falls in some alternative range of values.

The value in the null hypothesis usually represents *no effect*. The value in the alternative hypothesis then represents an effect of some type.

The symbol $\mathbf{H_0}$ denotes **null hypothesis** and the symbol $\mathbf{H_a}$ denotes **alternative hypothesis**.

</td></tr>
</table>

For the astrology experiment in Example 1, consider the hypothesis, "Based on any person's horoscope, the probability p that an astrologer can correctly predict which of three personality charts applies to that person equals 1/3." This hypothesis states that there is *no effect* in the sense that an astrologer's predictive power is no better than random guessing. This is a *null hypothesis*, symbolized by $H_0: p = 1/3$. If it is true, any difference that we observe between the sample proportion of correct guesses and 1/3 is due merely to ordinary sampling variability. The *alternative hypothesis* states that there *is* an effect, in the sense that an astrologer's predictions are *better* than random guessing. This is $H_a : p > 1/3$.

A null hypothesis has a *single* parameter value, such as $H_0: p = 1/3$. An alternative hypothesis has a *range* of values that are alternatives to the one in H_0, such as $H_a : p > 1/3$ or $H_a : p \neq 1/3$. You formulate the hypotheses for a significance test *before* viewing or analyzing the data.

EXAMPLE 2: HYPOTHESES FOR A THERAPEUTIC TOUCH STUDY

Picture the Scenario

Therapeutic touch (TT) practitioners claim to improve or heal many medical conditions by using their hands to manipulate a "human energy field" they perceive above the patient's skin. (The patient does not have to be touched.) A test investigating this claim used the following experiment[2]: A TT practitioner was blind-folded. In each trial, the researcher placed her hand over either the right or left

---

[2] Rosa, L, et al., *JAMA*, vol. 279, pp. 1005-1010, 1998.

hand of the TT practitioner, the choice being determined by flipping a coin. The TT practitioner was asked to identify whether his or her right or left hands was closer to the hand of the researcher. Let p denote the probability of a correct prediction by a TT practitioner. With random guessing, $p = \frac{1}{2}$. However, the TT practitioners claimed that they could do better than random guessing. They claimed that $p > \frac{1}{2}$.

Questions to Explore

Consider the hypothesis, "In any given trial, the probability p of guessing the correct hand is larger than 1/2."

a. Is this a null or an alternative hypothesis?

b. How can we express the hypothesis that being a TT practitioner has "no effect" on the probability of a correct guess?

Think It Through

a. The hypothesis states that $p > 1/2$. It has a range of parameter values, so it is an alternative hypothesis, symbolized by $H_a : p > 0.50$.

b. The quoted hypothesis is an alternative to the null hypothesis that a TT practitioner's predictions are equivalent to random guessing. This "no effect" hypothesis is $H_0 : p = 0.50$.

♦

To practice this concept, try Exercises 8.1 and 8.2.

In a significance test, the null hypothesis is presumed to be true unless the data give strong evidence against it. The "burden of proof" falls on the alternative hypothesis. In the TT study, we assume that $p = 0.50$ unless the data provide strong evidence against it and in favor of the TT practitioners' claim that $p > 0.50$. An analogy may be found in a legal trial in a courtroom, in which a jury must decide the guilt or innocence of a defendant. The null hypothesis, corresponding to "no effect," is that the defendant is innocent. The alternative hypothesis is that the defendant is guilty. The jury presumes that the defendant is innocent unless the prosecutor can provide strong evidence that the defendant is guilty "beyond a reasonable doubt." The "burden of proof" is on the prosecutor to convince the jury that the defendant is guilty.

Step 3: Test Statistic

The parameter to which the hypotheses refer has a point estimate. A **test statistic** describes how far that estimate falls from the parameter value given in the null hypothesis. Usually this is measured by the number of standard errors between the two.

RECALL

The **standard error** of an estimate describes the spread of the sampling distribution of that estimate (Sections 6.4 and 7.2). Denoted by *se*, it is used both in confidence intervals and in significance tests.

For instance, consider the null hypothesis $H_0 : p = 1/3$ that, based on a person's horoscope, the probability p that an astrologer can correctly predict which of three personality charts applies to that person equals 1/3. For the experiment described in Example 1, 40 of 116 predictions were correct. The estimate of the probability p is the sample proportion, $\hat{p} = 40/116 = 0.345$. The test statistic compares this to the value in the null hypothesis ($p = 1/3$), using a z-score that measures the number of standard errors that the estimate falls from 1/3. The next section shows the details.

Step 4: P-Value

To interpret a test statistic value, we use a probability summary of the evidence against the null hypothesis, H_0. Here's how we get it: We presume that H_0 is true, since the "burden of proof" is on the alternative, H_a. Then we consider the sorts of values we'd expect to get for the test statistic, according to its sampling distribution. If the sample test statistic falls well out in a tail of the sampling distribution, it is far from what H_0 predicts. If H_0 were true, such a value would be unlikely. We summarize how far out in the tail the test statistic falls by the tail probability of that value and values even more extreme. See Figure 8.1. This probability is called a **P-value**. The smaller the P-value, the stronger the evidence is against H_0.

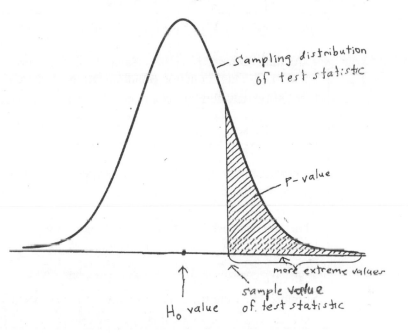

Figure 8.1: Suppose H_0 Were True. The P-Value is the Probability of a Test Statistic Value like the Observed One or Even More Extreme. This is the shaded area in the tail of the sampling distribution. **Question**: Which gives stronger evidence against the null hypothesis, a P-value of 0.20 or of 0.01? Why?

IN WORDS

The **P-value** is a tail probability beyond the observed test statistic value. Smaller P-values provide stronger evidence against the null hypothesis.

P-Value

The **P-value** is the probability that the test statistic equals the observed value or a value even more extreme. It is calculated by presuming that the null hypothesis H_0 is true.

In the astrology study, suppose a P-value is small, such as 0.01. This means that if H_0 were true (that an astrologer's predictions correspond to random guessing), it would be unusual to get sample data such as we observed. Such a P-value provides strong evidence against the null hypothesis of random guessing and in support of the astrologers' claim. On the other hand, if the P-value is not near 0, the data are consistent with H_0. For instance, a P-value such as 0.26 or 0.63 indicates that if the astrologer were actually randomly guessing, the observed data would not be unusual.

Step 5: Conclusion

The conclusion of a significance test reports the P-value and *interprets* what it says about the question that motivated the test. Sometimes this includes a decision about the validity of the null hypothesis H_0. For instance, based on the P-value, can we reject H_0 and conclude that astrologers' predictions are better than random guessing? As we'll discuss in the next section, we can reject H_0 in favor of H_a when the P-value is very small, such as 0.05 or less.

Summary: The Five Steps of a Significance Test

1. **Assumptions**
Method of data production (randomization), sample size condition, shape of population distribution.

2. **Hypotheses**
Null hypothesis, H_0 (a single parameter value, usually "no effect")

Alternative hypothesis, H_a (a set of alternative parameter values).

3. **Test statistic**
Measures distance between point estimate of parameter and its null hypothesis value, usually by the number of standard errors between them.

4. **P-value**
Suppose H_0 were true. The P-value is the probability the test statistic takes the observed value or a value more extreme. Smaller P-values represent stronger evidence against H_0.

5. **Conclusion**
Report and interpret the P-value in the context of the study. Based on the P-value, make a decision about H_0 if one is needed.

SECTION 8.1: PRACTICING THE BASICS

8.1 H_0 or H_a ?: For (a) and (b), is it a null hypothesis, or an alternative hypothesis?
a. In Canada, the proportion of adults who favor legalized gambling equals 0.50.
b. The proportion of all Canadian college students who are regular smokers is less than 0.24, the value it was ten years ago.
c. Introducing notation for a parameter, state the hypotheses in (a) and (b) in terms of the parameter values.

8.2 H$_0$ or H$_a$?: For each of the following, is it a null hypothesis, or an alternative hypothesis? Why?
a. The mean IQ of all students at Lake Wobegon High School is larger than 100.
b. The probability of rolling a 6 with a particular die equals 1/6.
c. The proportion of all adults in the European Union who have a favorable view of current American foreign policy is less than 0.50.

8.3 Burden of proof: For a new pesticide, should the Environmental Protection Agency (EPA) have the burden of proof to show that it is harmful to the environment, or should the producer of the pesticide have the burden of proof to show that it is not harmful to the environment? Give the analog of the null hypothesis and the alternative hypothesis if the burden of proof is on the EPA to show the new pesticide is harmful.

8.4 Iowa GPA: Suppose the mean GPA of all students graduating from the University of Iowa in 1975 was 2.80. The registrar plans to look at records of students graduating in 2005 to see if mean GPA has changed. Define notation and state the null and alternative hypotheses for this investigation.

8.5 Low-carbohydrate diet: A study plans to have a sample of obese adults follow a proposed low-carbohydrate diet for three months. The diet imposes limited eating of starches (such as bread and pasta) and sweets, but otherwise no limit on calorie intake. Consider the hypothesis,

"The population mean of the values of

weight change = weight at start of study – weight at end of study

is a positive number."

a. Is this a null or an alternative hypothesis? Explain your reasoning.
b. Define a relevant parameter, and express the hypothesis that the diet has "no effect" in terms of that parameter.

8.6 Examples of hypotheses: Give an example of a null hypothesis and an alternative hypothesis about a (a) population proportion, (b) population mean.

8.7 z test statistic: To test H$_0$: $p = 0.50$ that a population proportion equals 0.50, the test statistic is a z-score that measures the number of standard errors between the sample proportion and the H$_0$ value of 0.50. If $z = 3.6$, do the data support the null hypothesis, or do they give strong evidence against it? Explain.

8.8 P-value: Indicate whether each of the following P-values gives strong evidence against the null hypothesis or not especially strong evidence.
a. 0.38
b. 0.001

8.2 SIGNIFICANCE TESTS ABOUT PROPORTIONS

For categorical variables, the parameters of interest are the population proportions in the categories. We'll use the astrology study to illustrate a significance test for population proportions.

EXAMPLE 3: ARE ASTROLOGERS' PREDICTIONS BETTER THAN GUESSING?

Picture the Scenario
Many people take astrological predictions seriously, but there has never been any scientific evidence that astrology works. One scientific "test of astrology" used the experiment mentioned in Example 1: For each of 116 adult volunteers, an astrologer prepared a horoscope based on the positions of the planets and the moon at the moment of the person's birth. Each adult subject also filled out a California Personality Index (CPI) survey. For a given adult, his or her birth data and horoscope were shown to an astrologer together with the results of the personality survey for that adult and for two other adults randomly selected from the experimental group. The astrologer was asked which personality chart of the three subjects was the correct one for that adult, based on their horoscope.

The 28 participating astrologers were randomly chosen from a list prepared by the National Council for Geocosmic Research (NCGR), an organization dealing with astrology and respected by astrologers worldwide. The NCGR sampling frame consisted of astrologers with some background in psychology who were familiar with the CPI and who were held in high esteem by their astrologer peers. The experiment was double-blind: Each subject was identified by a random number, and neither the astrologers nor the experimenter knew which number corresponded to which subject. The chapter of the NCGR that recommended the participating astrologers claimed that the probability of a correct guess on any given trial in the experiment was larger than 1/3, the value for random guessing. In fact, they felt that it would exceed ½.

Question to Explore
Put this investigation in the context of a significance test by stating null and alternative hypotheses.

Think It Through
The outcome of each trial in the experiment is categorical. A prediction falls in the category "correct prediction" or the category "incorrect prediction." For each person, let p denote the probability of a correct prediction by the astrologer. We can regard this as the population proportion of correct guesses for the population of people and population of astrologers from which the study participants were sampled.

Hypotheses refer to the probability p of a correct prediction on any given trial. With random guessing, $p = 1/3$. To test the hypothesis of random guessing against the astrologers' claim that $p > 1/3$, we would test $H_0 : p = 1/3$ against $H_a : p > 1/3$.

Insight
In the actual experiment, the astrologers were correct with 40 of their 116 predictions. We'll see how to use these data to test these hypotheses as we work through the five steps of a significance test for a proportion in the next subsection.

◆

To practice this concept, try Exercises 8.9 and 8.10.

What Are the Steps of a Significance Test about a Population Proportion?

Step 1: Assumptions

A large-sample significance test about a population proportion makes three assumptions:

> **RECALL**
>
> Section 6.4 introduced the sampling distribution of a sample proportion. It has exact standard error
>
> $$\sqrt{p(1-p)}.$$
>
> It is approximately normal when
>
> $np \geq 15$ and $n(1-p) \geq 15$.

- The variable is categorical.
- The data are obtained using randomization (such as a random sample).
- The sample size is sufficiently large that the sampling distribution of the sample proportion $\hat{p}$ is approximately normal. The approximate normality happens when the *expected numbers of successes and failures are both at least 15, at the null hypothesis value for p.*

The sample size guideline is the one we used in Chapter 6 for judging when the normal distribution approximates the binomial distribution well. For the astrology experiment with $n = 116$ trials, when H_0 is true that $p = 1/3$, we expect $116(1/3) = 38.7 \approx 39$ correct guesses and $116(2/3) = 77.3 \approx 77$ incorrect ones, both above 15.

As with other statistical inference methods, without randomization the validity of the results is questionable. A survey should use random sampling. An experiment should use principles of randomization and blinding with the study subjects, as was done in the astrology study. In that study, the astrologers were randomly selected, but the subjects evaluated were people (mainly students) who volunteered for the study. Because of this, any inference applies to the population of astrologers but only to the particular subjects in the study. If the study could have randomly chosen the subjects as well, then the inference would extend more broadly to *all* people.

Step 2: Hypotheses

The null hypothesis of a test about a proportion has the form

$$H_0 : p = p_0 ,$$

> **IN WORDS**
>
> p_0 is read as "p-zero," or "the null-hypothesized proportion value."

where p_0 represents a particular proportion value between 0 and 1. In Example 3, the null hypothesis of no effect states that the astrologers' predictions correspond to random guessing. This is $H_0 : p = 1/3$. The null hypothesis value p_0 is 1/3.

The alternative hypothesis refers to alternative parameter values from the number in the null hypothesis. One possible alternative hypothesis has the form

$$H_a : p > p_0 .$$

One-Sided H$_a$: $p >$ 1/3

H$_0$ H$_a$

↓   ↓

Two-Sided H$_a$: p ≠ 1/3

H$_a$ H$_0$ H$_a$

↓   ↓   ↓

This is used when a test is designed to detect whether p is *larger* than the number in the null hypothesis. In the astrology experiment, the astrologers claimed they could predict *better* than by random guessing. Their claim corresponds to H$_a$: $p > 1/3$. This is called a **one-sided** alternative hypothesis, because it has values falling only on one side of the null hypothesis value. We'll use this H$_a$ below. The other possible one-sided alternative hypothesis is H$_a$: $p < p_0$, such as H$_a$: $p < 1/3$.

A **two-sided** alternative hypothesis has the form

$$H_a : p \neq p_0.$$

It includes *all* the other possible values, both below and above the value p_0 in H$_0$. It states that the population proportion *differs* from the number in the null hypothesis. An example is H$_a$: $p \neq 1/3$. This states that the population proportion equals some number other than 1/3.

In summary, for Example 3 we'll use H$_0$: $p = 1/3$ and the one-sided H$_a$: $p > 1/3$.

Step 3: Test Statistic

The test statistic measures how far the sample proportion $\hat{p}$ falls from the null hypothesis value p_0, relative to what we'd expect if H$_0$ were true. The sampling distribution of the sample proportion has mean equal to the population proportion p and exact standard error equal to $\sqrt{p(1-p)/n}$. When H$_0$ is true, $p = p_0$, so the sampling distribution has mean p_0 and exact standard error $se_0 = \sqrt{p_0(1-p_0)/n}$. (We use the zero subscript here on *se* to reflect using the value the exact *se* takes when H$_0$ is true. We substitute p_0 rather than $\hat{p}$ for p in the standard error because we're presuming H$_0$ to be true in conducting the test.) The test statistic is

> **RECALL**
>
> The standard error describes the spread (standard deviation) of the sampling distribution of $\hat{p}$, which is approximately a normal distribution.

$$z = \frac{\hat{p} - p_0}{se_0} = \frac{\hat{p} - p_0}{\sqrt{\dfrac{p_0(1-p_0)}{n}}} = \frac{\text{sample proportion - null hypothesis proportion}}{\text{standard error when null hypothesis is true}}.$$

This z-score measures the number of standard errors between the sample proportion $\hat{p}$ and the null hypothesis value p_0.

In testing H$_0$: $p = 1/3$ for the astrology experiment with $n = 116$ trials, the standard error is $\sqrt{p_0(1-p_0)/n} = \sqrt{[(1/3)(2/3)]/116} = 0.0438$. The astrologers were correct with 40 of their 116 predictions, a sample proportion of $\hat{p} = 0.345$. The test statistic is

$$z = \frac{\hat{p} - p_0}{se} = \frac{\hat{p} - p_0}{\sqrt{\dfrac{p_0(1 - p_0)}{n}}} = \frac{0.345 - 1/3}{0.0438} = 0.26.$$

The sample proportion of 0.345 is only 0.26 standard errors above the null hypothesis value of 1/3.

Step 4: P-Value

Does $z = 0.26$ give much evidence against $H_0 : p = 1/3$ and in support of $H_a : p >$ 1/3? The P-value summarizes the evidence. It describes how unusual the data would be if H_0 were true, that is, if the probability of a correct prediction were 1/3. The P-value is the probability that the test statistic takes a value like the observed test statistic or even more extreme, if actually $p = 1/3$.

<table>
<tr><td>

RECALL

The **standard normal** is the normal distribution with mean = 0 and standard deviation = 1. See the end of Section 6.3.

</td><td>

Figure 8.2 shows the approximate sampling distribution of the z test statistic when H_0 is true. This is the **standard normal distribution**. For the astrology study, $z = 0.26$. Values even farther out in the right tail, above 0.26, are even more extreme. *With a probability distribution for a continuous random variable, such as the normal distribution, the probability of a single value (such as $z = 0.26$) is zero. So, the P-value is the probability of the more extreme values, $z > 0.26$.* From Table A or using software, the right-tail probability above $z = 0.26$ is 0.40. This P-value of 0.40 tells us that if H_0 were true, the probability would be 0.40 that the test statistic would be more extreme than the observed value.

</td></tr>
</table>

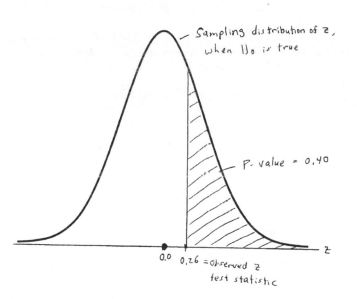

Figure 8.2: Calculation of P-value, when $z = 0.26$, for Testing $H_0 : p = 1/3$ against $H_a : p$ > 1/3. The P-value is the right-tail probability (if H_0 were true) of a test statistic value even more extreme than observed. **Question**: Logically, why are the *right-tail* z-scores considered to be the *more extreme* values for testing H_0 against $H_a : p > 1/3$?

Step 5: Conclusion

We summarize the test by reporting and interpreting the P-value. In the astrology study, the P-value of 0.40 is not especially small. It does not provide strong evidence against $H_0: p = 1/3$. The sample data are the sort we'd expect to see if $p = 1/3$ (that is, if astrologers were randomly guessing the personality type). There is not strong evidence that astrologers have special predictive powers.

♦

To practice this concept, try Exercise 8.15.

How Do We Interpret the P-value?

A significance test analyzes the strength of the evidence against the null hypothesis, H_0. We start by presuming that H_0 is true, putting the *burden of proof* on H_a. The approach taken is the indirect one of *proof by contradiction*. To convince ourselves that H_a is true, we must show the data contradict H_0, by showing they'd be unusual if H_0 were true. We analyze whether the data would be unusual if H_0 were true by finding the P-value. If the P-value is small, the data contradict H_0 and support H_a.

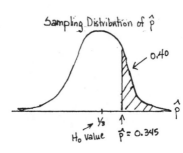

Sampling Distribution of $\hat{p}$

0.40

$\frac{1}{3}$ $\hat{p}$

H_0 value $\hat{p} = 0.345$

In the astrology study, the P-value for $H_a: p > 1/3$ is the right-tail probability of 0.40 from the sampling distribution of the z statistic. This P-value also approximates the probability that the sample proportion $\hat{p}$ takes a value that is as least as far above the null hypothesis value of 1/3 as the observed value of $\hat{p} = 0.345$ (See the margin figure). Since the P-value is not small, if truly $p = 1/3$, it would not be unusual to observe $\hat{p} = 0.345$. The data do not contradict H_0 that the astrologers' predictions correspond to random guessing.

Why do we find the total probability in the *right tail*? Because the alternative hypothesis $H_a: p > 1/3$ has values *above* (that is, to the right of) the null hypothesis value of 1/3. It's the relatively *large* values of $\hat{p}$ that support this alternative hypothesis.

Smaller P-values indicate stronger evidence against H_0, because the data would then be more unusual if H_0 were true. For instance, if we got a P-value of 0.01, then we might be more impressed by the astrologers' claims. When H_0 is true, the P-value is roughly equally likely to fall anywhere between 0 and 1. By contrast, when H_0 is false, the P-value is more likely to be near 0 than near 1.

Two-Sided Significance Tests

Sometimes we're interested in investigating whether a proportion falls above or below some point. For instance, can we conclude whether the population proportion who voted for a particular candidate is above ½, or below ½? We then use a two-sided alternative hypothesis. This has the form $H_a : p \neq p_0$, such as $H_a : p \neq 1/2$.

For two-sided tests, the values that are more extreme than the observed test statistic value are ones that fall farther in the tail in *either* direction. The P-value is the *two-tail* probability under the standard normal curve, because these are the test statistic values that provide even stronger evidence in favor of $H_a : p \neq p_0$ than the observed value. We calculate this by finding the tail probability in a single tail and then doubling it, since the distribution is symmetric. See Figure 8.3.

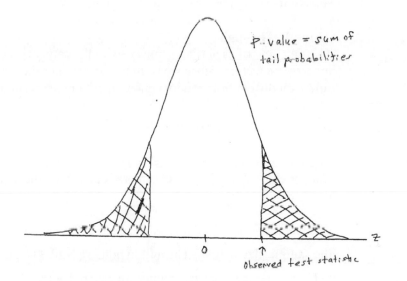

Figure 8.3: For the Two-Sided Alternative $H_a : p \neq p_0$, the P-value Is a Two-Tail Probability. Question: Logically, why are both the left-tail values and the right-tail values the more extreme values for $H_a : p \neq p_0$?

EXAMPLE 4: DR. DOG: CAN DOGS DETECT CANCER BY SMELL?

Picture the Scenario
A recent study investigated whether dogs can be trained to distinguish a patient with bladder cancer by smelling certain compounds released in the patient's urine.[3] Six dogs of varying breeds were trained to discriminate between urine from patients with bladder cancer and urine from control patients without it. The dogs were taught to indicate which among several specimens was from the bladder cancer patient by lying beside it.

---

[3] Article by C. M. Willis et al., *British Medical Journal*, vol. 329, September 25, 2004.

An experiment was conducted to analyze how the dogs' ability to detect the correct urine specimen compared to what would be expected with random guessing. Each of the six dogs was tested with nine trials. In each trial, one urine sample from a bladder cancer patient was randomly placed among six control urine samples. In the total of 54 trials with the six dogs, the dogs made the correct selection 22 times.

Let p denote the probability that a dog makes the correct selection on a given trial. Since the urine from the bladder cancer patient was one of seven specimens, with random guessing $p = 1/7$.

Question to Explore

Did this study provide strong evidence that the dogs' predictions were better or worse than with random guessing? Specifically, is there strong evidence that $p > 1/7$, with dogs able to select better than with random guessing, or that $p < 1/7$, with dogs' selections being poorer than random guessing?

Think It Through

The outcome of each trial is binary, with the categories correct selection and incorrect selection. Since we want to test whether the probability of a correct selection differs from random guessing, the hypotheses are:

$$H_0 : p = 1/7 \text{ and } H_a : p \neq 1/7.$$

The null hypothesis represents *no effect*, the selections being like random guessing. The alternative hypothesis says there is an effect, the selections differing from random guessing.

The sample proportion of correct selections by the dogs was $\hat{p} = 22/54 = 0.407$, for the sample size $n = 54$. The null hypothesis value is $p_0 = 1/7 = 0.143$. When $H_0 : p = 1/7$ is true, the expected counts are $np_0 = 54(1/7) = 7.7$ correct selections and $n(1 - p_0) = 54(6/7) = 46.3$ incorrect selections. The first of these is not larger than 15, so according to the guidelines given for step 1 of the test, n is not large enough to use the large-sample test. Later in this section, we'll see that the *two-sided* test is robust when this assumption is not satisfied. So we'll use the large-sample test, under the realization that the P-value is an approximation for the P-value of a small-sample test mentioned later.

> **RECALL**
>
> From Section 7.3, a method is **robust** with respect to a particular assumption if it works well even when that assumption is violated.

The standard error is $se_0 = \sqrt{p_0(1 - p_0)/n} = \sqrt{(1/7)(6/7)/54} = 0.0476$. The test statistic equals

$$z = \frac{\hat{p} - p_0}{se_0} = \frac{0.407 - 0.143}{0.0476} = 5.6.$$

The sample proportion, 0.407, falls more than five standard errors above the null hypothesis value of 0.143.

Figure 8.4 shows the approximate sampling distribution of the z test statistic when H_0 is true. The test statistic value of 5.6 is well out in the right tail. Values farther out in the tail, above 5.6, are even more extreme. The P-value is the total two-tail probability of the more extreme outcomes, above 5.6 or below -5.6. From software, the cumulative probability in the left tail below $z = -5.6$ is 0.000 to three decimal places (in fact, it is 0.00000001), and the probability in the two tails equals 2(0.000) = 0.000. This P-value provides extremely strong evidence against $H_0 : p = 1/7$.

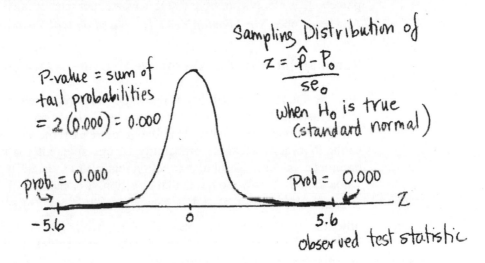

Figure 8.4: Calculation of P-value, when $z = 5.6$, for testing $H_0 : p = 1/7$ against $H_a : p \neq 1/7$. Presume H_0 is true. The P-value is the two-tail probability of a test statistic value even more extreme than observed. **Question**: Is P-value = 0.000 strong evidence supporting H_0, or strong evidence against it?

When the P-value in a two-sided test is small, the point estimate tells us the direction in which the parameter appears to differ from the null hypothesis value. In summary, since the P-value is very small and $\hat{p} > 1/7$, the evidence strongly suggests that the dogs' selections are *better* than random guessing.

Insight

Recall that one assumption for this significance test is randomization for obtaining the data. This study is like most medical studies in that its subjects were a *convenience sample* rather than a random sample from some population. It is not practical that a study can identify the population of all people who have bladder cancer and then randomly sample them for an experiment. Likewise, the dogs were not randomly sampled. Because of this, any inferential predictions are highly tentative. They are valid only to the extent that the patients and the dogs in the experiment are representative of their populations. The predictions become more conclusive if similar results occur in other studies with other samples. In medical studies, even though the sample is not random, it is important to employ randomization in any experimentation, for instance in the placement of the bladder cancer patient's urine specimen among the six control urine specimens.

> **RECALL**
>
> See Section 4.2 for potential difficulties with convenience samples, which use subjects who are conveniently available.

To practice this concept, try Exercise 8.17.

In Example 4, you might instead have been inclined to use $H_a : p > 1/7$, if you expected the dogs' predictions to be better than random guessing. However, most medical studies use two-sided alternative hypotheses. This recognizes that if an effect exists, it could be negative rather than positive. For instance, many medical studies compare a drug to a placebo. Although we may hope the drug is better than placebo, it's always possible that it is worse, perhaps because of bad side effects. And remember, *you shouldn't pick H_a based on looking at the data.*

Summary of How the Alternative Hypothesis Determines the P-value

The P-value is the probability of the values that are more extreme than the observed test statistic value. What is "more extreme" depends on the alternative hypothesis. For the two-sided alternative hypothesis, more extreme values fall in each direction, so the P-value is a two-tail probability. A one-sided alternative states that the parameter falls in a particular direction relative to the null hypothesis value. The P-value then uses only the tail in that direction. For instance, in the astrology study, the alternative $H_a : p > 1/3$ used only the *right* tail. For the alternative $H_a : p < 1/3$ in the other direction, we would have used the probability to the *left* of the observed test statistic value. See Figure 8.5. Notice that the P-values for the two one-sided alternative hypotheses add up to 1.0.

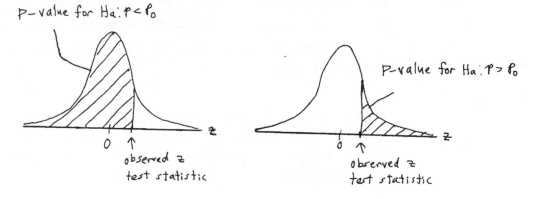

Figure 8.5: Calculation of P-value for One-Sided Alternative Hypotheses.   Question: For a given one-sided H_a, how do we know which tail to use for finding the tail probability?

Here's a summary of how we find the P-value:

Summary of P-Values for Different Alternative Hypotheses	
Alternative Hypothesis	*P-Value*
$H_a : p > p_0$	Right-tail probability
$H_a : p < p_0$	Left-tail probability
$H_a : p \neq p_0$	Two-tail probability

The Significance Level Tells Us How Strong the Evidence Must Be

Sometimes we need to make a decision about whether the data provide sufficient evidence to reject H_0. Before seeing the data, we decide how small the P-value would need to be to reject H_0. For example, we might decide that that we will reject H_0 if the P-value ≤ 0.05. The cutoff point of 0.05 is called the **significance level**. It is shown in Figure 8.6.

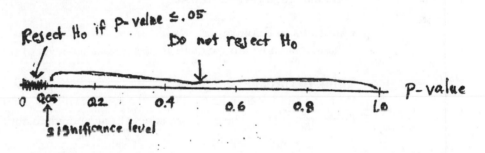

Figure 8.6: The Decision in a Significance Test. Reject H_0 if the P-value is less than or equal to a chosen **significance level**, usually 0.05.

Definition: Significance Level
The **significance level** is a number such that we reject H_0 if the P-value is less than or equal to that number. In practice, the most common significance level is 0.05.

Table 8.1 summarizes the two possible outcomes for a test decision when the significance level is 0.05. We either "reject H_0" or "do not reject H_0." If the P-value if larger than 0.05, the data do not sufficiently contradict H_0 for us to reject it. Then, H_0 is still plausible to us, and we "do not reject H_0." If the P-value is ≤ 0.05, we do feel the data provide enough evidence to "reject H_0." Recall that H_a had the

533

"burden of proof," and in this case we feel that the proof is sufficient. When we reject H_0, we say the results are **statistically significant**.

Table 8.1: Possible Decisions in a Test with Significance Level = 0.05

P-VALUE	DECISION ABOUT H_0
≤ 0.05	Reject H_0
> 0.05	Do not reject H_0

EXAMPLE 5: WHAT'S THE DECISION FOR THE ASTROLOGY STUDY?

Picture the Scenario

Let's continue our analysis of the astrology study from Example 3. The parameter p is the probability that an astrologer picks the correct one of three personality charts, based on the horoscope and birth data provided. We tested $H_0: p = 1/3$ against $H_a: p > 1/3$ and got a P-value of 0.40.

Questions to Explore

What decision would we make for a significance level of (a) 0.05? (b) 0.50?

Think It Through

(a) For a significance level of 0.05, the P-value of 0.40 is not less than 0.05. So we do not reject H_0. The evidence is not strong enough to conclude that the astrologers' predictions are better than random guessing.

(b) For a significance level of 0.50, the P-value of 0.40 *is* less than 0.50. So we reject H_0 in favor of H_a. We conclude that this result provides sufficient evidence to support $p > 1/3$. The results are statistically significant at the 0.50 significance level.

Insight

In practice, significance levels are normally close to 0, such as 0.05 or 0.01. The reason is that the significance level is also a type of error probability. We'll see in Section 8.4 that when H_0 is true, it's the probability of making an error by rejecting H_0. We used the significance level of 0.50 in (b) above for illustrative purposes, but the value of 0.05 used in part (a) is much more typical. The astrologers' predictions are consistent with random guessing, and we would not reject H_0 for these data.

◆

To practice this concept, try Exercise 8.18.

IN PRACTICE

Report the P-value, rather than merely indicating whether the results are statistically significant. Learning the actual P-value is more informative than learning only whether the test is "statistically significant at the 0.05 level." The P-values of 0.01 and 0.049 are both statistically significant in this sense, but the first P-value provides much stronger evidence than the second.

Now that we've studied all five steps of a significance test about a proportion, let's use them in a new example, much like those you'll see in the exercises. In this example and in the exercises, it may help you to refer to the following summary box:

Summary: Steps of a Significance Test for a Population Proportion *p*

1. Assumptions
- Categorical variable, with population proportion p defined in context
- Randomization, such as a simple random sample, for gathering data
- n large enough to expect at least 15 successes and 15 failures under H_0 (that is, $n\,p_0 \geq 15$ and $n(1 - p_0) \geq 15$). This is mainly important for one-sided tests.

2. Hypotheses

Null: $H_0 : p = p_0$, where p_0 is the hypothesized value, such as $H_0 : p = 0.50$

Alternative: $H_a : p \neq p_0$, such as $H_a : p \neq 0.50$ (two-sided)

or $H_a : p < p_0$ (one-sided) or $H_a : p > p_0$ (one-sided)

3. Test statistic

$$z = \frac{\hat{p} - p_0}{se_0} \text{ with } se_0 = \sqrt{p_0(1 - p_0)/n}$$

4. P-value

Alternative hypothesis	*P-value*
$H_a : p > p_0$	Right-tail probability
$H_a : p < p_0$	Left-tail probability
$H_a : p \neq p_0$	Two-tail probability

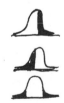

5. Conclusion

Smaller P-values give stronger evidence against H_0. If a decision is needed, reject H_0 if P-value is less than or equal to the preselected significance level (such as 0.05). Relate the conclusion to the context of the study.

EXAMPLE 6: CAN TT PRACTITIONERS DETECT A HUMAN ENERGY FIELD?

Picture the Scenario

Let's revisit the Therapeutic Touch (TT) experiment from Example 2. Each trial investigated whether a TT practitioner could correctly identify (while blind-folded) which of their hands was closer to the hand of a researcher. The researcher determined hand placement by flipping a coin. In a set of 150 trials with 15 TT practitioners (10 trials each), the TT practitioners were correct with 70 of their 150 predictions.

Questions to Explore

How strong is the evidence to support the TT practitioners' claim that they can predict the correct hand better than with random guessing? What decision would be made for a 0.05 significance level?

Think It Through

Let's follow the five steps of a significance test to organize our response to these questions:

1. *Assumptions*:
   - The response is categorical, with outcomes correct and incorrect for a TT practitioner's prediction. Let p denote the probability of a correct prediction.
   - We'll treat the 150 trials as a random sample of the possible trials that could occur in such an experiment with these TT practitioners. The inference will apply to the 15 particular TT practitioners used in this experiment. If the study randomly selected them from the population of TT practitioners, then the inference would apply to the entire population of TT practitioners.
   - We'll discuss the sample size requirement in the next step.

2. *Hypotheses*:

If a TT practitioner merely guesses, then $p = 0.50$. However, the TT practitioners claimed that $p > 0.50$. To check their claim, we'll test $H_0: p = 0.50$ against $H_a: p > 0.50$. Supposing H_0 is true, of 150 trials we expect $150(0.50) = 75$ correct predictions and $150(0.50) = 75$ incorrect ones. These both exceed 15. The sample size is large enough to use the large-sample significance test.

3. *Test statistic*:

The sample estimate of the probability p of a correct decision is $\hat{p} = 70/150 = 0.467$. The standard error of $\hat{p}$ when $H_0: p = 0.50$ is true is

$$se_0 = \sqrt{p_0(1-p_0)/n} = \sqrt{0.50(0.50)/150} = 0.0408.$$

The value of the test statistic is

$$z = \frac{\hat{p} - p_0}{se_0} = \frac{0.467 - 0.50}{0.0408} = \text{-}0.82.$$

DID YOU KNOW?

The TT experiment was a science fair project of a 9-year old girl. Her project was published in a major medical journal, *Journal of the American Medical Association (JAMA)*. What inspiration for future science fair researchers!

RECALL

H_0 contains a single number, whereas H_a has a range of values.

The sample proportion is 0.82 standard errors below the null hypothesis value.

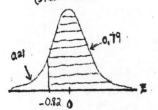

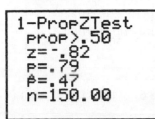

4. P-value:

For testing $H_0 : p = 0.50$ against $H_a : p > 0.50$, the P-value is the *right-tail* probability above $z = -0.82$ in the standard normal distribution (See the figure in the margin). From software or a normal table, this is $1 - 0.21 = 0.79$. TI-83+/84 output is also provided in the margin

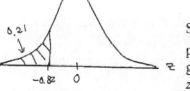

TI-83+/84 output

5. Conclusion:

The P-value of 0.79 is not small. If the null hypothesis $H_0 : p = 0.50$ were true, the data we observed would not be unusual. With a 0.05 significance level, the evidence is not strong enough to reject H_0. It seems plausible that $p = 0.50$. In summary, this experiment does not suggest that the TT practitioners can predict better than random guessing.

Insight

The P-value is larger than 1/2 because the data do not provide evidence against $H_0 : p = 0.50$ in favor of $H_a : p > 0.50$. The sample proportion of 0.467 actually falls in the direction of the *other* tail, $p < 0.50$. In this experiment, the TT practitioners predicted *worse* than what we'd expect for random guessing.

◆

To practice this concept, try Exercise 8.16.

Suppose we had instead chosen the one-sided alternative, $H_a : p < 0.50$. This predicts that the probability of a correct prediction is *worse* than with random guessing. Then, the P-value is a *left-tail* probability, everything to the left of $z = -0.82$ under the standard normal curve. The P-value is then 0.21. See the margin figure. The sum of the P-values for the one-sided alternatives always equals 1.0. The P-value = 0.79 for $H_a : p > 0.50$, the P-value = 0.21 for $H_a : p < 0.50$, and $0.79 + 0.21 = 1.0$.

IN PRACTICE

In practice, you should pick H_a before seeing the data, and you can use software to do the test. For the one-sided test with $H_a : p > 0.50$, Minitab reports the results shown in Table 8.2. Unless requested otherwise, by default software reports the two-sided P-value.

Some software reports the P-value to several decimals, such as 0.792892. We recommend rounding it to two or three decimal places, for instance, to 0.79, before reporting it. Reporting a P-value as 0.792892 suggests greater accuracy than actually exists, since the normal sampling distribution is only *approximate*.

Table 8.2: MINITAB Output for One-Sided Test in Example 6.

```
Test of p = 0.5 vs p > 0.5

       X          N        Sample p      Z-Value      P-Value
      70         150       0.466667        -0.82        0.793
```

Annotation of Output →

$\uparrow$ *Sample Size* $\uparrow$ *Test Statistic*

Category Count   *Sample Proportion*

"Do Not Reject H_0" Does Not Mean "Accept H_0"

A small P-value means that the sample data would be unusual if H_0 were true. If the P-value is not small, such as 0.79 for the one-sided test in Example 6 on TT practitioners, the null hypothesis is plausible. In this case, the conclusion is reported as "Do not reject H_0," since the data do not contradict H_0.

"Do not reject H_0" is not the same as saying "accept H_0." The population proportion has many plausible values besides the number in the null hypothesis. For instance, consider Example 6, with p the probability of a correct prediction by a TT practitioner. We did not reject $H_0 : p = 0.50$. Thus, p may equal 0.50, but other values are also plausible. A 95% confidence interval for p is

$$\hat{p} \pm 1.96\sqrt{\hat{p}(1-\hat{p})/n}, \text{ or } 0.467 \pm 1.96\sqrt{(0.467)(0.533)/150},$$

which equals (0.39, 0.55). Even though insufficient evidence exists to reject H_0, it is improper to accept it and conclude that $p = 0.50$. It is plausible that p is as low as 0.39 or as high as 0.55.

An analogy here is that of a legal trial in a courtroom. The null hypothesis is that the defendant is innocent. The alternative hypothesis is that the defendant is guilty. If the jury acquits the defendant, this does not mean that it *accepts* the defendant's claim of innocence. It merely means that innocence is plausible because guilt has not been established *beyond a reasonable doubt*.

We can "Accept H$_a$" when it contains all the plausible values.

The null hypothesis contains a single possible value for the parameter. Saying "Do not reject H$_0$" instead of "Accept H$_0$" emphasizes that that value is merely one of many plausible ones. Because of sampling error, there is a range of plausible values besides the H$_0$ value. "Accept H$_a$" *is* permissible for the alternative hypothesis. When the P-value is sufficiently small, a confidence interval would show that the entire range of plausible values falls within the range of numbers contained in H$_a$.

Use p_0 instead of $\hat{p}$ in the standard error for significance tests

In calculating the standard error in the test statistic, we substituted the null hypothesis value p_0 for the population proportion p in the formula $\sqrt{p(1-p)/n}$ for the exact standard error. *The parameter values for sampling distributions in significance tests are those from H$_0$, since the P-value is calculated presuming that H$_0$ is true.* So we use p_0 in standard errors for tests. This practice differs from confidence intervals, in which the sample proportion $\hat{p}$ substitutes for p (which is unknown) in the exact standard error. The confidence interval method[4] does not have a hypothesized value for p, so that method substitutes the point estimate $\hat{p}$ for p. To differentiate between the two cases, we've denoted $se_0 = \sqrt{p_0(1-p_0)/n}$ and $se = \sqrt{\hat{p}(1-\hat{p})/n}$.

How Do We Decide Between a One-Sided and a Two-Sided Test?

In practice, two-sided tests are more common than one-sided tests. Even if we think we know the direction of an effect, two-sided tests can also detect an effect that falls in the opposite direction. For example, in a medical study, even if we think a drug will perform better than a placebo, using a two-sided alternative allows us to detect if the drug is actually worse, perhaps because of bad side effects. However, as we saw with the astrology and TT examples, there are scenarios for which a one-sided test is natural.

Here are some things to consider in deciding on the alternative hypothesis:

- In deciding between one-sided and two-sided alternative hypotheses in a particular exercise or in practice, *consider the context of the real problem.*

For instance, in the astrology experiment, to test whether someone can guess *better* than with random guessing, we used the values $p > 1/3$ in the alternative hypothesis corresponding to that possibility. An exercise that says "Test whether the population

---

[4] If we instead conduct the test by substituting the sample proportion in the standard error, the standard normal approximation for the sampling distribution of z is much poorer.

proportion of ... *differs* from 0.50" suggests a two-sided alternative, $H_a : p \neq 0.50$, to allow for p to be larger or smaller that 0.50. "Test whether the population proportion of ... is *larger* than 0.50" suggests the one-sided alternative, $H_a : p > 0.50$.

- *In most research articles, significance tests use two-sided P-values.*

Partly this reflects an ethical approach to research that allows an effect to go in either direction. Using an alternative hypothesis that allows the effect to go in either direction is regarded as the most even-handed way to perform the test. In using two-sided P-values, researchers avoid the suspicion that they chose H_a when they saw the direction in which the data occurred. That would be cheating. (The P-value would then need to be based on a different sampling distribution that is conditioned on that extra information.)

- *Confidence intervals are two-sided.*

The practice of using a two-sided test coincides with the ordinary approach for confidence intervals. They are two-sided, obtained by adding and subtracting some quantity from the point estimate. There is a way to construct one-sided confidence intervals, for instance, concluding that a population proportion is *at least* equal to 0.70. In practice, though, two-sided confidence intervals are much more common.

IN PRACTICE

For the reasons just discussed, two-sided tests are more common than one-sided tests. Most examples and exercises in this book reflect the way tests are most often used and employ two-sided alternatives. *In practice, you should use a two-sided test unless you have a well-justified reason for a one-sided test.*

As you specify the hypotheses, also keep in mind two basic facts:

- The null hypothesis always has an equal sign (such as $H_0 : p = 0.5$), but the alternative hypothesis does not.

- The hypotheses always refer to population parameters, not sample statistics.

So *never* express a hypothesis using sample statistic notation, such as $H_0 : \hat{p} = 0.5$. There is no need to conduct statistical inference about statistics such as the sample proportion $\hat{p}$ since you can find their values exactly from the data.

The Binomial Test for Small Samples

The test about a proportion applies normal sampling distributions for the sample proportion $\hat{p}$ and the z-test statistic. Therefore, it is a *large-sample* test, because the central limit theorem implies approximate normality of the sampling distribution for large random samples. The guideline is that the expected numbers of successes and failures should be at least 15, when H_0 is true; that is, $n\,p_0 \geq 15$ and $n(1 - p_0) \geq 15$.

In practice, the large-sample z test still performs quite well for *two-sided* alternatives even for small samples. When p_0 is below 0.3 or above 0.7 and n is small, the sampling distribution is quite skewed. However, a tail probability that is smaller than the normal probability in one tail is compensated by a tail probability that is larger than the normal probability in the other tail. Because of this, the P-value from the two-sided test using the normal table gives a decent approximation to a P-value from a small-sample test.[5]

For one-sided tests, when p_0 differs from 0.5, the large-sample test does not work well when the sample size guideline ($n\,p_0 \geq 15$ and $n(1 - p_0) \geq 15$) is violated. In that case, you should use a small-sample test. This test uses the binomial distribution with parameter value p_0 to find the exact probability of the observed value and all the more extreme values, according to the direction in H_a. Since one-sided tests and small n are less common in practice, we will not study the binomial test here. Exercises 8.27 and 8.28 show how to do it.

SECTION 8.2: PRACTICING THE BASICS

8.9 Psychic: A person who claims to be psychic says that the probability p that he can correctly predict the outcome of the roll of a die in another room is greater than 1/6, the value that applies with random guessing. If we want to test this claim, we could use the data from an experiment in which he predicts the outcomes for n rolls of the die. State hypotheses for a significance test, letting the alternative hypothesis reflect the psychic's claim.

8.10 Believe in astrology?: You plan to apply significance testing to your own experiment for testing astrology, in which astrologers have to guess which of 4 personality profiles is the correct one for someone who has a particular horoscope. Define notation and state hypotheses, letting one hypothesis reflect the possibility that the astrologers' predictions could be better than random guessing.

8.11 Get P-value from z: For a test of $H_0 : p = 0.50$, the z test statistic equals 1.04.

a. Find the P-value for $H_a : p > 0.50$.

b. Find the P-value for $H_a : p \neq 0.50$.

c. Find the P-value for $H_a : p < 0.50$.

d. Do any of the P-values in (a), (b), or (c) give strong evidence against H_0? Explain.

---

[5] With small samples, this test works much better than the confidence interval of Section 6.2. Having an exact, rather than estimated, standard error, makes a great difference.

8.12 Get more P-values from z: Refer to the previous exercise. Suppose $z = 2.50$ instead of 1.04.

a. Find the P-value for (i) H$_a$: $p > 0.50$, (ii) H$_a$: $p \neq 0.50$, (iii) H$_a$: $p < 0.50$.

b. Do any of the P-values in (a) provide strong evidence against H$_0$? Explain

8.13 Find test statistic and P-value: For a test of H$_0$: $p = 0.50$, the sample proportion is 0.35 based on a sample size of 100.

a. Show that the test statistic is $z = -3.0$

b. Find the P-value for H$_a$: $p < 0.50$.

c. Does the P-value in (b) give much evidence against H$_0$? Explain.

8.14 Which z **has P-value = 0.05?**: The P-value for a test about a proportion is 0.05.

a. If the alternative hypothesis was H$_a$: $p > p_0$, report the value of the test statistic.

b. If the alternative hypothesis was H$_a$: $p < p_0$, report the value of the test statistic.

c. If H$_a$ was H$_a$: $p \neq p_0$, report the two possible test statistic values.

8.15 Another test of astrology: Examples 1, 3, and 5 referred to a study about astrology. Another part of the study used the following experiment: Professional astrologers prepared horoscopes for 83 adults. Each adult was shown three horoscopes, one of which was the one an astrologer prepared for them and the other two were randomly chosen from ones prepared for other subjects in the study. Each adult had to guess which of the three was theirs. Of the 83 subjects, 28 guessed correctly.

a. Defining notation, set up hypotheses to test that the probability of a correct prediction is 1/3 against the astrologers' claim that it exceeds 1/3.

b. Show that the sample proportion = 0.337, the standard error of the sample proportion for the test is 0.052, and the test statistic is $z = 0.08$.

c. Find the P-value. Would you conclude that people are more likely to select their "correct" horoscope than if they were randomly guessing, or are results consistent with random guessing?

d. In order for inference using the P-value in (c) to apply to all astrologers and to all subjects for whom they might make predictions, what do you need to assume?

8.16 Another test of therapeutic touch: Examples 1, 2, and 6 described a study about therapeutic touch (TT). A second run of the same experiment in the study used 13 TT practitioners who had to predict the correct hand in each of 10 trials.

a. Defining notation, set up hypotheses to test that the probability of a correct guess is 0.50 against the TT practitioners' claim that it exceeds 0.50.

b. Of the 130 trials, there were 53 correct guesses. Find and interpret the test statistic.

c. Report the P-value. (*Hint*: Since the sample proportion is in the opposite direction from the practitioners' claim, you should get a P-value > 0.50.) Indicate your decision, in the context of this experiment, using a 0.05 significance level.

d. Check whether the sample size was large enough to make the inference in (c). Indicate what assumptions you would need to make for your inferences to apply to all TT practitioners and to all subjects.

RECALL

Crossover designs were introduced in Section 4.3.

8.17 Testing a headache remedy: Studies that compare treatments for chronic medical conditions such as headaches can use the same subjects for each treatment. With a crossover design, each person crosses over from using one treatment to another during the study. One such study considered a drug (a pill called Sumatriptan) for treating migraine headaches in children.[6] The study observed each of 30 children at two times when he or she had a migraine headache. The child received the drug at one time and a placebo at the other time. The order of treatment was randomized and the study was double-blind. For each child, the response was whether the drug or the placebo provided better pain relief. Let p denote the proportion of children having better pain relief with the drug, in the population of children who suffer periodically from migraine headaches. Can you conclude that $p > 0.50$, with more than half of the population getting better pain relief with the drug, or that $p < 0.50$, with less than half getting better pain relief with the drug (i.e., the placebo being better)? Of the 30 children, 22 had more pain relief with the drug and 8 had more pain relief with the placebo.
a. Show that the test statistic equals 2.56.
b. Show that the P-value is 0.0104. Interpret.
c. Check the assumptions needed for this test, and discuss the limitations due to using a convenience sample.

8.18 Gender bias in selecting managers: For a large supermarket chain in Florida, a women's group claimed that female employees were passed over for management training in favor of their male colleagues. The company denied this claim, saying they picked the employees from the eligible pool at random to receive this training. Statewide, the large pool of more than 1000 eligible employees who can be tapped for management training is 40% female and 60% male. Since this program began, 28 of the 40 employees chosen for management training were male and 12 were female.
a. In conducting a significance test, the random sampling assumption is the claim of the company. In defining hypotheses, explain why this company's claim of a lack of gender bias is a "no effect" hypothesis. State the null and alternative hypotheses for a test to investigate whether there is sufficient evidence to support the women's claim.
b. The table shows results of using software to do a large-sample analysis. Explain why the large-sample analysis is justified, and show how software obtained the test statistic value.

Test of p = 0.60 vs not = 0.60

X	N	Sample p	95.0% CI	Z-Value	P-Value
28	40	0.70000	(0.558, 0.842)	1.29	0.1967

----

c. To what alternative hypothesis does the P-value in the table refer? Use it to find the P-value for the alternative hypothesis you specified in (a), and interpret it.
d. What decision would you make for a 0.05 significance level? Interpret.

---

[6] Data based on those in a study by M. L. Hamalainen et al., reported in *Neurology*, vol. 48, pp. 1100-1103, 1997.

8.19 Gender discrimination: Refer to the previous exercise.

a.   Explain why the alternative hypothesis of bias against males is $H_a : p < 0.60$.

b.   Show that the P-value for testing $H_0 : p = 0.60$ against $H_a : p < 0.60$ equals 0.90.
Interpret. Why is it large?

8.20 Protecting the environment?: When the 2000 General Social Survey asked,
"Would you be willing to pay much higher taxes in order to protect the
environment?" (variable GRNTAXES), 369 people answered yes and 483 answered
no. (We exclude those who made other responses.) Let p denote the population
proportion who would answer yes. Minitab software shows the following results of a
significance test to analyze whether a majority or minority of Americans would
answer yes:

```
-----------------------------------------------------------------------------
Test of p = 0.5 vs p not = 0.5


  X     N     Sample p          95% CI            Z-Value    P-Value
 369   852    0.433099   (0.399827, 0.466370)      -3.91      0.000
-----------------------------------------------------------------------------
```

a. What are the assumptions for the test? Do they seem to be satisfied for this
application? Explain.
b. Specify the null and alternative hypotheses that are tested on this printout.
c. Report the point estimate of p and the value of the test statistic.
d. Report and interpret the P-value.
e. According to the P-value, is it plausible that $p = 0.50$? Explain, in context.

8.21   Garlic to repel ticks: A recent study (*J. Amer. Med. Assoc.*, vol. 284, p.
831, 2000) considered whether daily consumption of 1200 mg of garlic could reduce
tick bites. The study used a crossover design with a sample of Swedish military
conscripts, half of whom used placebo first and garlic second and half the reverse.
The authors did not present the data, but the effect they described is consistent with
garlic being more effective with 37 subjects and placebo being more effective with
29 subjects. Does this suggest a real difference between garlic and placebo, or are the
results consistent with random variation? Answer by:
a. Setting up notation and checking sample size requirements for a large-sample test.
b. Stating hypotheses for a two-sided test.
c. Finding the test statistic value.
d. Finding and interpreting a P-value and stating the conclusion in context.

8.22 Exit poll predictions: According to an exit poll in the 2000 New York
senatorial election, 55.7% of the sample of size 2232 reported voting for Hillary
Clinton. Is this enough evidence to predict who won? Test that the population
proportion who voted for Clinton was 0.50 against the alternative that it differed from
0.50. Answer by:
a. Stating and checking assumptions.
b. Defining notation and stating hypotheses.
c. Finding the test statistic.
d. Reporting the P-value.
e. Explaining how to make a decision for the significance level of 0.05.

8.23 Guessing true or false: An exam has 100 true-or-false questions. The instructor suspects that one of her students, who has never attended class or handed in a homework assignment, is completely unprepared and will do no better than if he randomly guessed the answer to each question. Let p denote the probability that the student gets the correct answer on a question.

a. State the hypotheses to test that the student randomly guesses the answers and to be able to detect whether his responses are better than random guessing.

b. The student correctly answers 56 of the 100 questions. Find the P-value, and interpret. Are his results consistent with random guessing?

c. Make a decision about H_0, using a significance level of 0.05. Based on this decision, what can you conclude about the parameter?

d. Based on the conclusion in (c), can you "accept H_0"? Explain.

8.24 Which cola?: The 49 students in a class at the University of Florida made blinded evaluations of pairs of cola drinks. For the 49 comparisons of Coke and Pepsi, Coke was preferred 29 times. In the population that this sample represents, is this strong evidence that a majority prefers one of the drinks? Answer by explaining how to interpret each item on the following Minitab printout.

Test of $p = 0.50$ vs not $= 0.50$

X	N	Sample p	95.0% CI	Z-Value	P-Value
29	49	0.5918	(0.454, 0.729)	1.286	0.1985

8.25 Doctors and alternative therapies: Let p denote the proportion of general practitioners in Australia who have ever studied alternative therapy, such as acupuncture. Can you conclude whether this is less than, or greater than, 0.50? The printout shows results of a significance test for survey data from the University of Western Australia, in which 132 GPs said they had studied alternative therapy.

Test of $p = 0.50$ vs not $= 0.50$

Sample	X	N	Sample p	95.0% CI	Z-Value	P-Value
1	132	282	0.4681	(0.4098, 0.5263)	-1.072	0.2838

a. Explain how to get "Sample p."
b. Explain how to get the test statistic value that MINITAB reports.
c. Explain how to get "P-value." Interpret it.
d. What does the 95% confidence interval tell you that the test does not?

8.26 How to sell a burger: A fast-food chain wants to compare two ways of promoting a new burger (a turkey burger). One way uses a coupon available in the store. The other way uses a poster display outside the store. Before the promotion, their marketing research group matches 50 pairs of stores. Each pair has two stores with similar sales volume and customer demographics. The store in a pair that uses coupons is randomly chosen, and after a month-long promotion, the increases in sales of the turkey burger are compared for the two stores. The increase was higher for 28 stores using coupons and higher for 22 stores using the poster. Is this strong evidence to support the coupon approach, or could this outcome be explained by chance? Answer by performing all five steps of a two-sided significance test.

8.27 A binomial headache: A null hypothesis states that the population proportion p of headache sufferers who have more pain relief with aspirin than with another pain reliever equals 0.50. For a crossover study with 10 subjects, all 10 have more relief with aspirin. If the null hypothesis were true, by the binomial distribution the probability of this sample result equals $(0.50)^{10} = 0.001$. In fact, this is the small-sample P-value for testing $H_0: p = 0.50$ against $H_a: p > 0.50$. Does this P-value give

(a) strong evidence in favor of H_0, or (b) strong evidence against H_0? Explain why.

8.28 P-value for small samples: Example 4 on whether dogs can detect bladder cancer by selecting the correct urine specimen (out of 7) used the normal sampling distribution to find the P-value. The normal P-value approximates a P-value using the binomial distribution. That binomial P-value is more appropriate when either expected count is less than 15. In Example 4, n was 54, and 22 of the 54 selections were correct.

a. Supposing $H_0: p = 1/7$ is true, X = number of correct selections has the binomial distribution with $n = 54$ and $p = 1/7$. Why?

b. For $H_a: p > 1/7$, with $x = 22$, explain why the P-value using the binomial is $p(22) + p(23) + \ldots + p(54)$, where $p(x)$ denotes the binomial probability of outcome x with $p = 1/7$. (This equals 0.0000019.)

8.3 SIGNIFICANCE TESTS ABOUT MEANS

For quantitative variables, significance tests usually refer to the population mean μ. We illustrate the significance test about means with the following example.

EXAMPLE 7: MEAN WEIGHT CHANGE IN ANOREXIC GIRLS

Picture the Scenario
A recent study compared different psychological therapies for teenage girls suffering from anorexia, an eating disorder that can cause them to become dangerously underweight.[7] Each girl's weight was measured before and after a period of therapy. The variable of interest was the weight change, defined as weight at the end of the study minus weight at the beginning of the study. The weight change was positive if the girl gained weight and negative if she lost weight. The therapies were designed to aid weight gain.

In this study, 29 girls received the cognitive behavioral therapy. This form of psychotherapy stresses identifying the thinking that causes the undesirable behavior and replacing it with thoughts designed to help to improve this behavior. Table 8.3

---

[7] The data are courtesy of Prof. Brian Everitt, Institute of Psychiatry, London.

shows the data. The weight changes for the 29 girls had a sample mean of $\bar{x} = 3.0$ pounds and standard deviation of $s = 7.3$ pounds.

Table 8.3: Weights of Anorexic Girls (in Pounds) Before and After Treatment.
Example 7 uses the weight change as the variable of interest.

Girl	Weight Before	After	Change	Girl	Weight Before	After	Change	Girl	Weight Before	After	Change
1	80.5	82.2	1.7	11	85.0	96.7	11.7	21	83.0	81.6	-1.4
2	84.9	85.6	0.7	12	89.2	95.3	6.1	22	76.5	75.7	-0.8
3	81.5	81.4	-0.1	13	81.3	82.4	1.1	23	80.2	82.6	2.4
4	82.6	81.9	-0.7	14	76.5	72.5	-4.0	24	87.8	100.4	12.6
5	79.9	76.4	-3.5	15	70.0	90.9	20.9	25	83.3	85.2	1.9
6	88.7	103.6	14.9	16	80.6	71.3	-9.3	26	79.7	83.6	3.9
7	94.9	98.4	3.5	17	83.3	85.4	2.1	27	84.5	84.6	0.1
8	76.3	93.4	17.1	18	87.7	89.1	1.4	28	80.8	96.2	15.4
9	81.0	73.4	-7.6	19	84.2	83.9	-0.3	29	87.4	86.7	-0.7
10	80.5	82.1	1.6	20	86.4	82.7	-3.7				

Question to Explore

How could we frame this investigation in the context of a significance test that can detect a positive or negative effect of the therapy? State the null and alternative hypotheses for that test.

Think It Through

The response variable, weight change, is quantitative. Hypotheses refer to the population mean weight change μ, namely whether it is 0 (the "no effect" value) or differs from 0. To use a significance test to analyze the strength of evidence about the therapy's effect, we'll test $H_0: \mu = 0$ against $H_a: \mu \neq 0$.

Insight

We'll see how to test these hypotheses as we work through the five steps of a test about means in the next subsection. We'll analyze whether the difference between the sample mean weight change of 3.0 and the no-effect value of 0 can be explained by random variability.

What Are the Steps of a Significance Test about a Population Mean?

A significance test about a mean has the same five steps as a test about a proportion: Assumptions, hypotheses, test statistic, P-value, and conclusion. We will mention modifications as they relate to the test for a mean. However, the reasoning is the same as that for the significance test for a proportion.

Step 1: Assumptions

The three basic assumptions of a test about a mean are:

- The variable is quantitative.
- The data production employed randomization.

- The *population distribution* is approximately normal. This is most crucial when n is small and H_a is one-sided, as discussed later in the section.

The anorexia study is like the 'dogs detecting cancer' study in Example 4 in that its subjects were a convenience sample rather than a random sample. Inferences are at best tentative. They are more convincing if researchers can argue that the girls in the sample are representative of the population of girls who suffer from anorexia and if other studies show similar results. The study did employ randomization in assigning girls to one of three therapies, only one of which (cognitive behavioral) is considered in this example.

Step 2: Hypotheses

The null hypothesis in a test about a population mean has form

$$H_0 : \mu = \mu_0 ,$$

where μ_0 denotes a particular value for the population mean. The two-sided alternative hypothesis

$$H_a : \mu \neq \mu_0$$

includes values both below and above the number μ_0 listed in H_0. Also possible are the one-sided alternative hypotheses,

$$H_a : \mu > \mu_0 \text{ or } H_a : \mu < \mu_0 .$$

For instance, let μ denote the mean weight change with the cognitive behavioral therapy for the population represented by the sample of anorexic girls. If the therapy has no effect, then $\mu = 0$. If the therapy has a beneficial effect on weight, as the study expected, then $\mu > 0$. To test that the therapy has no effect against the alternative that it has a beneficial effect, we test $H_0 : \mu = 0$ against $H_a : \mu > 0$. In practice, the two-sided alternative $H_a : \mu \neq 0$ is more common, to take an objective approach that can detect either a positive or a negative effect of the therapy.

Step 3: Test Statistic

The test statistic is the distance between the sample mean $\bar{x}$ and the null hypothesis value μ_0, as measured by the number of standard errors between them. This is measured by

$$\frac{(\bar{x} - \mu_0)}{se} = \frac{\text{sample mean - null hypothesis mean}}{\text{standard error of sample mean}} .$$

In practice, as in forming a confidence interval for a mean (Section 7.3), the standard error is $se = s/\sqrt{n}$. The test statistic is

> **RECALL**
>
> From Section 6.5, the exact standard error of the sample mean is $\sigma/\sqrt{n}$, where σ = population standard deviation. In practice, σ is unknown, so we estimate the standard error by
>
> $se = s/\sqrt{n}$.

$$t = \frac{(\overline{x} - \mu_0)}{se} = \frac{(\overline{x} - \mu_0)}{s/\sqrt{n}}.$$

In the anorexia study, the sample mean $\overline{x} = 3.0$ and the sample standard deviation $s = 7.32$. The standard error $se = s/\sqrt{n} = 7.32/\sqrt{29} = 1.36$. The test statistic equals

$$t = (\overline{x} - \mu_0)/se = (3.0 - 0)/1.36 = 2.21.$$

RECALL

You can review the t-distribution in Section 7.3. We used it there to form a confidence interval for a population mean u.

We use the symbol t rather than z for the test statistic because (as in forming a confidence interval) using s to estimate σ introduces additional error: The sampling distribution is more spread out than the standard normal. When H_0 is true, the t test statistic has approximately the *t-distribution*. The t-distribution is specified by its degrees of freedom, which equal $n - 1$ for inference about a mean. This test statistic is called a **t-statistic**.

Figure 8.7 shows the t sampling distribution. The farther $\overline{x}$ falls from the null hypothesis mean μ_0, the farther out in a tail the t test statistic falls, and the stronger the evidence is against H_0.

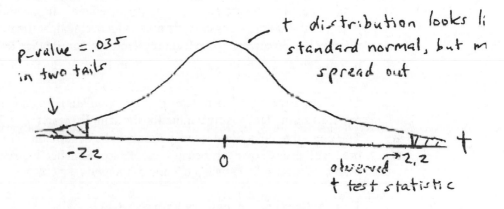

Figure 8.7: The t-Distribution of the t Test Statistic. There is stronger evidence against H_0 when the t test statistic falls farther out in a tail. The P-value for a two-sided H_a is a two-tail probability (shaded in figure). **Question**: Why is it that t-scores farther out in the tails provide stronger evidence against H_0?

Step 4: P-Value
The P-value is a single tail or a two-tail probability depending on whether the alternative hypothesis is one-sided or two-sided.

Alternative Hypothesis	P-Value
$H_a : \mu \neq \mu_0$	Two-tail probability from t distribution
$H_a : \mu > \mu_0$	Right-tail probability from t distribution
$H_a : \mu < \mu_0$	Left-tail probability from t distribution

For the anorexia study with $H_a : \mu \neq 0$, the P-value is the two-tail probability of a test statistic value farther out in either tail than the observed value of 2.21. See Figure 8.7. This probability is double the single-tail probability. Table 8.4 shows the way MINITAB software reports results. Since $n = 29$, $df = n-1 = 28$. The P-value is 0.036, or 0.04 rounded to two decimal places. This is the two-tail probability of t test statistic values below -2.21 and above +2.21 when df = 28.

Table 8.4: MINITAB Output for Analyzing Data from Anorexia Study

```
One-Sample T: wt_change
Test of mu = 0 vs not = 0

Variable    N   Mean   StDev   SE Mean      95% CI         T     P
wt_change  29  3.000  7.3204  1.3594  (0.2155, 5.7845)  2.21  0.036
```

Step 5: Conclusion

The conclusion of a significance test reports the P-value and *interprets* what it says about the question that motivated the test. Sometimes this includes a decision about the validity of H_0. In that case, we reject the null hypothesis when the P-value is less than or equal to the pre-selected significance level. In the anorexia study, the small P-value of 0.04 provides considerable evidence against the null hypothesis that the therapy has no effect. If we had preselected a significance level of 0.05, this would be enough evidence to reject $H_0 : \mu = 0$ in favor of $H_a : \mu \neq 0$.

The alternative hypothesis states that the population mean weight change μ is not equal to zero. The positive value for the sample mean ($\overline{x} = 3.0$) suggests that $\mu > 0$. The cognitive behavioral therapy seems to be beneficial. The effect may be small in practical terms, however, because the 95% confidence interval shown in Table 8.4 predicts that μ falls between 0.2 and 5.8 pounds.

♦

To practice this concept, try Exercise 8.34.

When research scientists conduct a significance test, their report would not show all results of a printout but would instead present a simple summary such as, "The diet had a statistically significant positive effect on weight (mean change = 3 pounds, $n = 29$, $t = 2.21$, P-value = 0.04)." Often results are summarized even further, such as "The diet had a significant positive effect on weight (P-value < 0.05)." This brief a summary is undesirable, because it does not show the estimated size of the effect (3 pounds) or the actual P-value. Don't condense your conclusions this much.

The *t*-Statistic and *z*-Statistic Have the Same Form

As you read the examples in this section, notice the parallel between each step of the test for a mean and the test for a proportion. For instance, the t test statistic for a mean has the same form as the z test statistic for a proportion, namely,

Form of Test Statistic
$$\frac{\text{Estimate of parameter - } H_0 \text{ value of parameter}}{\text{standard error of estimate}}$$

For the test about a mean, the estimate $\bar{x}$ of the population mean μ replaces the estimate $\hat{p}$ of the population proportion p, the H_0 mean μ_0 replaces the H_0 proportion p_0, and the standard error of the sample mean replaces the standard error of the sample proportion. The box summarizes tests for population means.

Summary: Steps of a Significance Test for Population Mean μ

1. Assumptions
- Quantitative variable, with population mean μ defined in context
- Data are obtained using randomization, such as a simple random sample
- Population distribution is approximately normal

2. Hypotheses

Null: $H_0 : \mu = \mu_0$, where μ_0 is the hypothesized value (such as $H_0 : \mu = 0$)

Alternative: $H_a : \mu \neq \mu_0$ (two-sided)

or $H_a : \mu < \mu_0$ (one-sided) or $H_a : \mu > \mu_0$ (one-sided)

3. Test statistic

$$t = \frac{(\bar{x} - \mu_0)}{se}, \text{ where } se = s/\sqrt{n}$$

4. P-value

Use *t*-distribution (Table B) with $df = n-1$

Alternative Hypothesis	*P-Value*
$H_a : \mu \neq \mu_0$	Two-tail probability
$H_a : \mu > \mu_0$	Right-tail probability
$H_a : \mu < \mu_0$	Left-tail probability

5. Conclusion

Smaller P-values give stronger evidence against H_0 and supporting H_a. If using a significance level to make a decision, reject H_0 if P-value $\leq$ significance level (such as 0.05). Relate the conclusion to the context of the study.

How Can We Perform a One-Sided Test about a Population Mean?

One-sided alternative hypotheses apply for a prediction that μ differs from the null hypothesis value in a certain direction. For example, $H_a : \mu > 0$ predicts that the true mean is *larger* than the null hypothesis value. Its P-value is the probability of a t value *larger* than the observed value, that is, in the *right* tail. Likewise, for $H_a : \mu < 0$ the P-value is the *left*-tail probability. In each case, again $df = n - 1$.

EXAMPLE 8: DOES A LOW-CARBOHYDRATE DIET WORK?

Picture the Scenario
In a recent study,[8] 41 overweight subjects were placed on a low carbohydrate diet but given no limit on caloric intake. The prediction was that subjects on such a diet would lose weight, on the average. After 16 weeks, their weight change averaged -9.7 kg (that is, a *decrease* in weight, on the average), with a standard deviation of 3.4 kg.

Question to Explore
Find and interpret the P-value for testing $H_0 : \mu = 0$ against $H_a : \mu < 0$.

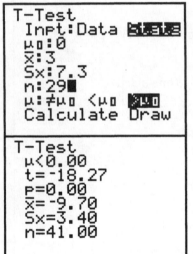

Think It Through
The standard error $se = s/\sqrt{n} = 3.4/\sqrt{41} = 0.53$. The test statistic equals

$$t = (\bar{x} - \mu_0)/se = (-9.7 - 0)/0.53 = -18.3,$$

with $df = n - 1 = 40$. Since H_a predicts that the mean is *below* 0, the P-value is the left-tail probability *below* the test statistic value of -18.3. This is zero to many decimal places. See Figure 8.8. The P-value of 0.000 (rounded to three decimal places) means that if $H_0 : \mu = 0$ were true, it would be practically impossible to observe a t statistic of -18.3 or even larger in the negative direction. A P-value of 0.000 is very strong evidence against $H_0 : \mu = 0$ and in favor of $H_a : \mu < 0$.

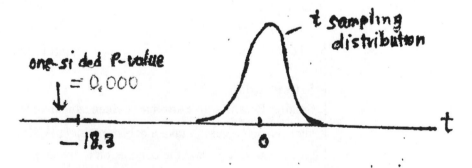

Figure 8.8: P-value for Testing $H_0 : \mu = 0$ against $H_a : \mu < 0$. Question: Why are large negative, rather than large positive, t test statistic values the ones that support $H_a : \mu < 0$?

---

[8] *Am. J. Med.*, vol. 113, pp. 30-36, 2002

In summary, on the average, people on this diet lose weight. This conclusion is tentative, since this study used a convenience sample rather than a random sample of subjects. No information was given about the shape of the distribution of weight changes, but with n this large ($n = 41$) the normal population assumption is not usually crucial because the sampling distribution will still be roughly bell shaped.

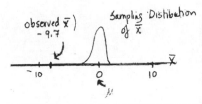

Insight

The P-value is the probability of t values below the observed t test statistic value of -18.3. Equivalently, it is the probability of a sample mean weight change $\bar{x} < -9.7$ (the observed value), if H_0 were true that $\mu = 0$. See the first figure in the margin.

◆

To practice this concept, try Exercise 8.32.

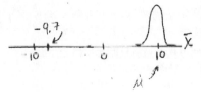

We've seen that if $\mu = 0$, it would be extremely unusual to get a sample mean of -9.7. If in fact μ were a number *greater than* 0, it would be even more unusual. For example, a sample value of $\bar{x} = -9.7$ is even more unusual when $\mu = 10$ than when $\mu = 0$, since -9.7 is farther out in the tail of the sampling distribution of $\bar{x}$ when $\mu = 10$ than when $\mu = 0$. See the second figure in the margin.

- Thus, when we reject $H_0: \mu = 0$ in favor of $H_a: \mu < 0$, we can also reject the broader null hypothesis of $H_0: \mu \geq 0$.

In other words, we conclude that $\mu = 0$ is false and that $\mu > 0$ is false. However, statements of null hypotheses use a *single* number in the null hypothesis, because a single number is entered in the test statistic to compare to the sample mean.

How Can We Use the *t*-Table to Approximate a P-Value?

Statistical software and many calculators can find the P-value for you, as illustrated in Table 8.4 for the anorexia study of Example 7. If you have the test statistic value but are not using software, you can use the t table (Table B in the Appendix). Table B is not detailed enough to provide an *exact* tail probability, but it provides enough information to determine whether a one-tail probability is greater than or less than 0.100, 0.050, 0.025, 0.010, 0.005, or 0.001.

Let's see how to do this for the t test statistic from the anorexia study (Example 7), $t = 2.21$ based on df = 28. Look at the row of Table B for df = 28. You will see:

df	$t_{.100}$	$t_{.050}$	$t_{.025}$	$t_{.010}$	$t_{.005}$	$t_{.001}$
28	1.313	1.701	2.048	2.467	2.763	3.408

Note that $t = 2.21$ falls between 2.048 and 2.467. Now, the value $2.048 = t_{.025}$ has a right-tail probability of 0.025 and the value $2.467 = t_{.010}$ has a right-tail probability of 0.010. So the right-tail probability for $t = 2.21$ falls between 0.010 and 0.025. Figure 8.9 illustrates.

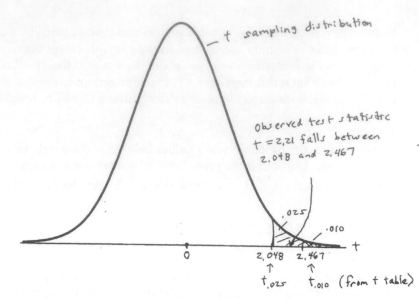

Figure 8.9: For *df* = 28, *t* = 2.21 has a Right-Tail Probability between 0.010 and 0.025.
The two-sided P-value falls between 2(0.010) = 0.02 and 2(0.025) = 0.05. **Question**: Using software or a calculator, can you show that the actual right-tail probability equals 0.017 and that the two-sided P-value equals 0.035?

To get the P-value for the two-sided H_a, double the bounds, because you want a two-tail probability. For these data, we double 0.01 and 0.025 to report 0.02 < P-value < 0.05. This is sufficient to tell us we can reject H_0 at the 0.05 significance level. In fact, software told us (Table 8.4) that the actual P-value = 0.036.

Results of Two-Sided Tests and Results of Confidence Intervals Agree

For the anorexia study, we got a P-value of 0.036 for testing H_0: $\mu = 0$ against H_a: $\mu \neq 0$ for the mean weight change with the cognitive behavioral therapy. With the 0.05 significance level, we would reject H_0. Table 8.4 showed that a 95% confidence interval for the population mean weight change μ is (0.2, 5.8) pounds. We conclude that the population mean weight change μ is positive, because all the numbers in this interval are greater than 0. However, μ is possibly small, because the lower endpoint is close to 0. The confidence interval shows just how different from 0 the population mean weight change is likely to be. It is estimated to fall between 0.2 and 5.8 pounds. The effect of the therapy may be very small.

Both the significance test and the confidence interval suggested that μ differs from 0. In fact, conclusions about means using two-sided significance tests are consistent with conclusions using confidence intervals. If a two-sided test says you can reject the hypothesis that $\mu = 0$, then 0 is not in the corresponding confidence interval.

If P-value ≤ 0.05 in a two-sided test, a 95% confidence interval does not contain the H_0 value.

This correspondence applies whenever the confidence level is the complement of the significance level. For example, if P-value $\leq$ 0.05 results from testing H_0: $\mu = 0$ against H_a: $\mu \neq 0$, then a 95% confidence interval for μ does not contain 0. In the anorexia study of Example 7, the two-sided test of H_0: $\mu = 0$ has P-value = 0.04 $\leq$ 0.05. This small P-value says we can reject H_0 at the 0.05 significance level. The 95% confidence interval for μ is (0.2, 5.8). The interval does not contain 0.0, so this method also suggests that μ is not exactly equal to 0.0.

By contrast, suppose that the P-value > 0.05 in a two-sided test of H_0: $\mu = 0$, so we cannot reject H_0 at the 0.05 significance level. Then, a 95% confidence interval for μ *does* contain 0. According to both methods, the value of 0 is a plausible one for μ.

Why are confidence intervals and two-sided tests about means consistent?

Figure 8.10 illustrates why decisions from two-sided tests about means are consistent with confidence intervals.

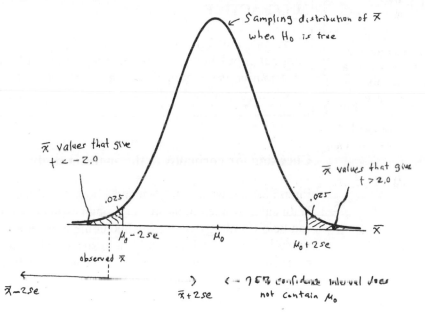

Figure 8.10: Relation between Confidence Interval and Significance Test. With large samples, if the sample mean falls more than about 2 standard errors from μ_0, then μ_0 does not fall in the 95% confidence interval and also μ_0 is rejected in a test at the 0.05 significance level. **Question**: Inference about proportions does not have an *exact* equivalence between these two methods. Why? (*Hint*: Are the exact same standard error values used in the two methods?)

With large samples, the *t*-score for a 95% confidence interval is approximately 2, so the confidence interval is roughly $\bar{x} \pm 2(se)$. If this interval does not contain a particular value μ_0, this means that the sample mean $\bar{x}$ falls more than 2 standard errors from μ_0. But this means that the test statistic $t = (\bar{x} - \mu_0)/se$ is larger than 2 in absolute value. But when this happens, the two-tail P-value is less than 0.05, so we would reject the hypothesis that $\mu = \mu_0$.

What If the Population Does Not Satisfy the Normality Assumption?

For the test about a mean, the third assumption states that the population distribution should be approximately normal. This ensures that the sampling distribution of the sample mean $\bar{x}$ is normal and the t test statistic has the t distribution. For large samples (roughly about 30 or higher), this assumption is not important, because an approximate normal sampling distribution for $\bar{x}$ occurs regardless of the population distribution. (Remember the central limit theorem?)

Chapter 14 presents a type of statistical method, called *nonparametric*, that does not require the normal assumption. However, the normal assumption needed with small samples for the t-test is not crucial when we use a two-sided H$_a$.

RECALL
From Section 6.5, with random sampling the sampling distribution of the sample mean is approximately normal for large n, by the central limit theorem.

RECALL
Section 7.3 defined a statistical method to be **robust** with respect to a particular assumption if it performs adequately even when that assumption is violated.

IN PRACTICE
Two-sided inferences using the t distribution are *robust* against violations of the normal population assumption. They still usually work well if the actual population distribution is not normal. The test does not work well for a one-sided test with small n when the population distribution is highly skewed.

Checking for normality in the anorexia study

Figure 8.8 shows a histogram and a box plot of the data from the anorexia study. With small n, such plots are very rough estimates of the population distribution. It can be difficult to determine whether the population distribution is approximately normal. However, Figure 8.11 does suggest skew to the right, with a small proportion of girls having considerable weight gains.

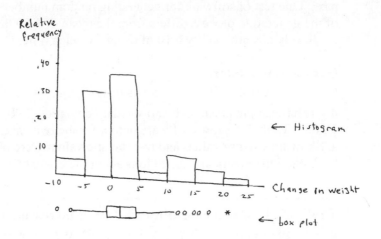

Figure 8.11: Histogram and Box Plot of Weight Change for Anorexia Sufferers.
Question: What do these plots suggest about the shape of the population distribution of weight change?

As just mentioned, a two-sided t test still works quite well even if the population distribution is skewed. So, we feel comfortable with the two-sided test in Example 7. However, this plot makes us a bit wary about using a one-sided test for these data. The sample size is not large ($n = 29$), and the histogram in Figure 8.11 shows substantial skew, with the box plot highlighting six quite large weight change values.

Regardless of robustness, look at the data

Whether n is small or large, you should look at the data to check for severe skew or for severe outliers that occur primarily in one direction. They could cause the sample mean to be a misleading measure. For the anorexia data, the median weight change is only 1.4 pounds, somewhat less than the mean of 3.0 because of the skew to the right. Even though the test showed significant evidence of a positive population mean weight change, the sample median is another indication that the size of the effect could be small. You also need to be cautious about any conclusion if it changes by removing an extreme outlier from the data set (Exercise 8.43).

What Effect Does the Sample Size Have on P-values?

The result of a significance test can depend strongly on how much data you have. For a given sample effect, the test statistic tends to be larger as the sample size increases, as the next example shows.

EXAMPLE 9: TESTING WHETHER SOFTWARE OPERATES PROPERLY

Picture the Scenario
Random numbers are used for conducting simulations and for identifying subjects to be chosen for a random sample. A difficult programming challenge is writing a computer program that can generate random numbers properly. Each digit must be equally likely to be 0, 1, 2, …, 9, and the digits in the sequence must be independent: The probability of each digit is 0.10, regardless of which digits were generated in the

past. One test of software for generating random numbers checks whether the mean of the generating process differs from the mean of 4.50 that holds if each digit 0, 1, ..., 9 truly has probability 0.10 of occurring each time.

Questions to Explore

Consider the test of $H_0: \mu = 4.50$ against $H_a: \mu \neq 4.50$ when a sequence of random digits has sample mean 4.40 and standard deviation 2.90, if (i) $n = 100$, (ii) $n = 10,000$. Table 8.5 shows software output for the two cases.
a. Show how the *se* values and test statistic values were obtained in the two cases.
b. Explain the practical implications about the effect of n on results of a test.

Table 8.5: Effect on a Significance Test of Increasing the Sample Size *n*. For a given size of effect (such as $\bar{x} - \mu_0 = 0.10$), as *n* increases, the test statistic increases and the P-value decreases.

Test of mu = 4.50 vs not = 4.50

(i) *n* = 100 →

Variable	N	Mean	StDev	SE Mean	T	P
RanDigit	100	4.40	2.90	0.290	-0.345	0.731

(ii) *n* = 10,000 →

Variable	N	Mean	StDev	SE Mean	T	P
RanDigit	10,000	4.40	2.90	0.029	-3.45	0.0006

Think It Through

a. In both cases, $\bar{x} = 4.40$, $s = 2.90$, and the sample effect as measured by $\bar{x} - \mu_0 = 4.40 - 4.50 = -0.10$ is the same. However, the standard error is smaller with a larger sample size. The *se* values are:

(i) $se = s/\sqrt{n} = 2.90/\sqrt{100} = 0.29$,   (ii) $se = s/\sqrt{n} = 2.90/\sqrt{10,000} = 0.029$.

The test statistics then are also quite different:

(i) $t = \dfrac{(\bar{x} - \mu_0)}{se} = \dfrac{4.40 - 4.50}{0.29} = -0.345$,   (ii) $t = \dfrac{(\bar{x} - \mu_0)}{se} = \dfrac{4.40 - 4.50}{0.029} = -3.45.$

From Table 8.5, the P-values are dramatically different, 0.73 compared to 0.0006. The same effect, $\bar{x} - \mu_0 = -0.10$, but based on a larger sample size, results in a much smaller P-value.

b. For a given sample mean, larger sample sizes produce larger test statistics. Why does this happen? As *n* increases, the standard error in the denominator of the *t* statistic decreases. So, the *t* statistic itself increases. The two-sided P-value is then smaller. This makes sense: For a given sample effect, we can be more certain that that effect reflects a true population effect if the sample size is large than if it is small.

Insight
An implication of this result is that, for large *n*, statistical significance may not imply an important result in practical terms. For instance, you can get a small P-value even if the sample mean falls quite near the null hypothesis value. We'll discuss this further in Section 8.5.

♦

To practice this concept, try Exercise 8.31.

SECTION 8.3: PRACTICING THE BASICS

8.29 Which *t* has P-value = 0.05?: A *t*-test for a mean uses a sample of 15 observations. Find the *t* test statistic value that has a P-value of 0.05 when the alternative hypothesis is (a) $H_a : \mu \neq 0$, (b) $H_a : \mu > 0$, (c) $H_a : \mu < 0$?

8.30   Practice mechanics of a test: A study has a random sample of 20 subjects. The test statistic for testing $H_0 : \mu = 100$ is $t = 2.40$. Find the approximate P-value for the alternative, (a) $H_a : \mu \neq 0$, (b) $H_a : \mu > 0$, (c) $H_a : \mu < 0$.

8.31   Effect of *n*: Refer to the previous exercise. If the same sample mean and standard deviation had been based on $n = 5$ instead of $n = 20$, the test statistic would have been $t = 1.20$.
a. Would the P-value for $H_a : \mu \neq 0$ be larger, or smaller, than when $t = 2.40$? Why?
b. Other things being equal, explain why larger sample sizes result in smaller P-values.

8.32 More on the anorexia study: Suppose the study about the anorexia therapy (Example 7) had predicted that the therapy induces a *positive* mean weight change. We could then test $H_0 : \mu = 0$ against $H_a : \mu > 0$. In Example 7, we saw that $\bar{x} = 3.0$ for $n = 29$ subjects, and $t = 2.21$, which has P-value = 0.035.
a. What is the P-value for testing $H_0 : \mu = 0$ against $H_a : \mu > 0$?
b. How is the P-value interpreted?
c. Explain why rejecting $H_0 : \mu = 0$ in favor of $H_a : \mu > 0$ also inherently rejects the broader null hypothesis of $H_0 : \mu \leq 0$.

8.33 Low-carb diet: Refer to the study about the low-carbohydrate diet in Example 8. The study made the inference that the diet results in a weight loss, on the average.
a. Can you justify this inference if weight change does not have a normal distribution? Explain.
b. What limitations are there from using a convenience sample in this study?

8.34   Weight change for controls: A disadvantage of the experimental design in Example 7 on weight change in anorexic girls is that girls could change weight merely from participating in a study. In fact, girls were randomly assigned to receive a therapy or to serve in a control group, so it was possible to compare weight change for the therapy group to the control group. For the 26 girls in the control group, the weight change had $\bar{x}$ = -0.5 and s = 8.0. Repeat all five steps of the test of H_0: μ = 0 against H_a: $\mu \neq 0$ for this group, and interpret the P-value.

8.35 **Crossover study**: A crossover study of 13 children suffering from asthma (*Clinical and Experimental Allergy*, vol. 20, pp. 429-432, 1990) compared single inhaled doses of formoterol (F) and salbutamol (S). The outcome measured was the child's peak expiratory flow (PEF) eight hours following treatment. The data on PEF follow:

Child	F	S	Child	F	S	Child	F	S	Child	F	S
1	310	270	5	410	380	9	330	365	13	220	90
2	385	370	6	370	300	10	250	210			
3	400	310	7	410	390	11	380	350			
4	310	260	8	320	290	12	340	260			

Let μ denote the population mean of the difference between the PEF values for the F and S treatments. Use a calculator or software for the following analyses:
a. Form the 13 difference scores, for instance 310 – 270 = 40 for child 1 and 330 – 365 = -35 for child 9, always taking F - S. Construct a dot plot or a box plot. Describe the sample data distribution.
b. Carry out the five steps of the significance test for a mean of the difference scores, using H_0: $\mu = 0$ and H_a: $\mu \neq 0$.
c. Discuss whether the assumptions seem valid for this example. What is the impact of using a convenience sample?

8.36 **Too little or too much wine?**: Hosmer Winery on Cayuga Lake, New York has a machine for dispensing their riesling wine into bottles, which are advertised to contain 750 ml. When the machine is in statistical control, the amount dispensed has a mean of 755 ml. Four observations are taken each day, to plot a daily mean over time on a control chart to check for irregularities. The most recent day's observations have a mean of 740 and a standard deviation of 5 *ml*. Could the difference between the sample mean and the target value be attributed to random variation, or is the true mean now different from 755? Answer by giving the five steps of a significance test.

8.37 **Selling a burger**: In Exercise 8.26, a fast-food chain compared two ways of promoting a turkey burger. In a separate experiment with 10 pairs of stores, the difference in the month's increased sales between the store that used coupons and the store with the outside poster had a mean of $3000 with a standard deviation of $4000. Does this provide strong evidence of a true difference between mean sales for the two advertising approaches? Answer by testing that the population mean difference is 0, carrying out the five steps of a significance test. Make a decision using a 0.05 significance level.

8.38 **Assumptions important?**: Refer to the previous exercise. Suppose you instead wanted to perform a one-sided test, because the study predicted that the increase in sales would be higher with coupons. Explain why the normal population assumption may possibly be problematic.

8.39   **Mother works**: In response to the statement, "A preschool child is likely to suffer if his or her mother works," the response categories (Strongly agree, Agree, Disagree, Strongly disagree) had counts (155, 611, 863, 180) for the 1809 responses in a recent General Social Survey. With scores (2, 1, -1, -2) for the four categories, software reported the following results:

```
----------------------------------------------------------------------
Test of mu = 0 vs mu not = 0

Variable        N          Mean        StDev         SE Mean
Opinion        1809       -0.1669      1.236          0.0291

Variable       95.0% CI                T              P
Opinion        (-0.224, -0.110)       -5.745         0.000
----------------------------------------------------------------------
```

a. Give an example of response counts that would have a sample mean of 0.

b. Define μ and set up null and alternative hypotheses to test whether the population mean differs from the neutral value, 0.

c. Carry out the significance test, discussing assumptions and showing how software got the results stated in the table for the standard error and the test statistic value. Interpret the P-value.

8.40 CI and test: Refer to the previous exercise, for which the P-value = 0.000.

a. Using a significance level 0.05, make a decision about H_0, and interpret.

b. Explain how the result of the 95% confidence interval shown in the table agrees with the test decision in (a).

8.41 Lake pollution: An industrial plant claims to discharge no more than 1000 gallons of wastewater per hour, on the average, into a neighboring lake. An environmental action group decides to monitor the plant, in case this limit is being exceeded. Doing so is expensive, and only a small sample is possible. A random sample of four hours is selected over a period of a week. The observations are:

$$2000, 1000, 3000, 2000.$$

a. Show that $\bar{x} = 2000$, $s = 816.5$, and standard error = 408.25.

b. To test H_0: $\mu = 1000$ against H_a: $\mu > 0$, show that the test statistic equals 2.45.

c. Using Table B or software, show that the P-value is less than 0.05, so there is enough evidence to reject the null hypothesis at the 0.05 significance level.

d. Explain how your one-sided analysis in (b) implicitly tests the broader null hypothesis that $\mu \leq 1000$.

8.42 Anorexia in teenage girls: Example 7 described a study about various therapies for teenage girls suffering from anorexia. For each of 17 girls who received the family therapy, the changes in weight were:

$$11, 11, 6, 9, 14, -3, 0, 7, 22, -5, -4, 13, 13, 9, 4, 6, 11.$$

a. Plot these with a dot plot or box plot, and summarize.

b. Verify that the weight changes have $\bar{x} = 7.29$, $s = 7.18$, and $se = 1.74$ pounds.

c. Give all steps of a significance test about whether the population mean was 0, against an alternative designed to see if there is any effect.

▣ 8.43 Sensitivity study: Ideally, results of a statistical analysis should not depend greatly on a single observation. To check this, it's a good idea to conduct a **sensitivity study**. This entails re-doing the analysis after deleting an outlier from the data set or changing its value to a more typical value and checking whether results change much. If results change little, this gives us more faith in the conclusions that the statistical analysis reports. For the data in Table 8.3 from the anorexia study of Example 7, the greatest reported weight change of 20.9 pounds was a severe outlier. Suppose this observation was actually 2.9 pounds but was incorrectly recorded. Re-do the two-sided test of that example, and summarize how the results differ.

8.44 Test and CI: Results of 99% confidence intervals are consistent with results of two-sided tests with which significance level? Explain the connection.

8.4 DECISIONS AND TYPES OF ERRORS IN SIGNIFICANCE TESTS

In significance tests, the P-value summarizes the evidence about H_0. A P-value such as 0.001 casts strong doubt on H_0 being true, because if it were true the observed data would be very unusual.

When we need to decide whether the evidence is strong enough to reject H_0, we've seen that the key is whether the P-value falls below a pre-specified **significance level**. The significance level is usually denoted by the Greek letter α (alpha). In practice, $\alpha = 0.05$ is most common: We reject H_0 if the P-value ≤ 0.05. We do not reject H_0 if the P-value > 0.05. The smaller α is, the stronger the evidence must be to reject H_0. To avoid bias, we select α *before* looking at the data.

Two Potential Types of Errors in Test Decisions

Because of sampling variability, decisions in significance tests always have some uncertainty. A decision can be in error. There are two types of potential errors, called **Type I** and **Type II errors**.

Definition: Type I and Type II Errors

A **Type I error** occurs when H_0 is rejected, even though H_0 is true.

A **Type II error** occurs when H_0 is not rejected, even though H_0 is false.

When we make a decision, there are four possible results. These refer to the two possible decisions combined with the two possibilities for whether H_0 is true. Table 8.6 summarizes these four possibilities.

Table 8.6: The Four Possible Results of a Decision in a Significance Test. Type I and Type II errors are the two possible incorrect decisions.

REALITY ABOUT H_0	DECISION	
	Do not reject H_0	Reject H_0
H_0 true	Correct decision	Type I error
H_0 false	Type II error	Correct decision

Consider the experiment about astrology in Example 5. In that study H_0: $p = 1/3$ corresponded to random guessing by the astrologers. We got a P-value of 0.40. With significance level = 0.05, we do not reject H_0. If truly $p = 1/3$, this is a correct decision. However, if astrologers actually can predict better than random guessing (so that $p > 1/3$), we've made a Type II error, failing to reject H_0 when it is false.

In the anorexia study of Example 7, we got a P-value of 0.04 for testing H_0: $\mu = 0$ of no weight change, on the average. With significance level = 0.05, we can reject H_0. If the therapy truly has an effect, this is a correct decision. But if actually there is no effect (that is, if $\mu = 0$), we've made a Type I error, rejecting H_0 when it is true.

An analogy: Decision errors in a legal trial

These two types of errors can occur with any decision having two options, one of which is incorrect. For instance, consider a decision in a legal trial. The null hypothesis tested is the defendant's claim of innocence. The alternative hypothesis is that the defendant is guilty. The jury rejects H_0 if it decides that the evidence is sufficient to convict. The defendant is then judged guilty. A Type I error, rejecting a true null hypothesis, occurs in convicting a defendant who is actually innocent. Not rejecting H_0 means the defendant is acquitted (judged not guilty). A Type II error, not rejecting H_0 even though it is false, occurs in acquitting a defendant who is actually guilty. See Table 8.7. Another analogy occurs with medical diagnostic testing, such as using a mammogram to test whether a woman has breast cancer (Exercise 8.49).

Table 8.7: Possible Results of a Legal Trial

	LEGAL DECISION	
DEFENDANT	Acquit	Convict
Innocent (H_0)	Correct	Type I error
Guilty (H_a)	Type II error	Correct

The Significance Level Is the Probability of Type I Error

When H_0 is actually true, let's see how to find the probability of a Type I error. This is the probability of rejecting H_0, even though it is actually true. We'll find this for the two-sided test about a proportion.

With the $\alpha = 0.05$ significance level, we reject H_0 if the P-value ≤ 0.05. For two-sided tests about a proportion, the two-tail probability that forms the P-value is ≤ 0.05 whenever the test statistic z satisfies $|z| \geq 1.96$. The collection of test statistic values for which a test rejects H_0 is called the **rejection region**. These are the z test statistic values that occur when the sample proportion falls at least 1.96 standard errors from the null hypothesis value. They are the values we'd least expect to observe if H_0 were true. Figure 8.12 illustrates.

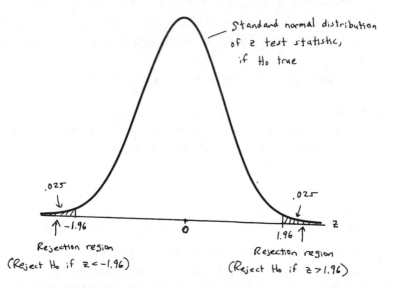

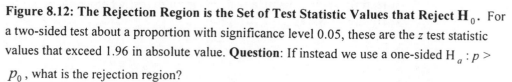

Figure 8.12: The Rejection Region is the Set of Test Statistic Values that Reject H_0. For a two-sided test about a proportion with significance level 0.05, these are the z test statistic values that exceed 1.96 in absolute value. **Question**: If instead we use a one-sided $H_a : p > p_0$, what is the rejection region?

Now, if H_0 is actually true, the sampling distribution of the z test statistic is the standard normal. Therefore, the probability of rejecting H_0, which is the probability that $|z| \geq 1.96$, is exactly 0.05. But this is precisely the significance level.

P(Type I error) = Significance level α

Suppose H_0 is true. The probability of rejecting H_0, thereby making a Type I error, equals the significance level for the test.

We can control the probability of a Type I error by our choice of the significance level. The more serious the consequences of a Type I error, the smaller α should be. In practice, $\alpha = 0.05$ is most common, just as a probability of error of 0.05 in interval estimation is most common (that is, 95% confidence intervals). However, this may be too high when a decision has serious implications. For instance, if a convicted defendant gets the death penalty, we would hope that the probability of convicting someone who is actually innocent is smaller than 0.05.

IN PRACTICE

In practice, we don't know whether a decision in a significance test is in error, just as we don't know whether a particular confidence interval truly contains an unknown parameter value. However, we can control the *probability* of an incorrect decision for either type of inference.

Although we don't know whether the decision in a particular test is correct, we justify the method in terms of the long-run proportions of Type I and Type II errors. We'll learn how to calculate P(Type II error) later in the chapter.

EXAMPLE 10: CAN WE ELIMINATE TYPE I ERRORS FROM LEGAL VERDICTS?

Picture the Scenario

In an ideal world, Type I or Type II errors would not occur. In practice, however, whether in significance tests or in applications such as courtroom trials or medical diagnoses, errors do happen. It can be surprising and disappointing how often they do, as we saw in the diagnostic testing examples of Chapter 5. Likewise, we've all read about defendants who were given the death penalty but later determined to be innocent, but we don't have reliable information about how often this occurs.

Question to Explore

When we make a decision, why don't we use an extremely small probability of Type I error, such as $\alpha = 0.000001$? For instance, why don't we make it almost impossible to convict someone who is really innocent?

PHOTO

Perhaps have photo here of a defendant in a courtroom.

Think It Through
When we make α smaller in a significance test, we need a smaller P-value to reject H_0. It then becomes harder to reject H_0. But this means that it will also be harder even if H_0 is false. The stronger the evidence that is required to convict someone, the more likely it becomes that we will fail to convict defendants who are actually guilty. In other words, the smaller we make the probability of Type I error, the *larger* the probability of Type II error becomes (failing to reject H_0 even though it is false).

If we tolerate only an extremely small chance of a Type I error (such as $\alpha = 0.000001$), then the test may be unlikely to reject the null hypothesis even if it is false --for instance, unlikely to convict someone even if they are guilty. Some of our laws are set up to make Type I errors very unlikely, and as a consequence some truly guilty individuals are not punished for their crimes.

Insight
This reasoning reflects a fundamental relation between the probabilities of the two types of errors, for a given sample size *n*:

P(Type II Error) Goes Up as P(Type I Error) Goes Down

The two probabilities are inversely related.

◆

To practice this concept, try Exercise 8.40.

Except in the final section of this chapter, we will not calculate the probability of Type II error. This calculation can be complex. In practice, to make a decision in a test, we only need to set the probability of Type I error, which is the significance level α.

These days, most research articles merely report the P-value rather than a decision about whether to reject H_0. From the P-value, readers can see the strength of evidence against H_0 and make their own decisions, if they want to.

ACTIVITY 1: WHY IS 0.05 COMMONLY USED AS A SIGNIFICANCE LEVEL?

A crossover study compares a drug with a placebo for children who suffer from migraine headaches. The study observed each child at two times when he or she had a migraine headache. The child received the drug at one time and a placebo at the other time. Let p denote the probability that the pain relief is better with the drug. You are going to decide whether you can reject $H_0 : p = 0.50$ in favor of $H_a : p \neq 0.50$. Ahead of time, you have no idea whether the drug will be better, or worse, than the placebo.

- The first child does better with the placebo. Would you reject H_0?

- The second child also does better with the placebo. Would you now reject H_0?

- The third child also does better with the placebo. Would you now reject H_0?

- The fourth child also does better with the placebo. Would you now reject H_0?

- The fifth child also does better with the placebo. Are you ready yet to reject H_0?

If you are like many people, by the time you see the fifth straight success for the placebo over the drug, you are willing to predict that the placebo is better. If the null hypothesis that $p = 0.50$ is actually true, then by the binomial distribution the probability this happens is $(0.50)^5 = 1/32 = 0.03$. For a two-sided test, this result gives a P-value $= 2(0.03) = 0.06$, close to 0.05. So, for many people, it takes a P-value near 0.05 before they feel there is enough evidence to reject a null hypothesis. This may be one reason the significance level of 0.05 has become common over the years in a wide variety of disciplines that use significance tests.

To practice this concept, try Exercise 8.138.

SECTION 8.4: PRACTICING THE BASICS

8.45 Error probability: A significance test about a proportion is conducted using a significance level of 0.05. The test statistic equals 2.58. The P-value is 0.01.
a. If H_0 were true, for what probability of Type I error was the test designed?
b. If this test resulted in a decision error, what type of error was it?

8.46 Astrology errors: Example 3, in testing $H_0 : p = 1/3$ against $H_a : p > 1/3$, analyzed whether astrologers could predict the correct personality chart (out of three possible ones) for a given horoscope better than by random guessing. In the words of that example, what would be a (a) Type I error, (b) Type II error?

8.47 Anorexia errors: Example 7 tested a therapy for anorexia, using hypotheses $H_0 : \mu = 0$ and $H_a : \mu \neq 0$ about the population mean weight change μ. In the words of that example, what would be a (a) Type I error, (b) Type II error?

8.48 Anorexia decision: Refer to the previous exercise. If we tested $H_0 : \mu = 0$ against $H_a : \mu > 0$, we would get a P-value of 0.02.

a. What would the decision be for a significance level of 0.05? Interpret in context.
b. If the decision in (a) is in error, what type of error is it?
c. Suppose the significance level were instead 0.01. What decision would you make, and if it is in error, what type of error is it?

8.49 Decision errors in medical diagnostic testing: Consider medical diagnostic testing, such as using a mammogram to detect whether a woman may have breast cancer. Identify the null hypothesis of no effect with no disease. Identify rejecting H_0 with a positive diagnostic test, which means that the test predicts that the person has the disease. See the table for a summary of the possible outcomes:

Medical Diagnostic Testing

	MEDICAL DIAGNOSIS	
DISEASE	Negative	Positive
No (H_0)	Correct	Type I error
Yes (H_a)	Type II error	Correct

----

a. When a radiologist interprets a mammogram, explain why a Type I error is a "false positive," predicting that a woman has breast cancer when actually she does not.
b. A Type II error is a "false negative." What does this mean?
c. A radiologist wants to decrease the chance of telling a woman that she may have breast cancer when actually she does not. To do this, a positive test result will be reported only when there is *extremely* strong evidence that breast cancer is present. What is the disadvantage of this approach?

8.50 Detecting prostate cancer: A *New York Times* article (Feb. 17, 1999) about the PSA blood test for detecting prostate cancer stated: "The test fails to detect prostate cancer in 1 in 4 men who have the disease."
a.   For the PSA test, explain what it means to make a Type I error.
b.   For the PSA test, explain what it means to make a Type II error.
c.   To which type of error does the probability of 1 in 4 refer?
d.   The article also stated that "as many as two-thirds of the men tested receive false-positive results." That is, given that you receive a positive result, the probability that you do not actually have prostate cancer is 2/3. Explain the difference between this and the conditional probability of a Type I error, given that you do not actually have prostate cancer.

8.51 Which error is worse?: Which error, Type I or Type II, would usually be considered more serious for decisions in the following tests? Explain why.
a. A trial to test a murder defendant's claimed innocence, when conviction results in the death penalty.
b. A medical diagnostic procedure, such as a mammogram.

8.5 LIMITATIONS OF SIGNIFICANCE TESTS

Chapters 7 and 8 have presented the two primary methods of statistical inference --- confidence intervals and significance testing. We will use both of these methods throughout the rest of the book. Of the two methods, confidence intervals can be more useful, for reasons we'll discuss in this section. Significance tests have more potential for misuse. We'll now summarize their major limitations.

Statistical Significance Does Not Mean Practical Significance

When we conduct a significance test, its main relevance is studying whether the true parameter value is

- above, or below, the value in H_0, and

- sufficiently different from the value in H_0 to be of practical importance.

A significance test does give us information about whether the parameter differs from the H_0 value and its direction from that value, but we'll see now that it does not tell us about practical importance.

There is an important distinction between *statistical significance* and *practical significance*. A small P-value, such as 0.001, is highly statistically significant, giving strong evidence against H_0. It does not, however, imply an *important* finding in any practical sense. The small P-value merely means that if H_0 were true, the observed data would be unusual. It does not mean that the true value of the parameter is far from the null hypothesis value in practical terms. In particular, whenever the sample size is large, small P-values can occur when the point estimate is quite near the parameter value in H_0, as the following example shows.

EXAMPLE 11: POLITICAL CONSERVATISM AND LIBERALISM IN AMERICA

Picture the Scenario

Where do Americans say they fall on the conservative-liberal political spectrum? The General Social Survey asks, "I'm going to show you a seven-point scale on which the political views that people might hold are arranged from extremely liberal, point 1, to extremely conservative, point 7. Where would you place yourself on this scale?" Table 8.8 shows the scale and the distribution of 2644 responses for the GSS in 2000.

569

Table 8.8: Responses of 2644 Subjects on a Seven-Point Scale of Political Views

CATEGORY	COUNT
1. Extremely liberal	107
2. Liberal	308
3. Slightly liberal	285
4. Moderate, middle of road	1054
5. Slightly conservative	390
6. Conservative	411
7. Extremely conservative	89

This categorical variable has seven categories that are ordered in terms of degree of liberalism or conservatism. Categorical variables that have *ordered* categories are called **ordinal variables**. Sometimes we treat an ordinal variable in a quantitative manner by assigning scores to the categories and summarizing the data by the mean. This summarizes whether observations gravitate toward the conservative or the liberal end of the scale. For the category scores of 1 to 7, as in Table 8.8, a mean of 4.0 corresponds to the moderate outcome. A mean below 4 shows a propensity toward liberalism, and a mean above 4 shows a propensity toward conservatism. The 2644 observations in Table 8.8 have a mean of 4.10 and a standard deviation of 1.41.

Questions to Explore

a. Do these data indicate that the population has a propensity toward liberalism or toward conservatism? Answer by conducting a significance test that compares the population mean to the moderate value of 4.0, by testing $H_0: \mu = 4.0$ against $H_a: \mu \neq 4.0$.

b. Does this test show (i) statistical significance? (ii) practical significance?

Think
It Through

a. For the sample data, the standard error $se = s/\sqrt{n} = 1.41/\sqrt{2644} = 0.0275$. The test statistic for $H_0: \mu = 4.0$ equals

$$t = (\bar{x} - \mu_0)/se = (4.10 - 4.00)/0.0275 = 3.53.$$

Its two-sided P-value is 0.0004, the probability of observing a sample mean farther from 4.0 than the observed value. There is *extremely* strong evidence that the true mean exceeds 4.0. We conclude that the true mean falls on the conservative side of moderate.

b. Although the P-value is very small, on a scale of 1 to 7, the sample mean of 4.10 is very close to the value of 4.0 in H_0. It is only one tenth of the distance from the moderate score of 4.0 to the slightly conservative score of 5.0. Although the difference of 0.10 between the sample mean of 4.10 and the null hypothesis mean of 4.0 is highly significant statistically, this difference is small in practical terms. In practice, we'd regard a mean of 4.10 as "moderate" on this 1 to 7 scale. In summary, there's statistical significance but not practical significance.

Insight

As we also saw in Example 9, with large samples P-values can be very small even when the sample estimate falls close to the parameter value in H_0. The P-value measures the extent of evidence about H_0, not how far the true parameter value happens to be from H_0. Always inspect the difference between the sample estimate and the null hypothesis value to gauge the practical implications of a test result.

◆

To practice this concept, try Exercise 8.44.

Significance Tests Are Less Useful than Confidence Intervals

Although significance tests can be useful, most statisticians believe that this method has been overemphasized in research.
- A significance test merely indicates whether the particular parameter value in H_0 (such as $\mu = 0$) is plausible.

When a P-value is small, the significance test indicates that that value is not plausible, but it tells us little about which potential parameter values *are* plausible.

- A confidence interval is more informative, because it displays the entire set of believable values.

A confidence interval shows how badly H_0 may be false by showing whether the values in the interval are far from the H_0 value. It helps us to determine whether the difference between the true value and the H_0 value has practical importance.

Let's illustrate with Example 11 and the 1 to 7 scale for political beliefs. A 95% confidence interval for μ is $\bar{x} \pm 1.96(se) = 4.10 \pm 1.96(0.0276)$, or $(4.05, 4.15)$. We can conclude that the difference between the population mean and the moderate score of 4.0 is small. Figure 8.13 illustrates. Although the P-value of 0.0004 provided extremely strong evidence against H_0: $\mu = 4.0$, in practical terms the confidence interval shows that H_0 is not wrong by much.

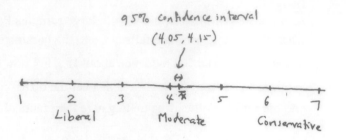

Figure 8.13: Statistical Significance but not Practical Significance. In testing $H_0 : \mu = 4.0$, the P-value = 0.0004, but the confidence interval of (4.05, 4.15) shows that μ is very close to the H_0 value of 4.0. **Question**: For $H_0 : \mu = 4.0$, does a sample mean of 6.10 and confidence interval of (6.05, 6.15) indicate (a) statistical significance? (b) practical significance?

By contrast, if $\bar{x}$ had been 6.10 (instead of 4.10), the 95% confidence interval would equal (6.05, 6.15). This indicates a substantial difference from 4.0, the mean response being near the conservative score rather than near the moderate score.

Misinterpretations of Results of Significance Tests

Unfortunately, results of significance tests are often misinterpreted. Here are common misinterpretations, some of which we've already discussed:

- **"Do not reject H_0" does not mean "Accept H_0."** If you get a P-value above 0.05 when the significance level is 0.05, you cannot conclude that H_0 is correct. We can never accept a single value, which H_0 contains, such as $p = 0.50$ or $\mu = 0$. A test merely indicates whether a particular parameter value is plausible. A confidence interval shows that there is a *range* of plausible values, not just a single one.

- **Statistical significance does not mean practical significance.** A small P-value does not tell us whether the parameter value differs by much in practical terms from the value in H_0.

- **The P-value cannot be interpreted as the probability that H_0 is true.** The P-value is

    P(test statistic takes observed value or beyond in tails | H_0 true),

    not P(H_0 true | observed test statistic value).

We've been calculating probabilities about test statistic values, not about parameters. It makes sense to find the probability that a test statistic takes a value in the tail, but the probability that a population mean = 0 does not make

RECALL

P(A | B) denotes the conditional probability of event A, given event B. To find a P-value, we condition on H_0 being true (that is, we suppose it is true), rather than find the probability it is true.

572

sense, because probabilities do not apply to parameters. The null hypothesis H_0 is either true or not true, and we simply do not know which.[9]

- **It is misleading to report results only if they are "statistically significant."** Some research journals have the policy of publishing results of a study only if the P-value ≤ 0.05. Here's a danger of this policy: If there truly is no effect, but 20 researchers independently conduct studies about it, we would expect about $20(0.05) = 1$ of them to obtain significance at the 0.05 level merely by chance. (When H_0 is true, about 5% of the time we get a P-value below 0.05 anyway.) If that researcher then submits results to a journal but the other 19 researchers do not, the article published will be a Type I error – reporting an effect when there really is not one. The popular media may then also report on this study, causing the general public to hear about an effect that does not actually exist.

- **Some tests may be statistically significant just by chance.** Related to the last comment, you should never scan pages and pages of computer output, looking for whatever results are statistically significant and reporting only those. If you run 100 tests, even if all the null hypotheses are correct, you would expect to get P-values of 0.05 or less about $100(0.05) = 5$ times. Keep this in mind and be skeptical of reports of significance that might merely reflect ordinary random variability. For instance, suppose an article reports an unusually high rate of a rare type of cancer in your town. It could be due to some cause such as air or water pollution. However, if researchers found this by looking at data for *all* towns and cities nationwide, it could also be due to random variability. Determining which is true may not be easy.

- **True effects may not be as large as initial estimates reported by the media.** Even if a statistically significant result is a true effect, the true effect may be smaller than suggested in the first article about it. For instance, often several researchers perform similar studies, but the results that get attention are the most extreme ones. This sensationalism may come about because the researcher who is the first to publicize the result is the one who got the most impressive sample result, perhaps way out in the tail of the sampling distribution of all the possible results. Then, the study's estimate of the effect may be greater than later research shows it to be. See the margin figure and the next example.

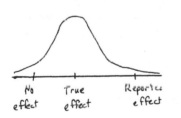

No effect   True effect   Reported effect

EXAMPLE 12: ARE MANY MEDICAL "DISCOVERIES" REALLY TYPE I ERRORS?

Picture the Scenario

What can be done with heart attack victims to increase their chance of survival? In 1992, trials of a clot-busting drug called anistreplase suggested that it doubled the chance of survival. Likewise, a 1993 study estimated that injections of magnesium could double the chance of survival. However, a much larger study in 1995 of heart attack survival rates among 58,000 patients indicated that this optimism was

---

[9] It is possible to find probabilities about parameter values using an approach called **Bayesian statistics**, but this requires extra assumptions and is beyond the scope of this text.

premature. The actual effectiveness of anistreplase seemed to be barely half that estimated by the original trial, and magnesium injections seemed to have no effect at all.[10] The anistreplase finding is apparently an example of a true effect not being as large as the initial estimate, and the report from the original magnesium study may well have been a Type I error.

Question to Explore

In medical studies, suppose that a true effect exists only 10% of the time. Suppose also that when an effect truly exists, there's a 50% chance of making a Type II error and failing to detect it. These were the hypothetical percentages used in a recent article in a medical journal[11]. The authors noted that many medical studies have a high Type II error rate because they are not able to use a large sample size. Assuming these rates, could a substantial percentage of medical "discoveries" actually be Type I errors? Approximate P(Type I error) in your answer by considering what you would expect to happen with 1000 medical studies that test various hypotheses.

Think It Through

Figure 8.14 is a tree diagram of what's expected. A true effect exists 10% of the time, or in 100 of the 1000 studies. We do not get a small enough P-value to detect this true effect 50% of the time, that is, in 50 of these 100 studies. An effect will be reported for the other 50 of the 100 that do truly have an effect. Now, for the 900 cases in which there truly is no effect, with the usual significance level of 0.05 we expect 5% of the 900 studies (that is, 45 studies) to incorrectly reject the null hypothesis and predict that there is an effect. So, of the 1000 studies, we expect 50 to report an effect that is truly there, but we also expect 45 to report an effect that does not really exist. If the assumptions are reasonable, then a proportion of $45/(45 + 50) = 0.47$ of medical studies that report effects (that is, reject H_0) are actually reporting Type I errors. In summary, nearly half the time when an effect is reported, there actually is no effect in the population!

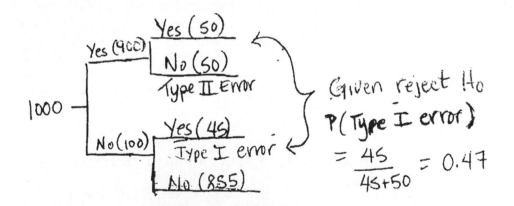

Figure 8.14: Tree Diagram of Results of 1000 Hypothetical Medical Studies. This assumes that a true effect exists 10% of the time and that there's a 50% chance of a Type II error when an effect truly does exist.

---

[10] "The great health hoax," by R. Matthews, in *The Sunday Telegraph*, Sept. 13, 1998.
[11] By J. Sterne, G. Smith, and D. R. Cox, *BMJ*, vol. 322, pp. 226-231 (2001).

Insight

Be skeptical when you hear reports of new medical advances. The true effect may be weaker than reported, or there may actually be no effect at all.

♦

To practice this concept, try Exercise 8.48.

Finally, remember that tests and confidence intervals use *sample* data to make inferences about *populations*. If you have data for an entire population, statistical inference is not relevant. For instance, if your college reports that the class of entering freshmen had a mean high school GPA of 3.60, there is no need to perform a test or construct a confidence interval. You already know the mean for that population.

On the Shoulders of ... Jerzy Neyman and the Pearsons

How can you build a framework for making decisions about everything from whether a roulette wheel is balanced to whether cloud-seeding results in rain?

The methods of confidence intervals and hypothesis testing were introduced in a series of articles beginning in 1928 by Jerzy Neyman (1894-1981) and Egon Pearson (1895-1980). Neyman emigrated from Poland to England and then to the U.S., where he established a top-notch Statistics Department at the University of California at Berkeley. He helped develop the theory of statistical inference, and he applied the theory to scientific questions in a variety of areas, such as agriculture, astronomy, biology, medicine, and weather modification. For instance, late in his career, Neyman's analysis of data from several randomized experiments showed that cloud-seeding can have a considerable effect on rainfall.

Much of Neyman's theoretical research was done with Egon Pearson, a professor at University College, London. Pearson's father, Karl Pearson, had developed one of the first statistical tests in 1900 to study various hypotheses, including whether the outcomes on a roulette wheel were equally likely. (We'll study his test, the **chi-squared test**, in Chapter 10.) Neyman and the younger Pearson developed the decision-making framework that introduced the two types of errors and the most powerful significance tests for various hypotheses.

PHOTO

Jerzy Neyman and Egon Pearson
((use picture from p. 239 of Neyman biography by C. Reid))

Figure 8.15 Jerzy Neyman and Egon Pearson. They developed statistical theory for making decisions about hypotheses.

SECTION 8.5: PRACTICING THE BASICS

8.52 Misleading summaries?: Two researchers conduct separate studies to test H_0: $p = 0.50$ against $H_a : p \neq 0.50$, each with $n = 400$. Researcher A gets 220 observations in the category of interest, and $\hat{p} = 220/400 = 0.550$. Researcher B gets 219 in the category of interest, and $\hat{p} = 219/400 = 0.5475$.

a. Show that $z = 2.00$ and P-value = 0.046 for Researcher A. Show that $z = 1.90$ and P-value = 0.057 for Researcher B.

b. Using $\alpha = 0.05$, indicate in each case whether the result is "statistically significant." Interpret.

c. From (a) and (b), explain why important information is lost by reporting the result of a test as "P-value ≤ 0.05" versus "P-value > 0.05," or as "reject H_0" versus "Do not reject H_0," instead of reporting the actual P-value.

d. Show that the 95% confidence interval for p is (0.501, 0.599) for Researcher A and (0.499, 0.596) for Researcher B. Explain how this method shows that, in practical terms, the two studies had very similar results.

8.53 Practical significance: A study considers whether the mean score on a college entrance exam for students in 2005 is any different from the mean score of 500 for students who took the same exam in 1975. Let μ represent the mean score for all students who took the exam in 2005. Find the P-value for testing $H_0 : \mu = 500$ against $H_a : \mu \neq 500$, using a nationwide random sample of 40,000 students who took the exam in 2005, for whom $\bar{x} = 498$ and $s = 100$. Explain why the test result is highly statistically significant, but not practically significant.

8.54 Effect of n: Example 11 analyzed political conservatism and liberalism in America. Suppose that the sample mean of 4.10 and sample standard deviation of 1.41 were from a sample size of only 25, rather than 2644.

a. Find the test statistic.

b. Find the P-value for testing $H_0 : \mu = 4.0$ against $H_a : \mu \neq 4.0$. Interpret.

c. Show that a 95% confidence interval for μ is (3.5, 4.7).

d. Together with the results of Example 11, explain what this illustrates about the effect of sample size on (i) the size of the P-value (for a given mean and standard deviation), (ii) the width of the confidence interval.

8.55 Fishing for significance: A research study conducts 60 significance tests. Of these, 3 are statistically significant at the 0.05 level. The study's final report stresses only the three tests with significant results, not mentioning the other 57 tests. Explain what is misleading about this.

8.56 Selective reporting: In 2004, New York Attorney General Eliot Spitzer filed a lawsuit against GlaxoSmithKline pharmaceutical company, claiming that the company failed to publish results of one of their studies that showed that an anti-depressant drug (Paxil) may make adolescents more likely to commit suicide. Partly as a consequence, editors of 11 medical journals agreed to a new policy to make researchers and companies register all clinical trials when they begin, so that negative results cannot later be covered up. The *International Journal of Medical Journal*

Editors wrote, "Unfortunately, selective reporting of trials does occur, and it distorts the body of evidence available for clinical decision-making." Explain why this controversy relates to the argument that it is misleading to report results only if they are "statistically significant." (*Hint*: See the subsection of this chapter on misinterpretations of significance tests.)

8.57 Is there really no difference?: An advertisement by Schering Corp. in 1999 for the allergy drug Claritin mentioned that in a clinical trial, the proportion who showed symptoms of nervousness was not significantly greater for patients taking Claritin than for patients taking placebo. Does this mean that the population proportion having nervous symptoms is exactly the same using Claritin and using placebo? How would you explain this to someone who has not studied statistics?

8.58 Does positive attitude lengthen life?: An article in the medical journal *BMJ* (by M. Petticrew et al., published 11/02) found no evidence to back the commonly held belief that a positive attitude can lengthen the lives of cancer patients. The authors noted that the studies that had indicated a benefit from some coping strategies tended to be smaller studies with weaker designs. Using this example and the text discussion, explain why you need to have some skepticism when you hear that new research suggests that some therapy or drug has an impact in treating a disease.

8.59 Inference using census data: The 2000 census of all Americans reported the proportion p of people who have a college education. Would it be correct to use this proportion in a significance test? For instance, could we test $H_0: p = 0.50$, and if we get a P-value of 0.0000001, reject the null hypothesis that exactly half of Americans have a college education? Explain.

8.6 HOW LIKELY IS A TYPE II ERROR (NOT REJECTING H_0 EVEN THOUGH IT'S FALSE)?

The probability of a Type I error is the significance level α of the test. When $\alpha = 0.05$, if H_0 is true then the probability of rejecting it equals 0.05.

A Type II error results from *not* rejecting H_0 even though it is false. This probability has more than one value, because H_a contains a range of possible values for the parameter. Each value in H_a has its own probability of Type II error. Let's see how to find the probability of a Type II error at a particular value.

EXAMPLE 13: FINDING P(TYPE II ERROR) AS PART OF A STUDY DESIGN

Picture the Scenario
Examples 1, 3, and 5 discussed an experiment to test astrologers' predictions. For each person's horoscope, an astrologer must predict which of three personality charts is the correct one. Let p denote the probability of a correct prediction. Consider the test of $H_0: p = 1/3$ (astrologers' predictions are like random guessing) against

$H_a : p > 1/3$ (better than random guessing), using the 0.05 significance level. Suppose an experiment plans to use $n = 116$ people, as in this experiment.

Questions to Explore

a. For what values of the sample proportion can we reject H_0?

b. For what values of the sample proportion would we make a Type II error?
c. The National Council for Geocosmic Research claimed that p would be 0.50 or higher. If truly $p = 0.50$, what is the probability that a significance test based on this experiment will make a Type II error, failing to reject H_0 even though it's false?

Think It Through

a. The standard error for the test statistic is

$$se_0 = \sqrt{p_0(1-p_0)/n} = \sqrt{[(1/3)(2/3)]/116} = 0.0438.$$

Now, for $H_a : p > 1/3$, a test statistic of $z = 1.645$ has a P-value (right-tail probability) of 0.05. If $z \geq 1.645$, the P-value is ≤ 0.05 and we can reject H_0. That is, we reject H_0 when $\hat{p}$ falls at least 1.645 standard errors above $p_0 = 1/3$,

$$\hat{p} \geq 1/3 + 1.645(se) = 1/3 + 1.645(0.0438) = 0.405.$$

We do not reject H_0 if $\hat{p} < 0.405$. Figure 8.16 shows the sampling distribution of $\hat{p}$ and these regions. The figure is centered at 1/3, because the test statistic is calculated and the rejection region is formed supposing that H_0 is correct.

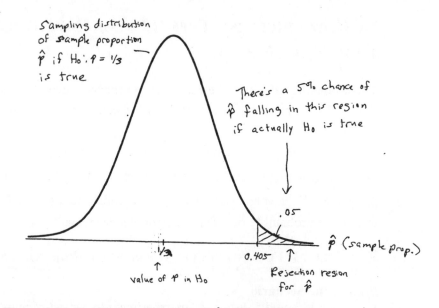

Figure 8.16: For Sample Proportion $\hat{p}$ above 0.405, Reject $H_0 : p = 1/3$ against $H_a : p >$ 1/3 at the 0.05 Significance Level. When the true $p > 1/3$, a Type II error occurs if $\hat{p} < .405$, since then the P-value > 0.05 and we do not reject H_0. **Question:** Why does each possible value of p from H_a have a separate probability of Type II error?

b. When H_0 is false, a Type II error occurs when we fail to reject H_0. From part (a) and Figure 8.16, we do not reject H_0 if $\hat{p} < 0.405$.

c. If the true value of p is 0.50, then the true sampling distribution of $\hat{p}$ is centered at 0.50, as Figure 8.17 shows. The probability of a Type II error is the probability that $\hat{p} < 0.405$ when $p = 0.50$. When $p = 0.50$, the standard error of $\hat{p}$ for a sample size of 116 is $\sqrt{0.50(0.50)/116} = 0.0464$. (Note that this differs a bit from the standard error for the test statistic, which uses 1/3 instead of 0.50 for p.)

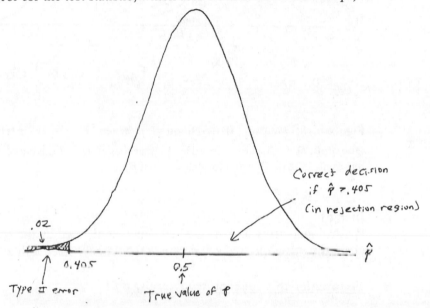

Figure 8.17: Calculation of P(Type II Error) when True Parameter $p = 0.50$. A Type II error occurs if the sample proportion $\hat{p} < 0.405$. **Question**: If the value of p decreases to 0.40, will the probability of Type II error decrease, or increase?

For a normal sampling distribution with mean 0.50 and standard error 0.0464, the $\hat{p}$ value of 0.405 has a z-score of

$$z = (0.405 - 0.50)/0.0464 = -2.04.$$

Using Table A or software, the left tail probability below -2.04 for the standard normal distribution equals 0.02. In summary, when $p = 0.50$ the probability of making a Type II error and failing to reject $H_0: p = 1/3$ is only 0.02.

Figure 8.18 shows the two figures together that we used in our reasoning. The normal distribution with mean 1/3 was used to find the rejection region, based on what we expect for $\hat{p}$ when $H_0: p = 1/3$ is true. The normal distribution with mean 0.50 was used to find the probability that $\hat{p}$ fails to fall in the rejection region even though $p = 0.50$ (that is, a Type II error occurs).

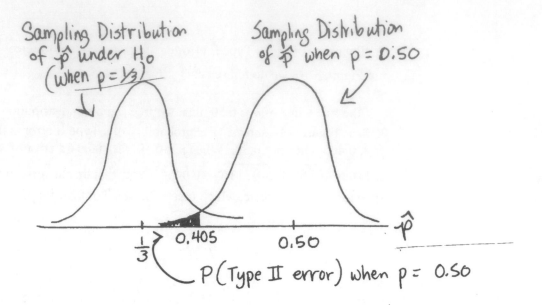

Figure 8.18: Sampling Distribution of $\hat{p}$ **when** $H_0 : p = 1/3$ **is true and when** $p = 0.50$.
Question: Why does the shaded area under the left tail of the curve for the case $p = 0.50$ represent P(Type II error) for that value of p?

Insight
If astrologers truly had the predictive power claimed by the National Council for Geocosmic Research, the experiment would have been unlikely to fail to detect it.

♦

To practice this concept, try Exercise 8.61.

The probability of Type II error increases when the true parameter value moves closer to H_0. To verify this, try to find the probability of Type II error when $p = 0.40$ instead of 0.50. You should get 0.54. (Remember to re-compute the exact standard error, now using 0.40 for p.)   If an astrologer can predict better than random guessing but not *much* better, we may not detect it with this experiment. Figure 8.19 plots the P(Type II error) for various values of p above 1/3. The farther the parameter value falls from the number in H_0, the less likely a Type II error.

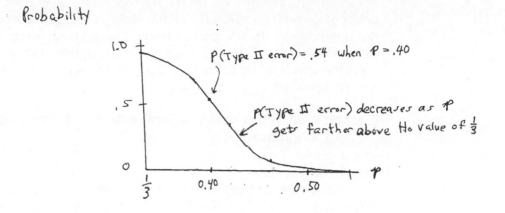

Probability

P(Type II error) = .54 when P = .40

P(Type II error) decreases as P gets farther above Ho value of $\frac{1}{3}$

Figure 8.19: Probability of Type II Error for Testing H$_0$: p = 1/3 against H$_a$: p > 1/3. This is plotted for the values of p in H$_a$, when the significance level = 0.05. **Question**: If the sample size n increases, how do you think this curve would change?

For a fixed significance level α, P(Type II error) decreases:

- as the parameter value moves farther into the H$_a$ values and away from the H$_0$ value.

- as the sample size increases.

Also, recall that P(Type II error) increases as α decreases. One reason that extremely small values, such as α = 0.001, are not common is that P(Type II error) is then too high. We may be unlikely to reject H$_0$, even though it is false.

Before conducting a study, researchers should find the probability of Type II error for the size of effect they want to be able to detect. If the probability is high, it may not be worth conducting the study unless they can use a larger sample size and lower it. They won't know the value of the parameter, so they won't know the actual P(Type II error). It may be large if n is small and if the true parameter value is not very far from the value in H$_0$. This may be the reason a particular significance test does not obtain a small P-value and reject H$_0$.

The Power of a Test

When H$_0$ is false, you want the probability of rejecting it to be high. The probability of rejecting H$_0$ is called the **power** of the test. For a particular value of the parameter from the range of alternative hypothesis values,

$$\boxed{\text{Power} = 1 - \text{P(Type II error)}}$$

In Example 13, for instance, P(Type II error) = 0.02 when $p = 0.50$. Therefore, the power of the test at $p = 0.50$ is 1 - 0.02 = 0.98. The higher the power, the better, so this is quite good. Before granting research support, many agencies (such as the National Institutes of Health) expect research scientists to show that for the planned study reasonable power (usually, at least 0.80) exists at values of the parameter that are considered practically significant.

EXAMPLE 14: WHAT WAS THE POWER OF THE TEST IN THE TT EXPERIMENT?

Picture the Scenario
In Example 6 on the therapeutic touch (TT) experiment, the data did not support the TT practitioners' claim to be able to detect a human energy field. The P-value was not small for testing $H_0 : p = 0.50$ against $H_a : p > 0.50$, where p is the probability of a correct prediction about which hand was near the researcher's hand. The medical journal article about the study stated, "The statistical power of this experiment was sufficient to conclude that if TT practitioners could reliably detect a human energy field, the study would have demonstrated this." For the test of $H_0 : p = 0.50$ with one of the sets of trials, the power was reported as 0.96 if actually $p = 2/3$.

Questions to Explore
a. How should the probability of 0.96 be interpreted?
b. In context, what is a Type II error for this experiment?
c. If $p = 2/3$, what is the probability of committing a Type II error?

Think It Through
a. The power of 0.96 is the probability of correctly rejecting H_0 when it is false. If the actual probability of correct predictions by TT practitioners was 2/3, there was a 96% chance of data such that the significance test performed would reject H_0.

b. A Type II error occurs if we do not reject $H_0 : p = 0.50$, when actually TT practitioners *can* predict correctly more than half the time. The consequence would be to question the truthfulness of what they have been practicing for over 30 years, when they actually do have some ability.

c. If $p = 2/3$, the value of P(Type II error) is 1- (Power at $p = 2/3$). This is $1 - 0.96 = 0.04$. A Type II error was unlikely if truly $p = 2/3$.

Insight
Based on the data, the researchers remained skeptical that TT practitioners could detect a human energy field. Because of the strong power at a value of p that TT practitioners claim is reasonable, the researchers felt justified in concluding that "TT claims are groundless and further use of TT by health professionals is unjustified."

♦

To practice this concept, try Exercise 8.65.

ACTIVITY 2: LET'S SIMULATE THE PERFORMANCE OF SIGNIFICANCE TESTS

Let's get a feel for the two possible errors in significance tests. To do this, for a given population proportion value you can simulate many samples and perform a significance test for each. You can then check how often the tests make an incorrect inference. You can conduct the simulation using statistical software (such as Minitab) or using a statistical applet that allows you to control the null hypothesis value, the true parameter value, the sample size, and the significance level.

Try this by going to the *significance test* applet at www.prenhall.com/???. Set the null hypothesis as $H_0: p = 1/3$ for a two-sided test using significance level $\alpha = 0.05$ with sample size 116, a case Example 3 on the astrology experiment considered. First, see what happens when you set the true $p = 1/3$ and repeatedly take samples of size 116 and conduct the test. At the menu, set the proportion value to 1/3, and select a random sample of size 116. When we did this, we got 43 outcomes of one type and 73 of the other. You will probably get something different, because the process is random. After generating the sample, the applet calculates the sample proportion (we got $\hat{p} = 0.37$) and the test statistic (we got $z = 0.80$). Our sample gave a two-sided P-value of 0.42. We did not reject H_0, since this is greater than 0.05. This is a correct inference, since H_0 was true in this simulation. When you perform the test, do you make the correct inference of not rejecting H_0? (About 5% of you *will* reject H_0, making a Type I error.)

To get a feel for what happens "in the long run," do this simulation 100 times. You will then perform 100 separate significance tests. Some will incorrectly reject $H_0: p = 1/3$. These are Type I errors. If you do this a much larger number of times (say, 10,000), you will see that close to 5% of the tests make a Type I error.

Next, change the value of p to 0.50, so H_0 is actually false. Again, perform a single test. Now if you get a P-value above 0.05 and fail to reject H_0, you have made a Type II error, since H_0 is actually false. Perform the test 100 times. What percentage of times did you make a Type II error? By Example 13, this should happen only about 2% of the time.

To practice this concept, try Exercise 8.141 and Exercise 8.56.

SECTION 8.6: PRACTICING THE BASICS

8.60 Two sampling distributions: A study is designed to test $H_0: p = 0.50$ against $H_a: p > 0.50$, taking a random sample of size $n = 100$, using significance level 0.05.
a. Show that the rejection region consists of values of $\hat{p} > 0.582$.
b. Sketch a single picture that shows (i) the sampling distribution of $\hat{p}$ when H_0 is true and (ii) the sampling distribution of $\hat{p}$ when $p = 0.60$. Label each sampling distribution with its mean and standard error, and highlight the rejection region.
c. Find P(Type II error) when $p = 0.60$.

8.61 Gender bias in selecting managers: Exercise 8.18 tested the claim that female employees were passed over for management training in favor of their male colleagues. Statewide, the large pool of more than 1000 eligible employees who can be tapped for management training is 40% female and 60% male. Let p be the probability of selecting a female for any given selection. For testing $H_0: p = 0.40$ against $H_a: p < 0.40$ based on random sample of 50 selections, using the 0.05 significance level, verify that:

a. A Type II error occurs if the sample proportion falls less than 1.645 standard errors below the null hypothesis value, which means that $\hat{p} > 0.286$.

b. When $p = 0.20$, a Type II error has probability 0.064.

8.62 Balancing Type I and Type II errors: Recall that the smaller the probability of Type I error, α, the larger the P(Type II error). Let's check this for Example 13. There we found P(Type II error) for testing $H_0: p = 1/3$ (astrologers randomly guessing) against $H_a: p > 1/3$ when actually $p = 0.50$, with $n = 116$. If we use $\alpha = 0.01$, verify that:

a. A Type II error occurs if the sample proportion falls less than 2.326 standard errors above the null hypothesis value, which means $\hat{p} < 0.435$.

b. When $p = 0.50$, a Type II error has probability 0.08. (By comparison, Example 13 found P(Type II error) = 0.02 when $\alpha = 0.05$, so we see that P(Type II error) increased when P(Type I error) decreased.)

8.63 P(Type II error) large when p close to H_0: For testing $H_0: p = 1/3$ (astrologers randomly guessing) against $H_a: p > 1/3$ with $n = 116$, Example 13 showed that P(Type II error) = 0.02 when $p = 0.50$. Now suppose that $p = 0.35$.

a. Show that P(Type II error) = 0.89.

b. Explain intuitively why P(Type II error) is large when the parameter value is close to the value in H_0 and decreases as it moves farther from that value.

8.64 Type II error with two-sided H_a: In Example 13 for testing $H_0: p = 1/3$ (astrologers randomly guessing) with $n = 116$ when actually $p = 0.50$, suppose we used $H_a: p \neq 1/3$. Then, show that:

a. A Type II error occurs if $0.248 < \hat{p} < 0.419$.

b. The probability is 0.00 that $\hat{p} < 0.248$ and 0.96 that $\hat{p} > 0.419$.

c. P(Type II error) = 0.04.

8.65 Power of TT: Consider Example 14 about the power of the test used in the TT experiment for testing $H_0: p = 0.50$ against $H_a: p > 0.50$, where p is the probability of a correct prediction about which hand was nearer the researcher's hand. In a significance test planned for a second set of trials, the power was reported as 0.98 at $p = 2/3$.

a. How should the probability of 0.98 be interpreted?

b. In context, what is a Type II error for this experiment?

c. What is the probability of committing a Type II error if $p = 2/3$?

⌨**8.65 Simulating Type II errors**: Refer to the simulation in Activity 2 at the end of the section.

a. Repeat the simulation, now assuming that actually $p = 0.45$. In a large number of simulations, what proportion of the time did you make a Type II error? What does theory predict for this proportion?

b. Do you think that the proportion of Type II errors will increase, or decrease, if actually $p = 0.50$? Check your intuition by conducting another simulation.

c. Refer to (a). Do you think that the proportion of Type II errors will increase, or decrease, if the sample size in each simulation is 50 instead of 116? Check your intuition by conducting another simulation.

Chapter Summary

A significance test helps us to judge whether a particular value for a parameter is plausible. Each significance test has five steps:

1. **Assumptions**: The most important assumption for any significance test is that the data result from a *random sample* or *randomized experiment*. Specific significance tests make other assumptions, such as summarized in the box below for significance tests about means and proportions.

2. **Null and alternative hypotheses** about the parameter: Null hypotheses have form $H_0: p = p_0$ for a proportion and $H_0: \mu = \mu_0$ for a mean, where p_0 and μ_0 denote particular values, such as $p_0 - 0.5$ and $\mu_0 = 0$. The most common alternative hypothesis is **two-sided**, such as $H_a: p \neq 0.5$. **One-sided** hypotheses such as $H_a: p > 0.5$ and $H_a: p < 0.5$ are also possible.

3. **Test statistic:** This measures how far the sample estimate of the parameter falls from the null hypothesis value. The z statistic for proportions and the t statistic for means have the form

$$\text{Test statistic} = \frac{\text{parameter estimate - null hypothesis value}}{\text{standard error}}.$$

   This measures the number of standard errors that the parameter estimate ($\hat{p}$ or $\bar{x}$) falls from the null hypothesis value (p_0 or μ_0).

4. **P-value:** This is a probability summary of the evidence that the data provide about the null hypothesis. It equals the probability that the test statistic takes a value like the observed one or even more extreme.

   - It is calculated by supposing that H_0 is true.

- The test statistic values that are "more extreme" depend on the alternative hypothesis. When H_a is two-sided, the P-value is a two-tail probability. When H_a is one-sided, the P-value is a one-tail probability.

- When the P-value is small, the observed data would be unusual if H_0 were true. The smaller the P-value, the stronger the evidence against H_0.

5. **Conclusion**: A test concludes by interpreting the P-value in the context of the study. Sometimes a decision is needed, using a fixed **significance level** α, usually $\alpha = 0.05$. Then we reject H_0 if the P-value $\leq \alpha$. Two types of error can occur:

- A **Type I error** results from rejecting H_0 when it is true. The significance level = P(Type I error), when H_0 is true.

- A **Type II error** results from failing to reject H_0 when it is false.

Summary: Significance Tests for Population Proportions and Means

	PARAMETER	
	PROPORTION	**MEAN**

1. Assumptions	Categorical variable
	Randomization
	Expected numbers of
	   successes and failures ≥ 15
		Quantitative variable
		Randomization
		Approximately normal
		   population

2. Hypotheses	$H_0 : p = p_0$ $H_a : p \neq p_0$ (two-sided) $H_a : p > p_0$ (one-sided) $H_a : p < p_0$ (one-sided)	$H_0 : \mu = \mu_0$ $H_a : \mu \neq \mu_0$ (two-sided) $H_a : \mu > \mu_0$ (one-sided) $H_a : \mu < \mu_0$ (one-sided)
3. Test statistic	$z = \dfrac{\hat{p} - p_0}{se}$ $(se = \sqrt{p_0(1 - p_0)/n}\,)$	$t = \dfrac{\bar{x} - \mu_0}{se}$ $(se = s/\sqrt{n}\,)$
4. P-value	Two-tail ($H_a : p \neq p_0$) or right tail ($H_a : p > p_0$) or left tail ($H_a : p < p_0$) probability from standard normal	Two-tail ($H_a : \mu \neq \mu_0$) or right tail ($H_a : \mu > \mu_0$) or left tail ($H_a : \mu < \mu_0$) probability from t dist ($df = n\text{-}1$)
5. Conclusion	Interpret P-value in context Reject H_0 if P-value $\leq \alpha$	Interpret P-value in context Reject H_0 if P-value $\leq \alpha$

Where We're Going

To introduce statistical inference, Chapters 7 and 8 presented methods for a single proportion or mean. In practice, inference is used more commonly to compare parameters for different groups (for example, females and males). The next chapter shows how to compare proportions and how to compare means for two groups.

SUMMARY OF NEW NOTATION IN CHAPTER 8

H_0 = null hypothesis, H_a = alternative hypothesis

p_0 = null hypothesis value of proportion, μ_0 = null hypothesis value of mean

α = significance level (usually 0.05; the P-value must be $\leq \alpha$ to reject H_0)
      = probability of Type I error

ANSWERS TO THE CHAPTER FIGURE QUESTIONS

Figure 8.1:   *A P-value of 0.01 gives stronger evidence against the null hypothesis. This smaller P-value (0.01 compared to 0.20) indicates it would be more unusual to get the observed sample data or a more extreme value if the null hypothesis is true.*

Figure 8.2:   *It is the relatively large values of $\hat{p}$ (with their corresponding right-tail z-scores) that support the alternative hypothesis of $p > 1/3$.*

Figure 8.3:   *Values more extreme than the observed test statistic value can be either relatively smaller or larger to support the alternative hypothesis $p \neq p_0$. These smaller or larger values are ones that fall farther in either the left or right tail.*

Figure 8.4:   *A P-value = 0.000 is strong evidence against H_0.*

Figure 8.5:   *The tail probability is found based on the direction stated in H_a. If H_a is $p > p_0$, the right tail is used. If H_a is $p < p_0$, the left tail is used.*

Figure 8.7:   *The t-score indicates the number of standard errors $\overline{x}$ falls from the null hypothesis mean. The farther out a t-score falls in the tails, the farther the sample mean falls from the null hypothesis mean, providing stronger evidence against H_0.*

Figure 8.8:   *For H_a: $\mu < 0$, the t-test statistics providing stronger evidence against H_0 will be t-statistics corresponding to the relatively small $\overline{x}$ values falling to the left of the hypothesized mean of 0. The t-test statistics for these values of $\overline{x}$ are negative.*

Figure 8.9: Answers will vary

Figure 8.10:   *The standard error for the significance test is computed using p_0, the hypothesized proportion. The standard error for the confidence interval is computed using $\hat{p}$, the sample proportion. The values p_0 and $\hat{p}$ are not necessarily the same; therefore, there is not an exact equivalence.*

Figure 8.11:   *The plots suggest that the population distribution of weight change is skewed to the right.*

Figure 8.12:   *Using a significance level of 0.05, the rejection region consists of values of z > 1.645.*

Figure 8.13:   *A sample mean of 6.10 indicates both statistical and practical significance.*

Figure 8.16:   *A Type II error results from not rejecting H_0 even though H_0 is false. The probability of a Type II error has more than one value because H_a contains a range of possible values for the parameter p.*

Figure 8.17:   *The probability of a Type II error will increase.*

Figure 8.18:   *A correct decision for the case p=0.50 is to reject $H_0 : p = 1/3$. The shaded area under the left tail of the curve for the case p = 0.50 represents the probability we would not reject $H_0 : p = 1/3$, which is the probability of an incorrect decision for this situation. This shaded area is the probability of committing a Type II error.*

Figure 8.19:   *By increasing the sample size n, the probability of a Type II error decreases. The curve would more quickly approach the horizontal axis, indicating that the Type II error probability more rapidly approaches a probability of 0 as p increases.*

CHAPTER PROBLEMS: PRACTICING THE BASICS

8.66 H_0 or H_a ?: For each of the following hypotheses, explain whether it is a null hypothesis, or an alternative hypothesis:
a. For females, the population mean on the political ideology scale is equal to 4.0.
b. For males, the population proportion who support the death penalty is larger than 0.50.
c. The diet has an effect, the population mean change in weight being less than 0.

8.67 Compare P-values: Does a P-value of 0.40 provide stronger, or weaker, evidence against the null hypothesis than a P-value of 0.01? Explain.

8.68 Income inequality: Do you think it should or should not be the government's responsibility to reduce income differences between the rich and poor? Let *p* denote the population proportion of American adults who believe it should be. Analyses using data from a recent General Social Survey are as follows:

```
------------------------------------------------------------------------------
Test of p = 0.50 vs p not = 0.50

X        N        Sample p      95.0% CI       Z-Value      P-Value
591      1227     0.4817        (0.454, 0.510) -1.286       0.1986
------------------------------------------------------------------------------
```

a. State the null and alternative hypotheses for the significance test reported here.
b. Report the value of the test statistic.

c. Report the P-value. What would your decision be for a 0.05 significance level? Interpret this in such a way that someone who had not taken a statistics course would understand.

d. In (c), explain why it does not make sense to "accept H_0." (Hint: Use the confidence interval reported.)

8.69 ESP: A person who claims to possess extrasensory perception (ESP) says she can guess more often than not the outcome of a flip of a balanced coin. Out of 20 flips, if she guesses correctly 12 times, would you conclude that she truly has ESP? Answer by reporting all five steps of a significance test of the hypothesis that each of her guesses has probability 0.50 of being correct against the alternative that corresponds to her having ESP.

8.70 Free throw accuracy: Consider all cases in which a pro basketball player shoots two free throws and makes one and misses one. Which do you think is more common: Making the first and missing the second, or missing the first and making the second? One of the best shooters was Larry Bird of the Boston Celtics. During 1980-1982 he made only the first free throw 34 times and made only the second 48 times (A. Tversky and T. Gilovich, *Chance*, vol. 2, pp. 16-21, 1989). Does this present much evidence that one sequence was truly more likely than the other sequence for Larry Bird? Answer by conducting a significance test, (a) defining notation and specifying assumptions and hypotheses, (b) finding the test statistic, and (c) finding the P-value and interpreting in this context.

8.71 Box or Draper?: A mayoral election in Madison, Wisconsin, has two candidates, Box and Draper.

a. For a random sample of 400 voters in an exit poll, 230 voted for Box and 170 for Draper. Conduct all five steps of a test of $H_0 : p = 0.50$ against $H_a : p \neq 0.50$, where p denotes the probability that a randomly selected voter prefers Box. Are you willing to predict the outcome of the election? Explain how to make a decision, using a significance level of 0.05.

b. Suppose the sample size had been 40 voters, of whom 23 voted for Box. Show that the sample proportion is the same as in (a), but the test statistic and P-value are very different. Are you now willing to predict the outcome of the election?

c. Using (a) and (b), explain how results of a significance test can depend on the sample size.

8.72 Environment: Let p denote the proportion of Floridians who think that government environmental regulations are too strict. A telephone poll of 834 people conducted by the Institute for Public Opinion Research at Florida International University is summarized on the following printout, where X is the number who said they thought regulations were too strict:

Test of p = 0.50 vs p not = 0.50

X	N	Sample p	95.0% CI	Z-Value	P-Value
222	834	0.2662	(0.236, 0.296)	13.50	0.0000

a. Specify the null and alternative hypotheses that are tested.

b. Explain how to interpret the P-value.

c. Using a significance level of 0.01, can you determine whether a majority or minority think that environmental regulations are too strict, or is it plausible that $p = 0.50$? Explain how you would summarize your conclusion.

d. Explain an advantage of the confidence interval shown over the significance test. What assumptions are needed for either of these methods to be valid?

8.73 Affirmative action: A Pew Research Center poll (May 14, 2003) asked the question, "All in all, do you think affirmative action programs designed to increase the number of black and minority students on college campuses are a good thing or a bad thing?" 60% said good, 30% said bad, and 10% said don't know. Let p denote the population proportion who said it is good. Conduct all five steps of a test of H_0: $p = 0.50$ against $H_a : p \neq 0.50$. Can you reject H_0 using a significance level of 0.05?

8.74 Plant inheritance: In an experiment on chlorophyll inheritance in maize (corn), of the 1103 seedlings of self-fertilized green plants, 854 seedlings were green and 249 were yellow. Theory predicts the ratio of green to yellow is 3 to 1. Show all five steps of a test of the hypothesis that 3 to 1 is the true ratio. Interpret the P-value in context.

8.75 Ellsberg paradox: You are told that a ball will be randomly drawn from one of two boxes (A and B), both of which contain black balls and red balls, and if a red ball is chosen, you will win $100. You are also told that Box A contains half black balls and half red balls, but you are not told the proportions in Box B. Which box would you pick?

a. Set up notation and specify hypotheses to test whether the population proportion who would pick Box A is 0.50.

b. For a random sample of 40 people, 36 pick Box A. Can you make a conclusion about whether the proportion for one option is higher in the population? Explain all steps of your reasoning. (Logically those who picked Box A would seem to think that Box B has greater chance of a black ball. However, a paradox first discussed by Daniel Ellsberg predicts that if they were now told that they would instead receive $100 if a black ball is chosen, they would overwhelmingly pick Box A again, because they prefer definite information over ambiguity.)

8.76 Traffic tickets: Let p denote the proportion of adult Americans who have ever received a ticket for a traffic violation (other than for illegal parking).

a. Can you conclude whether this is less than, or greater than, 0.50, based on GSS data in which 812 of 1466 people said they had received a ticket? Explain your reasoning by carrying out the five steps of a significance test.

b. Find a 95% confidence interval for p. Explain why it is more informative than the significance test.

8.77 Nonrandom sample?: A fraternity at a university lobbies the administration to start a hockey team. To bolster its case, it reports that of a simple random sample of 100 students, 83% support starting the team. Upon further investigation, their sample has 80 males and 20 females. Should you be skeptical of whether the sample was random, if you know that 55% of the student body population was male? Answer this by performing all steps of a two-sided significance test.

8.78 Interest charges on credit card: A bank wants to evaluate which credit card would be more attractive to its customers: One with a high interest rate for unpaid balances but no annual cost, or one with a low interest rate for unpaid balances but an annual cost of $40. For a random sample of 100 of its 52,000 customers, 40 say they prefer the one that has an annual cost. Software reports:

```
-----------------------------------------------------------------------------------
Test of p = 0.50 vs p not = 0.50


Sample   X    N    Sample p     95.0% CI      Z-Value    P-Value
1        40   100  0.40000     (0.304, 0.496)  -2.00      0.04550
-----------------------------------------------------------------------------------
```

Explain how to interpret all results on the printout. What would you tell the company about what the majority of its customers prefer?

8.79 Jurors and gender: A jury list contains the names of all individuals who may be called for jury duty. The proportion of the available jurors on the list who are women is 0.53. If 40 people are selected to serve as candidates for being picked on the jury, show all steps of a significance test of the hypothesis that the selections are random with respect to gender.
a. Set up notation and hypotheses, and specify assumptions.
b. 5 of the 40 selected were women. Find the test statistic.
c. Report the P-value, and interpret.
d. Explain how to make a decision using a significance level of 0.01.
e. If you made an error with the decision in (d), is it a Type I or a Type II error?

8.80 Errors: Explain what Type I and Type II errors mean in the context of the previous exercise.

8.81 Levine = author?: The authorship of an old document is in doubt. A historian hypothesizes that the author was a journalist named Jacalyn Levine. Upon a thorough investigation of Levine's known works, it is observed that one unusual feature of her writing was that she consistently began 10% of her sentences with the word "whereas". To test the historian's hypothesis, it is decided to count the number of sentences in the disputed document that begin with the word *whereas*. Out of the 300 sentences in the document, none begin with that word. Let p denote the probability that any one sentence written by the unknown author of the document begins with the word "whereas".
a. Conduct a test of the hypothesis $H_0 : p = 0.10$ against $H_a : p \neq 0.10$. What conclusion can you make, using significance level of 0.05?
b. What assumptions are needed for that conclusion to be valid? (F. Mosteller and D. Wallace conducted an investigation similar to this to determine whether Alexander Hamilton or James Madison was the author of 12 of the *Federalist Papers*. See *Inference and Disputed Authorship: The Federalist*, Addison-Wesley, 1964.)

8.82 Useful advertising?: The owner of a department store in Rochester, New York, initiates a week-long newspaper advertising campaign to increase awareness of the store. Before investing more money in advertising, the owner takes a phone survey to check whether potential customers are more inclined or less inclined to shop in this store after seeing the advertising. Consider the population of residents who have seen the ad and are either more inclined or less inclined to shop there than before seeing the ad. Of these people, let p denote the proportion more inclined. For a random sample of 100 names selected from the telephone book, 40 express an opinion. Of these 40 subjects, 28 say they are more inclined to shop at the store, and 12 say they are less inclined. Show all five steps of a test of $H_0 : p = 0.50$ against $H_a : p \neq 0.50$. Make a decision using the 0.05 significance level. Can the owner conclude that, of those with an opinion, a majority is more likely to shop at the store?

⌨**8.83 Practice steps of test for mean**: For a quantitative variable, you want to test $H_0 : \mu = 0$ against $H_a : \mu \neq 0$. The 10 observations are

$$3, 7, 3, 3, 0, 8, 1, 12, 5, 8.$$

a. Show that (i) $\bar{x} = 5.0$, (ii) $s = 3.71$, (iii) standard error $= 1.17$, (iv) test statistic $= 4.26$, (v) df $= 9$.
b. The P-value is 0.002. Interpret, and make a decision using a significance level of 0.05. Interpret.
c. If you had instead used $H_a : \mu > 0$, what would the P-value be? Interpret it.

8.84 Two ideal children?: Is the ideal number of children equal to 2, or higher or lower than that? For testing that the mean response from a recent GSS equals 2.0 for the question, "What do you think is the ideal number of children to have?," software shows results:

```
------------------------------------------------------------------------------
Test of mu = 2.0 vs mu not = 2.0

Variable    N      Mean    StDev    SE Mean      T          P
Children   1302    2.490   0.850    0.0236     20.80     0.0000
------------------------------------------------------------------------------
```

a. Report the test statistic value, and show how it was obtained from other values reported in the table.
b. Explain what the P-value represents, in terms of a t distribution, and interpret its value.

8.85 Hours at work: When the subjects in the 2002 GSS who were working full or part-time were asked how many hours they worked in the previous week at all jobs (variable HRS1), software produced the following analyses:

```
-------------------------------------------------------------------------
Test of mu = 40 vs not = 40

  N      Mean     StDev    SE Mean       95% CI             T       P
1729   41.7770   14.6230   0.3517   (41.0873, 42.4667)    5.05   0.000
-------------------------------------------------------------------------
```

For this printout, (a) state the hypotheses and (b) explain how to interpret the values of (i) SE Mean, (ii) T, (iii) P.

8.86 Females liberal or conservative?: Example 11 compared mean political beliefs (on a 1 to 7 point scale) to the moderate value of 4.0, using GSS data. Test whether the population mean equals 4.00 for females, for whom the sample mean was 4.02 and standard deviation was 1.40 for a sample of size 1504. Carry out the five steps of a significance test, reporting and interpreting the P-value in context.

8.87 Infant heart rate: Recent findings have suggested that neonatal sex differences exist in behavioral and physiological reactions to stress. One study (M. Davis and E. Emory, *Child Development*, Vol. 66, 1995, pp. 14-27) evaluated changes in the heart rate for a sample of infants placed in a stressful situation. For the 15 female infants, the following table is a printout for the data on the change in heart rate. State (a) hypotheses, (b) test statistic, (c) P-value, and (d) interpret the P-value in context.

Variable	Number of Cases	Mean	SD	SE of Mean	t-value	df	2 Tail Sig
CHANGE	15	10.70	17.70	4.570	2.341	14	.0346

8.88 Canada SAT: The mean score for all U.S. high school seniors taking the SAT college entrance exam equals 500. A study is conducted to see whether a different mean applies to Canadian seniors. For a random sample of 100 Canadian seniors, suppose the mean and standard deviation on this exam equal 508 and 100.
a. Set up hypotheses for a significance test, and compute the test statistic.
b. The P-value is 0.43. Interpret it, and make a decision about H_0, using a significance level of 0.05.
c. If the decision in (b) was in error, what type of error is it?
d. A 95% confidence interval for μ is (488.2, 527.8). Show the correspondence between the decision in the test and whether 500 falls in this confidence interval.

8.89 One-sided SAT: Refer to the previous exercise.
a. Suppose you had predicted that the mean would be *higher* in Canada. State the alternative hypothesis for this prediction, and report the P-value based on the result in (b) of the previous exercise.
b. Suppose you had predicted that the mean would be lower in Canada. State the alternative hypothesis for this prediction, and report the P-value.

8.90 Blood pressure: When Vincenzo De Cerce's blood pressure is in control, his systolic blood pressure reading has a mean of 130.   For the last six times he has monitored his blood pressure, he has obtained the values

140, 150, 155, 155, 160, 140.

a. Does this provide strong evidence that his true mean has changed? Carry out the five steps of the significance test, interpreting the P-value.
b. Review the assumptions that this method makes. For each assumption, discuss it in context.

8.91 Tennis balls in control?: When it is operating correctly, a machine for manufacturing tennis balls produces balls with a mean weight of 57.6 grams. A test of whether the process is in control using the last four balls manufactured shows:

```
Test of mu = 57.6 vs mu not = 57.6

Variable    N      Mean      StDev    SE Mean       T         P
Weight      4      57.297    0.602    0.301     -1.007    0.388
```

Since the sample mean is not exactly equal to 57.6, do the results suggest that the process no longer has a mean of 57.6? Why or why not?

8.91 Catalog sales: A company that sells its products through mail-order catalogs wants to evaluate whether the mean sales for their most recent catalog were higher or lower than the mean of $15 from past catalogs. For a random sample of 100 customers from their files, the mean sales were $10, with a standard deviation of $10. Report a P-value to provide the extent of evidence that the mean differed with this catalog. Interpret.

8.92 Wage claim false?: Management claims that the mean income for all senior-level assembly-line workers in a large company equals $500 per week. An employee decides to test this claim, believing that it is actually less than $500. For a random sample of nine employees, the incomes are

$$430, 450, 450, 440, 460, 420, 430, 450, 440.$$

Conduct a significance test of whether the population mean income equals $500 per week against the alternative that it is less. Include all assumptions, the hypotheses, test statistic, and P-value, and interpret the result in context.

8.93 CI and test: Refer to the previous exercise.
a. For which significance levels can you reject H_0? (i) 0.10, (ii) 0.05, (iii) 0.01.
b. Based on the answers in (a), for which confidence coefficients would the confidence interval contain 500? (i) 0.90, (ii) 0.95, (iii) 0.99.
c. Use (a) and (b) to illustrate the correspondence between results of significance tests and results of confidence intervals.

8.94 Mean IQ = 100?: A social psychologist plans to conduct an experiment with a random sample of 49 children from a school district. Before conducting the experiment, the psychologist checks how this sample compares to national norms on several variables. The IQ scores for the 49 children have $\bar{x} = 103$ and $s = 14$. Nationally, the population mean IQ equals 100. Is it plausible that the mean μ of the population of children in the school district from which these students were sampled equals 100?
a. Show all five steps of a test of $H_0: \mu = 100$ against $H_a: \mu \neq 100$. Interpret the P-value, and make a decision about H_0 using a significance level of 0.05.
b. If the decision in (a) is an error, what type of error is it, Type I or Type II? Why?
c. What conclusion applies for each of the following significance levels: (i) $\alpha = 0.20$, (ii) $\alpha = 0.10$, (iii) $\alpha = 0.01$. Why is $\alpha = 0.20$ rare in practice?

8.95 CI and test connection: The P-value for testing $H_0 : \mu = 100$ against $H_a : \mu \neq 100$ is 0.043.
a. What decision is made using a 0.05 significance level?
b. If the decision in (a) is in error, what type of error is it?
c. Does a 95% confidence interval for μ contain 100? Explain.

8.96 Religious beliefs: A journal article that deals with changes in religious beliefs over time states, "For these subjects, the difference in their responses on the scale of religiosity between age 16 and the current survey was statistically significant (P-value < 0.05)."
a. How would you explain to someone who has never taken a statistics course what it means for the result to be "statistically significant" with a P-value < 0.05?
b. Explain why it would have been more informative if the authors provided the actual P-value rather than merely indicating that it is below 0.05.
c. Can you conclude that a practically *important* change in religiosity has occurred between age 16 and the time of the current survey? Why or why not?

8.97 How to reduce chance of error?: In making a decision in a significance test, a researcher worries about rejecting H_0 when it may actually be true.
a. Explain how the researcher can control the probability of this type of error.
b. Why should the researcher probably not set this probability equal to 0.00001?

8.98 Legal trial errors: Consider the analogy (discussed in Section 8.4) between making a decision about a null hypothesis in a significance test and making a decision about the innocence or guilt of a defendant in a criminal trial.
a. Explain the difference between Type I and Type II errors in the trial setting.
b. In this setting, explain intuitively why decreasing the chance of Type I error increases the chance of Type II error.

8.99 P(Type II error) with smaller n: Consider Example 13 about testing $H_0 : p = 1/3$ against $H_a : p > 1/3$ for the astrology study, with $n = 116$. Find P(Type II error) for testing $H_0 : p = 1/3$ against $H_a : p > 1/3$ when actually $p = 0.50$, if the sample size is 60 instead of 116. Do this by showing that:
a. The standard error is 0.061 when H_0 is true.
b. The rejection region consists of $\hat{p}$ values above 0.433.
c. When $p = 0.50$, the probability that $\hat{p}$ falls below 0.433 is the left-tail probability below -1.03 under a standard normal curve. What is the answer? Why would you expect P(Type II error) to be larger when n is smaller?

CHAPTER PROBLEMS: CONCEPTS AND INVESTIGATIONS

⌨**8.100 Student data**: Refer to the "Florida student survey" data file on the text CD. Test whether the (a) population mean political ideology (on a scale of 1 to 7, where 4 = moderate) equals or differs from 4.0, (b) population proportion favoring affirmative action equals or differs from 0.50. For each part, write a one-page report showing all five steps of the test, including what you must assume for each inference to be valid.

⌨**8.101 Class data**: Refer to the data file your class created in Activity 3 at the end of Chapter 1. For a variable chosen by your instructor, conduct inferential statistical analyses. Prepare a report, summarizing and interpreting your findings. In this report, also use graphical and numerical methods presented earlier in this text to describe the data.

8.102 Gender of best friend: A GSS question asked the gender of your best friend. Of 1381 people interviewed, 147 said their best friend had the opposite gender, and 1234 said their best friend had the same gender. Prepare a short report in which you analyze these data using a confidence interval and a significance test. Which do you think is more informative? Why?

8.103 Lottery paradox: In an experiment, each student was given a lottery ticket for a lottery in which the winner would receive a large payment. Later, the students were asked if they would be willing to exchange their ticket for another one, plus a small monetary incentive. Of the 26 students, only 7 agreed to the exchange (M. Bar-Hillel and E. Neter, *J. Personality and Social Psychology*, Vol. 70, 1996, pp. 17-27). In a related experiment, 31 students were given a new pen and then later asked to exchange it for another pen and a small monetary incentive. All 31 agreed. Analyze these data using methods of the previous two chapters, preparing a one-page summary of your analyses and conclusions. Why do you think students behaved so differently in the two cases?

8.104 Baseball home team advantage: In major league baseball's 2004 season, the home team won 1299 games and the away team won 1129 (www.mlb.com).
a. Although these games are not a random sample, explain how you could think of these data as giving you information about some long-run phenomenon.
b. Analyze these data using a (i) significance test and (ii) confidence interval. Which method is more informative? Why?

8.105 Statistics and scientific objectivity: The President of the American Statistical Association recently stated, "Statistics has become the modern-day enforcer of scientific objectivity. Terms like randomization, blinding, and 0.05 significance wield a no-doubt effective objectivity nightstick." He also discussed how learning what effects *aren't* in the data is as important as learning what effects *are* significant. In this vein, explain how statistics provides an objective framework for testing the claims of what many believe to be "quack science," such as astrology and therapeutic touch. (*Source*: Bradley Efron, *Amstat News*, July 2004, p.3.)

8.106 Effect of choice of scores: For the seven categories of political beliefs in Example 11, we used the scores 1 through 7. Suppose we instead use the scores (-3, -2, -1, 0, 1, 2, 3), subtracting 4 from each of the original scores. We then test H_0:

$\mu = 0$ instead of $H_0: \mu = 4$. Explain what, if any, effect this change of scores would have on the sample mean, the standard deviation, the test statistic, the P-value, and the interpretation.

8.107 Alternative hypotheses: Explain how you would decide in practice whether to use a one-sided or two-sided alternative hypothesis. Explain how your choice affects calculation of the P-value.

8.108 Two-sided or one-sided?: A medical researcher gets a P-value of 0.056 for testing $H_0: \mu = 0$ against $H_a: \mu \neq 0$. Since he believes that the true mean is positive and is worried that his favorite journal will not publish the results because they are not "significant at the 0.05 level," he instead reports in his article the P-value of 0.028 for $H_a: \mu > 0$. Explain what is wrong with:
a. Reporting the one-sided P-value after seeing the data.
b. The journal's guideline of publishing results only if they are statistically significant.

8.109 Interpret P-value: An article in a political science journal states that "no statistically significant difference was found between men and women in their voting rates (P-value = 0.63)." In practical terms, how would you explain what this means to someone who has not studied statistics?

8.110 Subgroup lack of significance: A crossover study on comparing a magnetic device to placebo for reducing pain in 54 people suffering from low back or knee pain (www.holcombhealthcare.com/reports/pub-bio.html) reported a significant result overall, the magnetic device being preferred to placebo. However, when the analysis was done separately for the 27 people with shorter illness duration and the 27 people with longer illness duration, results were not significant. Explain how it might be possible that an analysis could give a P-value below 0.05 using an entire sample but not with subgroups, even if the subgroups have the same descriptive statistics (for instance, the same sample mean and standard deviation, or the same sample proportion).

8.111 Birth control pills and breast cancer: A medical study in 1996 concluded that women who take birth control pills have a higher chance of getting breast cancer. A later study by scientists at the NIH and CDC of 9200 women reported no effect (*New England Journal of Medicine*, 6/27/2002). Discuss the factors that can cause different medical studies to come to different conclusions.

8.112 Overestimated effect: When medical stories in the mass media report large dangers of certain agents (e.g., coffee drinking), later research often suggests that the effects may not exist or are weaker than first believed. Explain how this could be because some journals tend to publish only statistically significant results.

8.113 $\alpha = 0.05$ too large?: An alternative hypothesis states that a newly developed drug is better than the one currently used to treat a serious illness. If we reject H_0, the new drug will be prescribed instead of the current one.
a. Explain why we might prefer to use a significance level smaller than 0.05, such as 0.01.

b. What is a disadvantage of using $\alpha = 0.01$ instead of 0.05?

8.114 Why not accept H_0?: Explain why the terminology "do not reject H_0" is preferable to "accept H_0."

8.115 Report P-value: It is more informative and potentially less misleading if you conclude a test by reporting and interpreting the P-value rather than by merely reporting whether or not you reject H_0 at the 0.05 significance level. One reason is that a reader can then tell whether the result is significant *at any significance level*. Give another reason.

8.116 Significance: Explain the difference between *statistical significance* and *practical significance*.

8.117 Effect of n: You are testing $H_0 : p = 0.50$ against $H_a : p \neq 0.50$. Which provides more evidence against H_0: Counts (10, 20) in the two categories with $n = 30$, or counts (100, 200) with $n = 300$? Answer by giving your reasoning, without going through the computational steps of the tests.

8.118 More doctors recommend: An advertisement by Company A says that more doctors recommend pain reliever A than all other brands combined.
a. If the company based this claim on interviewing a random sample of doctors, explain how they could use a significance test to back up the claim.
b. The company later claims that 3 of 4 doctors recommend pain reliever A over all other brands combined. How would you explain to someone who has never studied statistics that this claim would be more impressive if it is based on a (i) random sample of 40 doctors than if it is based on a random sample of 4 doctors, (ii) random sample of 40 doctors nationwide than the sample of all 40 doctors who work in a particular hospital?

8.119 Medical diagnosis error: Consider the medical diagnosis of breast cancer with mammograms. An AP story (Sept. 19, 2002) said that a woman has about a 50% chance of having a false positive diagnosis over the course of 10 annual mammography tests. Relate this result to the chance of eventually making a Type I error if you do lots of significance tests.

8.120 Bad P-value interpretations: A random sample of size 1000 has $\bar{x} = 104$. The P-value for testing $H_0 : \mu = 100$ against $H_a : \mu \neq 100$ is 0.057. Explain what is incorrect about each of the following interpretations of this P-value, and provide a proper interpretation.
a. The probability that the null hypothesis is correct equals 0.057.
b. The probability that $\bar{x} = 104$ if H_0 is true equals 0.057.
c. If in fact $\mu \neq 100$ so H_0 is false, the probability equals 0.057 that the data would show at least as much evidence against H_0 as the observed data.
d. The probability of Type I error equals 0.057.
e. We can accept H_0 at the $\alpha = 0.05$ level.

f. We can reject H_0 at the $\alpha = 0.05$ level.

8.121 Interpret P-value: One interpretation for the P-value is that it is the smallest value for the significance level α for which we can reject H_0. Illustrate using the P-value of 0.057 from the previous exercise.

8.122 Incorrectly posed hypotheses: What is wrong with expressing hypotheses about proportions and means in a form such as H_0: $\hat{p} = 0.50$ and H_0: $\bar{x} = 0$?

8.123 Crossover designs: Exercise 8.17 explained how with crossover experiments it's possible to compare two treatments by having the same subjects use each, at two different times. An alternative to a crossover experiment randomly selects half the study subjects to receive one treatment and half to receive the other, then making comparisons.
a. What do you think is the advantage of a crossover design over this approach?
b. A crossover design is often not possible for comparing two treatments, such as when a medical condition can be treated only once. Give an example.

8.124 Simulating Type I errors: Refer to the simulation at the end of Section 8.6.
a. Repeat the simulation, assuming that actually $p = 1/3$. In 100 simulations, what proportion of the time did you make a Type I error? What does theory predict for this proportion?
b. Do you think that the proportion of Type I errors will increase, or decrease, if you use a significance level of 0.01 instead? Check your intuition by conducting another simulation. Interpret.
c. Refer to (a). Do you think that the proportion of Type I errors will increase, or decrease, if the sample size in each simulation is 60 instead of 116? Check your intuition by conducting another simulation.

Select the correct response in multiple-choice exercises 8.125 – 8.128.

8.125 Small P-value: The P-value for testing H_0: $\mu = 100$ against H_a: $\mu \neq 100$ is 0.001. This indicates that:
a. There is strong evidence that $\mu = 100$.
b. There is strong evidence that $\mu \neq 100$, since if μ were equal to 100, it would be unusual to obtain data such as those observed.
c. The probability that $\mu = 100$ is 0.001.
d. The probability that $\mu = 100$ is the significance level, usually taken to be 0.05.

8.126 Probability of P-value: When H_0 is true in a t test with significance level 0.05, the probability that the P-value falls ≤ 0.05
a. equals 0.05.
b. equals 0.95.
c. equals 0.05 for a 1-sided test and 0.10 for a two-sided test.
d. can't be specified, because it depends also on P(Type II error).

8.127 Pollution: Exercise 8.41 concerned an industrial plant that may be exceeding pollution limits. An environmental action group took four readings to analyze whether the true mean discharge of wastewater per hour exceeded the company claim of 1000 gallons. When we make a decision in the one-sided test using $\alpha = 0.05$:

a. If the plant is not exceeding the limit, but actually $\mu = 1000$, there is only a 5% chance that we will conclude that they are exceeding the limit.

b. If the plant is exceeding the limit, there is only a 5% chance that we will conclude that they are not exceeding the limit.

c. The probability that the sample mean equals exactly the observed value would equal 0.05 if H_0 were true.

d. If we reject H_0, the probability that it is actually true is 0.05.

e. All of the above.

8.128 Interpret P(Type II error): Let β denote the probability of Type II error. For a test of $H_0 : \mu = 0$ against $H_a : \mu > 0$ based on $n = 30$ observations and using $\alpha = 0.05$ significance level, $\beta = 0.36$ at $\mu = 4$. Identify the response that is *incorrect*.

a. At $\mu = 5$, $\beta < 0.36$.

b. If $\alpha = 0.01$, then at $\mu = 4$, $\beta > 0.36$.

c. If $n = 50$, then at $\mu = 4$, $\beta > 0.36$.

d. The power of the test is 0.64 at $\mu = 4$.

8.129 True or false: It is always the case that P(Type II error) = 1 - P(Type I error).

8.130 True or false 2: If we reject $H_0 : \mu = 0$ in a study about change in weight on a new diet using $\alpha = 0.01$, then we also reject it using $\alpha = 0.05$.

8.131 True or false 3: A study about the change in weight on a new diet reports P-value = 0.043 for testing $H_0 : \mu = 0$ against $H_a : \mu \neq 0$. If the authors had instead reported a 95% confidence interval for μ, then the interval would have contained 0.

8.132 True or false 4: A 95% confidence interval for μ = population mean IQ is (96, 110). So, in the test of $H_0 : \mu = 100$ against $H_a : \mu \neq 100$, the P-value > 0.05.

8.133 True or false 5: For a fixed significance level α, the probability of Type II error increases when the sample size increases.

8.134 True or false 6: The P-value is the probability that H_0 is true.

◆◆**8.135 Standard error formulas**: Suppose you wanted to test $H_0 : p = 0.50$, but you had 0 successes in n trials. If you had found the test statistic using the $se = \sqrt{\hat{p}(1-\hat{p})/n}$ designed for confidence intervals, show what happens to the test statistic. Explain why $se = \sqrt{p_0(1-p_0)/n}$ is a more appropriate se for tests.

◆◆ **8.136 Rejecting true H_0 ?**: A medical researcher conducts a significance test every time she analyzes a new data set. Over time, she conducts 100 independent tests.

a. Suppose the null hypothesis is true in every case. What is the distribution of the number of times she rejects the null hypothesis at the 0.05 level?

b. Suppose she rejects the null hypothesis in five of the tests. Is it plausible that the null hypothesis is correct in every case? Explain.

♦♦**8.137 Mid P-value**: For discrete variables, the **mid P-value** takes the probabilities of values that are *farther* out in the tail than observed (in the direction of H_a) and adds *half* the probability of values that are *as far* out as observed. Suppose $n = 10$, and suppose $x = 9$. With $H_a : p > p_0$ the mid P-value is $P(10) + P(9)/2$.

a. For $H_a : p < p_0$, explain why the mid P-value is $[P(0) + P(1) + \ldots + P(8)] + P(9)/2$.

b. With a sketched figure, explain why the sum of the two one-sided mid P-values equals 1.0. (The mid P-value has similar properties as the P-value for continuous variables, such as an expected value of 0.50 when H_0 is true. For two-sided tests, and for one-sided tests with $p_0 = 0.5$, the P-value using the normal distribution is designed to approximate well the binomial mid P-value, even when n is small.)

CHAPTER PROBLEMS: CLASS EXPLORATIONS

8.138 Refer to the activity at the end of Section 8.4. Each student should indicate how many successive "wins" by the (a) placebo over the drug would be necessary before they would feel comfortable rejecting $H_0 : p = 0.5$ in favor of $H_a : p \neq 0.5$ and concluding that $p < 0.5$, (b) drug over the placebo would be necessary before they would feel comfortable rejecting $H_0 : p = 0.5$ in favor of $H_a : p \neq 0.5$ and concluding that $p > 0.5$. The instructor will compile a "distribution of significance levels" for the two cases. Are they the same? In principle, should they be?

8.139 Refer to Exercise 7.141, "Randomized response" in Chapter 7. Before carrying out the method described there, the class was asked to hypothesize or predict what they believe is the value for the population proportion of students that have had alcohol at a party. Use the class estimate for $\hat{p}$ to carry out the significance test for testing this hypothesized value. Discuss whether to use a one sided or two sided alternative hypothesis. Describe how the confidence interval formed in Exercise 7.140 relates to the significance test results.

BIBLIOGRAPHY

Reid, C. (1997). *Neyman*. Berlin: Springer-Verlag.

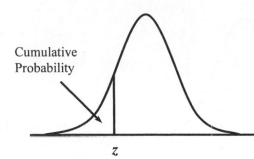

Cumulative
Probability

Cumulative probability for z is the area under
the standard normal curve to the left of z

z

TABLE A   Standard Normal Cumulative Probabilities

z	.00
-5.0	.000000287
-4.5	.00000340
-4.0	.0000317
-3.5	.000233

z	.00	.01	.02	.03	.04	.05	.06	.07	.08	.09
-3.4	.0003	.0003	.0003	.0003	.0003	.0003	.0003	.0003	.0003	.0002
-3.3	.0005	.0005	.0005	.0004	.0004	.0004	.0004	.0004	.0004	.0003
-3.2	.0007	.0007	.0006	.0006	.0006	.0006	.0006	.0005	.0005	.0005
-3.1	.0010	.0009	.0009	.0009	.0008	.0008	.0008	.0008	.0007	.0007
-3.0	.0013	.0013	.0013	.0012	.0012	.0011	.0011	.0011	.0010	.0010
-2.9	.0019	.0018	.0018	.0017	.0016	.0016	.0015	.0015	.0014	.0014
-2.8	.0026	.0025	.0024	.0023	.0023	.0022	.0021	.0021	.0020	.0019
-2.7	.0035	.0034	.0033	.0032	.0031	.0030	.0029	.0028	.0027	.0026
-2.6	.0047	.0045	.0044	.0043	.0041	.0040	.0039	.0038	.0037	.0036
-2.5	.0062	.0060	.0059	.0057	.0055	.0054	.0052	.0051	.0049	.0048
-2.4	.0082	.0080	.0078	.0075	.0073	.0071	.0069	.0068	.0066	.0064
-2.3	.0107	.0104	.0102	.0099	.0096	.0094	.0091	.0089	.0087	.0084
-2.2	.0139	.0136	.0132	.0129	.0125	.0122	.0119	.0116	.0113	.0110
-2.1	.0179	.0174	.0170	.0166	.0162	.0158	.0154	.0150	.0146	.0143
-2.0	.0228	.0222	.0217	.0212	.0207	.0202	.0197	.0192	.0188	.0183
-1.9	.0287	.0281	.0274	.0268	.0262	.0256	.0250	.0244	.0239	.0233
-1.8	.0359	.0351	.0344	.0336	.0329	.0322	.0314	.0307	.0301	.0294
-1.7	.0446	.0436	.0427	.0418	.0409	.0401	.0392	.0384	.0375	.0367
-1.6	.0548	.0537	.0526	.0516	.0505	.0495	.0485	.0475	.0465	.0455
-1.5	.0668	.0655	.0643	.0630	.0618	.0606	.0594	.0582	.0571	.0559
-1.4	.0808	.0793	.0778	.0764	.0749	.0735	.0721	.0708	.0694	.0681
-1.3	.0968	.0951	.0934	.0918	.0901	.0885	.0869	.0853	.0838	.0823
-1.2	.1151	.1131	.1112	.1093	.1075	.1056	.1038	.1020	.1003	.0985
-1.1	.1357	.1335	.1314	.1292	.1271	.1251	.1230	.1210	.1190	.1170
-1.0	.1587	.1562	.1539	.1515	.1492	.1469	.1446	.1423	.1401	.1379
-0.9	.1841	.1814	.1788	.1762	.1736	.1711	.1685	.1660	.1635	.1611
-0.8	.2119	.2090	.2061	.2033	.2005	.1977	.1949	.1922	.1894	.1867
-0.7	.2420	.2389	.2358	.2327	.2296	.2266	.2236	.2206	.2177	.2148
-0.6	.2743	.2709	.2676	.2643	.2611	.2578	.2546	.2514	.2483	.2451
-0.5	.3085	.3050	.3015	.2981	.2946	.2912	.2877	.2843	.2810	.2776
-0.4	.3446	.3409	.3372	.3336	.3300	.3264	.3228	.3192	.3156	.3121
-0.3	.3821	.3783	.3745	.3707	.3669	.3632	.3594	.3557	.3520	.3483
-0.2	.4207	.4168	.4129	.4090	.4052	.4013	.3974	.3936	.3897	.3859
-0.1	.4602	.4562	.4522	.4483	.4443	.4404	.4364	.4325	.4286	.4247
-0.0	.5000	.4960	.4920	.4880	.4840	.4801	.4761	.4721	.4681	.4641

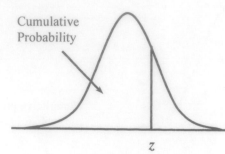

Cumulative Probability

Cumulative probability for z is the area under the standard normal curve to the left of z

TABLE A   Standard Normal Cumulative Probabilities *(continued)*

z	.00	.01	.02	.03	.04	.05	.06	.07	.08	.09
0.0	.5000	.5040	.5080	.5120	.5160	.5199	.5239	.5279	.5319	.5359
0.1	.5398	.5438	.5478	.5517	.5557	.5596	.5636	.5675	.5714	.5753
0.2	.5793	.5832	.5871	.5910	.5948	.5987	.6026	.6064	.6103	.6141
0.3	.6179	.6217	.6255	.6293	.6331	.6368	.6406	.6443	.6480	.6517
0.4	.6554	.6591	.6628	.6664	.6700	.6736	.6772	.6808	.6844	.6879
0.5	.6915	.6950	.6985	.7019	.7054	.7088	.7123	.7157	.7190	.7224
0.6	.7257	.7291	.7324	.7357	.7389	.7422	.7454	.7486	.7517	.7549
0.7	.7580	.7611	.7642	.7673	.7704	.7734	.7764	.7794	.7823	.7852
0.8	.7881	.7910	.7939	.7967	.7995	.8023	.8051	.8078	.8106	.8133
0.9	.8159	.8186	.8212	.8238	.8264	.8289	.8315	.8340	.8365	.8389
1.0	.8413	.8438	.8461	.8485	.8508	.8531	.8554	.8577	.8599	.8621
1.1	.8643	.8665	.8686	.8708	.8729	.8749	.8770	.8790	.8810	.8830
1.2	.8849	.8869	.8888	.8907	.8925	.8944	.8962	.8980	.8997	.9015
1.3	.9032	.9049	.9066	.9082	.9099	.9115	.9131	.9147	.9162	.9177
1.4	.9192	.9207	.9222	.9236	.9251	.9265	.9279	.9292	.9306	.9319
1.5	.9332	.9345	.9357	.9370	.9382	.9394	.9406	.9418	.9429	.9441
1.6	.9452	.9463	.9474	.9484	.9495	.9505	.9515	.9525	.9535	.9545
1.7	.9554	.9564	.9573	.9582	.9591	.9599	.9608	.9616	.9625	.9633
1.8	.9641	.9649	.9656	.9664	.9671	.9678	.9686	.9693	.9699	.9706
1.9	.9713	.9719	.9726	.9732	.9738	.9744	.9750	.9756	.9761	.9767
2.0	.9772	.9778	.9783	.9788	.9793	.9798	.9803	.9808	.9812	.9817
2.1	.9821	.9826	.9830	.9834	.9838	.9842	.9846	.9850	.9854	.9857
2.2	.9861	.9864	.9868	.9871	.9875	.9878	.9881	.9884	.9887	.9890
2.3	.9893	.9896	.9898	.9901	.9904	.9906	.9909	.9911	.9913	.9916
2.4	.9918	.9920	.9922	.9925	.9927	.9929	.9931	.9932	.9934	.9936
2.5	.9938	.9940	.9941	.9943	.9945	.9946	.9948	.9949	.9951	.9952
2.6	.9953	.9955	.9956	.9957	.9959	.9960	.9961	.9962	.9963	.9964
2.7	.9965	.9966	.9967	.9968	.9969	.9970	.9971	.9972	.9973	.9974
2.8	.9974	.9975	.9976	.9977	.9977	.9978	.9979	.9979	.9980	.9981
2.9	.9981	.9982	.9982	.9983	.9984	.9984	.9985	.9985	.9986	.9986
3.0	.9987	.9987	.9987	.9988	.9988	.9989	.9989	.9989	.9990	.9990
3.1	.9990	.9991	.9991	.9991	.9992	.9992	.9992	.9992	.9993	.9993
3.2	.9993	.9993	.9994	.9994	.9994	.9994	.9994	.9995	.9995	.9995
3.3	.9995	.9995	.9995	.9996	.9996	.9996	.9996	.9996	.9996	.9997
3.4	.9997	.9997	.9997	.9997	.9997	.9997	.9997	.9997	.9997	.9998

z	.00
3.5	.999767
4.0	.9999683
4.5	.9999966
5.0	.999999713

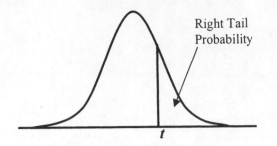

Right Tail
Probability

TABLE B *t*-Distribution critical values

			Confidence Level		
	80%	90%	95%	98%	99%
		Right Tail Probability			
df	$t_{.100}$	$t_{.050}$	$t_{.025}$	$t_{.010}$	$t_{.005}$
1	3.078	6.314	12.706	31.821	63.657
2	1.886	2.920	4.303	6.965	9.925
3	1.638	2.353	3.182	4.541	5.841
4	1.533	2.132	2.776	3.747	4.604
5	1.476	2.015	2.571	3.365	4.032
6	1.440	1.943	2.447	3.143	3.707
7	1.415	1.895	2.365	2.998	3.499
8	1.397	1.860	2.306	2.896	3.355
9	1.383	1.833	2.262	2.821	3.250
10	1.372	1.812	2.228	2.764	3.169
11	1.363	1.796	2.201	2.718	3.106
12	1.356	1.782	2.179	2.681	3.055
13	1.350	1.771	2.160	2.650	3.012
14	1.345	1.761	2.145	2.624	2.977
15	1.341	1.753	2.131	2.602	2.947
16	1.337	1.746	2.120	2.583	2.921
17	1.333	1.740	2.110	2.567	2.898
18	1.330	1.734	2.101	2.552	2.878
19	1.328	1.729	2.093	2.539	2.861
20	1.325	1.725	2.086	2.528	2.845
21	1.323	1.721	2.080	2.518	2.831
22	1.321	1.717	2.074	2.508	2.819
23	1.319	1.714	2.069	2.500	2.807
24	1.318	1.711	2.064	2.492	2.797
25	1.316	1.708	2.060	2.485	2.787
26	1.315	1.706	2.056	2.479	2.779
27	1.314	1.703	2.052	2.473	2.771
28	1.313	1.701	2.048	2.467	2.763
29	1.311	1.699	2.045	2.462	2.756
30	1.310	1.697	2.042	2.457	2.750
40	1.303	1.684	2.021	2.423	2.704
50	1.299	1.676	2.009	2.403	2.678
60	1.296	1.671	2.000	2.390	2.660
80	1.292	1.664	1.990	2.374	2.639
1000	1.282	1.646	1.962	2.330	2.581
∞	1.282	1.645	1.960	2.326	2.576

Statistics

Index

607